Free Student Aid.

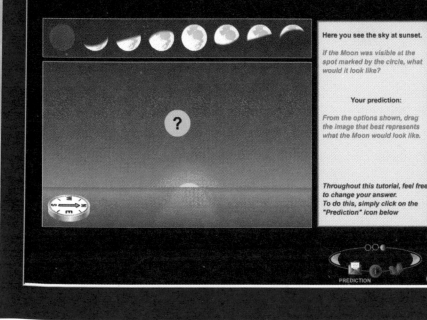

Log on.

Tune in.

Succeed.

To help you succeed in astronomy, your professor has arranged for you to enjoy access to great media resources, including planetarium software, Voyager: Skygazer, College Edition CD-ROM and The Astronomy Place web site. You'll find that these resources that accompany your textbook will enhance your course materials.

Here's your personal ticket to success:

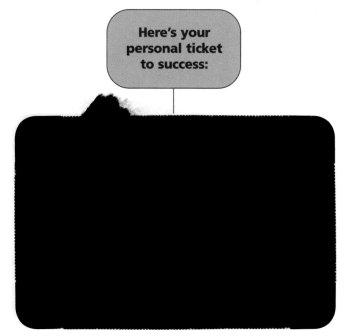

Detach this card and keep it handy. It's your ticket to valuable information.

Got technical questions?

For technical support, please visit www.aw.com/techsupport, send an email to online.support@pearsoned.com (for web site questions) or send an email to media.support@pearsoned.com (for CD-ROM questions) with a detailed description of your computer system and the technical problem. You can also call our tech support hotline at 1-800-677-6337 Monday-Friday, 8 a.m. to 5 p.m. CST.

What your system needs to use these media resources:

WINDOWS
250 MHz Intel Pentium processor or greater
Windows 95, 98 NT4, 2000
32 MB RAM installed, 64 MB preferred
800 X 600 screen resolution
4x CD-ROM drive
Browser: Internet Explorer 5.0 or Netscape
    Communicator 4.7
Plug-Ins: Shockwave Player 8, Flash Player 5,
    QuickTime 4.1

MACINTOSH
233 MHz G3 or higher
OS 8.1 or higher
32 MB RAM available, 64 MB preferred
800 x 600 screen resolution, thousands of colors
4x CD-ROM Drive
Browser: Internet Explorer 5.0 or Netscape
    Communicator 4.7
Plug-Ins: Shockwave Player 8, Flash Player 5,
    QuickTime 4.1

0-8053-8584-3

Cosmic resources for you and your students.

Included with all texts.

Available to supplement your course.

Help students develop a cosmic context with *The Cosmic Perspective.*

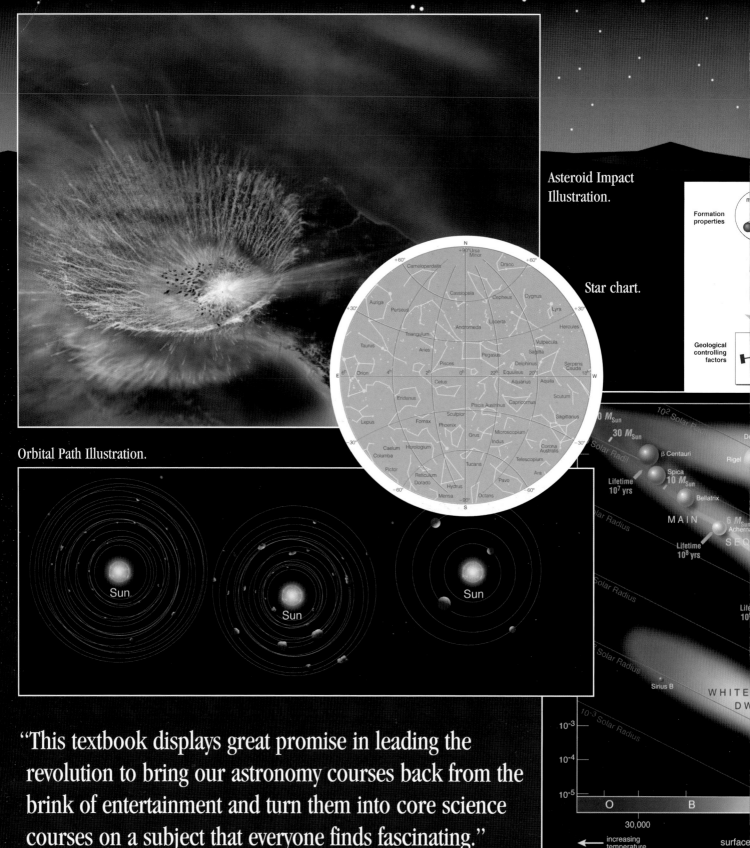

Asteroid Impact Illustration.

Star chart.

Formation properties

Geological controlling factors

Orbital Path Illustration.

Sun

Sun

Sun

30 M_{Sun}

β Centauri

Rigel

Spica

Lifetime 10^7 yrs

10 M_{Sun}

Bellatrix

MAIN

Achernar

SEQ

Lifetime 10^8 yrs

Sirius B

WHITE DW

10^{-3} Solar Radius

10^{-3}

10^{-4}

10^{-5}

O

B

30,000

increasing temperature

surface

"This textbook displays great promise in leading the revolution to bring our astronomy courses back from the brink of entertainment and turn them into core science courses on a subject that everyone finds fascinating."

— *James Schombert, University of Oregon*

New and favorite features of *The Cosmic Perspective*

REVISED! Thoroughly revised with current findings in astronomy. Updates include discoveries such as the accelerating universe/cosmological constant, the detection of more planets around other stars, the potential of water flows on Mars, and the latest theories on the very early universe.

NEW! Star charts. These visual tools are now included at the end of the book.

NEW! Additional end-of-chapter problems. Includes problems of varying difficulty level to allow for diverse learning levels.

NEW! New appendices are offered on constants and key equations.

NEW! Upgraded art program. Beautiful illustrations provide a new level of integration between scientific content and art that helps students better understand astronomy.

Planet Flow Diagram.

"Thinking About…" box.

"Common Misconceptions" box.

THINKING ABOUT . . .

What If Light Can't Catch You?

If you're like most students learning about relativity for the first time, you're probably already looking for loopholes in the logic of our thought experiments. For example, confronted with Thought Experiment 6, you might be tempted to ask, "What happens if you're traveling away from some planet faster than light, so the light from the planet can't catch you?" While it's surely true that light couldn't catch you if you were going faster than the speed of light, it also makes the question somewhat moot: If you can't see light from the planet, there is no way for you to know that the planet even exists.

In fact, what relativity really tells us about the speed of light is that it is a limit on the speed at which *information* can be transmitted. There are numerous circumstances in the universe in which, on philosophical grounds, an object may *seem* to be exceeding the speed of light. However, these circumstances do not provide any means of sending information or objects at speeds faster than the speed of light.

As an example, consider the implications of the fact that the universe is expanding [Section 1.3]. Recall that the more distant a galaxy, the faster the expansion of the universe is carrying it away from us. In principle, somewhere

far away, there could be ... sion is carrying galaxies ... speed of light. However, ... because their light cann ... that such galaxies are b ... universe, and the observ ... that we can study.

Other examples of t ... than the speed of light a ... world of quantum mech ... to quantum principles, ... can (in certain specific ... another place *instantane* ... light-years away. In fact, this process has been observed in laboratories over short distances. This instantaneous effect of one particle on another may at first seem to violate relativity, but it does not. The built-in randomness of quantum mechanics prevents this technique from being used to transmit useful information to a distant point. Moreover, if we wish to confirm that the second particle really was affected, we will have to receive a signal carrying information about the particle—and that signal can travel no faster than the speed of light.

Common Misconceptions: A Flat Earth

A widespread myth holds that Columbus proved the Earth to be round rather than flat. In fact, knowledge of the round Earth predated Columbus by nearly 2,000 years. However, it is probably true that most people in Columbus's day believed the Earth to be flat, largely because of the poor state of education. The vast majority of the public was illiterate and unaware of the scholarly evidence for a spherical Earth. Interestingly, Columbus's primary argument with other scholars concerned the distance from Europe to Asia going westward—and it was Columbus who was wrong. He underestimated the true distance and as a result was woefully unprepared for the voyage to Asia that he thought he was undertaking. Indeed, his voyages would almost certainly have ended in disaster had it not been for the presence of the Americas, which offered a safe landing well to the east of Asia.

NEW! Electronic Textbook CD-ROM includes the entire text and illustration program, browsable, and hotlinked in XML. Interactive tutorials, worksheets, and end-of-chapter problems from **The Astronomy Place** are also provided on this accompanying free CD-ROM. You and your students can now easily access both text and media resources on any subject covered in the book in one convenient location.

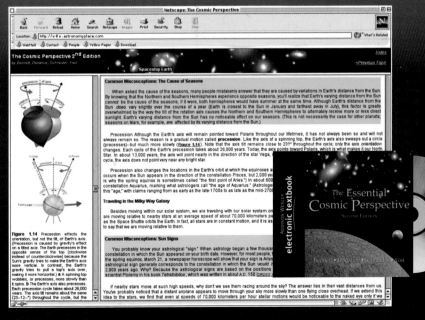

Hertzsprung-Russell Diagram.

The Astronomy Place

www.astronomyplace.com

An extensive media supplement to the text, The Astronomy Place includes a wealth of resources for you and your students. At the heart of the site are a suite of interactive tutorials and mini-documentaries linked to the online textbook.

The Astronomy Place also includes:

- Browse the Book feature (in XML)
- Activity worksheets for Voyager: SkyGazer, College Edition
- Self-assessment quizzes, with hints and rejoinders
- Chapter links and summaries
- Interactive periodic table and calculator
- Online glossary/appendix
- Chat/discussion area
- Online syllabus builder

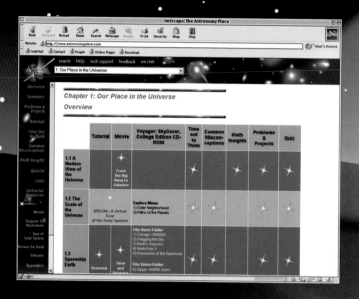

What students are saying about astronomyplace.com:

"The Astronomy Place helps to explain concepts that are difficult to illustrate in a 2-dimensional textbook. Sometimes the tutorials answered questions that I didn't even realize I had!" — *Amy Lindsay*

"These tutorials greatly increased my knowledge and confidence... I like that you learn something and then you are able to apply it through questions and visual sequences." — *James Tomlinson*

"The [interactive] depictions help to give a much clearer understanding for students like me who are visual learners ... any question I might have about a homework problem or clarification of a lecture can almost always be found [on The Astronomy Place]." — *Ashley Radovich*

"I like the fact that the people who wrote the book also developed the web site. It makes it easier to learn because the [book and site] are cohesive." — *Richard Murray*

Tutorials

A comprehensive collection of extensive interactive and animated tutorials that focus on important and hard-to-grasp concepts. These tutorials carefully build understanding through five structured steps: Introduction, Objectives, Lessons, Exercises, and Summary.

Tutorials include:

- Scale of the Universe
- Seasons
- Phases of the Moon
- Eclipses
- Orbits and Kepler's Laws
- Light and Spectroscopy
- Formation of the Solar System
- Extrasolar Planet Detection
- Shaping Planetary Surfaces
- Surface Temperature of Terrestrial Planets
- Hertzsprung-Russell Diagram
- Stellar Evolution
- Measuring Cosmic Distances
- Detecting Dark Matter in Spiral Galaxies
- Hubble's Law
- Fate of the Universe

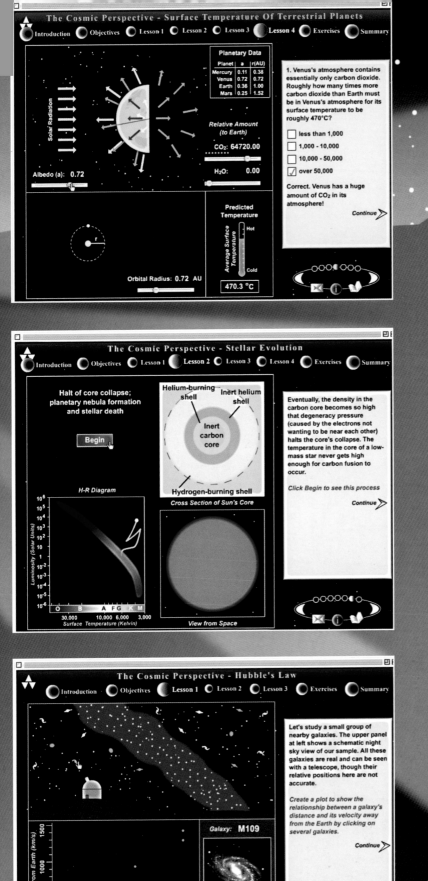

Mini-documentaries

In addition, The Astronomy Place offers ten 15-minute, animated and narrated mini-documentaries that provide engaging summaries of key topics:

- From the Big Bang to Galaxies
- Time and Season
- The Celestial Sphere
- Orbits in the Solar System
- History of the Solar System
- The Sun
- Lives of Stars
- Double Stars
- The Milky Way Galaxy
- Search for Extraterrestrial Life

"The tutorials really help. When you read [a] book you aren't really asked to apply that knowledge. The tutorials help you make the connections."

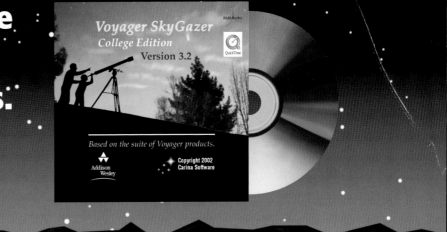

A spectacular resource for a visual science.

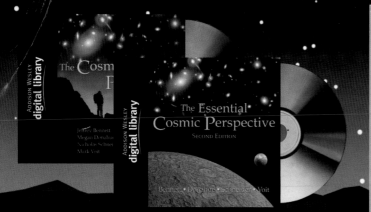

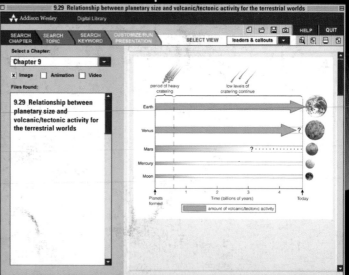

9.29 Relationship between planetary size and volcanic/tectonic activity for the terrestrial worlds

The Addison Wesley Digital Library for:

The Essential Cosmic Perspective, **Second Edition** ISBN 0-8053-8590-8
The Cosmic Perspective, **Second Edition** ISBN 0-8053-8569-X

This cross-platform CD-ROM features most of the illustrations, tables, and star charts from the book which can be customized for lecture presentation. Edit labels, import illustrations and photos from other sources, and export figures into other programs including PowerPoint or for use on the Web.

"[*The Cosmic Perspective*] distinguishes itself from the competition by a perspective that reaches far beyond the typical introductory text.... *The Cosmic Perspective* has excellent discussions of space and time, as well as spacetime and gravity..."
— *Marc Davis, University of California, Berkeley*

Flexible formats to suit your course.

The text is now available in four versions:

The Essential Cosmic Perspective, **Second Edition** ISBN 0-8053-8584-3

By **Jeffery Bennett, Megan Donahue, Nicholas Schneider, and Mark Voit**

The Cosmic Perspective, **Second Edition** ISBN 0-8053-8041-8

By **Jeffery Bennett, Megan Donahue, Nicholas Schneider, and Mark Voit**

The Solar System: The Cosmic Perspective, **Second Edition** ISBN 0-8053-8553-3

Chapters 1-13

Stars, Galaxies, and Cosmology: The Cosmic Perspective, **Second Edition** ISBN 0-8053-8556-8

Chapters 1, 4–7, 14–22, S2–S5

"Most long-standing astronomy textbooks incorporate new material... by adding pictures, or rewriting certain sections. Few reorganize the entire text to reflect our new understanding... The authors of *The Cosmic Perspective* have clearly thought very hard about the broad new themes in astronomy... and have incorporated these themes into their textbook. It is the most refreshing astronomy text I have seen in years."
— *Richard Gray, Appalachian State University*

"Student questions have been anticipated by an exceptionally thorough writing style. The explanations of complex concepts are artfully accomplished... The consistency is markedly superior... The questions are especially good."
— *John Stolar, West Chester University*

The Essential
Cosmic Perspective

SECOND EDITION

Jeffrey Bennett
University of Colorado at Boulder

Megan Donahue
Space Telescope Science Institute

Nicholas Schneider
University of Colorado at Boulder

Mark Voit
Space Telescope Science Institute

Addison
Wesley

San Francisco, California   Reading, Massachusetts   New York
Harlow, England   Don Mills, Ontario   Sydney   Mexico City   Madrid   Amsterdam

Acquisitions Editor: *Adam Black*
Editorial Director: *Frank Ruggirello*
Market Developer: *Chalon Bridges*
Marketing Manager: *Christy Lawrence*
Publishing Associate: *Liana Allday*
Developmental Editor: *Laura Maria Bonazzoli*
Senior Producer: *Claire Masson*
Production Coordination: *Joan Marsh*
Production: *Mary Douglas, Rogue Valley Publications*
Photo Research: *Myrna Engler*
Graphic Artists: *Joe Bergeron, John Goshorn/Techarts,*
    Blakeley Kim, Emiko Paul/fiVth.com, Quade
    Paul/fiVth.com
Copyeditors: *Mary Roybal, Karen Stough*
Text Designers: *Blakeley Kim, Andrew Ogus*
Partial Cover Photo: *Quade Paul/fiVth.com; PhotoDisk;*
    and Hubble Space Telescope
Cover Designers: Quade Paul/fiVth.com and Emiko
    Paul/fiVth.com
Manufacturing Supervisor: *Vivian McDougal*
Composition: *Thompson Type/Alma Bell, Lori Shranko,*
    Janice Adamski
Cover Printer: *Coral Graphics*
Prepress Services: *H&S Graphics, Inc./Tom Andersen,*
    Mike Tambellini, Lori Jewell, Dave Frank
Printer and Binder: *Von Hoffmann Press*

For permission to use copyrighted material, grateful acknowledgment is made to the copyright holders on pp. C-1–C-3, which are hereby made part of this copyright page.

Library of Congress Cataloging-in-Publication Data--2nd ed.
The essential cosmic perspective / Jeffrey O. Bennett . . . [et al.].
    p. cm.
  Includes index.
  ISBN 0-8053-8584-3
  1. Astronomy.   I. Bennett, Jeffrey O.
QB43.3.C68 2003
520—dc21   2001045214

1 2 3 4 5 6 7 8 9 10—VHP—05 04 03 02

We shall not cease from exploration
And the end of all our exploring
Will be to arrive where we started
And know the place for the first time.
T. S. ELIOT

Dedication

TO ALL WHO HAVE EVER WONDERED about the mysteries of the universe. We hope this book will answer some of your questions—and that it will also raise new questions in your mind that will keep you curious and interested in the ongoing human adventure of astronomy.

And, especially, to the members of the "baby boom" that has occurred among the authors and editors during the writing of this book: Michaela, Emily, Rachel, Sebastian, Elizabeth, Nathan, Grant, Georgia, Brooke, Brian, and Nolan. The study of the universe begins at birth, and we hope that you will grow up in a world with far less poverty, hatred, and war so that all people will have the opportunity to contemplate the mysteries of the universe into which they are born.

BRIEF CONTENTS

DETAILED CONTENTS

PREFACE

We humans have gazed into the sky for countless generations, wondering how our lives are connected to the Sun, Moon, planets, and stars that adorn the heavens. Today, through the science of astronomy, we know that these connections go far deeper than our ancestors ever imagined. This book tells the story of modern astronomy and the new perspective—*The Cosmic Perspective*—with which it allows us to view ourselves and our planet. It is written for anyone who is curious about the universe, but it is designed primarily as a textbook for college students not planning to major in mathematics or science.

The Essential Cosmic Perspective, Second Edition, is designed primarily for one-term courses in introductory astronomy. The essential edition focuses on the major conceptual ideas of the cosmic perspective and minimizes the use of mathematics.

Approach

This book grew out of our experience teaching astronomy to both college students and the general public over the past 20 years. During this time, a flood of new discoveries fueled a revolution in our understanding of the cosmos, but the basic organization and approach of most astronomy textbooks remained unchanged. We felt the time had come to rethink how to organize and teach the major concepts in astronomy. This book is the result. Several major innovations in organization set this book apart.

A "Big Picture" Context Throughout the Book.
Many traditional textbooks begin with the sky as viewed by the ancients and gradually work through to the expanding universe and modern ideas of cosmology at the very end of the book. The potential advantage to this traditional organization is that, if students fully absorb it, they recreate for themselves the thought patterns and paradigm shifts that led

from ancient superstitions about the sky to modern understanding of the universe. Unfortunately, this potential advantage is very difficult to realize in practice, because it means that students must learn a sequence of successive paradigms, why each was rejected, and why the next one followed, before finally arriving at what we now consider the correct understanding. That is, students must essentially recreate the thinking produced over thousands of years by some of the greatest minds in history, all with just a few hours of study per week over one or two course terms. Not only is such a process unrealistic for most students, but it also tends to lead to confusion among the various paradigms, thereby preventing students from understanding either the modern view or how we arrived at it. We have therefore chosen a different approach, in which we begin the book with a broad overview of modern scientific understanding of the cosmos and then use the rest of the book to help students understand how and why we have reached this understanding and what gaps remain in our knowledge. We believe that, for the vast majority of students, this approach enables them to understand the scientific process much better than they would with the historical approach, because knowing the end goal helps them focus their studies. We maintain this "big picture" context throughout the book by always telling students where we are going when we introduce new topics and by reviewing the context at the end of each chapter (in the end-of-chapter sections called *The Big Picture*).

Stronger Emphasis on the Scientific Process.
Nearly all astronomy textbooks emphasize the scientific process, but we believe that our approach allows us to go beyond what is possible with a more traditional, historical approach to astronomy. In particular, by starting with the context of modern astronomy, we

are able to focus clearly on how science works (Chapter 3) and on key physical principles that apply universally (Chapters 4 through 6). For example, we use the concept of conservation of energy over and over throughout the text to show how it helps explain everything from whether planets are geologically active to the processes that govern the lives of stars.

True Comparative Planetology. We offer a true comparative planetology approach, in which the discussion emphasizes general principles that apply to planetary geology and planetary atmospheres as opposed to simply describing one planet at a time. This approach has several crucial advantages over the more common, planet-by-planet approach. For example, by focusing on planetary processes rather than simply on morphological features, we strengthen students' grasp of how a few key physical principles apply throughout the universe, thereby furthering our goal of helping students understand the scientific process. In addition, long after the course is over, students are much more likely to retain the understanding of physical processes built in this way than they are of any particular planetary features that they memorize through a planet-by-planet approach. Perhaps most important, the comparative planetology approach helps students see the relevance of planetary science to their own lives by enabling them to gain a far deeper appreciation of our own unique world.

An Evolutionary Approach to Galaxies and Dark Matter. Over the past few decades, a revolution has occurred in our understanding of galactic evolution. For example, whereas quasars were once considered a distinct class of objects, the evidence now strongly contradicts this idea; instead, it now appears that quasars represent an unusually active stage of life in the centers of some young galaxies. Despite this revolution in scientific understanding, most introductory textbooks still present galaxies in the order that made sense when early galaxy classification schemes were first developed a half-century ago. We have taken an approach that teaches about galaxies in a way that closely parallels how nearly all textbooks (including ours) teach about stars: we begin with our own galaxy as a paradigm (Chapter 16) and then discuss the variety of galaxies, how we determine key parameters such as distances, and how we've arrived at the current state of knowledge regarding galaxy evolution (Chapter 17). Because dark matter plays such an important role in this evolutionary understanding, we then devote an entire chapter (Chapter 18) to its study.

An Integrated Approach to Cosmology. Ultimately, one of the primary goals of astronomy is to understand our cosmic origins, which means understanding cosmology. Most introductory textbooks present cosmology only in a final chapter. We integrate the study of cosmology throughout much of the text. Students first encounter modern cosmological ideas, including the notion of the Big Bang and the idea of an expanding universe, as part of our big picture overview in Chapter 1. Our entire presentation of galaxies is steeped in cosmology, since the study of galactic evolution cannot be effectively separated from cosmological ideas. Thus, we do not have a single cosmology chapter in the traditional sense, but instead we have a chapter devoted to the Big Bang theory (Chapter 19) that builds upon the understanding students develop throughout the text.

Pedagogical Features

Besides the main narrative, the book includes a number of pedagogical features designed to enhance student learning:

End-of-Chapter Questions, Problems, and Projects. The following four sets of questions appear at the end of each chapter:

- *Review Questions.* The Review Questions replace a standard chapter summary. Like a summary, they review the important concepts from the chapter. However, they are more effective as a study tool than a standard summary because they require students to think rather than enabling mindless use of a highlight pen. While it is possible to assign review questions as homework, these questions generally can be answered with little more than a quick re-reading of the relevant sections of the text.

- *Discussion Questions.* These questions are meant to be particularly thought provoking and generally do not have objective answers. As such, they are ideal for in-class discussion.

- *Problems.* Each chapter includes several problems of varying levels of difficulty that are designed to be assigned as written homework. Problems are generally arranged from easier to harder and to the extent possible follow the order of presentation in the chapter.

- *Web Projects.* Each chapter ends with one or more Web Projects, which are designed for independent research. The projects typically ask students to learn more about a topic of relevance to the chapter, such as a current or planned mission. Students can find useful links for all the Web Projects at The Astronomy Place (www.astronomyplace.com).

Time Out to Think. This feature gives students the opportunity to reflect on important new concepts. It also serves as an excellent starting point for classroom discussions. Answers or discussion points for all the Time Out to Think questions can be found in the Instructor's Guide.

Common Misconceptions. These boxes address and correct popularly held but incorrect ideas related to the chapter material.

Thinking About. These boxes contain supplementary discussion of topics related to the chapter material but not prerequisite to the continuing discussion.

The Big Picture. This end-of-chapter feature helps students put what they've learned into the context of the overall goal of gaining a new perspective on ourselves and our planet.

Cross-References. When we discuss a concept that is covered in greater detail elsewhere in the book, we include a cross-reference in brackets. For example, [Section 5.2] means that the concept being discussed is covered in greater detail in Section 5.2.

Glossary. A detailed glossary makes it easy for students to look up unfamiliar terms.

Appendixes. The appendixes include a number of useful references, including summaries of key constants (Appendix A), key formulas (Appendix B), and a brief review of key mathematical skills such as working with scientific notation and operating with units (Appendix C).

Star Charts (Appendixes I, J). The list of constellations and the star charts help interested students learn their way around the night sky.

Alternate Versions of the Book

The Cosmic Perspective is available in four versions tailored to particular course types:

- *The Cosmic Perspective, Second Edition* (full version, ISBN 0-8053-8044-2) This is the most complete version of the book, containing sufficient material for astronomy courses ranging in length from one quarter to a full year.

- *The Solar System: The Cosmic Perspective, Second Edition, Volume 1* (ISBN 0-8053-8544-1) This version contains Chapters 1–13 and Chapter S1 of the full version of the book. It is designed for courses that focus on the solar system rather than on stars, galaxies, and cosmology.

- *Stars, Galaxies, and Cosmology: The Cosmic Perspective, Second Edition, Volume 2* (ISBN 0-8053-8557-6) This version contains Chapters 1, 4–7, 14–22, and S2–S5 of the full version of the book. It is designed for courses that focus on stars, galaxies, and cosmology.

- *The Essential Cosmic Perspective, Second Edition* (ISBN 8053-8584-3) This is a condensed version of the full book, designed primarily for use in courses that provide an overview of all of astronomy but at a lower level of depth than offered in the full version of the book.

Supplements and Resources

The Cosmic Perspective is much more than just a textbook; it is a complete package of resources designed to help instructors and students. Here, we've broken the resources into two categories, the first designed to help students directly and the second designed to help instructors prepare course materials and lectures.

Resources to Help Students Learn

The following resources are designed to reinforce the content knowledge students will learn from the textbook.

The Astronomy Place (www.astronomyplace.com). This Web site is designed specifically to accompany *The Cosmic Perspective* textbooks. Among its constantly growing list of features, instructors and students will find:

- A complete set of student study resources for each chapter in the textbook; the resources include chapter summaries, additional information on Common Misconceptions and Mathematical Insights, self-assessment quizzes, and chapter-specific links (including links for the end-of-chapter Web Projects). In addition, all Time Out to Think questions and all end-of-chapter problems are digitized for on-line assignments. The entire textbook is available on-line in XML and can be browsed by topic.

- A suite of interactive tutorials that build understanding through a carefully constructed, step-by-step process based on the type of interaction and discussion that typically would occur during office hours.

- Ten 5-minute animated and narrated mini-movies that provide engaging summaries of key topics.

- Activity worksheets to be used with the *SkyGazer, College Edition* software packaged with the textbook.

- On-line versions of the textbook glossary and appendixes.

- Updates on the latest astronomy news.

***Carl Sagan's* Cosmos Series.** The newly released *Best of Cosmos* (ISBN: 0-8053-8571-1) and complete, revised, enhanced, and updated *Cosmos* series (Video ISBN: 0-8053-8570-3, DVD ISBN: 0-8053-8572-X) are available free to qualified adopters of *The Cosmic Perspective*.

SkyGazer, College Edition. Based on *Voyager III,* one of the world's most popular planetarium programs, *SkyGazer, College Edition* makes it easy for students to learn constellations and explore the wonders of the sky through interactive exercises. The *SkyGazer* CD is packaged free with all new copies of the textbook. It can also be ordered separately (CD-ROM ISBN: 0-8053-8022-1).

The Astronomy Tutor Center. This center provides one-on-one tutoring by qualified college instructors in any of four ways—phone, fax, email, and the Internet—during afternoon/evening hours and on weekends. The tutor center instructors will answer questions and provide help with examples, exercises, and other content found in *The Cosmic Perspective*. Students who register to use the astronomy tutor center can receive as much tutoring as they wish for the duration of their course. Registration is free with purchase of a new textbook or can be purchased separately; see www.aw.com/tutorcenter for more information.

Course Preparation Resources

Instructor's Resource Guide. (ISBN: 0-8053-8591-6) Includes an overview of the philosophy of the text, discussion of strategies for teaching astronomy, suggestions on preparing an astronomy course, sample syllabi for courses of different length and emphasis, a reference guide for integrating the *Cosmos* video series and The Astronomy Place Web tutorials and movies with this textbook, a chapter-by-chapter commentary with notes and teaching hints, answers or discussion points for all the Time Out to Think questions in the book, and solutions to end-of-chapter problems.

The Astronomy Place Instructor Resources (*www.astronomyplace.com*). The Astronomy Place has numerous resources designed specifically for instructors, including:

- A digitized version of the Instructor's Resource Guide (not including solutions to problems).

- Syllabus Manager, an easy to use on-line course management system. It can be used to build and maintain one or more syllabi on the Web. The system also allows you to receive and grade student assignments on-line.

- Customization resources that allow you to use the content of The Astronomy Place in WebCT, BlackBoard, or CourseCompass. If you are interested in these customized options, please contact your local Addison Wesley sales representative.

Printed Test Bank. (ISBN: 0-8053-8592-4) Includes over 1500 multiple-choice, true/false, and free-response questions for each chapter.

TestGen EQ CD-ROM. (ISBN: 0-8053-8589-4) Provides identical questions from the Printed Test Bank on one CD-ROM (Windows/Mac compatible). Allows the instructor to view and edit questions, add questions, and create and print different versions of tests. A built-in question editor creates graphs, imports graphics, and inserts text.

Transparency Acetates. (ISBN: 0-8053-8546-0) A selection of over 200 figures from the text printed on full-color transparency acetates.

Addison Wesley Science Digital Library. (ISBN: 0-8053-8590-8) This cross-platform CD-ROM includes illustrations, tables, and star charts from the book and many animations/video clips from the Web site that can be customized for lecture presentation.

Acknowledgments

A textbook may carry author names, but it is the result of hard work by a long list of committed individuals. We could not possibly list everyone who has helped, but we would like to call attention to a few people who have played particularly important roles. First, we thank our editors and friends at Addison Wesley Longman who have stuck with us through thick and thin, including Ben Roberts, Adam Black, Robin Heyden, Sami Iwata, Bill Poole, Linda Davis, Joan Marsh, Stacy Treco, Christy Lawrence, Nancy Gee, Liana Allday, and Tony Asaro. Special thanks to our production team, especially Mary Douglas, Myrna Engler, Mary Roybal, and Karen Stough; our art team, led by Blakeley Kim and Emiko Paul; and our Web team, led by Claire Masson, Jim Dove, and Ian Shakeshaft.

We've also been fortunate to have an outstanding group of reviewers whose extensive comments and suggestions helped us shape the book. We thank all those who have reviewed drafts of the book in various stages, including:

Christopher M. Anderson, University of Wisconsin
Peter S. Anderson, Oakland Community College
John Beaver, University of Wisconsin at Fox Valley
Priscilla J. Benson, Wellesley College
Eric Carlson, Wake Forest University
Supriya Chakrabarti, Boston University
Dipak Chowdhury, Indiana University–Purdue University at Fort Wayne
Robert Egler, North Carolina State University at Raleigh
Robert A. Fesen, Dartmouth College
Sidney Freudenstein, Metropolitan State College of Denver
Richard Gelderman, Western Kentucky University
Richard Gray, Appalachian State University
David Griffiths, Oregon State University
David Grinspoon, University of Colorado
Jim Hamm, Big Bend Community College
Charles Hartley, Hartwick College
Joe Heafner, Catawba Valley Community College
Richard Ignace, University of Iowa
Bruce Jakosky, University of Colorado
Kurtis Koll, Cameron University
Kristine Larsen, Central Connecticut State University
Larry Lebofsky, University of Arizona
Patrick Lestrade, Mississippi State University
David M. Lind, Florida State University
Michael LoPresto, Henry Ford Community College
William R. Luebke, Modesto Junior College
Marie Machacek, Massachusetts Institute of Technology

Marles McCurdy, Tarrant County College
Steven Majewski, University of Virginia
Phil Matheson, Salt Lake Community College
Barry Metz, Delaware County Community College
John P. Oliver, University of Florida
Jorge Piekarewicz, Florida State University
Harrison B. Prosper, Florida State University
Christina Reeves-Shull, Richland College
Elizabeth Roettger, DePaul University
John Safko, University of South Carolina
James A. Scarborough, Delta State University
Joslyn Schoemer, Denver Museum of Nature and Science
James Schombert, University of Oregon
Gregory Seab, University of New Orleans
Mark H. Slovak, Louisiana State University
Dale Smith, Bowling Green State University
John Spencer, Lowell Observatory
Darryl Stanford, City College of San Francisco
John Stolar, West Chester University
Jack Sulentic, University of Alabama
C. Sean Sutton, Mount Holyoke College
Beverley A. P. Taylor, Miami University
Donald M. Terndrup, Ohio State University
David Trott, Metro State College
Darryl Walke, Raritan Valley Community College
Fred Walter, State University of New York, Stony Brook
Mark Whittle, University of Virginia
Jonathan Williams, University of Florida
J. Wayne Wooten, Pensacola Junior College
Arthur Young, San Diego State University
Dennis Zaritsky, University of California, Santa Cruz

Historical Accuracy Reviewer—Owen Gingerich, Harvard–Smithsonian

In addition, we thank the following colleagues who helped us clarify technical points or checked the accuracy of technical discussions in the book:

Thomas Ayres, University of Colorado
Cecilia Barnbaum, Valdosta State University
Rick Binzel, Massachusetts Institute of Technology
Howard Bond, Space Telescope Science Institute
Humberto Campins, University of Florida
Robin Canup, Southwest Research Institute
Mark Dickinson, Space Telescope Science Institute
Jim Dove, Metropolitan State College of Denver
Harry Ferguson, Space Telescope Science Institute
Andrew Hamilton, University of Colorado
Todd Henry, Georgia State University
Dave Jewitt, University of Hawaii
Hal Levison, Southwest Research Institute
Mario Livio, Space Telescope Science Institute

Mark Marley, New Mexico State University
Kevin McLin, University of Colorado, Boulder
Rachel Osten, University of Colorado, Boulder
Bob Pappalardo, Brown University
Michael Shara, American Museum of Natural History
Glen Stewart, University of Colorado
John Stolar, West Chester University
Dave Tholen, University of Hawaii
Nick Thomas, MPI/Lindau (Germany)
John Weiss, University of Colorado, Boulder
Don Yeomans, Jet Propulsion Laboratory

Finally, we thank the many people who have greatly influenced our outlook on education and our perspective on the universe over the years, including Tom Ayres, Fran Bagenal, Forrest Boley, Robert A. Brown, George Dulk, Erica Ellingson, Katy Garmany, Jeff Goldstein, David Grinspoon, Don Hunten, Catherine McCord, Dick McCray, Dee Mook, Cheri Morrow, Charlie Pellerin, Carl Sagan, Mike Shull, John Spencer, and John Stocke.

Jeff Bennett
Megan Donahue
Nick Schneider
Mark Voit

HOW TO SUCCEED IN YOUR ASTRONOMY COURSE

Most readers of this book are enrolled in a college course in introductory astronomy. If you are one of these readers, we offer you the following hints to help you succeed in your astronomy course.

Using This Book

Before we address general strategies for studying, here are a few guidelines that will help you use *this* book most effectively.

- Read assigned material twice.

  - Make your first pass *before* the material is covered in class. Use this pass to get a "feel" for all the material and to identify concepts that you may want to ask about in class.

  - Read the material for the second time shortly after it is covered in class. This will help solidify your understanding and allow you to make notes that will help you study for exams later.

- Take advantage of the features that will help you study.

  - Always read the *Time Out to Think* features, and use them to help you absorb the material and to identify areas where you may have questions.

  - Use the *cross-references* indicated in brackets in the text and the *glossary* at the back of the book to find more information about terms or concepts that you don't recall.

  - Go to The Astronomy Place at **www.astronomyplace.com** to find additional study aids, including tutorials and quizzes with answers.

- It's your book, so don't be afraid to make notes in it that will help you study later.

  - Don't highlight—<u>underline!</u> Using a pen or pencil to <u>underline</u> material requires greater care than highlighting and therefore helps keep you alert as you study. And be selective in your underlining—for purposes of studying later, it won't help if you underlined everything.

  - There's plenty of "white space" in the margins and elsewhere, so use it to make notes as you read. Your own notes will later be very valuable when you are doing homework or studying for exams.

- After you complete the reading, and again when you study for exams, make sure you can answer the *Review Questions* at the end of each chapter. If you are having difficulty with a review question, reread the relevant portions of the chapter until the answer becomes clear.

Budgeting Your Time

One of the easiest ways to ensure success in any college course is to make sure you budget enough time for studying. A general rule of thumb for college classes is that you should expect to study about 2 to 3 hours per week *outside* of class for each unit of credit. For example, based on this rule of thumb, a student taking 15 credit hours should expect to spend 30 to 45 hours each week studying outside of class. Combined with time in class, this works out to a total of 45 to 60 hours spent on academic work—not much more than the time a typical job requires, and you get to choose your own hours. Of course, if you are working while you attend school, you will need to budget your time carefully.

As a rough guideline, your studying time in astronomy might be divided as shown in the table at the top of p. xix. If you find that you are spending

If Your Course Is:	Time for Reading the Assigned Text (per week)	Time for Homework Assignments (per week)	Time for Review and Test Preparation (average per week)	Total Study Time (per week)
3 credits	2 to 4 hours	2 to 3 hours	2 hours	6 to 9 hours
4 credits	3 to 5 hours	2 to 4 hours	3 hours	8 to 12 hours
5 credits	3 to 5 hours	3 to 6 hours	4 hours	10 to 15 hours

fewer hours than these guidelines suggest, you can probably improve your grade by studying more. If you are spending more hours than these guidelines suggest, you may be studying inefficiently; in that case, you should talk to your instructor about how to study more effectively.

General Strategies for Studying

- Don't miss class. Listening to lectures and participating in discussions is much more effective than reading someone else's notes. Active participation will help you retain what you are learning.

- Budget your time effectively. An hour or two each day is more effective, and far less painful, than studying all night before homework is due or before exams.

- If a concept gives you trouble, do additional reading or studying beyond what has been assigned. And if you still have trouble, ask for help: You surely can find friends, colleagues, or teachers who will be glad to help you learn.

- Working together with friends can be valuable in helping you understand difficult concepts. However, be sure that you learn *with* your friends and do not become dependent on them.

- Be sure that any work you turn in is of *collegiate quality*: neat and easy to read, well organized, and demonstrating mastery of the subject matter. Although it takes extra effort to make your work look this good, the effort will help you solidify your learning and is also good practice for the expectations that future professors and employers will have.

Preparing for Exams

- Study the review questions, and rework problems and other assignments; try additional questions to be sure you understand the concepts. Study your performance on assignments, quizzes, or exams from earlier in the term.

- Try the quizzes available on the text Web site at **www.astronomyplace.com**

- Study your notes from lectures and discussions. Pay attention to what your instructor expects you to know for an exam.

- Reread the relevant sections in the textbook, paying special attention to notes you have made on the pages.

- Study individually *before* joining a study group with friends. Study groups are effective only if every individual comes prepared to contribute.

- Don't stay up too late before an exam. Don't eat a big meal within an hour of the exam (thinking is more difficult when blood is being diverted to the digestive system).

- Try to relax before and during the exam. If you have studied effectively, you are capable of doing well. Staying relaxed will help you think clearly.

ABOUT THE AUTHORS

Jeffrey Bennett received a B.A. in biophysics from the University of California at San Diego in 1981 and a Ph.D. in astrophysics from the University of Colorado in 1987. His thesis research focused on Sun-like stars, but he now specializes in mathematics and science education. He has extensive teaching experience at the elementary and secondary levels and has taught more than fifty college courses in astronomy, physics, mathematics, and education. He led the development of the Colorado Scale Model Solar System and proposed and served as a co-principal investigator on the Voyage scale model solar system (pictured here) for the National Mall. Besides his astronomy textbooks, he has written textbooks in mathematics (Bennett and Briggs, *Using and Understanding Mathematics,* Addison Wesley, 2002) and statistics (Bennett, Briggs, & Triola, *Statistical Reasoning for Everyday Life,* Addison Wesley, 2001), and a popular book (*On the Cosmic Horizon,* Addison Wesley, 2001). He is currently working on a children's book called *Max Goes to the Moon.* When not working, he enjoys participating in masters swimming and hiking the trails of Boulder, Colorado, with his family.

Megan Donahue is an astronomer at the Space Telescope Science Institute in Baltimore, Maryland. After growing up in rural Nebraska, she obtained an S.B. in physics from the Massachusetts Institute of Technology in 1985 and a Ph.D. in astrophysics from the University of Colorado in 1990. Her thesis on intergalactic gas and clusters of galaxies won the Robert J. Trumpler Award (1993). She continued her research as a Carnegie Fellow at the Observatories of the Carnegie Institution in Pasadena, California, and later was an Institute Fellow at the Space Telescope Science Institute. She is an active observer, using ground-based telescopes, the Hubble Space Telescope, and orbiting X-ray telescopes. Her research focuses on questions of galaxy evolution, the nature of intergalactic space, large-scale structure formation, and dark matter and the fate of the universe. She married Mark Voit while in graduate school and they are the parents of two children, Michaela and Sebastian.

Nicholas Schneider is an associate professor in the Department of Astrophysical and Planetary Sciences at the University of Colorado and a researcher in the Laboratory for Atmospheric and Space Physics. He received his B.A. in physics and astronomy from Dartmouth College in 1979 and his Ph.D. in planetary science from the University of Arizona in 1988. In 1991, he received the National Science Foundation's Presidential Young Investigator Award. His research interests include planetary atmospheres and planetary astronomy, with a focus on the odd case of Jupiter's moon Io. He enjoys teaching at all levels and is active in efforts to improve undergraduate astronomy education. Off the job, he enjoys exploring the outdoors with his family and figuring out how things work.

Mark Voit is an astronomer in the Office of Public Outreach at the Space Telescope Science Institute. He earned his A.B. in physics at Princeton University in 1983 and his Ph.D. in astrophysics at the University of Colorado in 1990. He continued his studies at the California Institute of Technology, where he was Research Fellow in theoretical astrophysics. NASA then awarded him a Hubble Fellowship, under which he conducted research at the Johns Hopkins University. His research interests range from interstellar processes in our own galaxy to the clustering of galaxies in the early universe. Occasionally he escapes to the outdoors, where he and his wife, Megan Donahue, enjoy running, hiking, orienteering, and playing with their children. Mark is also author of the popular book *Hubble Space Telescope: New Views of the Universe*.

PART I
Developing Perspective

CHAPTER 1

Our Place in the Universe

Far from city lights on a clear night, you can gaze upward at a sky filled with
stars. If you lie back and watch for a few hours, you will observe the stars
marching steadily across the sky. Confronted by the seemingly infinite heavens,
you might wonder how the Earth and the universe came to be. With these
thoughts, you will be sharing an experience common to humans around the
world and in thousands of generations past.

Remarkably, modern science offers answers to many fundamental
questions about the universe and our place within it. We now know the basic
content and scale of the universe. We know the age of the Earth and the
approximate age of the universe. And, although much remains to be discovered,
we are rapidly learning how the simple constituents of the early universe developed into the incredible diversity of life on Earth.

In this first chapter, we will survey the content and history of the
universe, the scale of the universe, and the motions of the Earth in our universe. We'll thereby develop a "big picture" perspective on our place in the
universe that will provide a base on which we can build a deeper understanding
in the rest of the book. You may wish to begin by studying the two paintings
that appear in the first few pages of the chapter, summarizing (respectively) our
cosmic address and cosmic origins.

the Local Supercluster

the Local Group

FIGURE 1.1 Our place in the universe. This painting illustrates our cosmic address: The Earth is one of nine planets in our solar system; our solar system is one among more than 100 billion star systems in the Milky Way Galaxy; the Milky Way is one of the two largest of about 30 galaxies in the Local Group; the Local Group lies near the outskirts of the Local Supercluster; and the Local Supercluster fades into the background painting of structure throughout the universe.

Earth

the Milky Way Galaxy

the solar system
(not to scale)

1.1 A Modern View of the Universe

If you observe the sky carefully, you can see why most of our ancestors believed that the heavens revolved about the Earth. The Sun, Moon, planets, and stars appear to circle around our sky each day, and we cannot feel the constant motion of the Earth as it rotates on its axis and orbits the Sun. Thus, it seems quite natural to assume that we live in an Earth-centered, or *geocentric,* universe.

Nevertheless, we now know that the Earth is a planet orbiting a rather average star in a vast cosmos. (In astronomy, the term *cosmos* is synonymous with *universe.*) The historical path to this knowledge was long and complex, involving the dedicated intellectual efforts of thousands of individuals. In later chapters, we'll encounter many of these individuals and explore how their discoveries changed human understanding of the universe. First, however, it's useful to have at least a general picture of the universe as we know it today.

Our Cosmic Address

Take a look at Figure 1.1 on the preceding pages. Going counterclockwise from Earth, this painting illustrates the basic levels of structure that describe what we might call our "cosmic address."

Earth is a planet in our **solar system**, which consists of the Sun and all the objects that orbit it: nine planets and their moons, the chunks of rock that we call asteroids, the balls of ice that we call comets, and countless tiny particles of interplanetary dust.

Our Sun is a star, just like the stars we see in our night sky. The Sun and all the stars we can see with the naked eye make up only a small part of a huge, disk-shaped collection of stars called the **Milky Way Galaxy**. A galaxy is a great island of stars in space, containing from a few hundred million to a trillion or more stars. The Milky Way Galaxy is relatively large, containing more than 100 billion stars. Our solar system is located a little over halfway from the galactic center to the edge of the galactic disk.

Many galaxies congregate in groups; groups that contain more than a few dozen galaxies are called **galaxy clusters**. Our Milky Way belongs to a group of 30 or so galaxies called the **Local Group**.

On the largest scale, the universe has a frothlike appearance in which galaxies and galaxy clusters are loosely arranged in giant chains and sheets. In some places the galaxies and galaxy clusters are more tightly packed than in others; we call these giant structures **superclusters**. The supercluster to which our Local Group belongs is called, not surprisingly, the **Local Supercluster**. Between these vast structures lie huge voids containing few, if any, galaxies.

Basic Astronomical Definitions

Star A large, glowing ball of gas that generates energy through nuclear fusion in its core. The term *star* is sometimes applied to objects that are in the process of becoming true stars (e.g., protostars) and to the remains of stars that have died (e.g., neutron stars).

Planet An object that orbits a star and that, while much smaller than a star, is relatively large in size. There is no "official" minimum size for a planet, but the nine planets in our solar system are all at least 2,000 kilometers in diameter. Planets may be rocky, icy, or gaseous in composition, and they shine primarily by reflecting light from their star.

Moon (or *satellite*) An object that orbits a planet. The term *satellite* is also used more generally to refer to any object orbiting another object.

Asteroid A relatively small, rocky object that orbits a star. Asteroids are sometimes called *minor planets* because they are similar to planets but smaller.

Comet A relatively small, icy object that orbits a star.

Star system One or more stars and any planets or other material that orbits them. The term *solar system* refers to our own star system (*solar* means "of the Sun"). Many star systems contain two or more stars that orbit each other; two-star systems are called *binary star systems.*

Star cluster A group of stars (ranging in number from a few hundred to a few million) that are closely associated in space. In general, all the stars in a star cluster were born at roughly the same time from the same cloud of interstellar gas.

Galaxy A great island of stars in space, containing from a few hundred million to a trillion or more stars.

Galaxy cluster A collection of galaxies held together by gravity. Small clusters are generally called *groups* (such as our Local Group), with the term *cluster* reserved for collections of hundreds or thousands of galaxies.

Supercluster A giant structure encompassing many groups and clusters of galaxies.

Universe (or *cosmos*) The sum total of all matter and energy. By current understanding, only part of the universe is observable to us even in principle; we call this our *observable universe.*

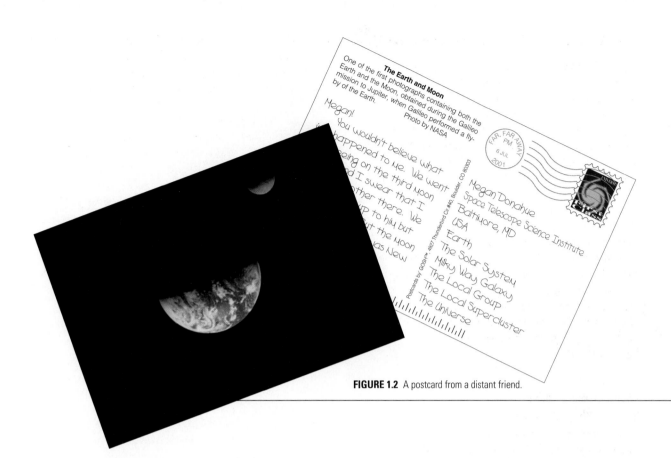

The Earth and Moon

One of the first photographs containing both the Earth and the Moon, obtained during the Galileo mission to Jupiter, when Galileo performed a fly-by of the Earth.
Photo by NASA

Megan!
You wouldn't believe what
happened to me. We went
seeing on the third moon
and I swear that I
other there. We
up to him but
the moon
was new

FAR, FAR AWAY
PM
6 JUL
2001

Megan Donahue
Space Telescope Science Institute
Baltimore, MD
USA
Earth
The Solar System
Milky Way Galaxy
The Local Group
The Local Supercluster
The Universe

Postcards by GOSH™ 4927 Thunderbird Cat #40 Boulder, CO 80303

FIGURE 1.2 A postcard from a distant friend.

Finally, the **universe** is the sum total of all matter and energy. That is, it encompasses the superclusters and voids, and everything within them. To review the different levels of structure in the universe, you might imagine how a faraway friend would address a postcard to Earth (Figure 1.2).

TIME OUT TO THINK *Some people think that our tiny physical size in the vast universe makes us insignificant. Others think that our ability to learn about the wonders of the universe gives us significance despite our small size. What do you think?*

Our Cosmic Origins

How did we come to be? Much of the rest of this text discusses the scientific evidence concerning our cosmic origins, and we'll see that humans are relative newcomers in an old universe. For now, let's look at a quick overview of the scientific story of creation, as summarized in Figure 1.3.

As we'll discuss shortly, telescopic observations of distant galaxies show that the entire universe is *expanding*. That is, average distances between galaxies are increasing with time. If the universe is expand-

ing, everything must have been closer together in the past. From the observed rate of expansion, astronomers estimate that the expansion started somewhere between 12 billion and 16 billion years ago. Astronomers call this beginning the **Big Bang**.

The universe as a whole has continued to expand ever since the Big Bang, but on smaller scales the force of gravity has drawn matter together. Structures such as galaxies and clusters of galaxies represent places where gravity has won out against the overall expansion. That is, while the universe as a whole continues to expand, individual galaxies and their contents do *not* expand.

Within galaxies, gravity drives the collapse of clouds of gas and dust, forming stars and planets. Stars have limited lifetimes, and when they die they may spew much of their content back into interstellar space. This material can then become part of new clouds of gas, which gravity collapses into new star systems. Thus, galaxies function as cosmic recycling plants, recycling material expelled from dying stars into subsequent generations of stars and planets.

Not only is material recycled, but its chemical nature also changes with time. All the matter in the universe was created in the Big Bang, but the early universe contained only the simplest chemical elements: hydrogen and helium (and trace amounts of lithium). All other elements were manufactured by

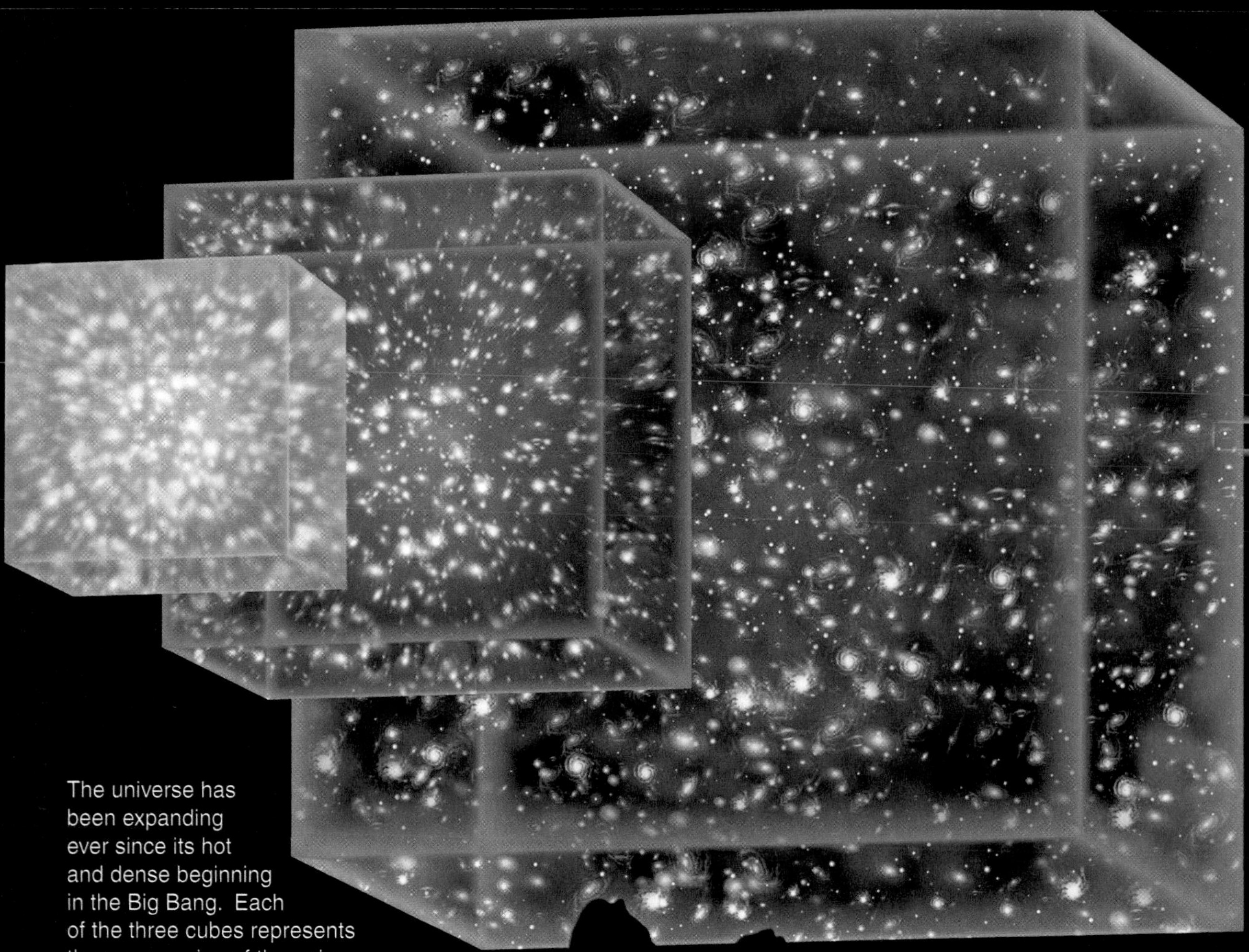

The universe has
been expanding
ever since its hot
and dense beginning
in the Big Bang. Each
of the three cubes represents
the same region of the universe,
showing how the region expands with time.

FIGURE 1.3 Our cosmic origins: All the matter and energy in the universe
was created in the Big Bang. This sequence of paintings shows the progression of that matter and energy from the Big Bang to human life. Note that the
elements from which we are made were produced in stars that shined long
ago, and these elements formed the Earth, thanks to the recycling role played
by our galaxy.

The Earth was built with elements
produced in stars that lived and died
in the Milky Way before our solar system
formed.

A few billion years after the Big Bang, caused local concentrations of matter apse into galaxies even while the universe hole continued to expand.

Galaxies like the Milky Way act as cosmic recyc Stars are made from the material in clouds of ga within the galaxy, and stars return material to int space when they die.

A star forms at the center of a c cloud of gas and dust, and plan form in the spinning disk that su the young star.

Massive stars explode when they die, scattering the elements they've roduced into space.

Stars shine with the energy produ by nuclear fusion in their cores; th fusion also creates heavier eleme

stars through **nuclear fusion**, in which lightweight elements fuse (i.e., their nuclei join together) to form heavier elements. The energy released by nuclear fusion, which occurs deep in stellar cores, is what makes stars shine. For most of their lives stars shine by fusing hydrogen into helium, but near the ends of their lives the more massive stars generate energy with advanced fusion reactions that produce elements such as carbon, oxygen, nitrogen, and iron. Many of these massive stars die in titanic explosions that release these heavy elements into space, where they mix with other gas and dust in the galaxy to be incorporated into later generations of star systems.

The processes of heavy-element production and cosmic recycling had already been taking place for several billion years by the time our solar system formed, about 4.6 billion years ago. The cloud that gave birth to our solar system was about 98% hydrogen and helium; the other 2% contained all the other chemical elements. The small rocky planets of our solar system, including the Earth, were made from a small part of this 2%. We do not know exactly how the elements on the Earth's surface developed into the first forms of life, but we do know that life was already flourishing on Earth more than 3.5 billion years ago. Biological evolution took over once life arose, leading to the great diversity of life on Earth today.

In summary, all the material from which we and the Earth are made (except hydrogen and most helium) was created inside stars that died before the birth of our Sun. We are intimately connected to the stars because we are products of stars. In the words of astronomer Carl Sagan (1934–96), we are "star stuff."

Images of Time

We study the universe by studying light from distant stars and galaxies. Light travels extremely fast by earthly standards: The speed of light is 300,000 kilometers per second, a speed at which it would be possible to circle the Earth nearly eight times in just 1 second. Nevertheless, even light takes a substantial amount of time to travel the vast distances in space. For example, light takes about 1 second to reach the Earth from the Moon and about 8 minutes to reach the Earth from the Sun. Light from the stars takes many years to reach us, so we measure distances to the stars in units called **light-years**. One light-year is the distance that light can travel in 1 year—about 10 trillion kilometers, or 6 trillion miles. (You can calculate this distance by multiplying the speed of light by the number of seconds in one

year.) Note that a light-year is a unit of *distance,* not of time.

The brightest star in the night sky, Sirius, is about 8 light-years from our solar system, which means that it takes light from Sirius about 8 years to reach us. Thus, when we look at Sirius, we see light that left the star about 8 years ago. The Orion Nebula, a star-forming region visible to the naked eye as a small, cloudy patch in the sword of the constellation Orion, lies about 1,500 light-years from Earth. Thus, we see the Orion Nebula as it looked about 1,500 years ago—about the time of the fall of the Roman Empire. If any major events have occurred in the Orion Nebula since that time, we cannot yet know about them because the light from these events would not yet have reached us.

Because it takes time for light to travel through space,

the farther away we look in distance, the further back we look in time.

If we look at a galaxy that lies 10 million light-years away, we see it as it was 10 million years ago. If we observe a distant cluster of galaxies that lies 1 billion light-years away, we see the cluster as it was 1 billion years ago.

Ultimately, the speed of light limits the portion of the universe that we can see. For example, if the universe is 12 billion years old, then light from galaxies more than 12 billion light-years away would not have had time to reach us. We would say that the **observable universe** extends 12 billion light-years in all directions from Earth. (As we'll discuss in Chapter 17, this explanation is somewhat oversimplified, but it captures the basic point that our *observable* universe may not be the *entire* universe.)

It is amazing to realize that any "snapshot" of a distant galaxy or cluster of galaxies is a picture of both space and time. For example, the Great Galaxy in Andromeda, also known as M31, lies about 2.5 million light-years from Earth. Figure 1.4, therefore, is a picture of how M31 looked about 2.5 million years ago, when early humans were first walking the Earth. Moreover, the diameter of M31 is about 100,000 light-years, so light from the far side of the galaxy required 100,000 years more to reach us than light from the near side. Thus, the picture of M31 shows 100,000 years of time. This single photograph captured light that left the near side of the galaxy some 100,000 years later than the light it captured from the far side. When we study the universe, it is impossible to separate space and time.

FIGURE 1.4 M31, the Great Galaxy in Andromeda, is about 2.5 million light-years away, so this photo captures light that traveled through space for 2.5 million years to reach us. Because the galaxy is 100,000 light-years in diameter, the photo also captures 100,000 years of time in M31: We see the galaxy's near side as it looked 100,000 years later than the time at which we see the far side.

1.2 The Scale of the Universe

We've given some numbers in our description of the size and age of the universe, but these numbers probably have little meaning for you—after all, they are literally astronomical. In this section, we will try to give meaning to incredible cosmic distances and times.

A Walking Tour of the Solar System

One of the best ways to develop perspective on cosmic sizes and distances is to start with a scale model of our solar system, such as the Voyage scale model solar system on the National Mall in Washington, DC (Figure 1.5). The model scale is 1 to 10 billion; that is, diameters and distances in the model

FIGURE 1.5 This photograph shows the monuments for the Sun (gold sphere on rightmost pedestal) and inner planets in the Voyage scale model solar system on the National Mall in Washington, DC. The building at the left is the National Air and Space Museum. The planets of the outer solar system can be found farther along the walkway. Voyage was created for the National Mall and other locations around the world by the Challenger Center for Space Science Education, the Smithsonian Institution, and NASA.

are *one ten-billionth* (10^{-10}) of actual diameters and distances in the solar system (Table 1.1). Figure 1.6 shows the sizes of the Sun and the planets on this scale.

Pages 12 through 23 take you on a virtual walk through the Voyage model solar system. Each page represents one stop, starting from the Sun and continuing to each of the planets. We also include stops for asteroids and comets. After you complete the 12-page tour, be sure that you have noticed the following key ideas about the scale of the solar system:

■ The Sun is roughly the size of a grapefruit, while the planets range in size from dust-speck-size Pluto to marble-size Jupiter. Earth is about the size of a pinhead.

■ The inner planets (Mercury, Venus, Earth, Mars) are all located within a couple of dozen steps of

the Sun, with the Earth located 15 meters from the Sun. The outer planets (Jupiter, Saturn, Uranus, Neptune, Pluto) are spread much farther apart, with Pluto located about 600 meters (just over $\frac{1}{3}$ mile) from the Sun. Walking the full length of the model would take about 10 minutes, not including stops.

■ Perhaps the most striking feature of the solar system is its *emptiness*. Although our model shows the planets laid out in a straight line from the Sun, a better model would show their orbits extending all the way around the Sun. Such a model would require an area measuring well over a kilometer (0.6 mile) on a side, equivalent to more than 300 football fields! Aside from the grapefruit-size Sun at the center, all we would find in the rest of this area would be the nine

Table 1.1 Solar System Sizes and Distances, 1-to-10-Billion Scale

Object	Real Diameter	Real Distance from Sun (average)	Model Diameter	Model Distance from Sun
Sun	1,392,500 km	—	139 mm = 13.9 cm	—
Mercury	4,880 km	57.9 million km	0.5 mm	6 m
Venus	12,100 km	108.2 million km	1.2 mm	11 m
Earth	12,760 km	149.6 million km	1.3 mm	15 m
Mars	6,790 km	227.9 million km	0.7 mm	23 m
Jupiter	143,000 km	778.3 million km	14.3 mm	78 m
Saturn	120,000 km	1,427 million km	12.0 mm	143 m
Uranus	52,000 km	2,870 million km	5.2 mm	287 m
Neptune	48,400 km	4,497 million km	4.8 mm	450 m
Pluto	2,260 km	5,900 million km	0.2 mm	590 m

planets and their moons, plus scattered asteroids and comets—no object on this scale is larger than marble-size Jupiter, and most are far smaller than pinhead-size Earth.

If you are having difficulty visualizing the model, it's easy to make your own model. Use a grapefruit or similar-size object to represent the Sun and a pinhead to represent the Earth. In a long hallway or outdoors, place your Sun on a chair or table and place your tiny Earth about 15 meters from your Sun. Once you have these objects in place, you may find it easier to visualize the sizes and locations for the rest of the planets (using the information in Table 1.1).

(Text continues on p. 24, following the 12-page virtual tour.)

FIGURE 1.6 The sizes of the Sun and the planets on the 1-to-10-billion scale.

The Sun

A visible light photograph of the Sun's surface. The dark splotches are sunspots—each large enough to swallow several Earths.

This photograph, from the *SOHO* spacecraft, shows a huge streamer of hot gas on the Sun; the image of the Earth was added for size comparison. The photograph was taken with a camera sensitive to ultraviolet light from the sun; the features shown would have been nearly invisible to our eyes.

We start our tour at the Sun, the central object of our solar system, which is about the size of a grapefruit in our model. The Sun is by far the largest object in our solar system and the only object easily visble anywhere within the model. In fact, the Sun contains about 99.9% of the solar system's total mass; that is, it outweighs everything else in the solar system combined by a factor of one thousand.

Although the Sun's surface looks solid in photographs, it is a roiling sea of hot (about 6,000°C, or 10,000°F) hydrogen and helium gas. The surface is speckled with sun spots that appear dark in photographs only because they are slightly cooler than their surroundings. Solar storms sometimes send streamers of hot gas soaring far above the surface and can disrupt radio communications on Earth and disable orbiting satellites.

The Sun is gaseous throughout. If you could plunge beneath the Sun's surface, you'd find ever higher temperatures as you went deeper. You'd find the ultimate source of the Sun's energy deep in its core, where the temperatures are so high that the Sun becomes a natural nuclear-fusion power plant. Each second, fusion transforms about 600 million tons of the Sun's hydrogen into 596 million tons of helium; the "missing" 4 million tons becomes energy in accord with Einstein's famous equation $E = mc^2$. Despite losing 4 million tons of mass each second, the Sun contains so much hydrogen that it has shone steadily for almost 5 billion years already and will continue to shine for some 5 billion years to come.

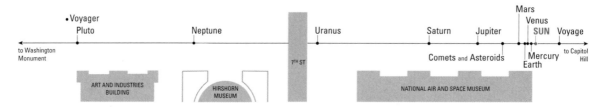

This map shows the Sun's location in the Voyage scale model solar system on the National Mall in Washington, DC. The Sun itself is about the size of a grapefruit on this scale.

This image shows what it would look like to be orbiting a few hundred kilometers above Mercury's surface with your back toward the Sun. Among the stars, you can see the Earth and the Moon as the blue speck and its tiny companion. The view was created using imagery from NASA's *Mariner 10* spacecraft but with computer manipulation to provide color and the orbital viewpoint. The inset (right) is a composite photograph of the full disk of Mercury by *Mariner 10;* the blank strip at the upper right was not photographed. (Image above from the Voyage scale model solar system, developed by Challenger Center for Space Science Education, the Smithsonian Institution, and NASA. Image created by ARC Science Simulations © 2001.)

The innermost planet, Mercury, lies just a few steps from the model Sun. Mercury is the smallest planet except for Pluto, and in our model it is only about as big as the period at the end of this sentence. Mercury is a desolate, cratered world with no active volcanoes, no earthquakes, no wind, no rain, and no life. Because there is virtually no air to scatter sunlight or color the sky, you could see stars even in the daytime if you stood on Mercury with your back toward the Sun. You wouldn't want to stay long, however, because the ground on Mercury's day side is nearly as hot as hot coals (about 425°C). Nighttime would not be much more comfortable: With no atmosphere to retain heat during the long nights (which last about 3 months), temperatures plummet below −150°C (about −240°F)—far colder than Antarctica in winter.

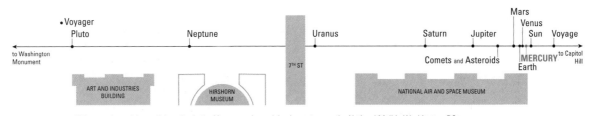

This map shows Mercury's location in the Voyage scale model solar system on the National Mall in Washington, DC.
The dot at the top of the page (next to title) shows Mercury's size on the scale.

Venus

This image shows a realistic picture of the surface of Venus. The surface topography is based on actual data from NASA's *Magellan* spacecraft, and the colors represent what scientists believe our eyes would see on Venus. The inset (left) shows the full disk of Venus as photographed by NASA's *Pioneer Venus Orbiter*. This photograph was taken with cameras sensitive to ultraviolet light; with visible light, cloud features cannot be distinguished from the general haze. (Image above from the Voyage scale model solar system, developed by Challenger Center for Space Science Education, the Smithsonian Institution, and NASA. Image by David P. Anderson, Southern Methodist University © 2001.)

We find the second planet from the Sun, Venus, just a few steps beyond Mercury in our model. Because Venus and the Earth are nearly identical in size—both pinheads in the model solar system—Venus is sometimes called our sister planet.

Venus is covered in dense clouds, and because it is not much closer to the Sun than the Earth science fiction writers once speculated that Venus might be a lush, tropical paradise. We now know that an extreme *greenhouse effect* bakes its surface to an incredible 450°C (about 850°F) and traps heat so

effectively that nighttime offers no relief—day and night, Venus is hotter than a pizza oven. All the while, the thick atmosphere bears down on the surface with a pressure equivalent to that nearly a kilometer (0.6 mile) beneath the ocean surface on Earth. Besides the crushing pressure and searing temperature, a visitor to Venus would feel the corrosive effects of sulfuric acid and other toxic chemicals in its atmosphere. Far from being a beautiful sister planet to Earth, Venus resembles a traditional view of hell.

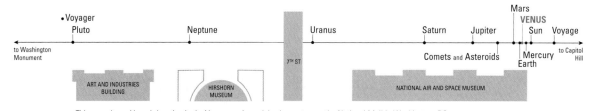

This map shows Venus's location in the Voyage scale model solar system on the National Mall in Washington, DC. The dot at the top of the page (next to title) shows Venus's size on the scale.

Earth

This image, computer generated from satellite data, shows the striking contrast between the daylight and nighttime hemispheres of Earth. The day side reveals little evidence of human presence, but at night our presence is revealed by the lights of human activity (mostly from cities as well as from agricultural, oil, and gas fires). (From the Voyage scale model solar system, developed by Challenger Center for Space Science Education, the Smithsonian Institution, and NASA. Image created by ARC Science Simulations © 2001.)

This famous photograph shows astronaut Buzz Aldrin standing on the Moon in July 1969 during the Apollo 11 mission. The reflection in his visor shows Neil Armstrong and the lunar module (spacecraft), along with some scientific equipment. Aldrin and Armstrong were the first of 12 humans who walked on the Moon; Armstrong took the first step, saying, "That's one small step for [a] man, one giant leap for mankind."

Beyond Venus and only about 15 meters from the model Sun, we find our home planet. Its tiny size in a big solar system reminds us that Earth is a rare and precious oasis of life. It is the only planet with oxygen for us to breathe and ozone to shield us from deadly solar radiation. Only on Earth does water flow freely to nurture life. And temperatures are ideal for life because Earth's atmosphere contains just enough carbon dioxide and water vapor to maintain a moderate greenhouse effect.

Despite its small size, Earth is striking in its beauty. Blue oceans cover nearly three-fourths of the surface, broken by the continental land masses and scattered islands. The polar caps are white with snow and ice, and white clouds are scattered above the surface. At night, the glow of artificial lights clearly reveals the presence of an intelligent civilization.

Earth is the first planet on our tour with a moon. On our model scale, the Moon is a barely visible speck that orbits the Earth at a distance of about 4 centimeters. Thus, your thumb could easily cover both the Earth and the Moon. The Moon is cratered and desolate, much like Mercury, though it has its own interesting history and geology. It is also the only world besides Earth on which humans have ever stepped. From July 1969 through December 1972, NASA's Apollo program successfully landed 12 people (six crews of two) on the lunar surface.

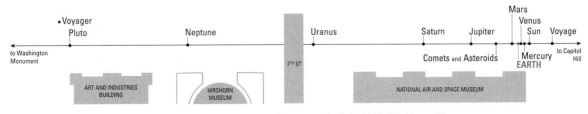

This map shows Earth's location in the Voyage scale model solar system on the National Mall in Washington, DC. The dot at the top of the page (next to title) shows Earth's size on the scale.

Mars •

A few more steps take us to the model Mars, which is about half the size (diameter) of Earth. Mars has two tiny moons, Phobos and Deimos, but they are so small as to be microscopic on our 1-to-10-billion scale.

Mars is a world of wonders, with extinct volcanoes that dwarf the largest mountains on Earth, a great canyon that runs nearly one-fifth of the way around the planet, and polar caps made of frozen carbon dioxide ("dry ice") and water ice. Although Mars is frozen today, the presence of dried-up riverbeds and rock-strewn floodplains offers clear evidence that Mars was warm and wet sometime in the distant past. Thus, Mars may once have been hospitable for life, though its wet era probably ended at least 3 billion years ago.

Mars looks almost Earth-like in photographs taken by spacecraft on its surface, but you wouldn't want to visit without a space suit. The air pressure is far less than that on top of Mount Everest, the temperature is usually well below freezing, the trace amounts of oxygen would not be nearly enough to

breathe, and the lack of atmospheric ozone would leave you exposed to deadly ultraviolet radiation from the Sun.

Mars is the most studied planet besides Earth. More than a dozen spacecraft have flown past, orbited, or landed on Mars, and plans are in the works for many more. We may even send humans to Mars within our lifetime. Overturning rocks in ancient riverbeds or chipping away at ice in the polar caps, explorers will search for fossil evidence of past life—and perhaps even find a few places where microbes survive today.

The surface of Mars, photographed from NASA's *Pathfinder* lander in 1997; part of the lander is visible in the foreground. The little rover called *Sojourner* is in the upper right, studying a rock that scientists dubbed Yogi. The inset (right) shows the full disk of Mars; the "gash" going horizontally across the center is the giant canyon known as Valles Marineris.

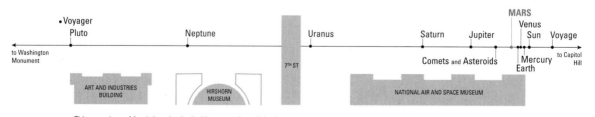

This map shows Mars's location in the Voyage scale model solar system on the National Mall in Washington, DC. The dot at the top of the page (next to title) shows Mars's size on the scale.

The Asteroid Belt

The asteroid Eros, photographed by the *NEAR* spacecraft that orbited it during 2000. At the end of its mission in February 2001, NASA engineers guided *NEAR* to a soft landing on Eros.

As we pass by Mars, we notice one of the most important characteristics of our solar system. In contrast to the few steps separating the four inner planets in the model solar system, the walk from Mars to Jupiter covers more than 50 meters—over half the length of a football field. As we look ahead, we see that increasingly large distances separate the remaining planets. Thus, the solar system has two major regions: the **inner solar system**, where Mercury, Venus, Earth, and Mars orbit the Sun relatively closely, and the **outer solar system**, where the remaining planets are widely separated. In between we find the **asteroid belt**, where thousands of asteroids orbit the Sun.

Asteroids are chunks of rock and metal left over from the formation of our solar system—bits and pieces that never became part of a moon or a planet. The largest asteroid, Ceres, is about 1,000 kilometers (600 miles) across. But most asteroids are far smaller and would be

microscopic on our 1-to-10-billion scale. Despite their large number, the total mass of the asteroids is less than that of any of the inner planets. The average distance between asteroids is millions of kilometers—so the real asteroid belt does not at all resemble the crowded rock fields depicted in many science fiction films.

Not all asteroids are confined to the asteroid belt, and those that orbit elsewhere in the solar system may collide with moons or planets; impact craters are the scars of such collisions. (Impact craters may also result from comet collisions.) Asteroid impacts may have played an important role in the development of life on Earth. For example, an asteroid impact may have been responsible for the extinction of the dinosaurs some 65 million years ago. Smaller pieces of asteroids fall to the Earth quite often; we call these fragments *meteorites*.

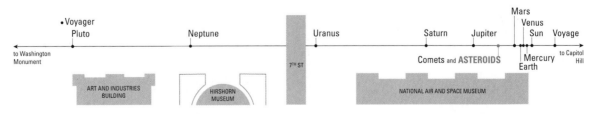

This map shows the Asteroid Belt's location in the Voyage scale model solar system on the National Mall in Washington, DC. Individual asteroids are too small to see on this scale.

17

Jupiter

This image shows a realistic view of what it would look like to be orbiting near Jupiter's moon Io as Jupiter comes into view displaying its Great Red Spot. The extraordinarily dark rings discovered in the Voyager missions are exaggerated to make them visible. This computer visualization was created using data from both NASA's Voyager and Galileo missions. (From the Voyage scale model solar system, developed by Challenger Center for Space Science Education, the Smithsonian Institution, and NASA. Image created by ARC Science Simulations © 2001.)

The model Jupiter is the size of a marble, making it a giant in comparison to the planets of the inner solar system. Indeed, Jupiter is so different from the planets of the inner solar system that we must create an entirely new mental image of the term *planet*. Its mass is more than 300 times that of the Earth, and its volume is more than 1,000 times that of the Earth. Its most famous feature—a long-lived storm called the Great Red Spot—is itself large enough to swallow two or three Earths. Like the Sun, Jupiter is made primarily of hydrogen and helium and has no solid surface. If you plunged deep into Jupiter, you would be crushed by the increasing gas pressure long before you ever reached its core.

Jupiter reigns over at least 28 moons and a thin set of rings (too faint to be seen in most photographs). The four largest moons—Io, Europa, Ganymede, and Callisto (often called the *Galilean moons* because they were discovered by Galileo)—are fascinating worlds in their own right and are easily visible on the scale of the model solar system. Io is the most volcanically active place in the solar system. Europa's icy crust probably hides a subsurface ocean of liquid water, making Europa a promising place to search for life. Indeed, the moons of Jupiter and the other outer planets are at least as interesting as the planets themselves.

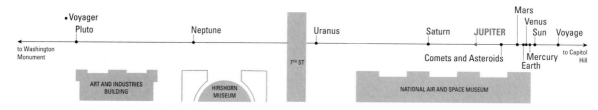

This map shows Jupiter's location in the Voyage scale model solar system on the National Mall in Washington, DC. The image at the top of the page (next to title) shows Jupiter's size on the scale.

18

Saturn

This computer simulation, built upon imagery from the *Voyager 1* mission, recreates the striking view from the spacecraft at somewhat higher detail. We see the shadow of the rings on Saturn's sunlit face, and the rings become lost in Saturn's shadow on the night side. (From the Voyage scale model solar system, developed by Challenger Center for Space Science Education, the Smithsonian Institution, and NASA. Image created by ARC Science Simulations © 2001.)

The distance from Jupiter to Saturn is about 65 meters in our model, or nearly three-fourths the length of a football field; notice how much longer this walk takes than the few steps between the planets of the inner solar system. Saturn is only slightly smaller than Jupiter in size but is considerably less massive (about one-third Jupiter's mass) because it is less dense. Like Jupiter, Saturn is made mostly of hydrogen and helium and has no solid surface.

Saturn is famous for its spectacular rings. Although all four of the giant outer planets have rings, only Saturn's can be seen easily through a small telescope on Earth. The rings may look solid from a distance, but this appearance is deceiving. If you could wander into the rings, you'd find yourself surrounded by countless individual particles of rock and ice, ranging in size from dust grains to city blocks. Each particle orbits Saturn like a tiny moon.

At least 30 moons orbit Saturn. Most are far too small to be visible in our model solar system (though they are still considerably larger than the ring particles). But Saturn's largest moon, Titan, is bigger than the planet Mercury and is blanketed by a thick atmosphere. On Titan's surface, you'd find an atmospheric pressure slightly greater than that on Earth, and you would inhale air with roughly the same nitrogen content as air on Earth. However, you'd need to bring your own oxygen, and you'd certainly want a warm spacesuit to protect yourself against frigid outside temperatures. NASA's *Cassini* spacecraft is currently on the way to Saturn; it will drop a probe to Titan's surface after it arrives in 2004.

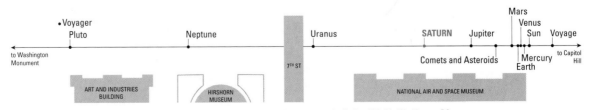

This map shows Saturn's location in the Voyage scale model solar system on the National Mall in Washington, DC. The image at the top of the page (next to title) shows Saturn's size on this scale.

Uranus

To go from Saturn to Uranus in our model, we must walk as far as we've walked in our entire tour so far, again illustrating the vast distances that separate planets in the outer solar system. Uranus is much smaller than either Jupiter or Saturn, but it is still much larger than Earth. It is made largely of hydrogen, helium, and hydrogen compounds such as methane (CH_4); the latter gives Uranus its pale blue-green color. Like the other giants of the outer solar system, it lacks a solid surface. At least 21 moons orbit Uranus, along with a set of rings similar to those of Saturn but much darker and more difficult to see.

The entire Uranus system—planet, rings, and moon orbits—is tipped on its side compared to the rest of the planets. This unusual orientation may be the result of a cataclysmic collision suffered by Uranus as it was forming some 4.6 billion years ago. It also is responsible for the most extreme pattern of seasons on any planet. If you lived on a platform floating in Uranus's atmosphere near the north pole, you'd have continuous daylight for half of each orbit, or 42 years. Then, after a very gradual sunset, you'd be plunged into a 42-year-long night.

Uranus was the first "discovered" planet—all the planets closer to the Sun are easily visible to the naked eye and thus were known to all ancient cultures. English astronomer William Herschel discovered Uranus in 1781. He originally suggested naming the planet *Georgium Sidus*, Latin for "George's star," in honor of his patron, King George III. Fortunately, the idea of "Planet George" never caught on. Instead, many eighteenth- and nineteenth-century astronomers referred to the new planet as "Herschel." The modern name Uranus, after the mythological father of Saturn, was first suggested by one of Herschel's contemporaries, astronomer Johann Bode, and was generally accepted by the mid-nineteenth century.

This image shows a view from a vantage point high above Uranus's moon Ariel. Several other moons are visible in the image, and the thin vertical line is the ring system (actually too dark to see from this vantage point). Computer simulation based upon data from NASA's *Voyager 2* mission. (From the Voyage scale model solar system, developed by Challenger Center for Space Science Education, the Smithsonian Institution, and NASA. Image created by ARC Science Simulations © 2001.)

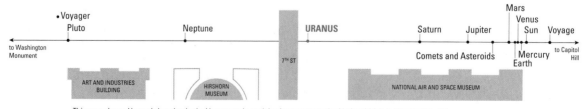

This map shows Uranus's location in the Voyage scale model solar system on the National Mall in Washington, DC. The image at the top of the page (next to title) shows Uranus's size on the scale.

Neptune

This image shows what it would look like to be orbiting Neptune's moon Triton as Neptune itself comes into view. The dark rings are exaggerated to make them visible in this computer simulation using data from NASA's *Voyager 2* mission. (From the Voyage scale model solar system, developed by Challenger Center for Space Science Education, the Smithsonian Institution, and NASA. Image created by ARC Science Simulations © 2001.)

From Uranus, we must cross a distance equivalent to more than one and a half football fields to reach Neptune in our model. Neptune looks nearly like a twin of Uranus, with very similar size and composition, although it is more strikingly blue. Neptune has rings and at least eight moons. Its largest moon, Triton, is unusual both in having a "backward" orbit around Neptune (orbiting in a direction opposite to Neptune's rotation) and in having active sites that spew plumes of gas into the sky. The backward orbit leads astronomers to believe that Triton originally may not have been a moon but instead was captured by Neptune sometime after it formed.

Neptune's discovery was a triumph for the theory of gravity, developed by Isaac Newton in the late 1600s. By the mid-nineteenth century, careful observations of Uranus had shown its orbit to be slightly inconsistent with that predicted by Newton's theory of gravity. In the early 1840s, Englishman John Couch Adams suggested that the inconsistency could be explained by a previously unseen "eighth planet" orbiting the Sun beyond Uranus. By making calculations based on Newton's theory, he even predicted the location of the planet and urged a telescopic search. Unfortunately, Adams was a student at the time and was unable to convince British astronomers to carry out the search.

In the summer of 1846, French astronomer Urbain Leverrier made similar calculations independently. He sent a letter to Johann Galle, of the Berlin Observatory, suggesting a search for the eighth planet. On the night of September 23, 1846, Galle pointed his telescope to the position suggested by Leverrier. There, within 1° of its predicted position, he saw the planet Neptune. Hence, Neptune's discovery truly was made by mathematics and physics and was merely confirmed with a telescope.

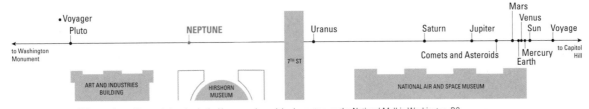

This map shows Neptune's location in the Voyage scale model solar system on the National Mall in Washington, DC. The image at the top of the page (next to title) shows Neptune's size on this scale.

Pluto ·

Pluto and Charon, as photographed by the Hubble Space Telescope. Because no spacecraft has been to Pluto, we do not yet have clear pictures of this tiny planet or its moon.

At its average distance from the Sun, the model Pluto lies another 140 meters beyond Neptune. The grapefruit-size Sun, more than a half kilometer away, appears tiny from here. It is easy to imagine that the world of Pluto must be cold and dark.

One look at the model Pluto, along with its moon Charon, shows it to be out of character with the rest of the planets. Pluto is neither large and gaseous like the other planets of the outer solar system nor rocky like the planets of the inner solar system. Instead, it is very small—the smallest planet by far—and is made mostly of ices. Pluto's orbit is also unusual. Whereas all the other planets orbit the Sun along nearly circular paths and in nearly the same plane, Pluto's orbit is highly elongated and substantially inclined relative to the orbits of the other planets. Pluto actually comes closer to the Sun

than Neptune for 20 years of each 284-year orbit. The last such period ended in 1999, so the next won't begin until 2263.

Although Pluto's characteristics make it a "misfit" among the planets, they give it much in common with the only other inhabitants of the outskirts of the solar system: the balls of ice and dust we call *comets*. In the past decade or so, astronomers have discovered many other Pluto-like objects, though none quite as large as Pluto. Indeed, it now seems likely that Pluto is merely the largest (or one of the largest) among thousands of "giant comets" that orbit the Sun beyond Neptune. Some astronomers have gone so far as to suggest demoting Pluto from its status as a planet, but most favor keeping its popular title as the ninth planet—a title it has held since its discovery in 1930 by Clyde Tombaugh.

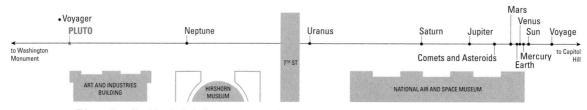

This map shows Pluto's location in the Voyage scale model solar system on the National Mall in Washington, DC. The dot at the top of the page (next to title) shows Pluto's size on the scale.

Comet Hale–Bopp, photographed over Boulder, Colorado, during its appearance in 1997.

After Pluto's discovery, many people continued searching the skies in hopes of discovering a tenth planet, sometimes called "Planet X." Given the sensitivity with which our telescopes now scan the skies, it is unlikely that another large planet orbits our Sun (though other Pluto-size objects are possible). Thus, Pluto's orbit marks the end of the realm of the planets. However, the most numerous objects in the solar system still lie ahead: comets.

You are probably familiar with the occasional appearance of a comet in the inner solar system, where the heat of the Sun evaporates some of its ice and it grows a long, beautiful tail. But with very few exceptions, nearly all comets spend nearly their entire lifetime in the extreme outer reaches of the solar system. Our present technology does not allow us to detect comets when they are very far from the Sun. However, to account for the frequency with which we see comets in the inner solar system, astronomers have calculated that there must be something like a trillion (10^{12}) comets inhabiting the outskirts of our solar system.

The realm of the comets probably consists of two vast regions of space. The region containing the Pluto-like "giant comets"—called the *Kuiper belt*—begins near the orbit of Neptune and probably continues out to several times this distance. The comets in this region have orbits that lie close to the plane of planetary orbits and go around the Sun in the same direction as the planets. The second and much larger region—called the *Oort cloud*—may extend more than one-fourth of the way to the nearest stars. Comets in this region have orbits that are inclined at all angles to the plane of planetary orbits. Thus, the Oort cloud would look roughly spherical in shape if we could see it. But it is so vast that, even if it has a trillion comets, each comet would typically be separated from the next by more than a billion kilometers.

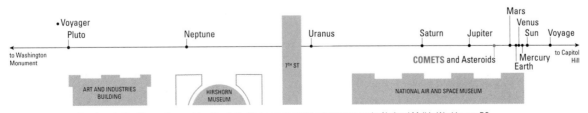

Comets and asteroids are discussed together in the Voyage scale model solar system on the National Mall in Washington, DC. Most comets actually reside well beyond the orbit of Pluto and are too small to see on this scale.

23

FIGURE 1.7 Stars of the constellation Centaurus, including Alpha Centauri, which is visible only from tropical and southern latitudes. Stars are so far away that they should appear as mere points of light; photographs make them seem to have a measurable size only because of the limitations of cameras (brighter stars appear larger than dimmer stars because they are overexposed). In fact, despite being separated by about the distance between the Sun and Uranus, the two largest stars in the Alpha Centauri system look like a single point of light in this photograph. (The nearest of the three stars, Proxima Centauri, is too dim to see with the naked eye.)

Onward to the Stars

The nearest star system to our own, called Alpha Centauri, contains three stars (Figure 1.7). The largest is about the same size as our Sun—a grapefruit on the 1-to-10-billion scale. After completing the short walk from the Sun to Pluto, how much farther would you have to walk to reach Alpha Centauri?

To answer this question, recall that a light-year is about 10 trillion kilometers, which becomes 1,000 kilometers on the 1-to-10-billion scale (10 trillion ÷ 10 billion = 1,000). Alpha Centauri's real distance, about 4.4 light-years, therefore becomes about 4,400 kilometers (2,700 miles) on this scale—roughly equivalent to the distance from New York to Los Angeles. In other words, looking at the largest star of Alpha Centauri from the Earth is equivalent to looking from New York at a very bright grapefruit in Los Angeles (neglecting the problems introduced by the curvature of the Earth). All other stars in our night sky are even farther away. No wonder the stars appear as mere points of light in the sky even when viewed through powerful telescopes.

Now, consider the difficulty of seeing *planets* orbiting other stars. Imagine looking from New York and trying to see a pinhead-size model Earth orbiting a grapefruit-size model Sun in Los Angeles. You probably won't be surprised to learn that we have not yet discovered Earth-size planets around other stars. Indeed, the bigger surprise may be that we *have* discovered dozens of extra-solar planets (planets around other stars), though most are closer in size to Jupiter than to Earth [Section 7.6]. With planet-detecting technology improving rapidly, we will probably know whether Earth-size planets orbit nearby stars within a couple of decades.

TIME OUT TO THINK *Consider what you have just learned about the scale of our solar system and the distances to the stars. Do these facts alter your perspective on the Earth or humanity in any way? Explain.*

The Milky Way Galaxy and Beyond

The vast separation between our solar system and Alpha Centauri is typical of the separations among star systems here in the outskirts of the Milky Way Galaxy. Thus, the 1-to-10-billion scale is useless for modeling even just a few dozen of the nearest stars, because they could not all be spaced properly on the surface of the Earth. Visualizing the entire galaxy requires a new scale.

Let's further reduce our solar system scale by a factor of 1 billion (making it a scale of 1 to 10^{19}). On this new scale, each light-year becomes 1 millimeter. For example, the 4.4-light-year separation between the Sun and Alpha Centauri becomes just 4.4 millimeters on this scale—smaller than the width of your little finger. Aside from the fact that it makes the stars themselves become microscopic (about the size of individual atoms), this new scale makes it easy to visualize the Milky Way Galaxy. Its 100,000-light-year diameter becomes 100 meters on this scale, or about the length of a football field.

Another way to put the galaxy into perspective is to consider its number of stars—more than 100 billion. Imagine that tonight you are having difficulty falling asleep (perhaps because you are contemplating the scale of the universe). Instead of counting sheep, you decide to count stars. If you are able to count about one star each second, on average, how long would it take you to count 100 billion stars in the Milky Way? Clearly, the answer is 100 billion (10^{11}) seconds. But how long is that? Amazingly, 100 billion seconds turns out to be more than 3,000 years. (You can confirm this by dividing 100 billion by the number of seconds in 1 year.) Thus, you would need thousands of years just to *count* the stars in the Milky Way Galaxy, and this assumes you never take a break—no sleeping, no eating, and absolutely no dying!

As incredible as the scale of our galaxy may seem, the Milky Way is only one of at least 100 billion galaxies in the observable universe. If we assume 100 billion stars per galaxy, the total number of stars in the observable universe is roughly

$$100 \text{ billion} \times 100 \text{ billion} =$$
$$10,000,000,000,000,000,000,000 \ (10^{22})$$

How big is this number?

Visit a beach. Run your hands through the fine-grained sand. Try to imagine counting every one of

FIGURE 1.8 The number of stars in the universe is larger than the total number of grains of sand on all the beaches on Earth.

the tiny grains of sand as they slip through your fingers (Figure 1.8). Then imagine counting every grain of sand on the beach where you are sitting. Next think about counting *all* the grains of dry sand on *all* the beaches everywhere on Earth. The number you would count would be less than the number of stars in the observable universe.

The Scale of Time

Now that we have developed some perspective on the scale of space, we can do the same for the scale of time. Imagine the entire history of the universe, from the Big Bang to the present, compressed into a single year. We can represent this history with a *cosmic calendar,* on which the Big Bang takes place at the first instant of January 1 and the present day is just before the stroke of midnight on December 31 (Figure 1.9). If we assume that the universe is about 12 billion years old (recall that we think the age is between 12 billion and 16 billion years), then each month on the cosmic calendar represents about 1 billion years.

On this scale, the Milky Way Galaxy may have formed sometime in February. Many generations of stars lived and died in the subsequent cosmic months, enriching the galaxy with the heavier elements from which we and the Earth are made. Not until about August 13, which represents a time 4.6 billion years ago, did our solar system form from a cloud of gas and dust in the Milky Way.

No one knows exactly when the earliest life arose on Earth, but it certainly was within the first billion years, or by mid-September on the cosmic calendar.

**Common Misconceptions:
Confusing Very Different Things**

Most people are familiar with the terms *solar system* and *galaxy,* but many people sometimes mix them up. Notice how incredibly different our solar system is from our galaxy. The solar system is a *single* star system consisting of our Sun and the various objects that orbit it, including Earth and eight other planets. The galaxy is a collection of some 100 billion star systems—so many that it would take thousands of years just to count them. Thus, confusing the terms *solar system* and *galaxy* represents a mistake by a factor of 100 billion—a fairly big mistake!

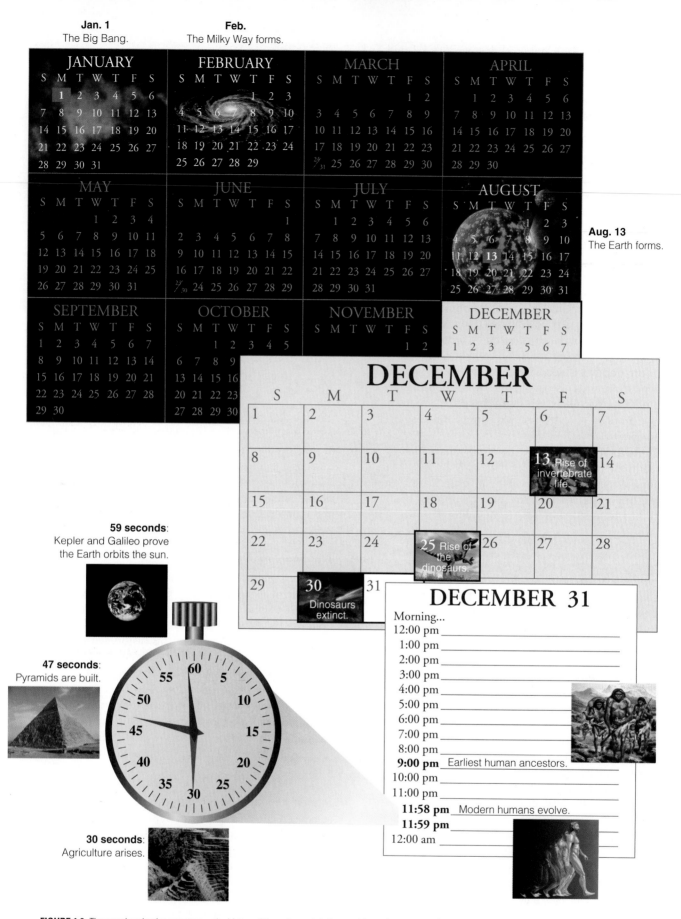

Jan. 1
The Big Bang.

Feb.
The Milky Way forms.

Aug. 13
The Earth forms.

59 seconds:
Kepler and Galileo prove the Earth orbits the sun.

47 seconds:
Pyramids are built.

30 seconds:
Agriculture arises.

DECEMBER

S	M	T	W	T	F	S
1	2	3	4	5	6	7
8	9	10	11	12	13 Rise of invertebrate life	14
15	16	17	18	19	20	21
22	23	24	25 Rise of the dinosaurs.	26	27	28
29	30 Dinosaurs extinct.	31				

DECEMBER 31

Morning...
12:00 pm _____
1:00 pm _____
2:00 pm _____
3:00 pm _____
4:00 pm _____
5:00 pm _____
6:00 pm _____
7:00 pm _____
8:00 pm _____
9:00 pm _Earliest human ancestors._
10:00 pm _____
11:00 pm _____
11:58 pm _Modern humans evolve._
11:59 pm _____
12:00 am _____

FIGURE 1.9 The cosmic calendar compresses the history of the universe into 1 year; this version assumes that the universe is 12 billion years old, so that each month represents about 1 billion years. It is only within the last few seconds of the last day that human civilization has taken shape. (This version of the cosmic calendar is adapted from a version created by Carl Sagan.)

For most of the Earth's history, living organisms remained relatively primitive and microscopic in size. Larger invertebrate life arose only about 600 million years ago, or about December 13 on the cosmic calendar. The age of the dinosaurs began shortly after midnight on Christmas (December 25) on the cosmic calendar. The dinosaurs vanished abruptly around 1:00 A.M. on December 30 (65 million years ago in real time), probably due to the impact of an asteroid or comet [Section 10.6].

With the dinosaurs gone, small furry mammals inherited the Earth. Some 60 million years later, or around 9:00 P.M. on December 31 of the cosmic calendar, the earliest hominids (human ancestors) walked upright. Most of the major events of human history took place within the final seconds of the final minute of the final day on the cosmic calendar. Agriculture arose about 30 seconds ago. The Egyptians built the pyramids about 13 seconds ago. It was only about 1 second ago on the cosmic calendar that Kepler and Galileo first proved that the Earth orbits the Sun. The average-age college student was born less than 0.1 second ago, around 11:59:59.9 P.M. on December 31. On this scale of cosmic time, the human species is the youngest of infants, and a human lifetime is a mere blink of an eye.

1.3 Spaceship Earth

Now that we have discussed the scale of space and of time, the next step in our "big picture" overview is understanding motion in the universe. Wherever you are as you read this book, you probably have the feeling that you're "just sitting here." Nothing could be further from the truth. In fact, you are being spun in circles as the Earth rotates, you are racing around the Sun in the Earth's orbit, and you are careening through the cosmos in the Milky Way galaxy. In the words of noted inventor and philosopher R. Buckminster Fuller (1895–1983), you are a traveler on *spaceship Earth*.

Rotation and Revolution

The Earth **rotates** on its axis once each day. (The axis is a line joining the North and South Poles and passing through the center of the Earth.) As a result, you are whirling around the Earth's axis at a speed of 1,000 kilometers per hour (600 miles per hour) or more—faster than most airplanes travel (Figure 1.10). It is this rotation that makes the Sun and the stars appear to make their daily circles around the sky. You see sunrise when the rotating Earth first allows you to glimpse the Sun in the east, noon when your location faces directly toward the Sun, and sunset just as your location rotates to the Earth's night side,

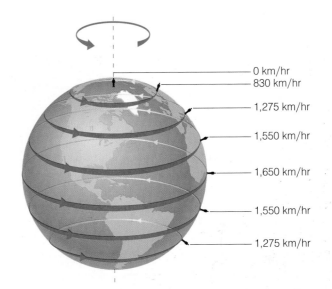

FIGURE 1.10 As the Earth rotates, your speed around the Earth's axis depends on your latitude. Unless you live at very high latitude, your speed is over 1,000 km/hr. Notice that the Earth rotates counterclockwise as viewed from above the North Pole, so you are always rotating from west to east (which is why the Sun rises in the east and sets in the west).

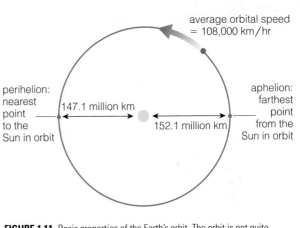
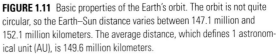

FIGURE 1.11 Basic properties of the Earth's orbit. The orbit is not quite circular, so the Earth–Sun distance varies between 147.1 million and 152.1 million kilometers. The average distance, which defines 1 astronomical unit (AU), is 149.6 million kilometers.

where the Sun is hidden from view. (Because the Earth orbits the Sun at the same time that it rotates, our 24-hour day is slightly longer—by about 4 minutes—than the Earth's actual rotation period.)

The Earth **revolves** around the Sun once each year, following an orbit that is not quite circular (Figure 1.11). The *average* distance of the Earth from the Sun is called an **astronomical unit**, or **AU** (more technically, an astronomical unit is the *semimajor axis* of the Earth's orbit):

$$1 \text{ AU} \approx 150 \text{ million km (93 million miles)}$$

The Earth travels slightly faster in its orbit when it is nearer to the Sun and slightly slower when it is farther

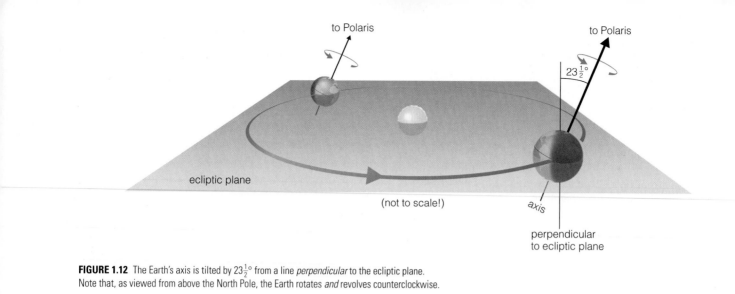

to Polaris

to Polaris

$23\frac{1}{2}°$

ecliptic plane

(not to scale!)

axis

perpendicular
to ecliptic plane

FIGURE 1.12 The Earth's axis is tilted by $23\frac{1}{2}°$ from a line *perpendicular* to the ecliptic plane. Note that, as viewed from above the North Pole, the Earth rotates *and* revolves counterclockwise.

from the Sun. But at all times the Earth is carrying you around the Sun at a speed in excess of 100,000 kilometers per hour (60,000 miles per hour).

The plane of the Earth's orbit around the Sun is called the **ecliptic plane** (Figure 1.12). The Earth's rotation axis happens to be tilted by $23\frac{1}{2}°$ from a line perpendicular to the ecliptic plane, pointing nearly in the direction of Polaris, the North Star. Viewed from above the North Pole, the Earth both rotates and revolves counterclockwise.

Seasons The combination of the Earth's rotation and its revolution around the Sun explains why we have seasons. Because the Earth's rotation axis remains pointed in the same direction (toward Polaris) throughout the year, its orientation *relative to the Sun* changes as it orbits the Sun, as shown in Figure 1.13. The familiar seasonal changes—longer and warmer days in summer, and shorter and cooler days in winter—arise because of changes in how directly the Sun's rays strike a particular location on Earth. When the rays are fairly direct, the Sun's path through the sky is long and high and the days are warm. (By analogy, think of how you sunburn more easily on parts of your body that face the Sun more directly.) When the Sun's rays strike less directly, the Sun's path is short and low through the sky and the days are cool. Notice that the Sun's rays strike the Northern Hemisphere more directly on the side of the orbit near point 2 in Figure 1.13, while they strike the Southern Hemisphere more directly on the side of the orbit near point 4. That is why the seasons are opposite in the Northern and Southern Hemispheres.

The four labeled points in Figure 1.13 have special names related to the seasons. The Earth is at

**Common Misconceptions:
The Cause of Seasons**

When asked the cause of the seasons, many people mistakenly answer that they are caused by variations in Earth's distance from the Sun. By knowing that the Northern and Southern Hemispheres experience opposite seasons, you'll realize that Earth's varying distance from the Sun *cannot* be the cause of the seasons; if it were, both hemispheres would have summer at the same time. Although Earth's distance from the Sun *does* vary slightly over the course of a year (Earth is closest to the Sun in January and farthest away in July), this factor is greatly overwhelmed by the way the tilt of the rotation axis causes the Northern and Southern Hemispheres to alternately receive more or less direct sunlight. Earth's varying distance from the Sun has no noticeable effect on our seasons. (This is not necessarily the case for other planets; seasons on Mars, for example, *are* affected by its varying distance from the Sun.)

point 1 on about March 21 each year, which represents the **spring equinox** (or *vernal equinox*). Both hemispheres receive equal amounts of sunlight at this time; it is the beginning of spring for the Northern Hemisphere but the beginning of fall for the Southern Hemisphere. The Earth reaches point 2, which represents the **summer solstice**, around June 21. This is the day on which the Northern Hemisphere receives its most direct sunlight and has the longest period of daylight of any day of the year; it is usually considered the first day of summer. Note that the Southern Hemisphere receives its least direct

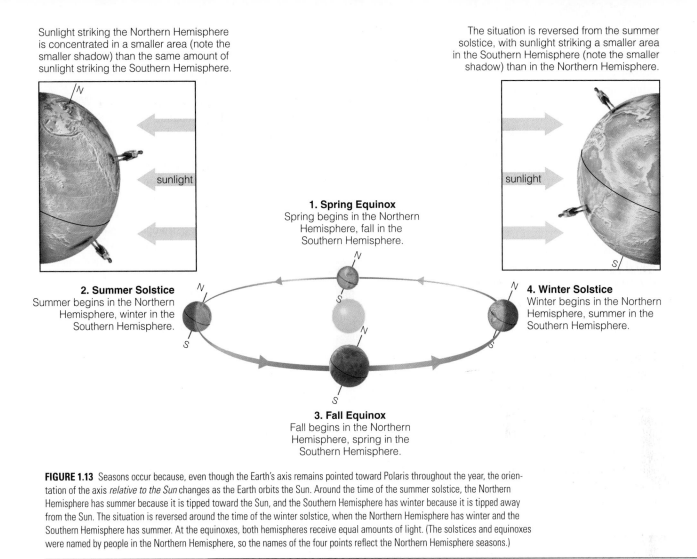

Sunlight striking the Northern Hemisphere is concentrated in a smaller area (note the smaller shadow) than the same amount of sunlight striking the Southern Hemisphere.

sunlight

The situation is reversed from the summer solstice, with sunlight striking a smaller area in the Southern Hemisphere (note the smaller shadow) than in the Northern Hemisphere.

sunlight

1. Spring Equinox
Spring begins in the Northern Hemisphere, fall in the Southern Hemisphere.

2. Summer Solstice
Summer begins in the Northern Hemisphere, winter in the Southern Hemisphere.

4. Winter Solstice
Winter begins in the Northern Hemisphere, summer in the Southern Hemisphere.

3. Fall Equinox
Fall begins in the Northern Hemisphere, spring in the Southern Hemisphere.

FIGURE 1.13 Seasons occur because, even though the Earth's axis remains pointed toward Polaris throughout the year, the orientation of the axis *relative to the Sun* changes as the Earth orbits the Sun. Around the time of the summer solstice, the Northern Hemisphere has summer because it is tipped toward the Sun, and the Southern Hemisphere has winter because it is tipped away from the Sun. The situation is reversed around the time of the winter solstice, when the Northern Hemisphere has winter and the Southern Hemisphere has summer. At the equinoxes, both hemispheres receive equal amounts of light. (The solstices and equinoxes were named by people in the Northern Hemisphere, so the names of the four points reflect the Northern Hemisphere seasons.)

sunlight and has its shortest daylight period on the summer solstice, so this is the first day of winter for the Southern Hemisphere. Point 3 represents the **fall equinox** (or *autumnal equinox*), which occurs around September 21; both hemispheres again receive the same amount of sunlight, but it is the beginning of fall in the Northern Hemisphere and the beginning of spring in the Southern Hemisphere. Finally, point 4 represents the **winter solstice**, which occurs around December 21. The situation here is the reverse of that on the summer solstice; it is usually considered the first day of winter for the Northern Hemisphere and the first day of summer for the Southern Hemisphere.

TIME OUT TO THINK *After studying Figure 1.13, explain why regions near the Earth's equator do not experience seasons in the same way as mid-latitudes such as North America, Europe, and Australia. When is the Sun's path in the sky highest for equatorial regions?*

Precession Although the Earth's axis will remain pointed toward Polaris throughout our lifetimes, it has not always been so and will not always remain so. The reason is a gradual motion called **precession**. Like the axis of a spinning top, the Earth's axis also sweeps out a circle (precesses)—but much more slowly (Figure 1.14). Note that the axis tilt remains close to $23\frac{1}{2}°$ throughout the cycle; only the axis *orientation* changes. Each cycle of the Earth's precession takes about 26,000 years. Today, the axis points toward Polaris, which is what makes it our North Star. In about 13,000 years, the axis will point nearly in the direction of the star Vega, making *it* our North Star. During most of the precession cycle, the axis does not point very near any bright star.

Precession also changes the locations in the Earth's orbit at which the equinoxes and solstices occur. For example, the spring equinox today occurs when the Sun appears in the direction of the constellation Pisces, but 2,000 years ago it occurred when the Sun appeared in Aries. (That is why the spring equinox is sometimes called "the first point of Aries.")

In about 600 years, precession will move the spring equinox into the constellation Aquarius, marking what some astrologers call "the age of Aquarius."

Traveling in the Milky Way Galaxy

Besides moving within our solar system, we are traveling with our solar system on a journey through the Milky Way Galaxy. For example, we are moving relative to nearby stars at an average speed of about 70,000 kilometers per hour (40,000 miles per hour)—about three times as fast as the Space Station orbits the Earth. In fact, all stars are in constant motion, and it is as valid to say that the stars are moving relative to us as it is to say that we are moving relative to them.

If nearby stars move at such high speeds, why don't we see them racing around the sky? The answer lies in their vast distances from us. You've probably noticed that a distant airplane appears to move through your sky more slowly than one flying close overhead. If we extend this idea to the stars, we find that even at speeds of 70,000 kilometers per hour stellar motions would be noticeable to the naked eye only if we watched for thousands of years. That is why the patterns in the constellations seem to remain fixed. Nevertheless, in 10,000 years the constellations will be noticeably different from those we see today. In 500,000 years they will be unrecognizable. If you could watch a time-lapse movie made over millions of years, you would see the stars of our *local solar neighborhood* (the Sun and nearby stars)

racing around in seemingly random directions (Figure 1.15).

However, if you widened your view beyond the local solar neighborhood, you would see that all the stars are moving in a simple and general way: The entire Milky Way Galaxy is rotating. Stars at different distances from the galactic center take different amounts of time to complete an orbit. Our solar system, located about 28,000 light-years from the galactic center, completes one orbit of the galaxy in

FIGURE 1.14 Precession affects the orientation, but not the tilt, of Earth's axis. (Precession is caused by gravity's effect on a tilted axis. The Earth precesses in the opposite sense of the top [clockwise instead of counterclockwise] because the Sun's gravity tries to make the Earth's axis more vertical. In contrast, the Earth's gravity tries to pull a top's axis over, making it more horizontal.)

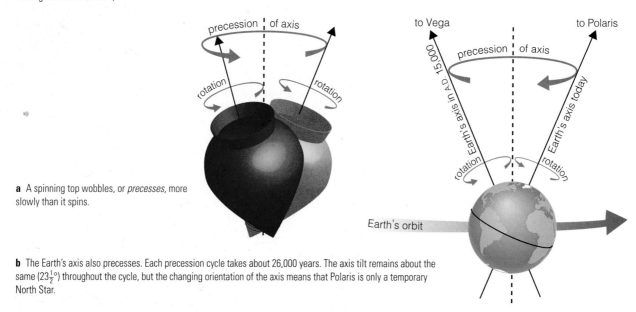

a A spinning top wobbles, or *precesses*, more slowly than it spins.

b The Earth's axis also precesses. Each precession cycle takes about 26,000 years. The axis tilt remains about the same $(23\frac{1}{2}°)$ throughout the cycle, but the changing orientation of the axis means that Polaris is only a temporary North Star.

The box represents stars and their motions in the local solar neighborhood.

FIGURE 1.15 As shown in this painting, the local solar neighborhood is only a tiny portion of the Milky Way Galaxy. The stars in the local solar neighborhood actually move quite fast relative to our solar system, but the enormous distances between stars make this motion barely detectable on human time scales. If you could watch a time-lapse movie made over a time period of millions of years, you would see the stars in the local solar neighborhood racing around in seemingly random directions.

about 230 million years (Figure 1.16). Even if you could watch from outside our galaxy, this motion would be unnoticeable to your naked eye. However, if you calculate the speed of our solar system as we orbit the center of the galaxy, you will find that it is close to 1 million kilometers per hour (600,000 miles per hour).

It is worth noting that the galactic rotation reveals one of the greatest mysteries in science—one that we will study in depth later in the book. The speeds at which stars orbit the galactic center depend on the strength of gravity, and the strength of gravity depends on how mass is distributed throughout the galaxy. Thus, careful study of the galactic rotation allows us to determine the distribution of mass in the galaxy. Remarkably, we find that the stars in the disk of the galaxy represent only the "tip of the iceberg" compared to the mass of the entire galaxy (Figure 1.17). That is, most of the mass of the galaxy is located outside the visible disk, in the galaxy's *halo*. We don't know the nature of this mass because we have not detected any light coming from it, and

FIGURE 1.16 This painting shows how the entire Milky Way Galaxy rotates (in a direction that would tend to wind up the spiral arms). Our Sun and solar system are located about 28,000 light-years from the galactic center. At this distance, each orbit around the galactic center takes about 230 million years.

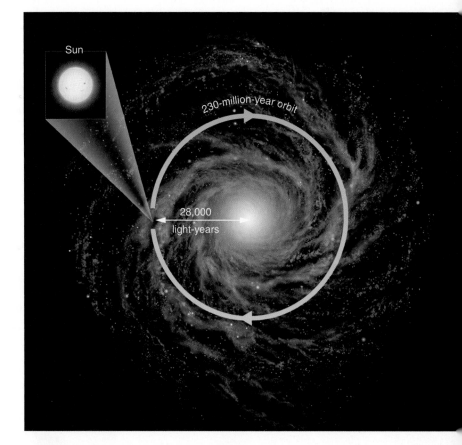

Sun

230-million-year orbit

28,000 light-years

Most of the galaxy's light comes from stars and gas in the galactic disk...

...but most of the galaxy's mass lies above and below the disk in the *halo*.

FIGURE 1.17 This painting shows an edge-on view of the Milky Way Galaxy. Most visible stars reside within the galaxy's thin *disk*, which runs horizontally across the page in this figure. But careful study of galactic rotation shows that most of the mass lies in the galactic *halo*—a large, spherical region that surrounds and encompasses the disk. Because this mass emits no light that we have detected, we call it dark matter. (The dark matter may extend quite far out; see Figure 18.1.)

we therefore call it **dark matter**. Studies of other galaxies show that they also are made mostly of dark matter. In fact, most of the mass in the universe appears to be made of this mysterious dark matter, but we do not yet know what it is.

The Expanding Universe

The billions of galaxies in the universe also move relative to one another. Within the Local Group (see Figure 1.1), the motions are about as we might expect. Some of the galaxies move toward the Milky Way Galaxy, some move away, and some move in more complex ways. (For example, two small galaxies, known as the Large and Small Magellanic Clouds, apparently orbit the Milky Way.) Again, the speeds are enormous by earthly standards—the Milky Way is moving toward the Great Galaxy in Andromeda (M31) at about 300,000 kilometers per hour (180,000 miles per hour)—but are unnoticeable to our eyes. We needn't worry about a collision any time soon. Even if the Milky Way and Andromeda Galaxies are approaching each other head-on (which they might not be), it will be nearly 10 billion years before the collision begins.

When we look outside the Local Group, however, we find two astonishing facts that were first discovered in the 1920s by Edwin Hubble, for whom the Hubble Space Telescope was named:

1. Virtually every galaxy outside the Local Group is moving *away* from us.

2. The more distant the galaxy, the faster it appears to be racing away from us.

Upon first becoming aware of these two facts, you might be tempted to conclude that our Local Group (which is held together by gravity) suffers a cosmic case of chicken pox. However, there is a much more natural explanation: *The entire universe is expanding.* We'll save details about this expansion for later in the book (Chapter 17), but you can understand the basic idea by thinking about a raisin cake baking in an oven.

Imagine that you make a raisin cake in which the distance between adjacent raisins is 1 centimeter. You place the cake in the oven, where it expands as it bakes. After 1 hour, you remove the cake, which has expanded so that the distance between adjacent raisins is 3 centimeters (Figure 1.18).

Pick any raisin (it doesn't matter which one), call it the Local Raisin, and identify it in the pictures of the cake both before and after baking; Figure 1.18 shows one possible choice for the Local Raisin, with several other nearby raisins labeled. Before baking, Raisin 1 is 1 centimeter away from the Local Raisin, Raisin 2 is 2 centimeters away, and Raisin 3 is 3 centimeters away. After baking, Raisin 1 is 3 centimeters away from the Local Raisin, Raisin 2 is 6 centimeters away, and Raisin 3 is 9 centimeters away. Nothing should seem surprising so far, because the new distances simply reflect the fact that the cake has expanded uniformly everywhere.

Now, suppose you live *inside* the Local Raisin. From your vantage point, the other raisins appear to move away from you as the cake expands. For example, Raisin 1 is 1 centimeter away before baking and 3 centimeters away after baking. It therefore appears to move 2 centimeters during the hour, so you'll see it moving away from you at a speed of 2 centimeters per hour. Raisin 2 is 2 centimeters away before baking and 6 centimeters away afterward; thus, it appears to move 4 centimeters during the hour of baking, giving it a speed of 4 centimeters per hour away from you. The following table lists the distances of other raisins before and after baking, along with their speeds as seen from the Local Raisin.

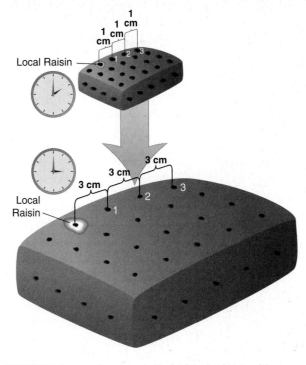

FIGURE 1.18 An expanding raisin cake illustrates basic principles of the expansion of the universe.

Distances and Speeds As Seen from the Local Raisin

Raisin Number	Distance Before Baking	Distance After Baking (1 hour later)	Speed
1	1 cm	3 cm	2 cm/hr
2	2 cm	6 cm	4 cm/hr
3	3 cm	9 cm	6 cm/hr
4	4 cm	12 cm	8 cm/hr
⋮	⋮	⋮	⋮
10	10 cm	30 cm	20 cm/hr
⋮	⋮	⋮	⋮

Note that, although the entire cake expanded uniformly, from the vantage point of the Local Raisin each subsequent raisin moved away at increasingly faster speeds. Also, because selection of the Local Raisin was arbitrary, you would find the same results no matter which raisin you chose to represent your Local Raisin.

If you now imagine the Local Raisin to represent our Local Group of galaxies and the other raisins to represent more distant galaxies or clusters of galaxies, you have a basic picture of the expansion of the universe. That is, *space* itself is growing, and more distant galaxies move away from us faster because they are carried along with this expansion like raisins in an expanding cake. And, just as the raisins themselves do not expand, the individual galaxies and clusters of galaxies do not expand (because they are bound together by gravity). The effects of expansion would appear basically the same viewed from any place in the universe.

Summary of Motion

Let's summarize the motions we have covered. We spin around the Earth's axis as we revolve around the Sun. Our solar system moves among the stars of the local solar neighborhood, while this entire neighborhood orbits the center of the Milky Way Galaxy. Our galaxy, in turn, moves among the other galaxies of the Local Group, while the Local Group is carried along with the overall expansion of the universe. Table 1.2 lists the motions and their associated speeds. Spaceship Earth is indeed carrying us on a remarkable journey!

Table 1.2 The Motions of Spaceship Earth

Motion	Typical Speed
rotation	1,000 km/hr or more around axis
revolution	100,000 km/hr around Sun
motion within local solar neighborhood	70,000 km/hr relative to nearby stars
rotation of the Milky Way Galaxy	1,000,000 km/hr around galactic center
motion within Local Group	300,000 km/hr toward Andromeda Galaxy
universal expansion	more distant galaxies moving away faster, with the most distant moving at speeds close to the speed of light

1.4 The Human Adventure of Astronomy

In a relatively few pages, we've laid out a fairly complete overview of our modern scientific view of the universe. But our goal in this book is not for you simply to be able to recite this view. Rather it is to help you understand the evidence supporting this view and the extraordinary story of how it developed. In many ways, this book is devoted to this goal.

Astronomy is a human adventure in the sense that it affects virtually everyone—even those who have never looked at the sky—because the development of astronomy has been so deeply intertwined with the development of civilization as a whole. Revolutions in astronomy have gone hand in hand with the revolutions in science and technology that have shaped modern life. Witness the repercussions of the Copernican revolution, which changed our view of the Earth from the center of the universe to just one planet orbiting the Sun. This revolution, which we will discuss further in Chapter 3, began when Copernicus published his idea of a Sun-centered solar system in 1543. Three subsequent figures—Tycho Brahe, Johannes Kepler, and Galileo—provided the key evidence that eventually led to wide acceptance of the Copernican idea, and the revolution culminated with Isaac Newton's uncovering of the laws of motion and gravity. Newton's work, in turn, became the foundation of physics that helped fuel the industrial revolution.

More recently, the development of space travel and the computer revolution have helped fuel tremendous progress in astronomy. We've learned a lot about our solar system by sending probes to the planets, and many of our most powerful observatories, including the Hubble Space Telescope, reside in space. On the ground, computer design and control have led to tremendous growth in the size and power of telescopes, particularly in the past decade.

Many of these efforts, along with the achievements they spawned, have led to profound social change. The most famous example involved Galileo, whom the Vatican put under house arrest in 1633 for his claims that the Earth orbits the Sun. Although the Church soon recognized that Galileo was right, he was formally vindicated only with a statement by Pope John Paul II in 1992. In the meantime, his case spurred great debate in religious circles and had a profound influence on both theological and scientific thinking.

As you progress through this book and learn about astronomical discovery, try to keep in mind the context of the human adventure. You will then be learning not just about a science, but about one of the great forces that have helped shape our modern world. This context will also lead you to think about how the many astronomical mysteries that remain—such as the makeup of dark matter, the events of the first instant of the Big Bang, and the question of life

FIGURE 1.19 This fanciful image imagines hikers looking at a sky filled with galaxies; the galaxies are actually part of a photograph taken by the Hubble Space Telescope (see Figure 18.9). Although such a view could not really exist anywhere in the universe, the image shows how our minds can connect the deepest mysteries of the cosmos to our lives here on Earth. Indeed, a primary goal of this book is to help you understand the extraordinary story of how we have reached our current scientific understanding of the universe and to appreciate the many mysteries that still remain.

beyond Earth—may influence our future. What would it mean to us if we were ever to learn the complete story of our cosmic origins? How would our view of the Earth be changed if we came to learn that Earth-like planets are common or exceedingly rare? Only time may answer these questions, but the chapters ahead give you the foundation you need to understand how we changed from primitive people looking at patterns in the night sky to a civilization capable of asking deep questions about our existence.

THE BIG PICTURE

In this first chapter, we developed a broad overview of our place in the universe. It is not yet necessary for you to remember the details—everything presented in this chapter will be covered in greater depth later in the book. However, you should understand enough so that all the following "big picture" ideas are clear:

- The Earth is not the center of the universe but instead is a planet orbiting a rather ordinary star in the Milky Way Galaxy. The Milky Way Galaxy, in turn, is one of billions of galaxies in our observable universe.

- We are "star stuff." The atoms from which we are made began as hydrogen and helium in the Big Bang and were later fused into heavier elements in massive stars. When these stars died, they released these atoms into space, where our galaxy recycled them. Our solar system formed from such recycled matter some 4.6 billion years ago.

- Cosmic distances are literally astronomical, but we can put them in perspective with the aid of scale models and other scaling techniques. When you think about these enormous scales, don't forget that every star is a sun and every planet is a unique world.

- We are latecomers on the scale of cosmic time. The universe was more than halfway through its history by the time our solar system formed, and then it took billions of years more before humans arrived on the scene.

- All of us are being carried through the cosmos on spaceship Earth. Although we cannot feel this motion in our everyday lives, the associated speeds are surprisingly high. Learning about the motions of spaceship Earth not only gives us a new perspective on the cosmos, but also helps us understand its nature and history.

- Throughout history, the development of astronomy has gone hand in hand with social and technological development. In this sense, astronomy touches everyone, making it a human adventure to be enjoyed by all.

Review Questions

1. What do we mean by a *geocentric* universe? In broad terms, contrast a geocentric universe with our modern view of the universe.

2. Describe the major levels of structure in the universe.

3. What do we mean when we say that the universe is *expanding*? Why does an expanding universe suggest a beginning in what we call the *Big Bang*?

4. Summarize our cosmic origins and explain what Carl Sagan meant when he said that we are "star stuff."

5. Explain the statement *The farther away we look in distance, the further back we look in time.*

6. What do we mean by the *observable universe?*

7. Describe key features of our solar system as they would appear on a scale of 1 to 10 billion.

8. How do the distances to the stars compare to distances within our solar system? Give a few examples to put the comparison in perspective.

9. Describe at least two ways to put the scale of the Milky Way Galaxy in perspective.

10. How many galaxies are in the observable universe? How many stars are in the observable universe? Put these numbers in perspective.

11. Imagine describing the cosmic calendar to friends. In your own words, give them a feel for how the human race fits into the scale of time.

12. Briefly summarize the major motions of the Earth through the universe.

13. What is an *astronomical unit?* What is the *ecliptic plane?*

14. Describe the cause of the Earth's seasons, and explain why the seasons are opposite in the Northern and Southern Hemispheres.

15. What is *precession,* and how does it affect the appearance of our sky?

16. Describe the raisin cake model of the universe, and explain how it shows that a uniform expansion would lead us to see more distant galaxies moving away from us at higher speeds.

Discussion Questions

1. *Vast Orbs.* The chapter-opening quotation from Christiaan Huygens suggests that humans might be less inclined to wage war if everyone appreciated the Earth's place in the universe. Do you agree? Defend your opinion.

2. *Infant Species.* In the last few tenths of a second before midnight on December 31 of the cosmic calendar, we have developed an incredible civilization and learned a great deal about the universe, but

we also have developed technology through which we could destroy ourselves. The midnight bell is striking, and the choice for the future is ours. How far into the next cosmic year do you think our civilization will survive? Defend your opinion.

3. *A Human Adventure.* How important do you think astronomical discoveries have been to our social development? Defend your opinion with examples drawn from your knowledge of history.

Problems

Sensible Statements? For **problems 1–8**, decide whether the statement is sensible and explain why it is or is not.

Example: I walked east from our base camp at the North Pole.

Solution: The statement does not make sense because *east* has no meaning at the North Pole—all directions are south from the North Pole.

1. The universe is between 12 billion and 16 billion years old.

2. The universe is between 12 billion and 16 billion light-years old.

3. It will take me light-years to complete this homework assignment!

4. Someday we may build spaceships capable of traveling at a speed of 1 light-minute per hour.

5. Astronomers recently discovered a moon that does not orbit a planet.

6. NASA plans soon to launch a spaceship that will leave the Milky Way Galaxy to take a photograph of the galaxy from the outside.

7. The observable universe is the same size today as it was a few billion years ago.

8. Photographs of distant galaxies show them as they were when they were much younger than they are today.

9. *No Axis Tilt.* Suppose the Earth's axis had no tilt. Would we still have seasons? Why or why not?

10. *Tour Report.* A friend asks you the following questions concerning your tour of a scale-model solar system. Answer each question in one paragraph.

    a. Is the Sun really much bigger than the Earth?

    b. Is it true that the Sun uses nuclear energy?

    c. Would it be much harder to send humans to Mars than to the Moon?

    d. In elementary school, I heard that Neptune is farther from the Sun than Pluto. Is this true?

    e. I read that Pluto is not really a planet. What's the story?

(continued)

f. Why didn't they have any stars besides the Sun in the scale model?

g. What was the most interesting thing you learned during your tour?

11. *Raisin Cake Universe.* Suppose that all the raisins in a cake are 1 centimeter apart before baking and after baking they are 4 centimeters apart.

    a. Draw diagrams to represent the cake before and after baking.

    b. Identify one raisin as the Local Raisin on your diagrams. Construct a table showing the distances and speeds of other raisins as seen from the Local Raisin.

    c. Briefly explain how your expanding cake is similar to the expansion of the universe.

12. *Light-Year.* Calculate the distance that light travels in 1 year. Show your work. (*Hint:* Multiply the speed of light, 300,000 km/s, by the number of seconds in a minute, the number of minutes in an hour, and so on until you have the distance light travels in a year.)

Web Projects

Find useful links for Web projects on the text Web site.

1. *Astronomy on the Web.* The Web contains a vast amount of astronomical information. Starting from the links on the textbook Web site, spend at least an hour exploring astronomy on the Web. Write two or three paragraphs summarizing what you learned from your Web surfing. What was your favorite astronomical Web site, and why?

2. *Voyage.* Visit the Web site for the Voyage scale model solar system on the National Mall (www. voyageonline.org). Write a one-page summary of what you learn from your virtual tour.

3. *NASA Missions.* Visit the NASA Web site to learn about upcoming missions that concern astronomy. Write a one-page summary of the mission you feel is most likely to give us new astronomical information during the time you are enrolled in your astronomy course.

CHAPTER 2

Discovering the Universe for Yourself

It is an exciting time in the history of astronomy. A new generation of telescopes is probing the depths of the universe. Increasingly sophisticated space probes are collecting new data about the planets and other objects in our solar system. Rapid advances in computing technology allow scientists to analyze the vast amount of new data and to model the processes that occur in planets, stars, galaxies, and the universe.

One of the goals of this book is to help *you* share in the ongoing adventure of astronomical discovery. One of the best ways to become a part of this adventure is to discover the universe for yourself by doing what other humans have done for thousands of generations: Go outside, observe the sky around you, and contemplate the awe-inspiring universe of which you are a part. In this chapter, we'll discuss a few key ideas that will help you understand what you see in the sky.

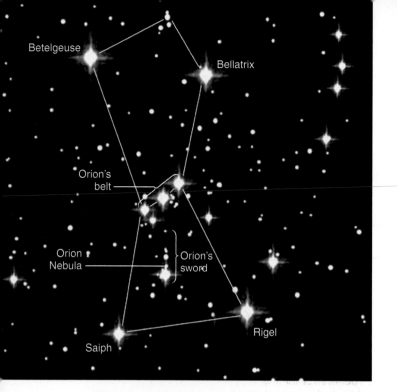

FIGURE 2.1 This photograph shows stars in the constellation Orion. Stars are so far away that they have no visible size; brighter stars appear larger in photographs only because they are overexposed. The crosses on bright stars are an artifact of the telescope used to take the photograph.

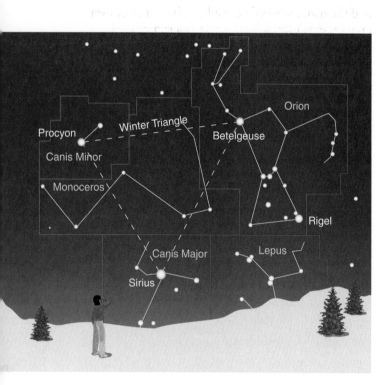

FIGURE 2.2 Red lines mark official borders of several constellations near Orion. Yellow lines connect recognizable patterns of stars within constellations. Sirius, Procyon, and Betelgeuse form a pattern spanning several constellations and called the Winter Triangle; it is easy to find on clear winter evenings.

2.1 Patterns in the Sky

Shortly after sunset, as daylight fades to darkness, the sky appears to fill slowly with stars. On clear, moonless nights far from city lights, as many as 2,000–3,000 stars may be visible to your naked eye. As you look at the stars, your mind might group them into many different patterns. If you observe the sky night after night or year after year, you will recognize the same patterns of stars.

People of nearly every culture gave names to patterns in the sky. The pattern that the Greeks named Orion, the hunter (Figure 2.1), was seen by the ancient Chinese as a supreme warrior called *Shen.* Hindus in ancient India also saw a warrior, called *Skanda,* who as the general of a great celestial army rode a peacock. The three stars of Orion's belt were seen as three fishermen in a canoe by Aborigines of northern Australia. As seen from southern California, these three stars climb almost straight up into the sky as they rise in the east, which may explain why the Chemehuevi Indians of the California desert saw them as a line of three sure-footed mountain sheep. These are but a few of the many names, each accompanied by a rich folklore, that have been given to the pattern of stars we call Orion.

Constellations

The patterns of stars seen in the sky are usually called constellations. Astronomers, however, use the term **constellation** to refer to a specific *region* of the sky. Any place in the sky belongs to some constellation; familiar patterns of stars merely help locate particular constellations. For example, the constellation Orion includes all the stars in the familiar pattern of the hunter, along with the region of the sky in which these stars are found (Figure 2.2).

The official borders of the constellations were set in 1928 by members of the International Astronomical Union (IAU), an association of astronomers from around the world. The IAU divided the sky into 88 constellations whose borders correspond roughly to the star patterns recognized by Europeans. Thus, despite the wide variety of names given to patterns of stars by different cultures, the "official" names of constellations visible from the Northern Hemisphere can be traced back to the ancient Greeks and to other cultures of southern Europe, the Middle East, and northern Africa. No one knows exactly when these constellations were first named, although some names probably go back at least 5,000 years. The official names of the constellations visible from the Southern Hemisphere are primarily those given by seventeenth-century European explorers.

Learning your way around the constellations is no more difficult than learning your way around your

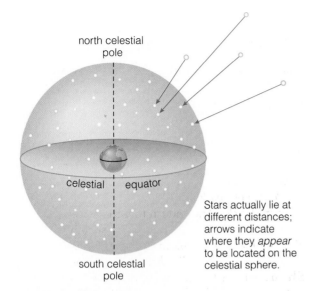

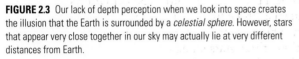

FIGURE 2.3 Our lack of depth perception when we look into space creates the illusion that the Earth is surrounded by a *celestial sphere*. However, stars that appear very close together in our sky may actually lie at very different distances from Earth.

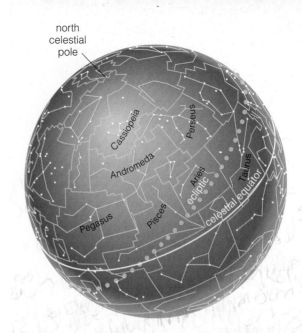

FIGURE 2.4 A model of the celestial sphere shows the patterns of the stars, the borders of the 88 official constellations, the ecliptic, and the celestial equator and poles. Because the celestial sphere shows the view from Earth, we imagine the Earth to reside in the center of the sphere. Thus, when we look into the sky, we see patterns of stars as they would appear from *inside* this imaginary sphere (which means the patterns appear left-right reversed compared to what we see when we look at the model from the outside).

neighborhood, and recognizing the patterns of just 20–40 constellations is enough to make the entire sky seem familiar. The best way to learn the constellations is to go out and view them, guided by the help of a few visits to a planetarium and the star charts in the back of this book. The *Skygazer* software that comes with this book can also help you learn constellations.

The Celestial Sphere

The stars in a particular constellation may appear to lie close to one another, but this is an illusion. All stars in a particular constellation lie in similar directions from Earth but not necessarily at similar distances (Figure 2.3). However, because we lack depth perception when we look into space, it looks like all the stars lie at the same distance. The ancient Greeks mistook this illusion for reality, imagining the Earth to be surrounded by a great **celestial sphere**.

Today, the concept of a celestial sphere is still useful for our learning about the sky, even though we know that the Earth does not really lie in the center of a giant ball. We give names to special locations on the imaginary celestial sphere. The point directly above the Earth's North Pole is called the **north celestial pole** (NCP), and the point directly above the Earth's South Pole is called the **south celestial pole**. The **celestial equator** represents an extension of the Earth's equator into space.

The stars form the fixed patterns of the constellations on the celestial sphere, while the Sun, Moon, and planets slowly wander among the stars. As we'll discuss shortly, the apparent motion of the Moon

and the planets is fairly complex. The Sun, however, appears to slowly circle the celestial sphere on a simple annual path called the **ecliptic**; this apparent path is the projection of the ecliptic plane (see Figure 1.12) into space. A model of the celestial sphere typically shows the patterns of the stars, the borders of the 88 official constellations, the ecliptic, and the celestial equator and poles (Figure 2.4).

Common Misconceptions: Stars in the Daytime

Because we don't see stars in the daytime, some people believe that the stars vanish in the daytime and "come out" at night. In fact, the stars are always present. The only reason you cannot see stars in the daytime is that their dim light is overwhelmed by the bright daytime sky. You *can* see bright stars in the daytime with the aid of a telescope, and you may see stars in the daytime if you are fortunate enough to observe a total eclipse of the Sun. Astronauts also see stars in the daytime—above the Earth's atmosphere, where no air is present to scatter sunlight through the sky, the Sun is a bright disk against a dark sky filled with stars. (However, because the Sun is so bright, astronauts must block its light and allow their eyes to adapt to darkness if they wish to *see* the stars.)

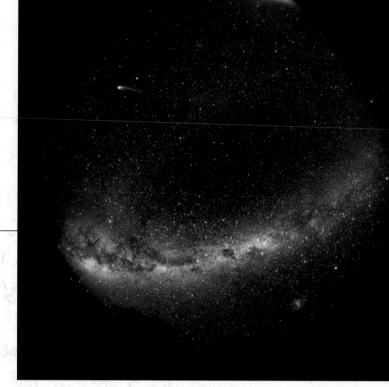

FIGURE 2.5 A "fish-eye" photograph of the Milky Way in the Australian sky. Near the upper left, Comet Hale–Bopp is visible in this 1997 photo.

The Milky Way

As your eyes adapt to darkness at a dark site, you'll begin to see the whitish band of light called the *Milky Way.* It is from this band of light that our Milky Way Galaxy gets its name. You can see only part of the Milky Way at any particular time, but it stretches all the way around the celestial sphere. If you look carefully, you will notice that the Milky Way varies in width and has dark fissures running through it. The widest and brightest parts of the Milky Way are most easily seen from the Southern Hemisphere (Figure 2.5), which probably explains why the Aborigines of Australia gave names to the patterns they saw within the Milky Way in the same way that other cultures gave names to patterns of stars.

The band of light called the Milky Way bears an important relationship to the Milky Way Galaxy: *It traces the galactic plane as it appears from our location in the outskirts of the galaxy* (Figure 2.6). Recall that the Milky Way Galaxy is shaped like a thin pancake with a bulge in the middle (see Figures 1.16 and 1.17). We view the universe from our location more than halfway outward from the center of this "pancake." When we look in any direction *within* the plane of the galaxy, we see countless stars, along with interstellar gas and dust. These stars and glowing

clouds of gas form the band of light we call the Milky Way. The dark fissures appear in regions where particularly dense interstellar clouds obscure our view of stars behind them. The central bulge of the galaxy makes the Milky Way wider in the direction of the galactic center, which is the direction of the constellation Sagittarius in our sky.

We see fewer stars when we look in directions pointing *away* from our location within the galactic plane, and there is relatively little gas and dust to obscure our view of more distant objects. Thus, we have a clear view to the far reaches of the universe, limited only by what our eyes or instruments allow. If you are fortunate enough to have a very dark sky, you may see a fuzzy patch in the constellation Andromeda (Figure 2.7). Although this patch may look like nothing more than a small cloud, you are actually seeing the Great Galaxy in Andromeda (M31)— some 2.5 million light-years away.

TIME OUT TO THINK *Contemplate the fact that light from M31 journeyed through space for some 2.5 million years to reach you and is the combined light of more than 100 billion stars. Do you think it possible that among some of those stars there are students like yourself gazing outward in amazement at the Milky Way Galaxy?*

FIGURE 2.6 Artist's conception of the Milky Way Galaxy from afar, showing how the galaxy's structure affects our view from Earth. When we look *into* the galactic plane in any direction, our view is blocked by stars, gas, and dust. Thus, we see the galactic plane as the band of light we call the Milky Way, stretching a full 360° around our sky (i.e., around the celestial sphere). We have a clear view to the distant universe only when we look away from the galactic plane.

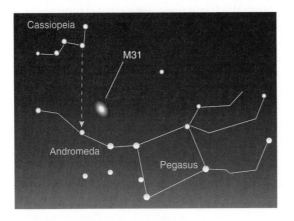

FIGURE 2.7 Location of M31 in the night sky.

2.2 The Circling Sky

If you spend a few hours out under a starry sky, you'll see stars (and the Moon and planets) rising and setting much like the Sun. In reality, it is we who are moving, not the stars. The Sun and stars appear to rise in the east and set in the west because the Earth rotates in the opposite direction, from west to east. In fact, the Earth's rotation makes the entire celestial sphere appear to rotate around us each day. If you could view the celestial sphere from outside, the daily circles of the stars would appear very simple (Figure 2.8).

However, because we live on the Earth, we see only *half* the celestial sphere at any one moment; the ground beneath us blocks our view of the other half. The particular half that we see depends on the time, the date, and our location on Earth. In this section, we'll discuss how to make sense of the sky as seen from Earth.

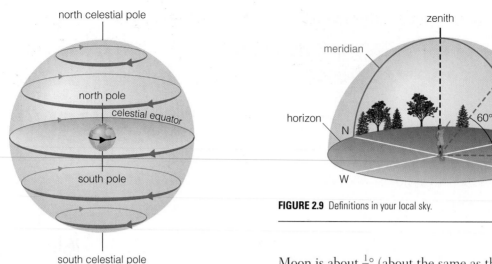

FIGURE 2.8 The Earth rotates from west to east (black arrows), making the celestial sphere *appear* to rotate around us from east to west. Viewed from outside, the stars (and the Sun, Moon, and planets) therefore appear to make simple daily circles around us. The red circles represent the apparent daily paths of a few selected stars.

FIGURE 2.9 Definitions in your local sky.

The Dome of the Sky

Picture yourself standing in a flat, open field. The sky appears to take the shape of a dome, making it easy to understand why people of many ancient cultures believed we live on a flat Earth lying under a great dome that encompasses the world. Today, we use the appearance of a dome to define the **local sky**—the sky as seen from wherever you happen to be standing (Figure 2.9). The boundary between Earth and sky is what we call the **horizon**. The point directly overhead is your **zenith**. Your **meridian** is an imaginary half-circle stretching from your horizon due south, through your zenith, to your horizon due north.

You can pinpoint the position of any object in your local sky by stating its **direction** along your horizon (more technically measured as *azimuth*) and its **altitude** above your horizon. For example, Figure 2.9 shows a person pointing to a star located in a southeasterly direction at an altitude of 60°. (The zenith has an altitude of 90° but no direction, since it is straight overhead.)

Because of our lack of depth perception in the sky, we cannot tell the true sizes of objects or the true distances between objects just by looking at them. For example, the Sun and the Moon look about the same size in our sky, but the Sun is actually about 400 times larger than the Moon in diameter. We can, however, measure *angles* in the sky. The **angular size** of an object like the Sun or the Moon is the angle it appears to span in your field of view. The **angular distance** between a pair of objects is the angle that appears to separate them. For example, Figure 2.10 shows that the angular size of the

Moon is about $\frac{1}{2}°$ (about the same as that of the Sun), while the angular distance between the "pointer stars" at the end of the Big Dipper's bowl is about 5°. You can use your outstretched hand to make rough estimates of angles in the sky (Figure 2.10c).

Variation with Latitude

If you stay in one place, you'll see the same set of stars following the same paths through your sky from one year to the next. But if you travel north or south, you'll notice somewhat different sets of stars moving through the sky on somewhat different paths. This observation convinced ancient scientists that the sky must be more than a simple dome.

According to historical records, the idea of the sky as a celestial sphere was first proposed by the Greek scientist Anaximander (c. 610–547 B.C.). Based on what he learned about how the sky changes with travel north or south, Anaximander also concluded that the Earth must not be flat. However, because *east–west* travel does *not* change the stars visible in the sky, he did not realize that the Earth is round and instead suggested that it might be a cylinder curved in the north–south direction. By about 500 B.C., the famous mathematician Pythagoras suggested that the Earth is a sphere. More than a century later, Aristotle (384–322 B.C.) cited observations of the Earth's curved shadow on the Moon during lunar eclipses as evidence for a spherical Earth.

To understand why the sky changes with north–south travel, we must first review how we locate points on the Earth (Figure 2.11). **Latitude** measures positions north or south; it is defined to be 0° at the equator, so the North Pole and the South Pole have latitude 90°N and 90°S, respectively. Note that "lines of latitude" actually are circles running parallel to the equator. **Longitude** measures east–west position, so "lines of longitude" are semicircles extending from the North Pole to the South Pole. The line of longitude passing through Greenwich, England, is

a The angular size of the Moon is about $\frac{1}{2}°$.

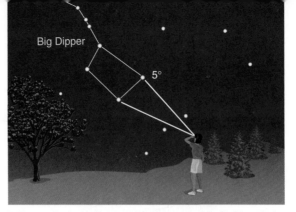

b The angular distance between the pointer stars of the Big Dipper is about 5°.

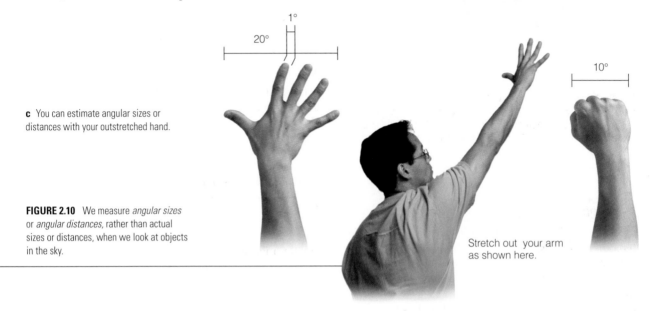

c You can estimate angular sizes or distances with your outstretched hand.

FIGURE 2.10 We measure *angular sizes* or *angular distances,* rather than actual sizes or distances, when we look at objects in the sky.

Stretch out your arm as shown here.

FIGURE 2.11 We can locate any place on Earth's surface by its latitude and longitude.

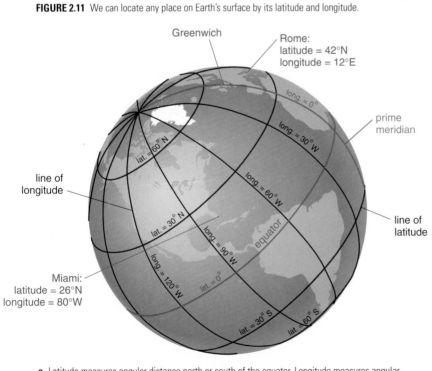

a Latitude measures angular distance north or south of the equator. Longitude measures angular distance east or west of the prime meridian, which passes through Greenwich, England.

b The entrance to the Old Royal Greenwich Observatory, near London. The line emerging from the door marks the prime meridian.

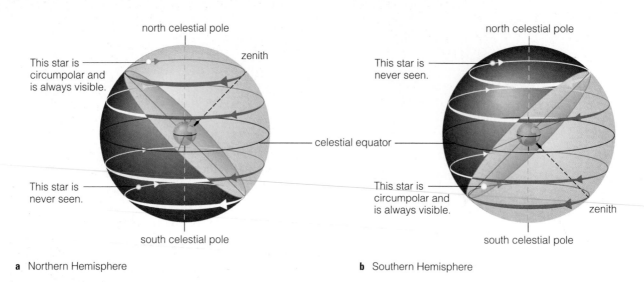

north celestial pole

zenith

This star is
circumpolar and
is always visible.

celestial equator

This star is
never seen.

south celestial pole

a Northern Hemisphere

north celestial pole

This star is
never seen.

celestial equator

This star is
circumpolar and
is always visible.

zenith

south celestial pole

b Southern Hemisphere

FIGURE 2.12 Sample local skies for (**a**) the Northern Hemisphere and (**b**) the Southern Hemisphere. Compare to Figure 2.8; note that your latitude determines your orientation relative to the celestial sphere, and the daily rotation determines the paths of stars through your local sky. To view the local skies more clearly, rotate the page so the zenith points up.

defined to be longitude 0° and is called the **prime meridian**. Stating a latitude and a longitude pinpoints a location. For example, Rome lies at about 42°N latitude and 12°E longitude and Miami lies at about 26°N latitude and 80°W longitude.

If we combine the idea of the sky's dome-shaped appearance when seen from the ground with the idea that stars make simple daily circles around the celestial sphere, we can see why the apparent paths of stars depend on *latitude*. Your latitude determines your orientation on Earth relative to the celestial sphere; "down" is toward the center of the Earth, and "up" (toward your zenith) is away from the center of the Earth. Figure 2.12 shows sample orientations relative to the celestial sphere for mid-latitudes in each hemisphere. If you study the diagrams carefully, you'll discover the following key features:

■ In the Northern Hemisphere, stars relatively near the north celestial pole remain constantly above the horizon. We say that such stars are **circumpolar**. They never rise or set but instead make daily counterclockwise circles around the north celestial pole. In the Southern Hemisphere, circumpolar stars make daily clockwise circles around the south celestial pole. Figure 2.13 shows a beautiful photograph of the daily circles of stars.

■ In the Northern Hemisphere, stars relatively near the south celestial pole remain constantly below the horizon and are never seen. In the Southern Hemisphere, stars near the north celestial pole are never visible. Thus, different sets of constellations are visible in northern and southern skies. Note that the set of constellations changes only with latitude, not with longitude.

Common Misconceptions: A Flat Earth

A widespread myth holds that Columbus proved the Earth to be round rather than flat. In fact, knowledge of the round Earth predated Columbus by nearly 2,000 years. However, it is probably true that most people in Columbus's day believed the Earth to be flat, largely because of the poor state of education. The vast majority of the public was illiterate and unaware of the scholarly evidence for a spherical Earth. Interestingly, Columbus's primary argument with other scholars concerned the distance from Europe to Asia going westward—and it was Columbus who was wrong. He underestimated the true distance and as a result was woefully unprepared for the voyage to Asia that he thought he was undertaking. Indeed, his voyages would almost certainly have ended in disaster had it not been for the presence of the Americas, which offered a safe landing well to the east of Asia.

■ All other stars (those that are neither circumpolar nor never visible) rise daily in the east and set daily in the west. Stars located north of the celestial equator on the celestial sphere rise north of due east and set north of due west, stars located on the celestial equator rise due east and set due west, and stars located south of the celestial equator rise south of due east and set south of due west.

FIGURE 2.13 This time-exposure photograph, taken at Arches National Park in Utah, shows how the Earth's rotation causes stars to trace daily circles around the sky. The north celestial pole lies at the center of the circles. Over the course of a full day, circumpolar stars trace complete circles, and stars that rise in the east and set in the west trace partial circles. Here we see only a portion of the full daily paths, because the time exposure lasted only a couple of hours.

■ If you study the geometry of the diagrams carefully, you'll see that in the Northern Hemisphere the altitude of the north celestial pole in your sky is equal to your latitude. For example, if the north celestial pole appears in your sky 40° above your horizon, your latitude is 40°N. Similarly, in the Southern Hemisphere the altitude of the south celestial pole in your sky is equal to your latitude.

The last feature listed above is very useful for navigation, since it allows you to determine your latitude just by finding the celestial pole in your sky. Finding the north celestial pole is fairly easy, because it lies very close to the star Polaris (Figure 2.14a). In the Southern Hemisphere, you can find the south celestial pole with the aid of the Southern Cross (Figure 2.14b).

TIME OUT TO THINK *Answer the following questions for your latitude: Where is the north (or south) celestial pole in your sky? Where should you look to see circumpolar stars? What portion of the celestial sphere is never visible in your sky?*

Common Misconceptions: What Makes the North Star Special?

Most people are aware that the North Star, Polaris, is a special star. Contrary to a relatively common belief, however, it is *not* the brightest star in the sky; more than 50 other stars are either considerably brighter or comparable in brightness. Polaris is special because it so closely marks the direction of due north and because its altitude in your sky is nearly equal to your latitude on Earth, making it very useful in navigation in the Northern Hemisphere.

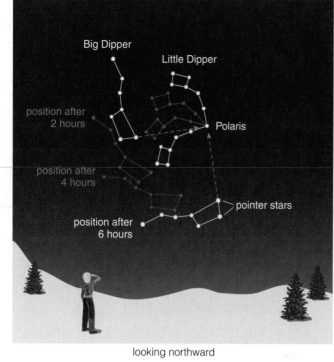

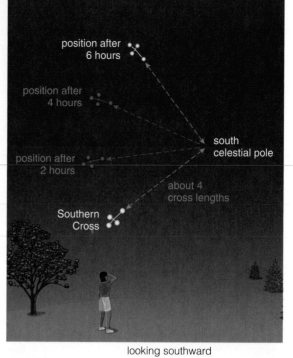

a In the Northern Hemisphere, the pointer stars of the Big Dipper point to Polaris, which lies within 1° of the north celestial pole. Note that the sky appears to turn *counterclockwise* around the north celestial pole.

b In the Southern Hemisphere, the Southern Cross points to the south celestial pole, which is not marked by any bright star. The sky appears to turn *clockwise* around the south celestial pole.

FIGURE 2.14 The altitude of the celestial pole in your sky is equal to your latitude.

Seasonal Changes in the Sky

The basic patterns of motion in the sky remain the same from one day to the next: The Sun, Moon, planets, and stars trace daily circles around the sky. However, if you observe the sky night after night, you will notice changes that cannot be seen over a single night. One pattern that becomes obvious after a few months is that the constellations visible at a particular time change with the seasons. For example, Orion is prominent in the evening sky in February, but by September it is visible only if you look shortly before dawn.

The nighttime constellations change with the seasons because of the Earth's orbit around the Sun. As shown in Figure 2.15, at different times of year we view the Sun from different places in our orbit. As a result, the Sun appears to move steadily *eastward* along a circular path through the constellations each year; we've already identified this path as the *ecliptic* (see Figure 2.4). The constellations along the ecliptic are called the constellations of the **zodiac**. (Tradition places 12 constellations along the zodiac, but the official borders include a wide swath of a thirteenth constellation, Ophiuchus.) The Sun's apparent location along the ecliptic determines which constellations you see at night. In late August, for example, the Sun appears to be in the constellation Leo, so you won't see the stars of Leo at all—they're above your horizon only during the daytime. Aquar-

ius, however, will be visible on your meridian at midnight because it is opposite Leo on the celestial sphere. Six months later, in February, it is Aquarius that cannot be seen and Leo that is on the meridian at midnight.

TIME OUT TO THINK *Based on Figure 2.15 and today's date, where among the zodiacal constellations does the Sun currently appear? What zodiacal constellation will be on your meridian at midnight? What zodiacal constellation will you see in the west shortly after sunset? Go outside at night to confirm your answers.*

The Sun's path through the daytime sky also changes with the seasons. Recall that the Earth's orientation *relative to the Sun* changes during the year because of the $23\frac{1}{2}°$ tilt of the Earth's axis (see Figure 1.13). This is also why the ecliptic crosses the celestial equator at angles of $23\frac{1}{2}°$ on the celestial sphere (see Figure 2.4). When the Sun appears north of the celestial equator, the Sun's daily path is long and high for the Northern Hemisphere (making the long, hot days of summer) but short and low for the Southern Hemisphere (making the short, cool days of winter). The situation for the two hemispheres is reversed when the Sun appears south of the celestial equator. At any particular location, you can notice this annual change by observing the Sun's position in your sky at the same time each day (Figure 2.16).

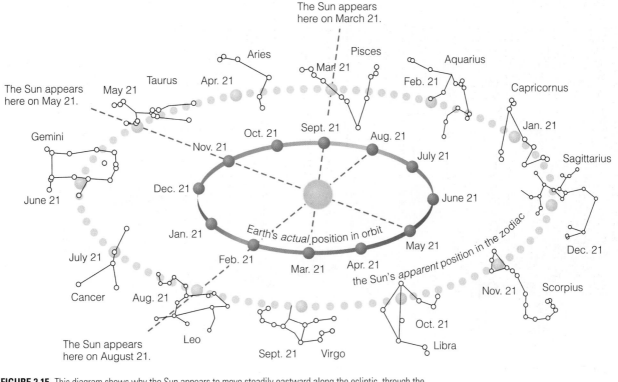

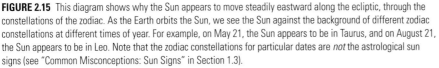

FIGURE 2.15 This diagram shows why the Sun appears to move steadily eastward along the ecliptic, through the constellations of the zodiac. As the Earth orbits the Sun, we see the Sun against the background of different zodiac constellations at different times of year. For example, on May 21, the Sun appears to be in Taurus, and on August 21, the Sun appears to be in Leo. Note that the zodiac constellations for particular dates are *not* the astrological sun signs (see "Common Misconceptions: Sun Signs" in Section 1.3).

FIGURE 2.16 This composite photograph shows images of the Sun, always at the same time of day (mean solar time), snapped at 10-day intervals over an entire year. The three bright streaks show the path of the Sun's rise on three particular dates. This photograph looks east, so north is to the left and south is to the right; it was taken in the Northern Hemisphere. Notice that the Sun's altitude varies considerably: It is high in the summer and low in the winter. The sunrise position also changes: The Sun rises north of due east in the summer and south of due east in the winter. (The "figure 8" pattern, which you'll find reproduced on many globes, is called an *analemma*.)

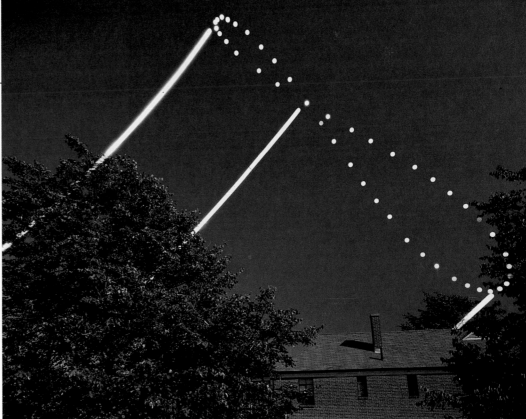

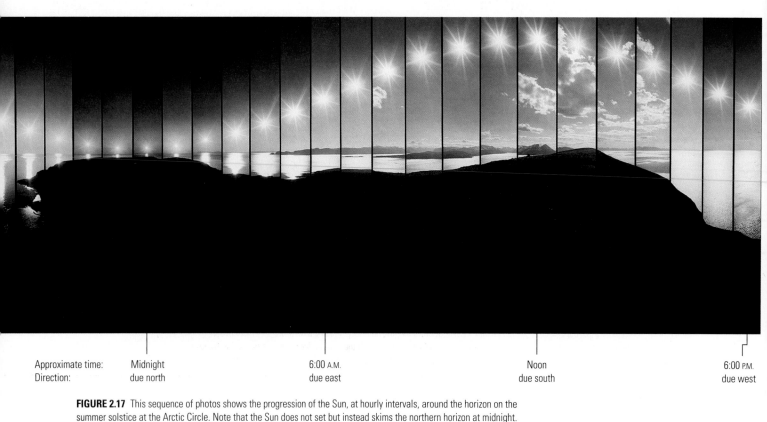

| Approximate time: | Midnight | 6:00 A.M. | Noon | 6:00 P.M. |
| Direction: | due north | due east | due south | due west |

FIGURE 2.17 This sequence of photos shows the progression of the Sun, at hourly intervals, around the horizon on the summer solstice at the Arctic Circle. Note that the Sun does not set but instead skims the northern horizon at midnight. It then gradually rises higher, reaching its highest point at noon, when it appears due south.

Moreover, just as the daily path of any star through your sky depends on your latitude, so does the daily path of the Sun at any time of year. During the summer, the Sun's daily path is longer at higher latitudes than at latitudes nearer the equator. The Sun's daily path is correspondingly shorter at higher latitudes in the winter. That is why higher latitudes have longer summer days and longer winter nights. In fact, the Sun becomes circumpolar at very high latitudes (within the *Arctic* or *Antarctic Circle*) in the summer. The Sun never sets during these summer days, giving these latitudes the name *land of the midnight Sun* (Figure 2.17). Of course, the name "land of noon darkness" is more appropriate in the winter, when the Sun never rises above the horizon at these high latitudes.

Common Misconceptions: High Noon

When is the Sun directly overhead in your sky? If you ask this question of friends and acquaintances, you'll probably find that many people answer "at noon." It's true that the Sun reaches its *highest* point each day when it crosses the meridian (hence the term "high noon," though the meridian crossing is rarely at precisely 12:00), but unless you live in the Tropics (between latitudes 23.5°S and 23.5°N), the Sun is *never* directly overhead. In fact, any time you can see the Sun as you walk around, you can be sure it is *not* at your zenith; unless you are lying down, seeing objects at the zenith requires tilting your head back into a very uncomfortable position.

2.3 The Moon, Our Constant Companion

Like all objects on the celestial sphere, each day the Moon rises in the east and sets in the west. Like the Sun, the Moon also moves gradually eastward through the constellations of the zodiac. However, it takes the Moon only about a month to make a complete circuit around the celestial sphere. If you carefully observe the Moon's position relative to bright stars over just a few hours, you can notice the Moon's drift among the constellations.

As the Moon moves through the sky, both its appearance and the time at which it rises and sets change with the cycle of **lunar phases**. Each complete cycle from one new moon to the next takes

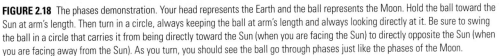

FIGURE 2.18 The phases demonstration. Your head represents the Earth and the ball represents the Moon. Hold the ball toward the Sun at arm's length. Then turn in a circle, always keeping the ball at arm's length and always looking directly at it. Be sure to swing the ball in a circle that carries it from being directly toward the Sun (when you are facing the Sun) to directly opposite the Sun (when you are facing away from the Sun). As you turn, you should see the ball go through phases just like the phases of the Moon.

about $29\frac{1}{2}$ days—hence the origin of the word *month* (think of "moonth"). The easiest way to understand the lunar phases is with a simple demonstration. Use a small ball to represent the Moon while your head represents the Earth. If it's daytime and the Sun is shining, take your ball outside and watch how you see phases as you move the ball around your head (Figure 2.18); if it's dark or cloudy, you can place a flashlight a few meters away to represent the Sun. As you hold your ball at various places in its "orbit" around your head, you'll observe that phases result from just two basic facts:

1. At any particular time, half of the ball faces the Sun (or flashlight) and therefore is bright, while the other half faces away from the Sun and therefore is dark.

2. As you look at the ball, you see some combination of its bright and dark faces. This combination is the phase of the ball.

We see lunar phases for the same reason. Half of the Moon is always illuminated by the Sun, but the amount of this illuminated half that we see from Earth depends on the Moon's position in its orbit (Figure 2.19). The time of day during which the Moon is visible also depends on its phase. For example, full moon occurs when the Moon is opposite the Sun in the sky, so the full moon rises around sunset, reaches the meridian at midnight, and sets around sunrise.

Common Misconceptions: Moon in the Daytime

In traditions and stories, night is so closely associated with the Moon that many people mistakenly believe that the Moon is visible only in the nighttime. In fact, the Moon is above the horizon as often in the daytime as at night, though it is easily visible only when its light is not drowned out by sunlight. For example, a first-quarter moon is easy to spot in the late afternoon as it rises through the eastern sky, and a third-quarter moon is visible in the morning as it heads toward the western horizon (see rise and set times in Figure 2.19). A related misconception appears in illustrations that show a star in the dark portion of the crescent moon (diagram below): A star in the dark portion appears to be in front of the Moon, which is impossible because the Moon is much closer to us than is any star.

This view, though common in art, can never occur because the star would have to be in front of the Moon.

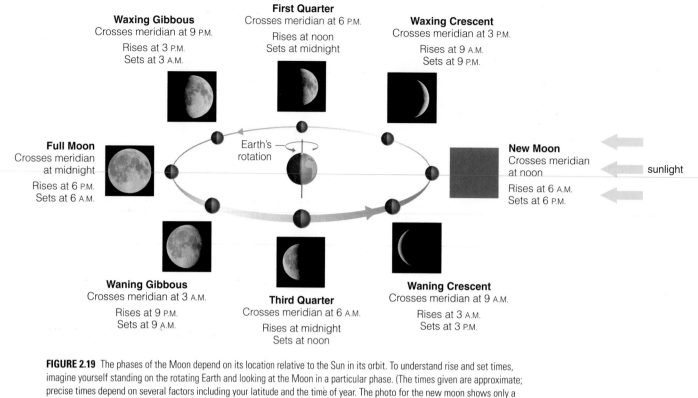

Waxing Gibbous
Crosses meridian at 9 P.M.
Rises at 3 P.M.
Sets at 3 A.M.

First Quarter
Crosses meridian at 6 P.M.
Rises at noon
Sets at midnight

Waxing Crescent
Crosses meridian at 3 P.M.
Rises at 9 A.M.
Sets at 9 P.M.

Earth's rotation

Full Moon
Crosses meridian at midnight
Rises at 6 P.M.
Sets at 6 A.M.

New Moon
Crosses meridian at noon
Rises at 6 A.M.
Sets at 6 P.M.

sunlight

Waning Gibbous
Crosses meridian at 3 A.M.
Rises at 9 P.M.
Sets at 9 A.M.

Third Quarter
Crosses meridian at 6 A.M.
Rises at midnight
Sets at noon

Waning Crescent
Crosses meridian at 9 A.M.
Rises at 3 A.M.
Sets at 3 P.M.

FIGURE 2.19 The phases of the Moon depend on its location relative to the Sun in its orbit. To understand rise and set times, imagine yourself standing on the rotating Earth and looking at the Moon in a particular phase. (The times given are approximate; precise times depend on several factors including your latitude and the time of year. The photo for the new moon shows only a blue sky, because a new moon is above the horizon in daylight but cannot be seen because it is very close to the Sun in the sky.)

TIME OUT TO THINK *Suppose you go outside in the morning and notice that the visible face of the Moon is half light and half dark. Is this a first-quarter or third-quarter moon? How do you know?* (Hint: *Study Figure 2.19.*)

Common Misconceptions: Moon on the Horizon

You've probably noticed that the full moon appears far larger when it is near the horizon than when it is high in your sky. However, this appearance is an illusion. (You can eliminate the illusion by viewing the Moon upside down between your legs when it is on the horizon.) If you measure the angular size of the full moon on a particular night, you'll find it is about the same whether it is near the horizon or high in the sky. (It actually appears slightly larger when overhead than when on the horizon, because you are viewing it from a position on Earth that is closer to the Moon by the radius of the Earth.) In fact, the Moon's angular size in the sky depends only on its distance from the Earth. Although this distance varies over the course of the Moon's monthly orbit, it does not change enough to cause a noticeable effect on a single night.

Although we see many *phases* of the Moon, we do not see many *faces*. In fact, from the Earth we always see (nearly) the same face of the Moon. This tells us that the Moon must rotate once on its axis in the same time that it makes a single orbit of the Earth, as you can observe with another simple demonstration. Place a ball on a table to represent the Earth, while you represent the Moon. Start by facing the ball. If you do not rotate as you walk around the ball, you'll be looking away from it by the time you are halfway around your orbit (Figure 2.20a). The only way you can face the ball at all times is by completing exactly one rotation while you complete one orbit (Figure 2.20b). (We'll learn why the Moon's periods of rotation and orbit are the same in Section 5.4.)

The View from the Moon

A good way to solidify your understanding of the lunar phases is to imagine that you live on the side of the Moon that faces the Earth. Look again at Figure 2.19. Note that at new moon you would be facing the day side of the Earth. Thus, you would see *full earth* when people on Earth see new moon. Similarly, at full moon you would be facing the night side of the Earth. Thus, you would see *new earth* when people on Earth see full moon. In general, you'd always see the Earth in a phase opposite the phase of the Moon that people on Earth see. Moreover, because the Moon always shows nearly the same face

a If you do not rotate while walking around the model, you will not always face it.

b You will face the model at all times only if you rotate exactly once during each orbit.

FIGURE 2.20 The fact that we always see the same face of the Moon means that the Moon must rotate once in the same amount of time that it takes to orbit the Earth once. You can see why by imagining that you are the Moon and walking around a model of the Earth.

to Earth, the Earth would appear to hang nearly stationary in your sky. In addition, because the Moon takes about a month to rotate, your "day" would last about a month; that is, you'd have about 2 weeks of daylight followed by about 2 weeks of darkness as you watched the Earth hanging in your sky and going through its cycle of phases.

Thinking about the view from the Moon clarifies another interesting feature of the lunar phases: The dark portion of the lunar face is not *totally* dark. Imagine that you are standing on the Moon when it is in a crescent phase. Because it's nearly new moon as seen from Earth, you would see nearly full earth in your sky. Just as we can see at night by the light of the Moon, the light of the Earth would illuminate your night moonscape. (In fact, because the Earth is much larger than the Moon, the full earth is much bigger and brighter in the lunar sky than the full moon is in Earth's sky.) This faint light illuminating the "dark" portion of the Moon's face is often called the *ashen light* or *earthshine*; it is the light that enables us to see the outline of the full face of the Moon even when the Moon is not full.

Eclipses

Look once more at Figure 2.19. If this figure told the whole story of the lunar phases, a new moon would always block our view of the Sun, and the Earth would always prevent sunlight from reaching a full moon. More precisely, this figure makes it look as if the Moon's shadow should fall on Earth during new moon and that Earth's shadow should fall on the Moon during full moon. Any time one astronomical object casts a shadow on another, we say that an **eclipse** is occurring. Thus, Figure 2.19 makes it look as if we should have an eclipse with every new moon and every full moon—but we don't.

Common Misconceptions: The "Dark Side" of the Moon

The term *dark side of the Moon* really should be used to mean the side facing away from the Sun. Unfortunately, *dark side* traditionally meant what would better be called the *far side*—the hemisphere that never can be seen from the Earth. Many people still refer to the far side as the "dark side," even though this side is not necessarily dark. For example, during new moon the far side faces the Sun and hence is completely sunlit; the only time the far side actually is completely dark is at full moon, when it faces away from both the Sun and the Earth.

The missing piece of the story in Figure 2.19 is that the Moon's orbit is inclined to the ecliptic plane by about 5°. An easy way to visualize this inclination is to imagine the ecliptic plane as the surface of a pond, as shown in Figure 2.21. Because of the inclination of its orbit, the Moon spends most of its time either above or below this surface. It crosses *through* this surface only twice during each orbit: once coming out and once going back in. The two points in each orbit at which the Moon crosses the surface are called the **nodes** of the Moon's orbit.

Figure 2.21 shows the position of the Moon's orbit at several different times of year. Note that the nodes are aligned the same way in each case (diagonally on the page). As a result, the nodes lie in a straight line with the Earth and the Sun only about twice each year. (For reasons we'll discuss shortly, it is not *exactly* twice each year.) Because an eclipse

FIGURE 2.21 This painting represents the ecliptic plane as the surface of a pond. The Moon's orbit is slightly tilted to the ecliptic plane. Thus, in this illustration, the Moon spends half of each orbit above the pond surface and half below the surface; the points at which the orbit crosses the surface represent the *nodes* of the Moon's orbit. Eclipses occur only when the Moon both is at a node (splashing through the pond surface) *and* has a phase of either new moon or full moon—as is the case with the lower left and top right orbits shown. At all other times, new moons and full moons occur above or below the ecliptic plane, so no eclipse is possible. (The figure is not to scale.)

can occur only when the Earth, Moon, and Sun lie along a straight line, two conditions must be met simultaneously for an eclipse to occur:

1. The nodes of the Moon's orbit must be nearly aligned with the Earth and the Sun.

2. The phase of the Moon must be either new or full.

There are two basic types of eclipse. A **lunar eclipse** occurs when the Moon passes through the Earth's shadow and therefore can occur only at *full moon*. A **solar eclipse** occurs when the Moon's shadow falls on the Earth and therefore can occur only at *new moon*. But the full story of eclipse types is more complex, because the shadow of the Moon or the Earth consists of two distinct regions: a central **umbra** where sunlight is completely blocked and a surrounding **penumbra** where sunlight is only partially blocked (Figure 2.22). Thus, an umbral shadow is totally dark, while a penumbral shadow is only slightly darker than no shadow.

Lunar Eclipses A lunar eclipse begins at the moment that the Moon's orbit first carries it into the Earth's penumbra. After that, we will see one of three types of lunar eclipse (Figure 2.23). If the Sun, the Earth, and the Moon are nearly perfectly aligned, the Moon will pass through the Earth's umbra, and we will see a **total lunar eclipse**. If the alignment is somewhat less perfect, only part of the full moon will pass through the umbra (with the rest in the penumbra), and we will see a **partial lunar eclipse**. If the Moon passes *only* through the Earth's penumbra, we will see a **penumbral lunar eclipse**.

Penumbral eclipses are the most common type of lunar eclipse, but they are difficult to notice because the full moon darkens only slightly. Partial lunar eclipses are easier to see because the Earth's umbral shadow always clearly darkens part of the Moon's face. Note that the Earth's umbra always casts a curved shadow on the Moon, demonstrating that

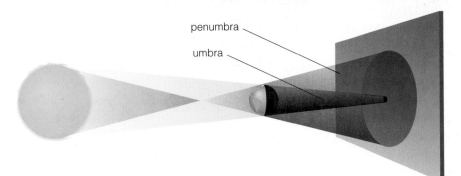

FIGURE 2.22 The shadow cast by an object in sunlight. Sunlight is fully blocked in the umbra and partially blocked in the penumbra.

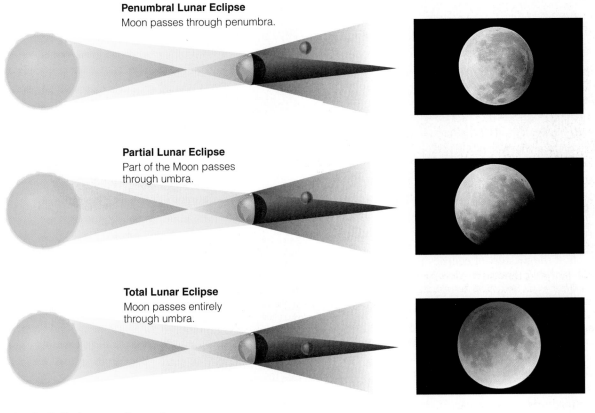

Penumbral Lunar Eclipse
Moon passes through penumbra.

Partial Lunar Eclipse
Part of the Moon passes through umbra.

Total Lunar Eclipse
Moon passes entirely through umbra.

FIGURE 2.23 The three types of lunar eclipse.

the Earth is round. A total lunar eclipse is particularly spectacular, because the Moon becomes dark and eerily red during **totality** (the time during which the Moon is entirely engulfed in the umbra)—dark because it is in shadow, and red because the Earth's atmosphere bends some of the red light from the Sun around the Earth and toward the Moon.

Solar Eclipses We can also see three types of solar eclipse (Figure 2.24). If a solar eclipse occurs when the Moon is relatively close to the Earth in its orbit, the Moon's umbra touches a small area of the Earth (no more than about 270 kilometers in diameter). Anyone within this area will see a **total solar**

eclipse. Surrounding this region of totality is a much larger area (typically about 7,000 kilometers in diameter) that falls within the Moon's penumbral shadow. Anyone within this region will see a **partial solar eclipse**, in which only part of the Sun is blocked from view. If the eclipse occurs when the Moon is relatively far from the Earth, the umbra may not reach the Earth at all. In that case, anyone in the small region of the Earth directly behind the umbra will see an **annular eclipse**, in which a ring of sunlight surrounds the disk of the Moon. (Again, anyone in the surrounding penumbral shadow will see a partial solar eclipse.) During any solar eclipse, the combination of the Earth's rotation and the orbital motion of the Moon causes the circular umbral and penumbral shadows to race across the face of the Earth at a typical speed of about 1,700 kilometers per hour (relative to the ground). As a result, the umbral (or annular) shadow traces a narrow path across the Earth, and totality (or annularity) never lasts more than a few minutes in any particular place.

A total solar eclipse is a spectacular sight. It begins when the disk of the Moon first appears to touch the Sun. Over the next couple of hours, the Moon appears to take a larger and larger "bite" out of the Sun. As totality approaches, the sky darkens and temperatures fall. Birds head back to their nests, and crickets begin their nighttime chirping. During the few minutes of totality, the Moon completely blocks the normally visible disk of the Sun, allowing the faint *corona* to be seen (Figure 2.25). The surrounding sky takes on a twilight glow, and planets and bright stars become visible in the daytime. As totality ends, the Sun slowly emerges from behind the Moon over the next couple of hours. However, because your eyes have adapted to the darkness, totality appears to end far more abruptly than it began.

Predicting Eclipses Few phenomena have so inspired and humbled humans throughout the ages as eclipses. For many cultures, eclipses were mystical events associated with fate or the gods, and countless stories and legends surround eclipses.

Much of the mystery of eclipses probably stems from the fact that they are relatively difficult to predict. Look again at Figure 2.21. The two periods each year when the nodes of the Moon's orbit are nearly aligned with the Sun are called **eclipse seasons**. Each eclipse season lasts a few weeks, so some type of lunar eclipse occurs during each eclipse season's full moon, and some type of solar eclipse occurs during its new moon. If Figure 2.21 told the whole story, eclipse seasons would occur every 6 months, and predicting eclipses would be easy. For example, if eclipse seasons always occurred in January and July, eclipses would occur only on the dates of new and full moons in those months. But Figure 2.21 does not show one important fact about the Moon's orbit: The nodes slowly precess around the orbit. As a result, eclipse seasons actually occur slightly less than 6 months apart (about 173 days apart) and therefore do not recur in the same months year after year.

The combination of the changing dates of eclipse seasons and the $29\frac{1}{2}$-day cycle of lunar phases makes eclipses recur in a cycle of about 18 years $11\frac{1}{3}$ days. For example, if a solar eclipse were to occur today, another would occur 18 years $11\frac{1}{3}$ days from now. This roughly 18-year cycle is called the **saros cycle**.

Astronomers in many ancient cultures identified the saros cycle and therefore could predict *when* eclipses would occur. However, the saros cycle does not account for all the complications involved in predicting eclipses. If a solar eclipse occurred today, the one that would occur 18 years $11\frac{1}{3}$ days from now would not be visible from the same places on the Earth and might not be of the same type (e.g., one might be total and the other only partial). As a result, no ancient culture achieved the ability to predict eclipses in every detail. Today, eclipses can be predicted because we know the precise details of the orbits of the Earth and the Moon; many astronomical software packages can do the necessary calculations. Table 2.1 lists upcoming total lunar eclipses, and Figure 2.26 shows paths of totality for total solar eclipses through 2022. It is well worth your while to plan a trip to see a total solar eclipse.

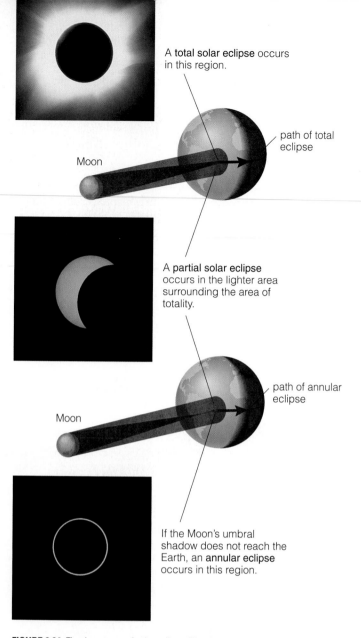

A **total solar eclipse** occurs in this region.

Moon

path of total eclipse

A **partial solar eclipse** occurs in the lighter area surrounding the area of totality.

Moon

path of annular eclipse

If the Moon's umbral shadow does not reach the Earth, an **annular eclipse** occurs in this region.

FIGURE 2.24 The three types of solar eclipse. (The photographs on the left are, from top to bottom, a total solar eclipse, a partial solar eclipse, and an annular eclipse.)

FIGURE 2.25 This multiple-exposure photograph shows the progression of a total solar eclipse. Totality (central image) lasts only a few minutes, during which time we can see the faint corona around the outline of the Sun.

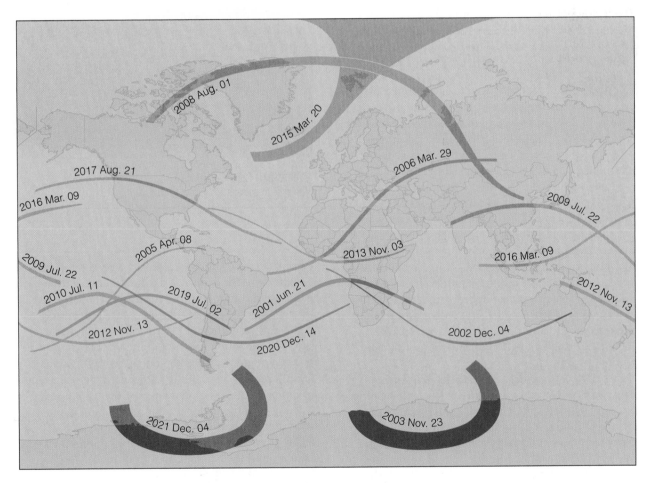

FIGURE 2.26 This map shows the paths of totality for solar eclipses from 2001 to 2022. Paths of the same color represent eclipses occurring in successive saros cycles, separated by 18 years 11$\frac{1}{3}$ days. For example, the 2019 eclipse occurs 18 years 11 days after the 2001 eclipse (both shown in green).

Table 2.1 Total Lunar Eclipses 2002–2010

May 16, 2003
November 9, 2003
May 4, 2004
October 28, 2004
March 3, 2007
August 28, 2007
February 21, 2008
December 21, 2010

2.4 The Ancient Mystery of the Planets

Five planets are easy to find with the naked eye: Mercury, Venus, Mars, Jupiter, and Saturn. Mercury can be seen only infrequently, and then only just after sunset or before sunrise because it is so close to the Sun. Venus often shines brightly in the early evening in the west or before dawn in the east; if you see a very bright "star" in the early evening or early morning, it probably is Venus. Jupiter, when it is visible at night, is the brightest object in the sky besides the Moon and Venus. Mars is recognizable by its red color, but be careful not to confuse it with a bright red star. Saturn is easy to see with the naked eye, but because many stars are just as bright as Saturn, it helps to know where to look. (Also, planets tend not to twinkle as much as stars.) Sometimes several planets may appear close together in the sky, offering a particularly beautiful sight (Figure 2.27).

Like the Sun and the Moon, the planets appear to move slowly through the constellations of the zodiac. (The word *planet* comes from the Greek for "wandering star.") However, although the Sun and the Moon always appear to move eastward relative to the stars, the planets occasionally reverse course and appear to move *westward* through the zodiac. A period during which a planet appears to move westward relative to the stars is called a period of **apparent retrograde motion** (*retrograde* means "backward"). Figure 2.28 shows a period of apparent retrograde motion for Jupiter.

Ancient astronomers could easily "explain" the daily paths of the stars through the sky by imagining that the celestial sphere was real and that it really rotated around the Earth each day. But the apparent retrograde motion of the planets posed a far greater mystery: What could cause the planets sometimes to go backward? As we'll discuss in Chapter 3, the ancient Greeks came up with some very clever ways to explain the occasional backward motion of the planets. But these ideas never allowed them to pre-

FIGURE 2.27 This photograph shows Mercury, Venus, Jupiter, and Saturn appearing close together in the sky on February 23, 1999.

dict planetary positions in the sky with great accuracy (though the accuracy was sufficient for their needs at that time), because they were wedded to the incorrect idea of an Earth-centered universe.

In contrast, apparent retrograde motion has a simple explanation in a Sun-centered solar system. You can demonstrate it for yourself with the help of a friend (Figure 2.29). Pick a spot in an open field to represent the Sun. You can represent the Earth, walking counterclockwise around the Sun, while your friend represents a more distant planet (e.g., Mars, Jupiter, or Saturn) by walking counterclockwise around the Sun at a greater distance. Your friend should walk more slowly than you because more distant planets orbit the Sun more slowly. As you walk, watch how your friend appears to move relative to buildings or trees in the distance. Although both of you always walk the same way around the Sun, your friend will *appear* to move backward against

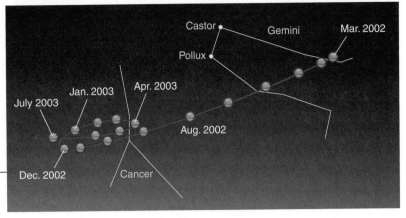

FIGURE 2.28 This diagram shows Jupiter's approximate position among the stars in our sky during 2002–2003. Jupiter generally appears to drift eastward among the stars, but for about 4 months each year it turns back toward the west. Here, this apparent retrograde motion occurs between about December 2002 and April 2003. Note that Jupiter moves about one-twelfth of the way through the zodiac each year because it takes 12 years to complete one orbit of the Sun.

Castor Gemini Mar. 2002
Pollux
July 2003 Jan. 2003 Apr. 2003
Aug. 2002
Dec. 2002 Cancer

Dots represent Jupiter's approximate position at 1-month intervals.
(Jupiter not to scale.)

the background during the part of your "orbit" at which you catch up to and pass him or her. (To understand the apparent retrograde motions of Mercury and Venus, which are closer to the Sun than is the Earth, simply switch places with your friend and repeat the demonstration.)

The apparent retrograde motion demonstration applies directly to the planets. For example, because Mars takes about 2 years to orbit the Sun (actually 1.88 years), it covers about half its orbit during the 1 year in which Earth makes a complete orbit. If you trace lines of sight from Earth to Mars from different points in their orbits, you will see that the line of sight usually moves eastward relative to the stars but moves westward during the time when Earth is passing Mars in its orbit (Figure 2.30). Like your friend in the demonstration, Mars never actually changes direction; it only *appears* to change direction from our perspective on Earth.

If the apparent retrograde motion of the planets is so readily explained by recognizing that the Earth is a planet, why wasn't this idea accepted in ancient times? In fact, the idea that the Earth goes around the Sun was suggested as early as 260 B.C. by the Greek astronomer Aristarchus but did not gain widespread acceptance until the seventeenth century. Although there were many reasons for the historic reluctance to abandon the idea of an Earth-centered universe, perhaps the most prominent involved the inability of ancient peoples to detect something called **stellar parallax**.

Extend your arm and hold up one finger. If you keep your finger still and alternately close your left eye and right eye, your finger will appear to jump back and forth against the background. This apparent shifting, called *parallax*, occurs simply because your two eyes view your finger from opposite sides of your nose. Note that if you move your finger closer to your face, the parallax increases. In contrast, if you look at a distant tree or flagpole instead of your finger, you probably cannot detect any parallax by alternately closing your left eye and right eye. Thus, parallax depends on distance, with nearer objects exhibiting greater parallax than more distant objects.

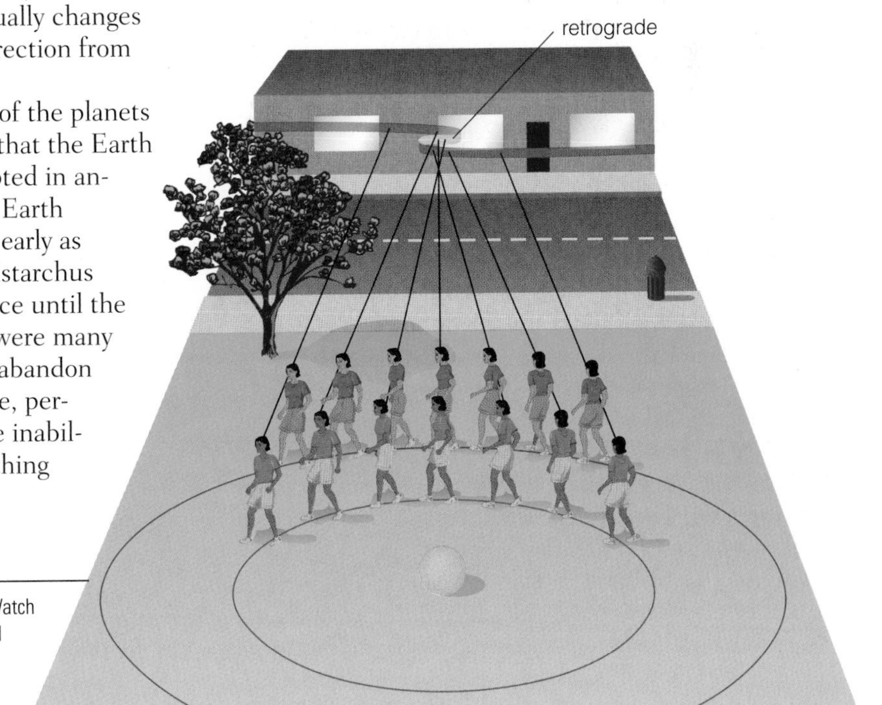

retrograde

FIGURE 2.29 The retrograde motion demonstration. Watch how your friend usually appears to you to move forward against the background of the building in the distance but appears to move backward as you catch up to and pass him or her in your "orbit."

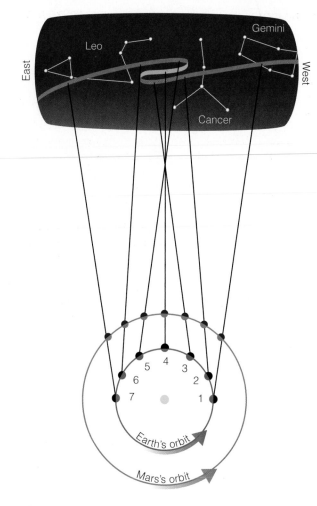

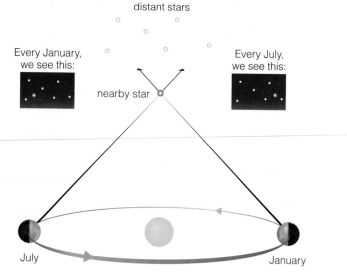

FIGURE 2.31 Stellar parallax is an apparent shift in the position of a nearby star as we look at it from different places in the Earth's orbit. This figure is greatly exaggerated; in reality, the amount of shift is far too small to detect with the naked eye.

FIGURE 2.30 The explanation for apparent retrograde motion. Follow the lines of sight from Earth to Mars in numerical order. The period during which the lines of sight shift *westward* relative to the distant stars is the period during which we observe apparent retrograde motion for Mars.

If you now imagine that your two eyes represent the Earth at opposite sides of its orbit around the Sun and that your finger represents a relatively nearby star, you have the idea of stellar parallax. That is, because we view the stars from different places in our orbit at different times of year, nearby stars should *appear* to shift back and forth against distant stars in the background during the course of the year (Figure 2.31). (The Greeks actually expected stellar parallax in a slightly different way, because they believed that all stars lay on the same celestial sphere: They assumed that, if Earth orbited the Sun, at different times of year we would be closer to different parts of the celestial sphere, which would change the angular separations of stars.) However, no matter how hard they searched, ancient astronomers could find no sign of stellar parallax. They therefore concluded that one of the following must be true:

1. The Earth orbits the Sun but the stars are so far away that stellar parallax is undetectable to the naked eye.

2. There is no stellar parallax because the Earth doesn't move; it is the center of the universe.

Unfortunately, with notable exceptions such as Aristarchus, ancient astronomers rejected the correct answer (1) because they could not imagine that the stars could be *that* far away. Today, we can detect stellar parallax with the aid of telescopes, thereby providing direct proof that the Earth really does orbit the Sun. Careful measurements of stellar parallax also provide the most reliable means of measuring distances to nearby stars [Section 13.2].

TIME OUT TO THINK *How far apart are opposite sides of the Earth's orbit? How far away are the nearest stars? Describe the challenge of detecting stellar parallax. It may help to visualize the Earth's orbit and the distance to stars on the 1-to-10-billion scale used in Chapter 1.*

Thus, the ancient mystery of the planets drove much of the historical debate over the Earth's place in the universe. In many ways, the modern technological society that we take for granted today can be traced directly back to the scientific revolution that began because of the quest to explain the slow wandering of the planets among the stars in our sky.

THINKING ABOUT . . .

Aristarchus

Until the early 1600s, nearly everyone believed that the Earth was the center of the universe. Yet Aristarchus (c. 310–230 B.C.) argued otherwise almost 2,000 years earlier.

Aristarchus proposed a Sun-centered system in about 260 B.C. Little of Aristarchus's work survives to the present day, so we cannot know why he made this proposal. It may have been an attempt to find a more natural explanation for the apparent retrograde motion of the planets. To account for the lack of detectable stellar parallax, Aristarchus suggested that the stars were extremely far away.

He further strengthened his argument by estimating the sizes of the Moon and the Sun. By observing the shadow of the Earth on the Moon during a lunar eclipse, Aristarchus estimated the Moon's diameter to be about one-third of the Earth's diameter—only slightly higher than the actual value. He then used a geometric argument, based on measuring the angle between the Moon and the Sun at first- and third-quarter phases, to conclude that the Sun must be larger than the Earth. (His measurements

were imprecise, so he estimated the Sun's diameter to be about 7 times the Earth's, rather than the correct value of about 100 times.) His conclusion that the Sun is larger than the Earth may have been another reason why he felt that the Earth should orbit the Sun, rather than vice versa.

Like most scientific work, Aristarchus's work built upon the work of others. In particular, Heracleides (c. 388–315 B.C.) had suggested that the Earth rotates; Aristarchus may have drawn on this idea to explain the apparent daily rotation of the stars in our sky. Heracleides also was the first to suggest that not all heavenly bodies circle the Earth; based on the fact that Mercury and Venus always are close to the Sun in the sky, he argued that these two planets must orbit the Sun. Thus, in suggesting that all the planets orbit the Sun, Aristarchus was extending the ideas of Heracleides and others before him. Alas, Aristarchus's arguments were not widely accepted in ancient times and were revived only with the work of Copernicus some 1,800 years later.

THE BIG PICTURE

In this chapter, we surveyed the phenomena of our sky. Keep the following "big picture" ideas in mind as you continue your study of astronomy:

- You can enhance your enjoyment of learning astronomy by spending time outside observing the sky. The more you learn about the appearance and apparent motions of the sky, the more you will appreciate what you can see in the universe.

- From our vantage point on Earth, it is convenient to imagine that we are at the center of a great celestial sphere—even though we really are on a planet orbiting a star in a vast universe. We can then understand what we see in the local sky by thinking about how the celestial sphere appears from our latitude.

- Most of the phenomena of the sky are relatively easy to observe and understand. But the more complex phenomena, particularly eclipses and apparent planetary motion, challenged our ancestors for thousands of years and helped drive the development of science and technology.

Review Questions

1. What is a *constellation?* How is a constellation related to a pattern of stars in the sky?

2. What is the *celestial sphere?* Describe the major features shown on a model of the celestial sphere.

3. What is the *ecliptic,* and how is it related to the ecliptic plane?

4. What is the *Milky Way* in our sky, and how is it related to the Milky Way Galaxy?

5. Why does the local sky look like a dome? Define *horizon, zenith,* and *meridian.* Describe how you can locate an object in the local sky by its altitude and its direction along the horizon.

6. Explain why we can measure only *angular* sizes and distances for objects in the sky.

7. Briefly describe how and why the sky varies with latitude.

8. Briefly describe how and why the sky changes with the seasons.

9. Describe the Moon's cycle of *phases* and explain why we see phases of the Moon.

10. Why don't we see an eclipse at every new and full moon? Describe the conditions that must be met to see a solar or lunar eclipse.

11. Why are eclipses so difficult to predict? Explain the importance of *eclipse seasons* and the *saros cycle* to eclipse prediction.

12. What is the *apparent retrograde motion* of the planets? Why was it difficult for ancient astronomers to explain but easy for us to explain?

13. What is *stellar parallax?* Describe the role it played in making ancient astronomers believe in an Earth-centered universe.

Discussion Questions

1. *Geocentric Language.* Many common phrases reflect the ancient Earth-centered view of our universe. For example, the phrase "the Sun rises each day" implies that the Sun is really moving over the Earth. In fact, the Sun only *appears* to rise as the rotation of the Earth carries us to a place where we can see the Sun in our sky. Identify other common phrases that imply an Earth-centered viewpoint.

2. *Flat Earth Society.* Believe it or not, there is an organization called the Flat Earth Society, whose members hold that the Earth is flat and that all indications to the contrary (such as pictures of the Earth from space) are fabrications made as part of a conspiracy to hide the truth from the public. Discuss the evidence for a round Earth and how you can check it for yourself. In light of the evidence, is it possible that the Flat Earth Society is correct? Defend your opinion.

Problems

Sensible Statements? For **problems 1–8**, decide whether the statement is sensible and explain why it is or is not. (For an example, see Chapter 1 problems.)

1. If you had a very fast spaceship, you could travel to the celestial sphere in about a month.

2. The constellation Orion didn't exist when my grandfather was a child.

3. When I looked into the dark fissure of the Milky Way with my binoculars, I saw what must have been a cluster of distant galaxies.

4. Last night the Moon was so big that it stretched for a mile across the sky.

5. I live in the United States, and during my first trip to Argentina I saw many constellations that I'd never seen before.

6. Last night I saw Jupiter right in the middle of the Big Dipper. (*Hint:* Is the Big Dipper part of the zodiac?)

7. Last night I saw Mars move westward through the sky in its apparent retrograde motion. (*Hint:* How long does it take to notice apparent retrograde motion?)

8. Although all the known stars appear to rise in the east and set in the west, we might someday discover a star that will appear to rise in the west and set in the east.

9. *View from Afar.* Describe how the Milky Way Galaxy would look in the sky of someone observing from a planet around a star in M31.

10. *Your View.*
    a. Find your latitude and longitude, and state the source of your information.
    b. Describe the altitude and direction in your sky at which the north or south celestial pole appears.
    c. Is Polaris a circumpolar star in your sky? Explain.
    d. Describe the path of the meridian in your sky.
    e. Describe the path of the celestial equator in your sky. (*Hint:* Study Figure 2.12.)

11. *View from the Moon.* Suppose you lived on the Moon, near the center of the face that we see from Earth.
    a. During the phase of full moon, what phase would you see for Earth? Would it be daylight or dark where you live?
    b. On Earth, we see the Moon rise and set in our sky each day. If you lived on the Moon, would you see the Earth rise and set? Why or why not? (*Hint:* Remember that the Moon always keeps the same face toward Earth.)
    c. What would you see when people on Earth were experiencing an eclipse? Answer for both solar and lunar eclipses.

12. *A Farther Moon.* Suppose the distance to the Moon was twice its actual value. Would it still be possible to have a total solar eclipse? An annular eclipse? A total lunar eclipse? Explain.

13. *A Smaller Earth.* Suppose the Earth was smaller in size. Would solar eclipses be any different? If so, how? What about lunar eclipses? Explain.

Web Projects

Find useful links for Web projects on the text Web site.

1. *Sky Information.* Search the Web for sources of daily information about sky phenomena (such as lunar phases, times of sunrise and sunset, and unusual sky events). Identify and briefly describe your favorite source.

2. *Constellations.* Search the Web for information about the constellations and their mythology. Write a short report about one or more constellations.

3. *Upcoming Eclipse.* Find information about an upcoming solar or lunar eclipse that you might have a chance to witness. Write a short report about how you could best witness the eclipse, including any necessary travel to a viewing site, and what you can expect to see. Bonus: Describe how you could photograph the eclipse.

PART II
Key Concepts for Astronomy

CHAPTER 3

The Science of Astronomy

Today we know that the Earth is a planet orbiting a rather ordinary star, in a galaxy of a hundred billion or more stars, in an incredibly vast universe. We know that the Earth, along with the entire cosmos, is in constant motion. We know that, on the scale of cosmic time, human civilization has existed only for the briefest moment. Yet we have acquired all this knowledge only recently in human history. How have we managed to learn these things?

It wasn't easy. Astronomy is the oldest of the sciences, with roots extending as far back as recorded history allows us to see. But while our current understanding of the universe rests on foundations laid long ago, the most impressive advances in knowledge have come in just the past few centuries.

In this chapter, we will trace how modern astronomy arose from its roots in ancient observations, including those of the Greeks. We'll then pay special attention to the unfolding of the Copernican revolution, which not only overturned the ancient belief in an Earth-centered universe but also laid the foundation for nearly all of modern science. Finally, we'll explore the nature of modern science and the scientific method.

3.1 Everyday Science

A common stereotype holds that scientists are men and women in white lab coats who somehow think differently than other people. In reality, scientific thinking is a fundamental part of human nature.

Think about how a baby behaves. By about a year of age, she notices that objects fall to the ground when she drops them. She lets go of a ball; it falls. She pushes a plate of food from her high chair; it falls too. She continues to drop all kinds of objects, and they all plummet to Earth. Through powers of observation, the baby learns about the physical world: Things fall when they are unsupported. Eventually, she becomes so certain of this fact that, to her parents' delight, she no longer needs to test it continually.

One day somebody gives the baby a helium balloon. She releases it; to her surprise, it rises to the ceiling! Her rudimentary understanding of physics must be revised. She now knows that the principle "all things fall" does not represent the whole truth, although it still serves her quite well in most situations. It will be years before she learns enough about the atmosphere, the force of gravity, and the concept of density to understand *why* the balloon rises when most other objects fall. For now, she is delighted to observe something new and unexpected.

The baby's experience with falling objects and balloons exemplifies scientific thinking, which, in essence, is a way of learning about nature through careful observation and trial-and-error experiments. Rather than thinking differently than other people, modern scientists simply are trained to organize this everyday thinking in a way that makes it easier for them to share their discoveries and employ their col-lective wisdom. This organizational scheme is what we call the *scientific method*; it is the method by which humanity has acquired its present physical knowledge of the universe.

TIME OUT TO THINK *When was the last time you used trial and error to learn something? Describe a few cases where you have learned by trial and error in cooking, participating in sports, fixing something, or any other situation.*

Just as it takes years for a child to learn to communicate through language, art, or music, it took humanity a long time to develop the principles of the scientific method. In its modern form, the scientific method requires painstaking attention to detail, relentless testing of each piece of information to ensure its reliability, and a willingness to give up old beliefs that are not consistent with observed facts about the physical world. For professional scientists, the demands of the scientific method are the "hard work" part of the job. At heart, professional scientists are like the baby with the balloon, delighted by the unexpected and motivated by those rare moments when they—and all of us—learn something new about the universe.

3.2 Ancient Observations

We will discuss the modern scientific method shortly, but first we will explore how it arose from the observations of ancient peoples. Our exploration begins in central Africa, where people of many indigenous societies predict the weather with reasonable accuracy

---

FIGURE 3.1 This diagram shows how central Africans used the Moon to predict the weather. The graph depicts the annual rainfall pattern in central Nigeria, characterized by a wet season and a dry season. The Moon diagrams represent the orientation of a waxing crescent moon relative to the western horizon at different times of year. The angle of the crescent "horns" allows observers to determine the time of year and hence the expected rainfall. (The orientation is measured in degrees, where 0° means that the crescent horns are parallel to the horizon.)

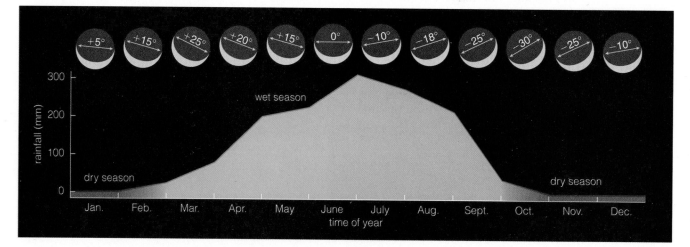

by making careful observations of the Moon. The Moon begins its monthly cycle as a crescent in the western sky just after sunset. Through long traditions of sky watching, central African societies learned that the orientation of the crescent "horns" relative to the horizon is closely tied to rainfall patterns (Figure 3.1).

No one knows when central Africans first developed the ability to predict weather using the lunar crescent. But the earliest-known written astronomical record comes from central Africa near the border of modern-day Congo and Uganda. It consists of an animal bone (known as the *Ishango* bone) etched with patterns that appear to be part of a lunar calendar, probably carved around 6500 B.C.

Why did ancient people bother to make such careful and detailed observations of the sky? In part, it was probably because they were just as curious and intelligent as we are today. In the daytime, they surely recognized the importance of the Sun to their lives. At night, without electric light, they were much more aware of the starry sky than we are today. Thus, it's not surprising that they paid attention to patterns of motion in the sky and developed ideas and stories to explain what they saw.

Astronomy played a practical role in ancient societies by enabling them to keep track of time and seasons, a crucial skill for people who depended on agriculture for survival. This ability may seem quaint today, when digital watches tell us the precise time and date, but it required considerable knowledge and skill in ancient times, when the only clocks and calendars were in the sky.

Modern measures of time still reflect their ancient astronomical roots. Our 24-hour day is the time it takes the Sun to circle our sky. The length of a month comes from the lunar cycle, and our calendar year is based on the cycle of the seasons. The days of the week are named after the seven naked-eye objects that appear to move among the constellations: the Sun, the Moon, and the five planets recognized in ancient times (Table 3.1).

FIGURE 3.2 This ancient Egyptian obelisk, which stands 83 feet tall and weighs 331 tons, resides in St. Peter's Square at the Vatican in Rome. It is one of 21 surviving obelisks from ancient Egypt, most of which are now scattered around the world. Shadows cast by the obelisks may have been used to tell time.

Determining the Time of Day

In the daytime, ancient peoples could tell time by observing the Sun's path through the sky. Many cultures probably used the shadows cast by sticks as simple sundials. The ancient Egyptians built huge obelisks, often inscribed or decorated in homage to the Sun, that probably also served as simple clocks (Figure 3.2).

At night, the Moon's position and phase give an indication of the time (see Figure 2.19). For example, a first-quarter moon sets around midnight, so it is not yet midnight if the first-quarter moon is still above the western horizon. The positions of the stars also indicate the time if you know the approximate date. For example, in December the constellation Orion rises around sunset, reaches the meridian around midnight, and sets around sunrise. Hence, if it is winter and Orion is setting, dawn must be approaching. Most ancient peoples probably were adept at estimating the time of night, although written evidence is sparse.

Table 3.1 The Seven Days of the Week and the Astronomical Objects They Honor

The correspondence between objects and days is easy to see in French and Spanish. In English, the correspondence becomes clear when we look at the names of the objects used by the Teutonic tribes who lived in the region of modern-day Germany.

Object	Teutonic Name	English	French	Spanish
Sun	Sun	Sunday	dimanche	domingo
Moon	Moon	Monday	lundi	lunes
Mars	Tiw	Tuesday	mardi	martes
Mercury	Woden	Wednesday	mercredi	miércoles
Jupiter	Thor	Thursday	jeudi	jueves
Venus	Fria	Friday	vendredi	viernes
Saturn	Saturn	Saturday	samedi	sábado

FIGURE 3.3 **(a)** Stonehenge today. **(b)** A sketch showing how archaeologists believe Stonehenge looked when construction was completed around 1550 B.C. Note, for example, that observers standing in the center would see the Sun rise directly over the Heel Stone on the summer solstice.

Determining the Time of Year

Many cultures built structures to help them mark the seasons. One of the oldest standing human-made structures served such a purpose: Stonehenge in southern England, which was constructed in stages from about 2750 B.C. to about 1550 B.C. (Figure 3.3). Observers standing in its center see the Sun rise directly over the Heel Stone only on the summer solstice. Stonehenge also served as a social gathering place and probably as a religious site; no one knows whether its original purpose was social or astronomical. Perhaps it was built for both—in ancient times, social rituals and practical astronomy probably were deeply intertwined.

One of the most spectacular structures used to mark the seasons was the Templo Mayor in the Aztec city of Tenochtitlán, located on the site of modern-day Mexico City (Figure 3.4). Twin temples surmounted a flat-topped, 150-foot-high pyramid. From the location of a royal observer watching from the opposite side of the plaza, the Sun rose directly through the notch between the twin temples on the equinoxes. Like Stonehenge, the Templo Mayor served important social and religious functions in addition to its astronomical role. Before it was destroyed by the Conquistadors, other Spanish visitors reported stories of elaborate rituals, sometimes including human sacrifice, that took place at the Templo Mayor at times determined by astronomical observations.

Many ancient cultures aligned their buildings with the cardinal directions (north, south, east, and west), enabling them to mark the rising and setting of the Sun relative to the building orientation. Some even created monuments that had a single special

astronomical purpose. Perhaps in a practical form of ancient art, someone among the ancient Anasazi people carved a spiral known as the Sun Dagger on a vertical cliff face near the top of a butte in Chaco Canyon, New Mexico (Figure 3.5). The Sun's rays form a dagger of sunlight that pierces the center of the carved spiral only once each year—at noon on the summer solstice.

Lunar Cycles

Many ancient civilizations paid particular attention to the lunar cycle, often using it as the basis for lunar calendars. The months on lunar calendars generally have either 29 or 30 days, chosen to make the average agree with the approximately $29\frac{1}{2}$-day lunar cycle. A 12-month lunar calendar has only 354 or 355 days, or about 11 days fewer than a calendar based on the tropical year. Such a calendar is still used in the Muslim religion, which is why the month-long fast of Ramadan (the ninth month) begins about 11 days earlier with each subsequent year.

Other lunar calendars take advantage of the fact that 19 years is almost precisely 235 lunar months. That is, every 19 years we get the same lunar phases on about the same dates. Because this fact was discovered in 432 B.C. by the Greek astronomer Meton, the 19-year period is called the **Metonic cycle**. A lunar calendar can be synchronized to the Metonic cycle by adding a thirteenth month to 7 of every 19 years (making exactly 235 months in each 19-year period), thereby ensuring that "new year" comes on approximately the same date every nineteenth year. The Jewish calendar follows the Metonic cycle,

FIGURE 3.4 This scale model shows the Templo Mayor and the surrounding plaza as they are thought to have looked before Aztec civilization was destroyed by the Conquistadors.

adding a thirteenth month in the third, sixth, eighth, eleventh, fourteenth, seventeenth, and nineteenth years of each cycle.

Some ancient cultures learned to predict eclipses by recognizing the 18-year saros cycle [Section 2.3]. In the Middle East, the ancient Babylonians achieved remarkable success in predicting eclipses more than 2,500 years ago. Perhaps the most successful eclipse predictions prior to modern times were made by the Mayans in Central America. The Mayan calendar featured a sacred cycle that almost certainly was re-

lated to eclipses. This Mayan cycle, called the *sacred round*, lasted 260 days—almost exactly $1\frac{1}{2}$ times the 173.32 days between successive eclipse seasons [Section 2.3]. Unfortunately, we know little more about the extent of Mayan knowledge, because the Spanish Conquistadors burned most Mayan writings.

The complexity of the Moon's orbit leads to other long-term patterns in the Moon's appearance. For example, the full moon rises at its most southerly point along the eastern horizon only once every 18.6 years, a phenomenon that may have been

FIGURE 3.5 The Sun Dagger is an arrangement of rocks that shape the Sun's light into a dagger, with a spiral carved on a rock face behind them. The dagger pierces the center of the carved spiral each year at noon on the summer solstice.

FIGURE 3.6 The Moon rising between two stones of the 4,000-year-old sacred stone circle at Callanish, Scotland (on the Isle of Lewis in the Scottish Hebrides). The full moon rises in this position only once every 18.6 years.

observed from the 4,000-year-old sacred stone circle at Callanish, Scotland (Figure 3.6).

Observations of Planets and Stars

Many cultures also made careful observations of the planets and stars. Mayan observatories in Central America, such as the one still standing at Chichén Itzá (Figure 3.7), had windows strategically placed for observations of Venus.

Observational aids could also mark the rising and setting of the stars. More than 800 lines, some stretching for miles, are etched in the dry desert sand of Peru between the Ingenio and Nazca Rivers. Many of these lines may simply have been well-traveled pathways, but others are aligned in directions that point to places where bright stars or the Sun rose at particular times of year. In addition to the many straight lines, the desert features many large figures of animals, which may be representations of constellations made by the Incas who lived in the region (Figure 3.8).

TIME OUT TO THINK *The animal figures show up clearly only when seen from above. As a result, some UFO enthusiasts argue that the patterns must have been created by aliens. What do you think of this argument? Defend your opinion.*

The great Incan cities of South America clearly were built with astronomy in mind. Entire cities

FIGURE 3.7 The ruins of the Mayan observatory at Chichén Itzá.

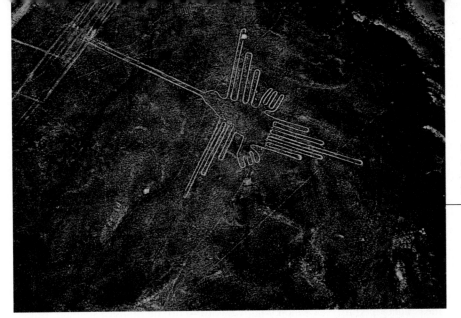

FIGURE 3.8 Hundreds of lines and patterns are etched in the sand of the Nazca desert in Peru. This aerial photo shows a large etched figure of a hummingbird.

appear to have been designed so that particular arrangements of roads, buildings, or other human-made structures would point to places where bright stars rose or set, or where the Sun rose and set at particular times of year.

Structures for astronomical observation also were popular in North America. Lodges built by the Pawnee people in Kansas featured strategically placed holes for observing the passage of constellations that figured prominently in their folklore. In the northern plains of the United States, Native American Medicine Wheels probably were designed for astronomical observations. The "spokes" of the Medicine Wheel at Big Horn, Wyoming, were aligned with the rising and setting of bright stars, as well as with the rising and setting of the Sun on the equinoxes and solstices (Figure 3.9). The 28 spokes of the Medicine Wheel probably relate to the month of the Native Americans, which they measured as 28 days (rather than 29 or 30 days) because they did not count the day of the new moon.

Perhaps the people most dependent on knowledge of the stars were the Polynesians, who lived and traveled among the many islands of the mid- and South Pacific. Because the next island in a journey usually was too distant to be seen, poor navigation meant becoming lost at sea. As a result, the most esteemed position in Polynesian culture was that of the Navigator, a person who had acquired the detailed knowledge necessary to navigate great distances among the islands. The Navigators employed a combination of detailed knowledge of astronomy and equally impressive knowledge of the patterns of waves and swells around different islands (Figure 3.10). The stars provided their broad navigational sense, pointing them in the correct direction of their intended destination. As they neared a destination,

FIGURE 3.9 The Big Horn Medicine Wheel in Wyoming. Note the 28 "spokes" radiating out from the center. These spokes probably relate to the month of the Native Americans.

FIGURE 3.10 A traditional Polynesian navigational instrument.

the wave and swell patterns guided them to their precise landing point. The Navigator memorized all his skills and passed them to the next generation through a well-developed program for training future Navigators. Unfortunately, with the advent of modern navigational technology, many of the skills of the Navigators have been lost.

From Observation to Science

Before a structure such as Stonehenge could be built, careful observations had to be made and repeated over and over to ensure their validity. Careful, repeatable observations also underlie the modern scientific method. To this extent, elements of modern science were present in many early human cultures.

The degree to which scientific ideas developed in different societies depended on practical needs, social and political customs, and interactions with other cultures. Because all of these factors can change, it should not be surprising that different cultures were more scientifically or technologically advanced than others at different times in history. The ancient Chinese kept remarkably detailed records of astronomical observations beginning at least 5,000 years ago (Figure 3.11).

Yet, by the end of the 1600s, Chinese science and technology had fallen behind that of Europe. Some historians argue that a primary reason for this decline was that the Chinese tended to regard their science and technology as state secrets. This secrecy may have slowed Chinese scientific development by preventing the broad-based collaborative science

that fueled the European advance beginning in the Renaissance.

In Central America, the ancient Mayans also were ahead of their time in many ways. For example, their system of numbers and mathematics looks distinctly modern; their invention of the concept of zero preceded by some 500 years its introduction in the Eurasian world (by Hindu mathematicians, around A.D. 600). The Aztecs, Incas, Anasazi, and other ancient peoples of the Americas may have been quite advanced in many other areas as well, but few written records survive to tell the tale.

It appears that virtually all cultures employed scientific thinking to varying degrees. Had the circumstances of history been different, any one of these many cultures might have been the first to develop what we consider to be modern science. But, in the end, history takes only one of countless possible paths, and the path that led to modern science emerged from the ancient civilizations of the Mediterranean and the Middle East.

3.3 The Modern Lineage

By 3000 B.C., civilization was well established in two major regions of the Middle East: Egypt and Mesopotamia (Figure 3.12). Their geographical location placed these civilizations at a crossroads for travelers, merchants, and armies of Europe, Asia, and Africa. This mixing of cultures fostered creativity, and the broad interactions among peoples ensured that new ideas spread throughout the region. Over the next 2,500 years, numerous great cultures arose. For example, the ancient Egyptians built the Great Pyramids between 2700 and 2100 B.C., using their astronomical knowledge to orient the Pyramids with the cardinal directions. They also invented papyrus scrolls and ink-based writing. The Babylonians invented methods of writing on clay tablets and developed arithmetic to serve in commerce and later in astronomical calculations. Many more of our modern principles of commerce, law, religion, and science originated with the cultures of Egypt and Mesopotamia.

The development of principles of modern science accelerated with the rise of Greece as a power in the Middle East, beginning around 500 B.C. In 330 B.C., Alexander the Great led the expansion of the Greek empire throughout the Middle East, absorbing all the former empires of Egypt and Mesopotamia. Alexander had a keen interest in science and education, perhaps fueled by his association with Aristotle, who was his personal tutor. He encouraged the pursuit of knowledge and respect for foreign cultures. On the Nile delta in Egypt, he founded the city

FIGURE 3.11 This photo shows a model of the celestial sphere and other instruments on the roof of the ancient astronomical observatory in Beijing. The observatory was built in the 1400s; the instruments shown here were built later and show the influence of Jesuit missionaries.

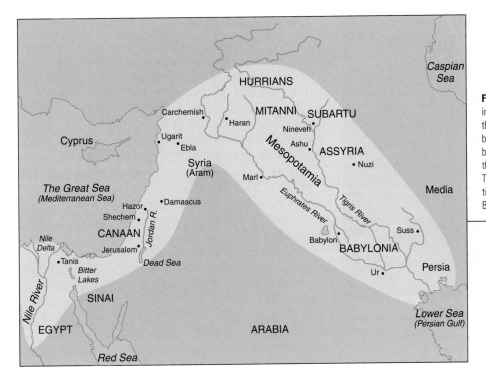

FIGURE 3.12 Map of the Middle East in ancient times. Mesopotamia was the ancient Greek name for the region between the Tigris and Euphrates Rivers, but modern historians use the name for the entire region today occupied by Iraq. Three major cultures thrived at various times in ancient Mesopotamia: Sumer, Babylonia, and Assyria.

THINKING ABOUT . . .

Eratosthenes Measures the Earth

In a remarkable ancient feat, the Greek astronomer and geographer Eratosthenes (c. 276–196 B.C.) estimated the size of the Earth in about 240 B.C. He did it by comparing the altitude of the Sun on the summer solstice in the Egyptian cities of Syene (modern-day Aswan) and Alexandria.

Eratosthenes knew that the Sun passed directly overhead in Syene on the summer solstice (it lies on the tropic of Cancer). He also knew that in the city of Alexandria to the north the Sun came within only 7° of the zenith on the summer solstice. He therefore concluded that Alexandria must be 7° of latitude to the north of Syene (Figure 3.13). Because 7° is $\frac{7}{360}$ of a circle, he concluded that the north-south distance between Alexandria and Syene must be $\frac{7}{360}$ of the circumference of the Earth.

Eratosthenes estimated the north-south distance between Syene and Alexandria to be 5,000 stadia (the *stadium* was a Greek unit of distance). Thus, he concluded that

$$\frac{7}{360} \times \text{circumference of Earth} = 5{,}000 \text{ stadia}$$

from which he found the Earth's circumference to be about 250,000 stadia.

Today, we don't know exactly what distance a stadium meant to Eratosthenes; however, based on the actual sizes of Greek stadiums, it must have been about $\frac{1}{6}$ kilometer. Thus, Eratosthenes estimated the circumference of the Earth to be about $\frac{250{,}000}{6} = 42{,}000$ kilometers—remarkably close to the modern value of just over 40,000 kilometers.

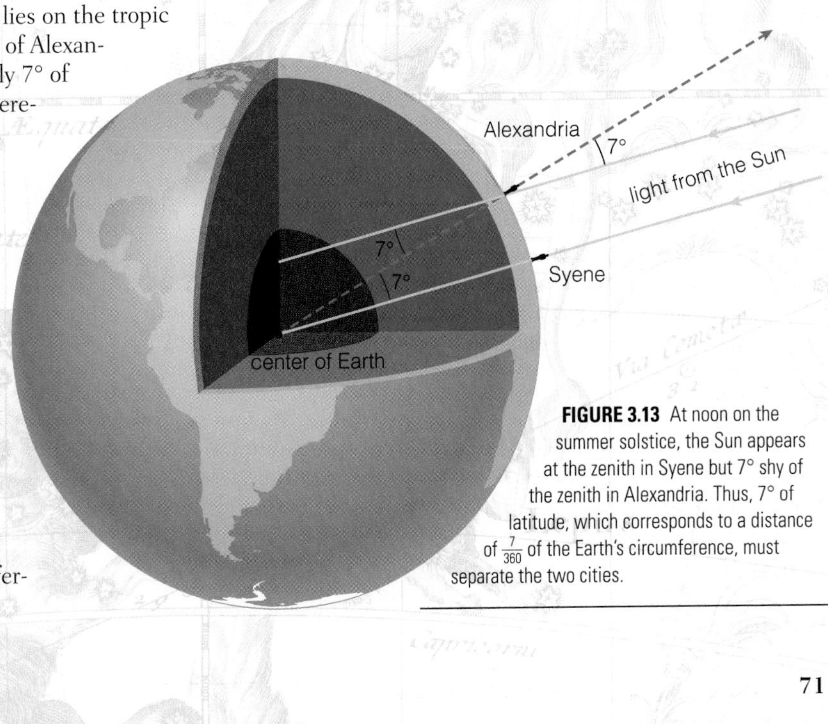

FIGURE 3.13 At noon on the summer solstice, the Sun appears at the zenith in Syene but 7° shy of the zenith in Alexandria. Thus, 7° of latitude, which corresponds to a distance of $\frac{7}{360}$ of the Earth's circumference, must separate the two cities.

FIGURE 3.14 This rendering, based on scholarly research, shows the Great Hall in the Library of Alexandria.

of Alexandria, which he hoped would become a center of world culture.

Although Alexander died at the age of 35, his dream came true. His successors continued to build Alexandria, including the construction of a great library and research center around 300 B.C. (Figure 3.14). The Library of Alexandria was the world's preeminent center of research for the next 700 years. Its end is associated with the death of Hypatia, perhaps the most prominent female scholar of the ancient world. Hypatia was a resident scholar of the Library of Alexandria, director of the observatory in Alexandria, and one of the leading mathematicians and astronomers of her time. Unfortunately, she lived during a time of rising sentiment against free inquiry and was murdered by anti-intellectual mobs in A.D. 415. The final destruction of the Library of Alexandria followed not long after her death.

At its peak, the Library of Alexandria held more than a half million books, handwritten on papyrus scrolls. As the invention of the printing press was far in the future, most of the scrolls probably were original manuscripts or the single copies of original manuscripts. When the library was destroyed, most of its storehouse of ancient wisdom was lost forever.

TIME OUT TO THINK *How old is the school that you attend? Compare its age to the 700 years that the Library of Alexandria survived. Estimate the number of books you've read in your life. Compare this number to the half million books once housed in the Library of Alexandria.*

The Ptolemaic Model of the Universe

Perhaps the most important scientific idea developed by the ancient Greeks was that of creating **models** of nature. Just as a model airplane is a representation of a real airplane, a scientific model seeks to represent some aspect of nature. However, scientific models usually are conceptual models rather than miniature representations. The purpose of a scientific model is to explain and predict real phenomena, without any need to invoke myth, magic, or the supernatural. The ancient Greek idea of a celestial sphere is an example of a scientific model: The celestial sphere is the model, and it can be used to explain and predict the apparent motions of stars in our sky.

By combining the concept of modeling with advances in logic and mathematics, the ancient Greeks practiced a pursuit of knowledge very much like that of modern science. They used their models to make predictions, such as predicting where the planets should appear among the constellations of the night sky. If the predictions proved inaccurate, they changed and refined their models. In this way, the Greeks soon learned that the simple idea of a celestial sphere could not explain all the phenomena they saw in the sky (particularly the apparent retrograde motion of the planets). They therefore added spheres for the Sun, the Moon, and each of the planets and continued to change their ideas of how the spheres moved in an ongoing attempt to make the model agree with reality (Figure 3.15).

The culmination of this ancient modeling came with the work of Claudius Ptolemy (c. A.D. 100–170),

FIGURE 3.15 This diagram depicts the heavenly spheres, an ancient Greek model of the universe (c. 200 B.C.).

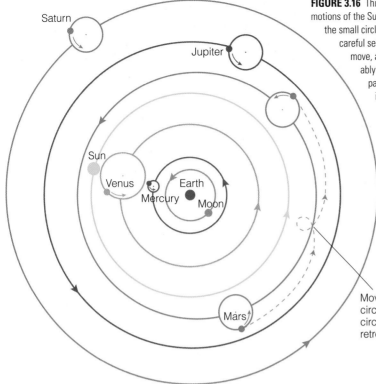

FIGURE 3.16 This diagram shows several key features of Ptolemy's model of the motions of the Sun, Moon, and planets around the Earth. The planets move around the small circles at the same time that these circles move around the Earth. By careful selection of the sizes of the circles and the rates at which planets move, along with a few other adjustments, this model made reasonably accurate predictions of planetary positions and showed their apparent retrograde motions. In order to show the circle-upon-circle idea clearly, this diagram exaggerates several features of the true Ptolemaic model; for example, for each retrograde loop, Mars moves much farther in the actual model than shown here.

Movement of small circles upon larger circles explained retrograde motion.

pronounced *tol-e-mee*. Ptolemy's model of the universe is a tribute to human ingenuity in that it made reasonably accurate predictions despite being built on the flawed premise of an Earth-centered universe. Following a Greek tradition dating back to Plato (428–348 B.C.), Ptolemy imagined that all heavenly motions must proceed in perfect circles. To explain the apparent retrograde motion of planets, Ptolemy used an idea first suggested by Apollonius (c. 240–190 B.C.) and further developed by Hipparchus (c. 190–120 B.C.). This idea held that each planet moved along a small circle that, in turn, moved around a larger circle (Figure 3.16). As seen from Earth, this "circle upon circle" motion meant that a planet usually moved eastward relative to the stars but sometimes moved westward in apparent retrograde motion.

Ptolemy selected the sizes of the circles and the rates of motion of the planets along the circles so that his model reproduced the apparent retrograde motion of all the known planets. He also incorporated more complex ideas into his model, such as allowing the Earth to be slightly off-center from the circles of a particular planet, so that it could forecast future planetary positions to within a few degrees—which was considered quite accurate at that time. Ptolemy's model remained the standard until the time of the Copernican revolution, some 1,400 years later.

The Islamic Role

Most of the great scholarly work of the ancient Greeks was lost with the fall of the Roman Empire and the destruction of the Library of Alexandria in the fifth century A.D. Much more would have been lost had it not been for the rise of a new center of intellectual achievement in Baghdad (present-day Iraq). While European civilization fell into the period of intellectual decline known as the Dark Ages, scholars of the new religion of Islam sought knowledge of mathematics and astronomy in hopes of better understanding the wisdom of Allah. During the eighth and ninth centuries A.D., scholars working in the Muslim empire centered in Baghdad translated and thereby saved many of the ancient Greek works.

Around A.D. 800, an Islamic leader named Al-Mamun (A.D. 786–833) established a "House of Wisdom" in Baghdad comparable in scope to the destroyed Library of Alexandria. Founded in a spirit of great openness and tolerance, the House of Wisdom employed Jews, Christians, and Muslims, all working together in scholarly pursuits. Using the translated Greek scientific manuscripts as building blocks, these scholars developed algebra and many new instruments and techniques for astronomical observation. Most of the official names of constellations and stars come from Arabic because of the work of the scholars at Baghdad. If you look at a star chart, you will see that the names of many bright stars begin with *al* (e.g., Aldebaran, Algol), which simply means "the" in Arabic. Ptolemy's book describing his astronomical system is also known by its Arabic name, *Almagest,* from Arabic words meaning "the greatest compilation."

The Islamic world of the Middle Ages was in frequent contact with Hindu scholars from India, who in turn brought knowledge of ideas and discoveries from China. Hence, the intellectual center in Baghdad achieved a synthesis of the surviving work of the ancient Greeks and that of the Indians and the Chinese. The accumulated knowledge of the Arabs spread throughout the Byzantine Empire (the eastern part of the former Roman Empire). When the Byzantine capital of Constantinople (modern-day Istanbul) fell to the Turks in 1453, many Eastern scholars headed west to Europe, carrying with them the knowledge that helped ignite the European Renaissance.

3.4 The Copernican Revolution

As we have seen, many different cultures developed scientific modes of thinking, and the ancient Greeks developed principles of modeling very similar to those underlying the modern scientific method. But there are important differences between these ancient ideas and modern science. For example, Ptolemy's writings suggest that he designed his model to aid in calculations, with little concern for whether it reflected reality. While modern scientists also employ purely computational models when necessary, understanding the underlying reality of nature is an important goal of modern science.

The principles that govern modern science coalesced during the European Renaissance. Within 100 years after the fall of Constantinople, Copernicus began the process of overturning the Earth-centered,

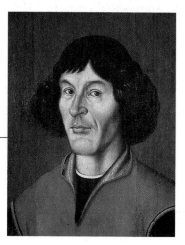

Copernicus (1473–1543)

Ptolemaic model, setting into motion a revolution that fundamentally changed the human view of the cosmos and spurred the development of the modern scientific method. In this section, we follow the dramatic story of the Copernican revolution.

Nicholas Copernicus: The Revolution Begins

Copernicus was born in Toruń, Poland, on February 19, 1473. His family was wealthy, and he received a first-class education, studying mathematics, medicine, and law. He began studying astronomy in his late teens. By that time, tables of planetary motion based on the Ptolemaic model of the universe were noticeably inaccurate. Copernicus concluded that planetary motion could be explained more simply in a Sun-centered solar system, as had been suggested by Aristarchus some 1,800 years earlier [Section 2.4], and he began developing a Sun-centered system for predicting planetary positions.

Copernicus was hesitant to publish his work for fear that his suggestion that the Earth moved would be considered absurd. Nevertheless, he discussed his system with other scholars and generated great interest in his work. At the urging of some of these scholars, including some high-ranking officials of the Catholic Church, he finally agreed to publish a book describing his system. The book, *De Revolutionibus Orbium Caelestium,* or "Concerning the Revolutions of the Heavenly Spheres," was published in 1543. Copernicus saw the first printed copy on the day he died—May 24, 1543.

In addition to its aesthetic advantages, the Sun-centered system of Copernicus allowed him to discover a mathematical relationship between a planet's true orbital period around the Sun and the time between successive appearances of the planet at a position opposite the Sun in the sky. He was also able to use geometrical techniques to estimate the distances of the planets from the Sun in terms of the Earth–Sun distance (i.e., distances in astronomical units). However, the model published by Copernicus did not predict planetary positions substantially more accurately than did the old Ptolemaic model. Moreover, it remained complex because Copernicus held to the ancient Greek belief that all heavenly motions must follow perfect circles. Because the true orbits of the planets are *not* circles, Copernicus found it necessary to add circles upon circles to his system, just as in the Ptolemaic system. As a result, the Copernican system won relatively few converts in the 50 years after it was published. After all, why throw out thousands of years of tradition for a new system that predicted planetary motion equally poorly?

Tycho Brahe (1546–1601)

Tycho Brahe: The Greatest Naked-Eye Observer of All Time

Part of the difficulty faced by astronomers who sought to improve either the Ptolemaic or the Copernican system was a lack of quality data. The telescope had not yet been invented, and existing naked-eye observations were not very accurate. In the late 1500s, a Danish nobleman named Tycho Brahe (1546–1601) set about correcting this problem.

When Tycho was a young boy, his family discouraged his interest in astronomy. He therefore hid his passion, learning the constellations from a miniature model of a celestial sphere that he kept hidden. As he grew older, Tycho was often arrogant about both his noble birth and his learned abilities. At age 20, he fought a duel with another student over which of them was the better mathematician. Part of his nose was cut off, so he designed a replacement piece made of silver and gold.

In 1563, Tycho decided to observe a widely anticipated conjunction of Jupiter and Saturn. To his surprise, the conjunction occurred nearly two days later than Copernicus had predicted. He resolved to improve the state of astronomical prediction and set about compiling careful observations of stellar and planetary positions in the sky.

Tycho's fame grew after he observed what he called a *nova*, meaning "new star," in 1572 and proved that it was at a distance much farther away than the Moon. (He compared observations made by other astronomers at other locations on Earth to his own, proving that the nova had no observable parallax and must be more distant than the Moon.) Today, we know that Tycho saw a *supernova*—the explosion of a distant star. In 1577, Tycho observed a comet and proved that it too lay in the realm of the heavens; others, notably Aristotle, had argued that comets were phenomena of the Earth's atmosphere. King

Frederick II of Denmark then decided to sponsor Tycho's ongoing work, providing him with money to build an unparalleled observatory for naked-eye observations. (After Frederick II died in 1588, Tycho moved to Prague, where his work was supported by German emperor Rudolf II.)

Over a period of three decades, Tycho and his assistants compiled naked-eye observations accurate to within less than 1 arcminute (an arcminute is $\frac{1}{60}$ of 1°). Because the telescope was invented shortly after his death, Tycho's data remain the best set of naked-eye observations ever made.

Despite the quality of his observations, Tycho never succeeded in coming up with a satisfying explanation for planetary motion. He did, however, succeed in finding someone who could: In 1600, he hired a young German astronomer named Johannes Kepler (1571–1630). Kepler and Tycho had a strained relationship while Tycho was living. But in 1601, as Tycho lay on his deathbed, he begged Kepler to find a system that would make sense of the observations so "that it may not appear I have lived in vain."

Tycho Brahe in his naked-eye observatory.

Kepler's Reformation: The Laws of Planetary Motion

Kepler was deeply religious and believed that understanding the geometry of the heavens would bring him closer to God. Kepler, like Copernicus, believed that Earth and the other planets traveled around the Sun in circular orbits, and he worked diligently to match circular motions to Tycho's data.

Kepler worked with particular intensity to find an orbit for Mars, which posed the greatest difficulties in matching the data to a circular orbit. After years of calculation, Kepler found a circular orbit that matched all of Tycho's observations of Mars's position along the ecliptic (east-west) to within 2 arcminutes. However, the same model did not correctly predict Mars's positions north or south of the ecliptic. Because Kepler sought a physically realistic orbit for Mars, he could not (as Ptolemy and Copernicus had done) tolerate one model for the east-west positions and another for the north-south positions. He attempted to find a unified model with a circular orbit, but in doing so he found that some of his predictions differed from Tycho's observations by as much as 8 arcminutes.

Kepler surely was tempted to ignore these discrepancies and attribute them to errors by Tycho. After all, 8 arcminutes is barely one-fourth the angular diameter of the full moon. But Kepler trusted Tycho's careful work, and the misses by 8 arcminutes finally led him to abandon the idea of circular orbits—

Kepler (1571–1630)

and to find the correct solution to the ancient riddle of planetary motion. About this event, Kepler wrote:

> *If I had believed that we could ignore these eight minutes [of arc], I would have patched up my hypothesis accordingly. But, since it was not permissible to ignore, those eight minutes pointed the road to a complete reformation in astronomy.*

FIGURE 3.17 Basic characteristics of circles and ellipses.

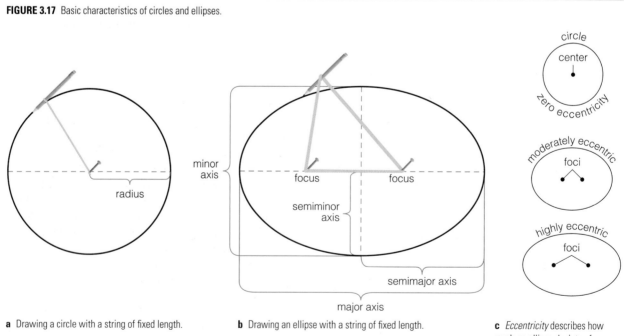

a Drawing a circle with a string of fixed length.

b Drawing an ellipse with a string of fixed length.

c *Eccentricity* describes how much an ellipse deviates from a perfect circle. (It is the center-to-focus distance divided by the length of the semimajor axis.)

Kepler summarized his discoveries with three simple laws that we now call **Kepler's laws of planetary motion**. He published the first two laws in 1610 and the third in 1618.

Kepler's key discovery was that planetary orbits are not circles but instead are a special type of oval called an **ellipse**. You probably know how to draw a circle by putting a pencil on the end of a string, tacking the string to a board, and pulling the pencil around (Figure 3.17a). Drawing an ellipse is similar, except that you must stretch the string around *two* tacks (Figure 3.17b). The locations of the two tacks are called the **foci** (singular, **focus**) of the ellipse. By altering the distance between the two foci while keeping the same length of string, you can draw ellipses of varying **eccentricity**, a quantity that describes how much an ellipse deviates from a perfect circle (Figure 3.17c): A circle has zero eccentricity, and greater eccentricity means a more elongated ellipse.

Kepler's first law states that *the orbit of each planet about the Sun is an ellipse with the Sun at one focus* (Figure 3.18). (There is nothing at the other focus.) This law tells us that a planet's distance from the Sun varies during its orbit: It is closest at the point called **perihelion**, and farthest at the point called **aphelion**. (*Helios* is Greek for the Sun, the prefix *peri* means "near," and the prefix *ap* (or *apo*) means "away." Thus, *perihelion* means "near the Sun" and *aphelion* means "away from the Sun.") The *average* of a planet's perihelion and aphelion distances is called its **semimajor axis**; we will refer to this simply as the planet's average distance from the Sun.

Kepler's second law states that *as a planet moves around its orbit, it sweeps out equal areas in equal times*. As shown in Figure 3.19, this means that the planet moves a greater distance when it is near perihelion than it does in the same amount of time near aphelion; that is, the planet travels faster when it is nearer to the Sun and slower when it is farther from the Sun.

Kepler's third law tells us that more-distant planets move more slowly in their orbits. More precisely, it describes how a planet's *orbital period* (the time it takes to complete one orbit of the Sun), measured in years, is related to its average distance from the Sun in astronomical units (1 AU ≈ 150 million km):

(orbital period in years)^2 = (average distance in AU)^3

This formula is often written more simply as $p^2 = a^3$, where p is the orbital period measured in years and a is the average distance measured in AU. Note that a planet's period does not depend on the eccentricity of its orbit: All orbits with the same semimajor axis have the same period. Nor does the period depend on the mass of the planet: Any object located an average of 1 AU from the Sun would orbit the Sun in

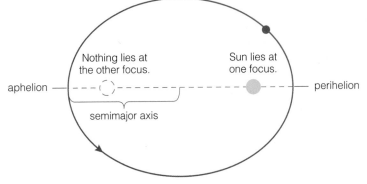

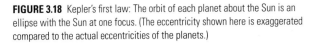

FIGURE 3.18 Kepler's first law: The orbit of each planet about the Sun is an ellipse with the Sun at one focus. (The eccentricity shown here is exaggerated compared to the actual eccentricities of the planets.)

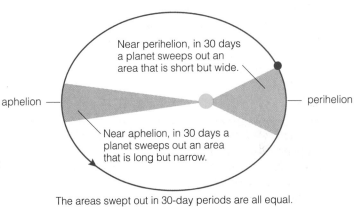

The areas swept out in 30-day periods are all equal.

FIGURE 3.19 Kepler's second law: As a planet moves around its orbit, it sweeps out equal areas in equal times.

a year (as long as the object's mass was small compared to the Sun's mass).

Galileo: The Death of the Earth-Centered Universe

The success of Kepler's laws in matching Tycho's data provided strong evidence in favor of Copernicus's placement of the Sun, rather than the Earth, at the center of the solar system. Nevertheless, many scientists still voiced reasonable objections to the Copernican view. There were three basic objections, all rooted in the 2,000-year-old beliefs of Aristotle and other ancient Greeks. First, Aristotle had held that the Earth could not be moving because, if it were, objects such as birds, falling stones, and clouds would be left behind as the Earth moved along its way. Second, the idea of noncircular orbits contradicted the ancient Greek belief that the heavens—the realm of the Sun, Moon, planets, and stars—must be perfect and unchanging. The third objection was the ancient argument that stellar parallax ought to be detectable if the Earth orbits the Sun [Section 2.4].

Galileo (1564–1642), a contemporary and correspondent of Kepler, answered all three objections.

Galileo defused the first objection with experiments that almost single-handedly overturned the Aristotelian view of physics. In particular, Galileo demonstrated that a moving object remains in motion *unless* a force acts to stop it (an idea now codified in Newton's first law of motion [Section 5.2]). This contradicted Aristotle's claim that the natural tendency of any moving object is to come to rest, and allowed Galileo to conclude that objects such as birds, falling stones, and clouds that are moving with the Earth should *stay* with the Earth unless some force knocks them away. This same idea explains why passengers in an airplane stay with the moving airplane even when they leave their seats.

Tycho's supernova and comet observations already had shown that the heavens could change, and Galileo shattered the idea of heavenly perfection after he built a telescope in late 1609. (Galileo did

Galileo (1564–1642)

not invent the telescope; it was invented in 1608 by Hans Lippershey. However, Galileo took what was little more than a toy and turned it into a scientific instrument.) Through his telescope, Galileo saw sunspots on the Sun, which were considered "imperfections" at the time. He also used his telescope to prove that the Moon has mountains and valleys like the "imperfect" Earth by noticing the shadows cast near the dividing line between the light and dark portions of the lunar face (Figure 3.20). If the heavens were not perfect, then the idea of elliptical (rather than circular) orbits was not so objectionable.

The absence of observable stellar parallax had been of particular concern to Tycho. Based on his estimates of the distances of stars, Tycho believed that his naked-eye observations were sufficiently precise to detect stellar parallax if the Earth did in fact orbit the Sun. (His planetary observations convinced him that the *planets* must orbit the Sun, so he advocated a model in which the Sun orbits the Earth while all other planets orbit the Sun. Few people took this model seriously, and Kepler's work soon made it a mere historical curiosity.) Refuting Tycho's argument required showing that the stars were more distant than Tycho had thought and therefore too distant for him to have observed stellar parallax. Although Galileo didn't actually prove this fact, he provided strong evidence in its favor. In particular, he saw with his telescope that the Milky Way resolved into countless individual stars, which helped him argue that the stars were far more numerous and more distant than Tycho had imagined.

The true death knell for an Earth-centered universe came with two of Galileo's earliest discoveries through the telescope. First, he observed four moons clearly orbiting Jupiter, *not* the Earth. (By itself, this observation still did not rule out a stationary, central Earth. However, it showed that moons can orbit a

FIGURE 3.20 The shadows cast by mountains and crater rims near the dividing line between the light and dark portions of the lunar face prove that the Moon's surface is not perfectly smooth.

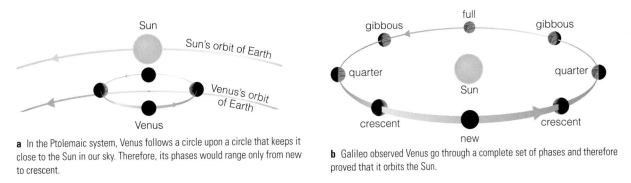

a In the Ptolemaic system, Venus follows a circle upon a circle that keeps it close to the Sun in our sky. Therefore, its phases would range only from new to crescent.

b Galileo observed Venus go through a complete set of phases and therefore proved that it orbits the Sun.

FIGURE 3.21 Galileo's telescopic observations of Venus proved that it orbits the Sun rather than the Earth.

moving planet like Jupiter, which overcame some critics' complaints that the Moon could not stay with a moving Earth.) Soon thereafter, he observed that Venus goes through phases like the Moon, proving that Venus must orbit the Sun and not the Earth (Figure 3.21). With Earth clearly removed from its position at the center of the universe, the scientific debate turned to the question of whether Kepler's laws were the correct model for our solar system. The most convincing evidence came in 1631, when astronomers observed a transit of Mercury across the Sun's face; Kepler's laws had predicted the transit with overwhelmingly better success than any competing model.

At last, the Copernican revolution took hold. Thanks in large part to Galileo, by the mid-1600s most astronomers accepted Kepler's model of planetary motion. Note that it took roughly a century from the time Copernicus published his ideas in 1543 to the time that a Sun-centered system became widely accepted. The dramatic shift in thought that occurred during this period offers an excellent example of the process of science. Prior to the start of this revolution, Ptolemy's model seemed an adequate way of explaining planetary motions in the sky. But as more and better observations were gathered, it became apparent that Ptolemy's model did not agree with the data. Copernicus offered an alternative model that had the right geneal idea in displacing the Earth from the center of the universe, but he still did not make predictions that agreed sufficiently with observations. As a result, the idea of an Earth that orbits the Sun remained controversial until Kepler created a model that successfully matched the observations and Galileo defused the objections raised on other grounds. The agreement between theory and observation became so strong that the debate finally turned away from how planets move to the question of *why* Kepler's laws hold true— a topic to which we will return in Chapter 5.

Although our historical hindsight shows clearly that Galileo won the day, the story was more complex in his own time, when Catholic Church doctrine still

THINKING ABOUT . . .

Aristotle

Aristotle (384–322 B.C.) is among the best-known philosophers of the ancient world. Both his parents died when he was a child, and he was raised by a family friend. In his 20s and 30s, he studied under Plato (427–347 B.C.) at Plato's Academy. He later founded his own school, called the Lyceum, where he studied and lectured on virtually every subject. Historical records tell us that his lectures were collected and published in 150 volumes. About 50 of these volumes survive to the present day.

Many of Aristotle's scientific discoveries involved the nature of plants and animals. He studied more than 500 animal species in detail, including dissecting specimens of nearly 50 species, and came up with a strikingly modern classification system. For example, he was the first person to recognize that dolphins should be classified with land mammals rather than with fish. In mathematics, he is known for laying the foundations of mathematical logic. Unfortunately, he was far less successful in physics and astronomy, areas in which many of his claims turned out to be wrong.

Interestingly, Aristotle's philosophies were not particularly influential until many centuries after his death. His books were preserved and valued by Islamic scholars, but they were unknown in Europe until they were translated into Latin in the twelfth and thirteenth centuries. Aristotle achieved his near-reverential status only after St. Thomas Aquinas (1225–1274) integrated Aristotle's philosophy into Christian theology. In the ancient world, Aristotle's greatest influence came indirectly, through his role as the tutor of Alexander the Great.

held the Earth to be the center of the universe. On June 22, 1633, Galileo was brought before a Church inquisition in Rome, where he was ordered to recant his claim that the Earth orbits the Sun. Already nearly 70 years old and fearing for his remaining life, Galileo did as ordered and his life was spared. However, legend has it that as he rose from his knees he whispered under his breath, *Eppur si muove*—Italian for "And yet it moves." (Given the likely consequences if Church officials had heard him say this, most historians doubt the veracity of this legend.) Although Galileo was not formally vindicated by the Church until 1992 [Section 1.4], the Church gave up the argument long before that. Galileo's book *Dialogue Concerning the Two Chief World Systems* was removed from the Church's index of banned books in 1824. Today, Catholic scientists are at the forefront of much astronomical research, and official Church teachings are compatible not only with the Earth's planetary status but also with the theories of the Big Bang and the subsequent evolution of the cosmos.

3.5 Modern Science and the Scientific Method

The success of the Copernican revolution demonstrated the value of careful observation and experiment to the advancement of human knowledge. As a result, the methods of modern science spread around the world, fueling a dramatic rise in human knowledge. It took humanity some 6,000 years to progress from the early lunar astronomy recorded in Africa to the ancient Greek idea of creating models of the universe, and it took humans another 2,000 years to finally recognize that the Earth orbits the Sun. In contrast, the past 400 years have seen discovery upon discovery, and we have come to know that the universe is filled with wonders far beyond the wildest imagination of our ancestors. In this section, we'll explore the general nature of modern science.

What Is Science?

The word **science** comes from the Latin *scientia*, meaning "knowledge"; it is also the root of the words *conscience* (related to the idea of self-knowledge or self-awareness) and *omniscience* (the quality of being all-knowing). Today, science connotes a special kind of knowledge: the kind that makes predictions that can be confirmed by rigorous observations and experiments. In brief, the modern **scientific method** is an organized approach to explaining observed facts with a model of nature, subject to the constraint that any proposed model must be testable and the provision that the model must be modified or discarded if it fails these tests.

In its most idealized form, the scientific method begins with a set of observed facts. A fact is supposed to be a statement that is objectively true. For example, we consider it a fact that the Sun rises each morning, that the planet Mars appeared in a particular place in our sky last night, and that the Earth revolves around the Sun. Facts are not always obvious, as illustrated by how long it took the latter fact to be discovered. In addition, our interpretations of facts often are based on beliefs about the world that others might not share. For example, when we say that the Sun rises each morning, we assume that it is the *same* Sun day after day—an idea that might not have been accepted by ancient Egyptians, whose mythology held that the Sun died with every sunset and was reborn with every sunrise. Nevertheless, facts are the raw material that scientific models seek to explain, so it is important that scientists agree on the facts. In the context of science, a fact must therefore be something that anyone can verify for himself or herself, at least in principle.

Once the facts have been collected, a model can be proposed to explain them. A useful model must also make predictions that can be tested through further observations or experiments. Kepler's model of planetary motion gained acceptance because it agreed so well with observations.

Common Misconceptions: Eggs on the Equinox

A central principle of science is that you needn't take scientific claims on faith; in principle, at least, you can always test them for yourself. Consider the claim, repeated in news reports every year, that the spring equinox is the only day on which you can balance an egg on its end. Many people believe this claim, but you'll be immediately skeptical if you think about the nature of the spring equinox. The equinox is merely a point in the Earth's orbit at which sunlight strikes both hemispheres equally (see Figure 1.13), and it's difficult to see how sunlight could affect an attempt to balance eggs (especially if the eggs are indoors). More important, you can test this claim directly. It's not easy to balance an egg on its end, but with practice you'll find that you can do it on *any* day of the year, not just on the spring equinox. Not all scientific claims are so easy to test for yourself, but the basic lesson should be clear: Before you accept any scientific claim, you should demand at least a reasonable explanation of the evidence that backs it up.

Astrology

Although the terms *astrology* and *astronomy* sound very similar, today they describe very different practices. In ancient times, however, astrology and astronomy often went hand in hand, and astrology played an important role in the historical development of astronomy. Indeed, astronomers and astrologers were usually one and the same.

In brief, the basic tenet of astrology is that human events are influenced by the apparent positions of the Sun, Moon, and planets among the stars in our sky. The origins of this idea are easy to understand. After all, there is no doubt that the position of the Sun in the sky influences our lives—it determines the seasons and hence the times of planting and harvesting, of warmth and cold, and of daylight and darkness. Similarly, the Moon determines the tides, and the cycle of lunar phases coincides with many biological cycles. Because the planets also appear to move among the stars, it seemed reasonable to imagine that planets also influence our lives, even if these influences were much more difficult to discover.

Ancient astrologers hoped that they might learn *how* the positions of the Sun, Moon, and planets influence our lives. They charted the skies, seeking correlations with events on Earth. For example, if an earthquake occurred when Saturn was entering the constellation of Leo, might Saturn's position have been the cause of the earthquake? If the king became ill when Mars appeared in the constellation Gemini and the first-quarter moon appeared in Scorpio, might it mean another tragedy for the king when this particular alignment of the Moon and Mars next recurred? Surely, the ancient astrologers thought, the patterns of influence eventually would become clear. Thus, the astrologers hoped that they might someday learn to forecast human events with the same reliability with which astronomical observations of the Sun could forecast the coming of spring.

This hope was never realized. Although many astrologers still attempt to predict future events, scientific tests have shown that their predictions come true no more often than would be expected by pure chance. Moreover, in light of our current understanding of the universe, the original ideas behind astrology no longer make sense. For example, today we use ideas of gravity and energy to explain the influences of the Sun and the Moon, and these same ideas tell us that the planets are too far from Earth to have a similar influence on our lives.

Of course, many people continue to practice astrology, perhaps because of its ancient and rich traditions. Scientifically, we cannot say anything about such traditions, because traditions are not testable predictions. But if you want to understand the latest discoveries about the cosmos, you'll need a science that can be tested and refined—and astrology can't meet these requirements.

In summary, the idealized scientific method proceeds as follows:

- *Observation:* The scientific method begins with the collection of a set of observed facts.

- *Hypothesis:* A model is proposed to explain the observed facts. A proposed model is often called a **hypothesis**, which essentially means an *educated guess.*

- *Further observations/experiments:* The model's predictions are tested through further observations or experiments. When a test is passed, we gain confidence that the model truly represents nature. When a test fails, we recognize that the model is flawed, and we therefore must refine or discard the model.

- *Theory:* A model must be continually challenged with new observations or experiments by many different scientists. A model achieves the status of a **scientific theory** only after passing a broad range of tests. Note that, while we can have great confidence that a scientific theory truly represents nature, we can never prove a theory to be true *beyond all doubt.* Therefore, even well-established theories must be subject to continuing challenges through further observations and experiments.

In reality, scientific discoveries rarely are made by a process as mechanical as the idealized scientific method described here. For example, we've seen that the Copernican revolution involved many twists and turns—and some moments of inspiration and luck—as it unfolded over more than a century. Nevertheless, with hindsight we can look back and see the steps of the scientific method. In that sense, the scientific method represents an ideal prescription for judging objectively whether a proposed model of nature is close to the truth.

Science, Pseudoscience, and Nonscience

People often seek knowledge in ways that do not follow the basic tenets of the scientific method and therefore are not science. When we leave the realm of science, claims of knowledge become far more subjective and difficult to verify. Nevertheless, it can be useful to categorize such claims of knowledge as either *pseudoscience* or *nonscience.*

By **pseudoscience** we mean attempts to search for knowledge in ways that may at first seem scientific but do not adhere to the testing and verification requirements of the scientific method; the prefix *pseudo* means "false." For example, at the beginning of each year you can find tabloid newspapers offering predictions made by people who claim to be able to "see" the future. Because they make specific predictions, we can test their claims by checking whether their predictions come true, and numerous studies have shown that their predictions come true no more often than would be expected by pure chance. But these seers seem unconcerned with the results of such studies. The fact that they make testable claims but then ignore the results of the tests marks their claimed ability to see the future as pseudoscience.

By **nonscience** we mean knowledge sought through pure intuition, societal traditions, ancient scriptures, and other processes that make no claim to adhere to the scientific method. In nonscience, experimental testing is irrelevant because nonscientific beliefs are based on things such as faith, political conviction, and tradition. Because nonscience makes no pretense of following the scientific method, science can say nothing about the validity of nonscience.

TIME OUT TO THINK *Can the scientific method be applied in any way to questions of faith, prayer, emotion, or values? Defend your view.*

Is Science Objective?

The boundaries between science, nonscience, and pseudoscience are sometimes blurry. In particular, because science is practiced by human beings, individual scientists carry their personal biases and beliefs with them in their scientific work. These biases can influence how a scientist proposes or tests a model, and in some cases scientists have been known to cheat—either deliberately or subconsciously—to obtain the results they desire. For example, in the late nineteenth and early twentieth centuries, some astronomers claimed to see networks of "canals" in their blurry telescopic images of Mars, and they hypothesized that Mars was home to a dying civilization that used the canals to transport water to thirsty cities [Section 8.6]. Because no such canals actually exist, these astronomers apparently were allowing their beliefs to influence how they interpreted blurry images. It was, in essence, a form of cheating—even though it certainly was not intentional.

Sometimes, bias can show up even in the thinking of the scientific community as a whole. Thus, some valid ideas may not be considered by any scientist because the ideas fall too far outside the general patterns of thought, or the **paradigm**, of the time. Einstein's theory of relativity provides an example. Many other scientists had gleaned hints of this theory in the decades before Einstein but did not investigate them, at least in part because they seemed too outlandish.

The beauty of the scientific method is that it encourages continued testing by many people. Even if personal biases affect some results, tests by others eventually will uncover the mistakes. Similarly, even when a new idea falls outside the accepted paradigm, sufficient testing and verification of the idea eventually will force a change in the paradigm. Thus, although individual scientists rarely follow the scientific method in its idealized form, the collective action of many scientists over many years generally *does* follow the basic tenets of the scientific method. That is why, despite the biases of individual scientists, science as a whole is usually objective.

THE BIG PICTURE

In this chapter, we focused on the scientific principles through which we have learned so much about the universe. Key "big picture" concepts from this chapter include the following:

- The basic ingredients of scientific thinking—careful observation and trial-and-error testing—are a part of everyone's experience. The modern scientific method simply provides a way of organizing this everyday thinking to facilitate the learning and sharing of new knowledge.

- Although knowledge about the universe is growing rapidly today, each new piece rests upon foundations of older discoveries. The foundations of astronomy reach far back into history and are intertwined with the general development of human culture and civilization.

- The Copernican revolution, which overthrew the ancient belief in an Earth-centered universe, did not occur instantaneously. It unfolded over a period of more than a century and involved careful observational, experimental, and theoretical work by many different people—especially Copernicus, Tycho Brahe, Kepler, and Galileo.

- The concept of making models of nature lies at the heart of modern science. Models are created by generalizing from many specific facts. They then yield further predictions that allow the models to be tested. Models that withstand repeated testing rise to the status of scientific theories, while models that fail must be modified or discarded.

Review Questions

1. In what ways is scientific thinking a fundamental part of human nature? What characteristics does the modern scientific method possess that go beyond this everyday type of thinking?

2. Explain how the people of central Africa predicted the weather by observing the Moon.

3. How are the names of the seven days of the week related to astronomical objects?

4. Briefly describe the astronomical uses of each of the following: Stonehenge, the Templo Mayor, the Sun Dagger, the Mayan observatory at Chichén Itzá, lines in the Nazca desert, Pawnee lodges, the Big Horn Medicine Wheel.

5. Explain why the Muslim fast of Ramadan occurs earlier with each subsequent year. What is the *Metonic cycle*? How does the Jewish calendar follow the Metonic cycle?

6. What role did the Navigator play in Polynesian society?

7. What was the Library of Alexandria? What role did it play in the ancient empires of Greece and Rome?

8. What do we mean by a *model* of nature? Explain how the celestial sphere represents such a model.

9. Who was Ptolemy? Briefly describe how the Ptolemaic model of the universe explains apparent retrograde motion while preserving the idea of an Earth-centered universe.

10. Briefly describe the role played by Islamic scholars of the Middle Ages in the development of modern science.

11. Was Copernicus the first person to suggest a Sun-centered solar system? What advantages did his model have over the Ptolemaic model? In what ways did it fail to improve on the Ptolemaic model?

12. How were Tycho Brahe's observations important to the development of modern astronomy?

13. State Kepler's three laws of motion, and explain the meaning of each law.

14. Describe how Galileo helped spur acceptance of Kepler's Sun-centered model of the solar system.

15. Describe the basic process of the *scientific method*. What is the role of a *model* in this method? What do we mean by *facts* when discussing science?

16. Briefly describe how science differs from *pseudoscience* and *nonscience*.

17. Briefly explain how science as a whole can be objective even while individual scientists have personal biases and beliefs.

Discussion Questions

1. *The Impact of Science.* The modern world is filled with ideas, knowledge, and technology that developed through science and application of the scientific method. Discuss some of these things and how they affect our lives. Which of these impacts do you think are positive? Which are negative? Overall, do you think science has benefited the human race? Defend your opinion.

2. *The Importance of Ancient Astronomy.* Why was astronomy important to people in ancient times? Discuss both the practical importance of astronomy and the importance it may have had for religious or other traditions. Which do you think was more important in the development of ancient astronomy, its practical or its philosophical role? Defend your opinion.

3. *Kepler's Choice.* Casting aside the idea that orbits must be perfect circles meant going against deeply entrenched beliefs, and Kepler said that it shook his deep religious faith. Given that only two of Tycho's observations disagreed with a perfectly circular orbit—and only by 8 arcminutes—do you think that most other people would have made the choice Kepler made to abandon perfect circles? Have you ever performed an experiment that disagreed with theory? Which did you question, the theory or your experiment? Why?

Problems

True Statements? For **problems 1–6**, decide whether the statements are true or false and clearly explain how you know.

1. If we defined hours as the ancient Egyptians did, we'd have the longest hours on the summer solstice and the shortest hours on the winter solstice.

2. The date of Christmas (December 25) is set each year according to a lunar calendar.

3. When navigating in the South Pacific, the Polynesians found their latitude with the aid of the pointer stars of the Big Dipper.

4. The Ptolemaic model reproduced apparent retrograde motion by having planets move sometimes counterclockwise and sometimes clockwise in their circles.

5. In science, saying that something is a theory means that it is really just a guess.

6. Upon its publication in 1543, the Copernican model was immediately accepted by most scientists because its predictions of planetary positions were essentially perfect.

7. *Cultural Astronomy.* Choose a particular culture of interest to you, and research the astronomical knowledge and accomplishments of that culture. Write a two- to three-page summary of your findings.

8. *Astronomical Structures.* Choose an ancient astronomical structure of interest to you (e.g., Stonehenge, Nazca lines, Pawnee lodges) and research its history. Write a two- to three-page summary of your findings. If possible, also build a scale model of the structure or create detailed diagrams to illustrate how the structure was used.

9. *The Copernican Revolution.* How did the Copernican revolution alter the course of human history? Write a one- to two-page essay in which you both summarize the unfolding of the revolution itself and describe/defend your opinion of how it changed human history.

10. *Scientific Test of Astrology.* Find out about at least one scientific test that has been conducted to test the validity of astrology. Write a short summary of how the test was conducted and what conclusions were reached.

Web Projects

Find useful links for Web projects on the text Web site.

1. *Greek Astronomers.* Many ancient Greek scientists had ideas that, in retrospect, seem well ahead of their time. Choose one or more of the following ancient Greek scientists, and learn enough about their work in science and astronomy to write a one- to two-page "scientific biography."

Thales	Anaximander	Pythagoras
Anaxagoras	Empedocles	Democritus
Meton	Plato	Eudoxus
Aristotle	Callipus	Aristarchus
Archimedes	Eratosthenes	Apollonius
Hipparchus	Seleucus	Ptolemy
Hypatia		

2. *The Ptolemaic Model.* This chapter gives only a very brief description of Ptolemy's model of the universe. Investigate the model in greater depth. Using diagrams and text as needed, give a two- to three-page description of the model.

The eternal mystery of the world is its comprehensibility.
The fact that it is comprehensible is a miracle.

ALBERT EINSTEIN

CHAPTER 4
A Universe of Matter and Energy

In this and the next two chapters, we turn our attention to the scientific concepts that lie at the heart of modern astronomy. We begin by investigating the nature of matter and energy, the fundamental stuff from which the universe is made.

The history of the universe essentially is a story about the interplay between matter and energy since the beginning of time. Interactions between matter and energy govern everything from the creation of hydrogen and helium in the Big Bang to the processes that make Earth a habitable planet. Understanding the universe therefore depends on familiarity with how matter responds to the ebb and flow of energy.

The concepts of matter and energy presented in this chapter will enable you to understand most of the topics in this book. Some of the concepts and terminology may already be familiar to you. If not, don't worry. We will go over them again as they arise in various contexts, and you can refer back to this chapter as needed during the remainder of your studies.

4.1 Matter and Energy in Everyday Life

The meaning of **matter** is obvious to most people, at least on a practical level. Matter is simply material, such as rocks, water, or air. You can hold matter in your hand or put it in a box. The meaning of **energy** is not quite as obvious, although we certainly talk a lot about it. We pay energy bills to the power companies, we use energy from gasoline to run our cars, and we argue about whether nuclear energy is a sensible alternative to fossil fuels. On a personal level, we often talk about how energetic we feel on a particular day. But what *is* energy?

Broadly speaking, energy is what makes matter move. For Americans, the most familiar way of measuring energy is in Calories, which we use to describe how much energy our bodies can draw from food. A typical adult uses about 2,500 Calories of energy each day. Among other things, this energy keeps our hearts beating and our lungs breathing, generates the heat that maintains our 37°C (98.6°F) body temperature, and allows us to walk and run.

Just as there are many different units for measuring height, such as inches, feet, and meters, there are many alternatives to Calories for measuring energy. If you look closely at an electric bill, you'll probably find that the power company charges you for electrical energy in units called *kilowatt-hours*. If you purchase a gas appliance, its energy requirements may be labeled in *British thermal units*, or *BTUs*. In science and internationally, the favored unit of energy is the **joule**, which is equivalent to $\frac{1}{4,184}$ of a Calorie. Thus, the 2,500 Calories used daily by a typical adult are equivalent to about 10 million joules. For purposes of comparison, some energies given in joules are listed in Table 4.1.

Although energy can always be measured in joules, it has many different forms. Already we have talked about food energy, electrical energy, the energy of a beating heart, the energy represented by our body temperature, and more. Fortunately, the many forms of energy can be grouped into three basic categories.

First, whenever matter is moving, it has energy of motion, or **kinetic energy** (*kinetic* comes from a Greek word meaning "motion"). Falling rocks, the moving blades on an electric mixer, a car driving down the highway, and the molecules moving in the air around us are all examples of objects with kinetic energy.

The second basic category of energy is **potential energy**, or energy being stored for later conversion

Table 4.1 Energy Comparisons

Item	Energy (joules)
Average daytime solar energy striking Earth, per m^2 per second	1.3×10^3
Energy released by metabolism of one average candy bar	1×10^6
Energy needed for 1 hour of walking (adult)	1×10^6
Kinetic energy of average car traveling at 60 mi/hr	1×10^6
Daily energy needs of average adult	1×10^7
Energy released by burning 1 liter of oil	1.2×10^7
Energy released by fission of 1 kg of uranium-235	5.6×10^{13}
Energy released by fusion of hydrogen in 1 liter of water	7×10^{13}
Energy released by 1-megaton H-bomb	5×10^{15}
Energy released by major earthquake (magnitude 8.0)	2.5×10^{16}
U.S. annual energy consumption	10^{20}
Annual energy generation of Sun	10^{34}
Energy released by supernova (explosion of a star)	$10^{44}–10^{46}$

into kinetic energy. A rock perched on a ledge has *gravitational* potential energy because it will fall if it slips off the edge. Gasoline contains *chemical* potential energy, which a car engine converts to the kinetic energy of the moving car. Power companies supply *electrical* potential energy, which we use to run dishwashers and other appliances.

The third basic category is energy carried by light, or **radiative energy** (the word *radiation* is often used as a synonym for *light*). Plants directly convert the radiative energy of sunlight into chemical potential energy through the process of *photosynthesis*. Radiative energy is fundamental to astronomy, because telescopes collect the radiative energy of light from distant stars.

TIME OUT TO THINK *We buy energy in many different forms. For example, we buy chemical potential energy in the form of food to fuel our bodies. Describe several other forms of energy that you commonly buy.*

Understanding how energy changes from one form to another helps us understand many common phenomena. For example, a diver standing on a 10-meter platform has gravitational potential energy owing to her height above the water and chemical potential energy stored in her body tissues. She uses the chemical potential energy to flex her muscles in such a way as to initiate her dive and then to help her execute graceful twists and spins (Figure 4.1). Meanwhile, her gravitational potential energy becomes kinetic energy of motion as she falls toward the water.

Following how energy changes from one form to another also helps us understand the universe. For example, the particles in a collapsing cloud of interstellar gas convert gravitational potential energy into kinetic energy as they fall inward, and their motion generates heat that can eventually ignite a star. But before we study astronomical phenomena, we need ways to describe energy quantitatively.

4.2 A Scientific View of Energy

In this section, we discuss a few ways to quantify kinetic and potential energy; we'll discuss radiative energy in Chapter 6, in which we study properties of light.

Kinetic Energy

We can calculate the kinetic energy of any moving object with a very simple formula:

$$\text{kinetic energy} = \tfrac{1}{2}mv^2$$

where m is the mass of the object and v is its speed (v for *velocity*). If we measure the mass in kilograms and the speed in meters per second, the resulting answer will be in joules. The kinetic energy formula is easy to interpret. The m in the formula tells us that kinetic energy is proportional to mass: A 5-ton truck has 5 times the kinetic energy of a 1-ton car moving at the same speed. The v^2 tells us that kinetic energy increases with the *square* of the velocity: If you double your speed (e.g., from 30 km/hr to 60 km/hr), your kinetic energy becomes $2^2 = 4$ times greater.

Thermal Energy Suppose we want to know about the kinetic energy of the countless tiny particles (atoms and molecules) inside a rock or of the countless particles in the air or in a distant star. Each of these tiny particles has its own motion relative to surrounding particles, and these motions constantly change as the particles jostle one another. The result is that the particles inside a substance appear to move randomly: Any individual particle may be moving in any direction with any of a wide range of speeds. Despite the seemingly random motion of particles within a substance, it's easy to measure the *average* kinetic energy of the particles—**temperature** is a measure of this average (Figure 4.2). A higher temperature simply means that, on average, the particles have more kinetic energy and hence are moving faster (Figure 4.3). (Kinetic energy depends on both mass and speed, but for a particular set of particles, greater kinetic energy means higher speeds.) The speeds of particles within a substance can be surprisingly fast. For example, the air molecules around you move at typical speeds of about 500 meters per second (about 1,000 miles per hour).

The energy contained *within* a substance as measured by its temperature is often called **thermal**

FIGURE 4.1 Understanding energy can help us understand both the graceful movements of a diver and the story of the universe.

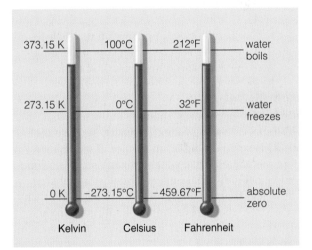

FIGURE 4.2 Temperature is a measure of the average kinetic energy of the particles in a substance. This diagram compares three common temperature scales. The Fahrenheit scale is used in the United States, but nearly all other countries use the Celsius scale. Scientists prefer the Kelvin scale because 0 K represents *absolute zero*, the coldest possible temperature. (The degree symbol ° is not usually used with the Kelvin scale.)

energy. Thus, thermal energy represents the collective kinetic energy of the many individual particles moving within a substance.

Temperature and Heat The concepts of *temperature* and *heat* are not quite the same. To understand the difference, imagine the following experiment (but don't try it!). Suppose you heat your oven to 500°F.

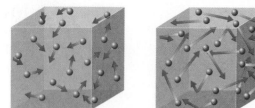

Longer arrows mean higher average speed.

FIGURE 4.3 The particles in the box on the right have a higher temperature because their average speeds are higher (assuming that both boxes contain particles of the same mass).

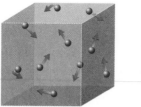

FIGURE 4.4 Both boxes have the same temperature, but the box on the right contains more *thermal energy* because it contains more particles.

Then you open the oven door, quickly thrust your arm inside (without touching anything), and immediately remove it. What will happen to your arm? Not much. Now suppose you boil a pot of water. Although the temperature of boiling water is only 212°F, you would be badly burned if you put your arm in the pot, even for an instant. Thus, the water in the pot transfers more thermal energy to your arm than does the air in the oven, even though it has a lower temperature. That is, your arm heats more quickly in the water. This is because thermal energy content depends on both the temperature and the total number of particles, which is much larger in a pot of water (Figure 4.4).

Let's look at what is happening on the molecular level. If air or water is hotter than your body, molecules striking your skin transfer some of their thermal energy to your arm. The high temperature in a 500°F oven means that the air molecules strike your skin harder, on average, than the molecules in a 212°F pot of boiling water. However, because the *density* is so much higher in the pot of water, many more molecules strike your skin each second. Thus, while each individual molecular collision transfers a little less thermal energy in the boiling water than in the oven, the sheer number of collisions in the water transfers so much thermal energy that your skin burns rapidly.

The environment in space provides another example of the difference between temperature and heat. Surprisingly, the temperature in low Earth orbit is several thousand degrees. However, astronauts working in Earth orbit (e.g., outside the Space Shuttle) tend to get very cold and therefore use heated space suits and gloves. The astronauts feel cold despite the high temperature because the extremely low density of space means that relatively few particles are available to transfer thermal energy to an astronaut. (You may wonder how the astronauts become cold given that the low density also means the astronauts cannot transfer much of their own thermal energy to the particles in space. It turns out that they lose their body heat by emitting *thermal radiation,* which we will discuss in Chapter 6.)

Potential Energy

Potential energy can be stored in many different forms and is not always easy to quantify. Fortunately, it is easy to describe two types of potential energy that we use frequently in astronomy.

Gravitational Potential Energy Gravitational potential energy is extremely important in astronomy. The conversion of gravitational potential energy into kinetic (or thermal) energy helps explain everything from the speed at which an object falls to the ground to the formation processes of stars and planets. The mathematical formula for gravitational potential energy can take a variety of forms, but in words the idea is simple: *The amount of gravitational potential energy released as an object falls depends on its mass, the strength of gravity, and the distance it falls.*

This statement explains why falling from a 10-story building hurts more than falling out of a chair. Your gravitational potential energy is much greater on top of the 10-story building than in your chair because you can fall much farther. Because your gravitational potential energy will be converted to kinetic energy as you fall, you'll have a lot more kinetic energy by the time you hit the ground after falling from the building than after falling from the chair, which means you'll hit the ground much harder.

Gravitational potential energy also helps us understand how the Sun became hot enough to sustain nuclear fusion. Before the Sun formed, its matter was contained in a large, cold, diffuse cloud of gas. Most of the individual gas particles were far from the center of this large cloud and therefore had considerable amounts of gravitational potential energy. As the cloud collapsed under its own gravity, the gravitational potential energy of these particles was converted to thermal energy, eventually making the center of the cloud hot enough to ignite nuclear fusion.

Mass-Energy Although matter and energy seem very different in daily life, they are intimately connected. Einstein showed that mass itself is a form of potential energy, often called **mass-energy**. The mass-energy of any piece of matter is given by the formula

$$E = mc^2$$

where E is the amount of potential energy, m is the mass of the object, and c is the speed of light. If the mass is measured in kilograms and the speed of light in meters per second, the resulting mass-energy has units of joules. Note that the speed of light is a large number ($c = 3 \times 10^8$ m/s) and the speed of light squared is much larger still ($c^2 = 9 \times 10^{16}$ m^2/s^2). Thus, Einstein's formula implies that a relatively small amount of mass represents a huge amount of mass-energy.

Mass-energy can be converted to other forms of energy, but noticeable amounts of mass become other forms of energy only under special but important circumstances. The process of nuclear fusion in the core of the Sun converts some of the Sun's mass into energy, ultimately generating the sunlight that sustains most life on Earth. On Earth, nuclear reactors and nuclear bombs also work in accord with Einstein's formula. In nuclear reactors, the splitting (fission) of elements such as uranium or plutonium converts some of the mass-energy of these materials into heat, which is then used to generate electrical power. In an H-bomb, nuclear fusion similar to that in the Sun uses a small amount of the mass-energy in hydrogen to devastating effect. Incredibly, a 1-megaton H-bomb that could destroy a major city requires the conversion of only about 0.1 kilogram of mass (about 3 ounces) into energy (Figure 4.5).

Just as $E = mc^2$ tells us that mass can be converted into other forms of energy, it also tells us that energy can be transformed into mass. In *particle accelerators*, scientists accelerate subatomic particles to extremely high speeds. When these particles collide with one another or with a barrier, some of the energy released in the collision spontaneously turns into mass, appearing as a shower of subatomic particles. These showers of particles allow scientists to test theories about how matter behaves at extremely high temperatures such as those that prevailed in the universe during the first fraction of a second after the Big Bang. Among the most powerful particle accelerators in the world are Fermilab in Illinois, the Stanford Linear Accelerator in California, and CERN in Switzerland.

TIME OUT TO THINK *Einstein's formula $E = mc^2$ is probably the most famous physics equation of all time. Considering its role as described in the preceding paragraphs, do you think its fame is well deserved?*

FIGURE 4.5 The energy from this H-bomb comes from converting only about 0.1 kg of mass into energy in accordance with $E = mc^2$.

Conservation of Energy

A fundamental principle in science is that, regardless of how we change the *form* of energy, the total *quantity* of energy never changes. This principle is called the **law of conservation of energy**. It has been carefully tested in many experiments, and it is a pillar upon which modern theories of the universe are built. Because of this law, the story of the universe is a story of the interplay of energy and matter: All actions in the universe involve exchanges of energy or the conversion of energy from one form to another.

For example, imagine that you've thrown a baseball so that it is moving and hence has kinetic energy. Where did this kinetic energy come from? The baseball got its kinetic energy from the motion of your arm as you threw it; that is, some of the kinetic energy of your moving arm was transferred to the baseball. Your arm, in turn, got its kinetic energy from the release of chemical potential energy stored in your muscle tissues. Your muscles got this energy from the chemical potential energy stored in the foods you ate. The energy stored in the foods came from sunlight, which plants convert into chemical potential energy through photosynthesis. The radiative energy of the Sun was generated through the process of nuclear fusion, which releases some of the mass-energy stored in the Sun's supply of hydrogen. Thus, the ultimate source of the energy of the moving baseball is the mass-energy stored in hydrogen—which was created in the Big Bang.

We have described where the baseball got its kinetic energy. Where will this energy go? As the baseball moves through the air, some of its energy is transferred to molecules in the air, generating heat or sound. If someone catches the baseball, its energy will cause his or her hand to recoil and perhaps will also generate some heat and sound. Ultimately, the energy of the moving baseball will be converted to a barely noticeable amount of heat (thermal energy) in the air, the ground, or a person's hand, making it extremely difficult to track. Nevertheless, the energy will never disappear. According to present understanding, the total energy content of the universe was determined in the Big Bang. It remains the same today and will stay the same forever into the future.

4.3 The Material World

Now that we have seen how energy animates the matter in the universe, it's time to consider matter itself in greater detail. You are familiar with two basic properties of matter on Earth from everyday experience. First, matter can exist in different **phases**: as a **solid**, such as ice or a rock; as a **liquid**, such as flowing water or oil; or as a **gas**, such as air. Second, even in a particular phase, matter exists in a great variety of different substances.

What is matter, and why does it have so many different forms? Today, we know that all ordinary matter is composed of **atoms** and that each different type of atom corresponds to a different chemical **element**. Among the most familiar chemical elements are hydrogen, helium, carbon, oxygen, silicon, iron, gold, silver, lead, and uranium. (Appendix D gives the periodic table of all the elements.)

The number of different material substances is far greater than the number of chemical elements because atoms can combine to form **molecules**. Some molecules consist of two or more atoms of the same element. For example, we breathe O_2, oxygen molecules made of two oxygen atoms. Other molecules, such as water (H_2O) and sulfuric acid (H_2SO_4), are made up of atoms of two or more different elements; such molecules are called **compounds**. The chemical properties of a molecule are different from those of its individual atoms. For example, water behaves very differently than pure hydrogen or pure oxygen, even though each water molecule is composed of two hydrogen atoms and one oxygen atom, as indicated by the familiar symbol H_2O.

Common Misconceptions: The Illusion of Solidity

Bang your hand on a table. Although the table feels solid, it is made almost entirely of empty space! Nearly all the mass of the table is contained in the nuclei of its atoms. But the volume of an atom is more than a trillion times the volume of its nucleus, so relatively speaking the nuclei of adjacent atoms are nowhere near to touching one another. The solidity of the table comes about from a combination of electrical interactions between the charged particles in its atoms and the strange quantum laws governing the behavior of electrons. If we could somehow pack all the table's nuclei together, the table's mass would fit into a microscopic speck. Although *we* cannot pack matter together in this way, nature can and does—in *neutron stars,* which we will study in Chapter 15.

Atomic Structure

Atoms are incredibly small: Millions could fit end to end across the period at the end of this sentence, and the number in a single drop of water (10^{22} to 10^{23} atoms) may exceed the number of stars in the observable universe. Yet atoms are composed of even smaller particles. A small, dense **nucleus** lies at the center of an atom; atomic nuclei (plural of nucleus) are made of particles called **protons** and **neutrons**. The nucleus is surrounded by particles called **electrons**.

The properties of an atom depend mainly on the amount of **electrical charge** in its nucleus. Electrical charge is a fundamental physical property that is always conserved, just as energy is always conserved. Electrons carry a negative electrical charge of -1, protons carry a positive electrical charge of $+1$, and neutrons are electrically neutral. Oppositely charged particles attract one another, and similarly charged particles repel one another. The attraction between the positively charged protons in the nucleus and the negatively charged electrons that surround it is what holds an atom together. Ordinary atoms have identical numbers of electrons and protons, making them electrically neutral overall.

Although electrons can be thought of as tiny particles, they are not quite like tiny grains of sand, and they don't really orbit the nucleus as planets orbit the Sun. Instead, the electrons in an atom are "smeared out," forming a kind of cloud that surrounds the nucleus and gives the atom its apparent size. The electrons aren't really cloudy; it's just that it is impossible to pinpoint their positions. An atomic nucleus

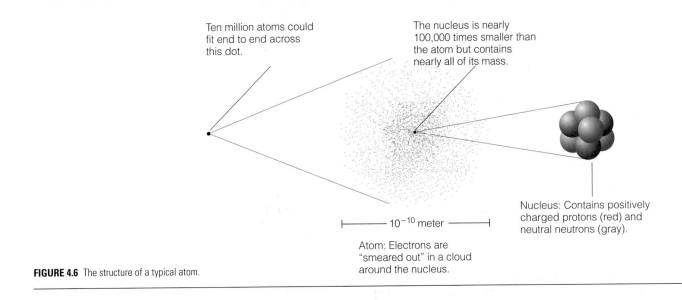

Ten million atoms could fit end to end across this dot.

The nucleus is nearly 100,000 times smaller than the atom but contains nearly all of its mass.

10^{-10} meter

Atom: Electrons are "smeared out" in a cloud around the nucleus.

Nucleus: Contains positively charged protons (red) and neutral neutrons (gray).

FIGURE 4.6 The structure of a typical atom.

is very tiny, even compared to the atom itself: If we imagine an atom on a scale on which its nucleus is the size of your fist, its electron cloud would be many miles wide. Nearly all the atom's mass resides in its nucleus, because protons and neutrons are each about 2,000 times more massive than an electron. Figure 4.6 shows the structure of a typical atom.

Each different chemical element contains a different number of protons in its nucleus called its **atomic number**. For example, a hydrogen nucleus contains just one proton, so its atomic number is 1; a helium nucleus contains two protons, so its atomic number is 2.

The *combined* number of protons and neutrons in an atom is called its **atomic mass** (or *atomic weight*). The atomic mass of ordinary hydrogen is 1 because its nucleus is just a single proton. Helium usually has two neutrons in addition to its two protons, giving it an atomic mass of 4. Carbon usually has six protons and six neutrons, giving it an atomic mass of 12.

Sometimes, the same element can have varying numbers of neutrons. For example, in rare cases a

hydrogen nucleus contains a neutron in addition to its proton, making its atomic mass 2; hydrogen with atomic mass 2 is often called *deuterium*. The different forms of hydrogen are called **isotopes** of one another. A third isotope of hydrogen, tritium (^{3}H), contains two neutrons in addition to its one proton. Each isotope of an element has the same number of protons but a different number of neutrons.

We usually denote different isotopes by showing the atomic mass to the upper left of the chemical symbol. For example, most carbon atoms contain six neutrons in addition to their six protons, making their atomic mass $6 + 6 = 12$; we write this isotope of carbon as ^{12}C and read it as carbon-12. The symbol ^{14}C (carbon-14) represents a rarer isotope of carbon with atomic mass 14 and hence eight neutrons rather than six. Figure 4.7 illustrates some atomic terminology.

TIME OUT TO THINK *The symbol ^{4}He represents helium with an atomic mass of 4; this is the most common form, containing two protons and two neutrons. What does the symbol ^{3}He represent?*

FIGURE 4.7 Terminology of atoms.

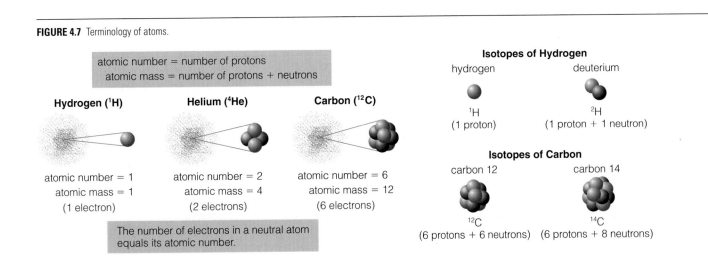

atomic number = number of protons
atomic mass = number of protons + neutrons

Hydrogen (^{1}H)
atomic number = 1
atomic mass = 1
(1 electron)

Helium (^{4}He)
atomic number = 2
atomic mass = 4
(2 electrons)

Carbon (^{12}C)
atomic number = 6
atomic mass = 12
(6 electrons)

The number of electrons in a neutral atom equals its atomic number.

Isotopes of Hydrogen
hydrogen
^{1}H
(1 proton)

deuterium
^{2}H
(1 proton + 1 neutron)

Isotopes of Carbon
carbon 12
^{12}C
(6 protons + 6 neutrons)

carbon 14
^{14}C
(6 protons + 8 neutrons)

Phases of Matter

Everyday experience tells us that, depending on the temperature, the same substance exists in different phases. The main difference between phases is how tightly neighboring particles are bound together. As a substance is heated, the average kinetic energy of its particles increases, enabling the particles to break the bonds holding them to their neighbors. Each change in phase corresponds to the breaking of a different kind of bond. Phase changes occur in all substances; let's consider what happens when we heat water, starting from its solid phase, ice (Figure 4.8):

- Below 0°C (32°F), water molecules have a relatively low average kinetic energy, and each molecule is bound tightly to its neighbors, making the *solid* structure of ice.

- As the temperature increases (but remains below freezing), the rigid arrangement of the molecules in ice vibrates more and more. At 0°C, the molecules have enough energy to break the solid bonds of ice. The molecules can then move relatively freely among one another, allowing the water to flow as a *liquid*. Even in liquid water, a type of bond between adjacent molecules still keeps them close together.

- When a water molecule breaks free of all bonds with its neighbors, we call it a molecule of water vapor, which is a *gas*. Molecules in the gas phase move independently of other molecules. Even at temperatures at which water is a solid or liquid, a few molecules will have enough energy to enter the gas phase. We call the process **evaporation** when molecules escape from a liquid and **sublimation** when they escape from a solid. As temperatures rise, the rates of sublimation and evaporation increase, eventually changing all the solid and liquid water into water vapor.

What happens if we continue to raise the temperature of water vapor? As the temperature rises, the molecules move faster, making collisions among them more violent. These collisions eventually split the water molecules into their component atoms of hydrogen and oxygen. The process by which the bonds that hold the atoms of a molecule together are broken is called **molecular dissociation**. At still

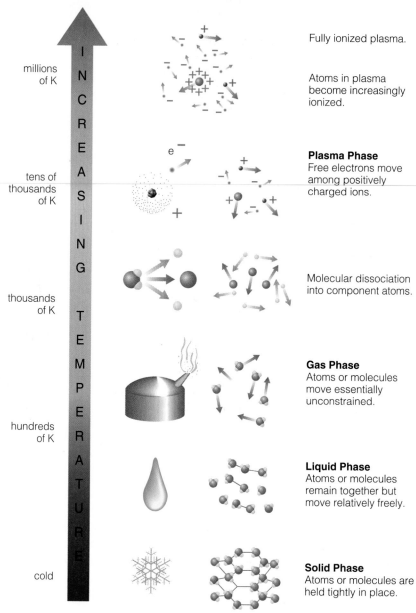

millions of K

tens of thousands of K

thousands of K

hundreds of K

cold

Fully ionized plasma.

Atoms in plasma become increasingly ionized.

Plasma Phase
Free electrons move among positively charged ions.

Molecular dissociation into component atoms.

Gas Phase
Atoms or molecules move essentially unconstrained.

Liquid Phase
Atoms or molecules remain together but move relatively freely.

Solid Phase
Atoms or molecules are held tightly in place.

FIGURE 4.8 The general progression of phase changes.

Common Misconceptions: One Phase at a Time?

In daily life, we usually think of H_2O as being in the phase of either solid ice, liquid water, or water vapor, with the phase depending on the temperature. In reality, two or even all three phases can exist at the same time. In particular, some sublimation *always* occurs over solid ice, and some evaporation *always* occurs over liquid water. Thus, the phases of solid ice and liquid water never occur alone but instead occur in conjunction with water vapor. The *amount* of water vapor increases with the temperature, which is why more and more steam rises from a kettle as you heat it.

higher temperatures, collisions can break the bonds holding electrons around the nuclei of individual atoms, allowing the electrons to go free. The loss of one or more negatively charged electrons leaves a remaining atom with a net positive charge. Such charged atoms are called **ions**, and the process of stripping electrons from atoms is called **ionization**. Thus, at high temperatures, what once was water becomes a hot gas consisting of freely moving electrons and positively charged ions of hydrogen and oxygen. This type of hot gas, in which atoms have become ionized, is called a **plasma**, sometimes referred to as "the fourth phase of matter." Other chemical substances go through similar phase changes, but the temperatures at which phase changes occur depend on the types of atom or molecules involved.

Note that neutral hydrogen contains only one electron, which balances the single positive charge of the one proton in its nucleus. Thus, hydrogen can be ionized only once, and the remaining hydrogen ion, designated H^+, is simply a proton. Oxygen, with atomic number 8, has eight electrons when it is neutral, so it can be ionized multiple times. *Singly ionized* oxygen is missing one electron, so it has a charge of $+1$ and is designated O^+. *Doubly ionized* oxygen, or O^{++}, is missing two electrons; *triply ionized* oxygen, or O^{+3}, is missing three electrons; and so on. At extremely high temperatures, oxygen can be *fully ionized*, in which case all eight electrons are stripped away and the remaining ion has a charge of $+8$.

4.4 Energy in Atoms

So far, we've seen two different ways in which atoms have energy. First, by virtue of their mass, they possess mass-energy in the amount mc^2. Second, they possess kinetic energy by virtue of their motion. Atoms also contain energy in a third way: as *electric potential energy* in the distribution of their electrons around their nuclei. The simplest case is that of hydrogen, which has only one electron. Remember that an electron tends to be "smeared out" into a cloud around the nucleus. When the electron is "smeared out" to the minimum extent that nature allows, the atom contains its smallest possible amount of electric potential energy; we say that the atom is in its **ground state** (Figure 4.9). If the electron somehow gains energy, then it becomes "smeared out" over a greater volume, and we say that the atom is in an **excited state**. If the electron gains enough energy, it can escape the atom completely, in which case the atom has been *ionized*.

Perhaps the most surprising aspect of atoms was discovered in the 1910s, when scientists realized that electrons in atoms can have only *particular* energies (which correspond to particular sizes and shapes of

ground state   excited state   ionization

FIGURE 4.9 In its ground state, an electron is "smeared out" to the minimum extent allowed by nature. Adding energy can raise the electron to an excited state that occupies a larger volume. Adding enough energy can ionize the atom.

FIGURE 4.10 (**a**) A window washer on an adjustable platform can be at any height. (**b**) A window washer on a ladder can be at only the particular heights of the steps. Similarly, electrons in an atom can have only particular energy levels.

the electron cloud). As a simple analogy, suppose you're washing windows on a building. If you use an adjustable platform to reach high windows, you can stop the platform at any height above the ground (Figure 4.10a). But if you use a ladder, you can stand only at *particular* heights—the heights of the rungs of the ladder—and not at any height in between (Figure 4.10b). The possible energies of electrons in atoms are like the possible heights on a ladder. Only a few particular energies are possible, and energies between these special few are not possible.

The possible energy levels of the electron in hydrogen are represented like the steps of a ladder

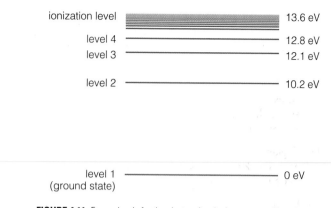

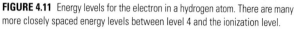

ionization level	13.6 eV
level 4	12.8 eV
level 3	12.1 eV
level 2	10.2 eV
level 1 (ground state)	0 eV

FIGURE 4.11 Energy levels for the electron in a hydrogen atom. There are many more closely spaced energy levels between level 4 and the ionization level.

in Figure 4.11. The ground state, or level 1, is the bottom rung of the ladder; its energy is labeled zero because the atom has no excess electrical potential energy to lose. Each subsequent rung of the ladder represents a possible excited state for the electron. Each level is labeled with the electron's energy above the ground state in units of **electron-volts**, or **eV** ($1 \text{ eV} = 1.60 \times 10^{-19}$ joule). For example, the energy of an electron in energy level 2 is 10.2 eV greater than that of an electron in the ground state. That is, an electron must gain 10.2 eV of energy to "jump" from level 1 to level 2. Similarly, jumping from level 1 to level 3 requires gaining 12.1 eV of energy.

Note that, unlike a ladder built for climbing, the rungs on the electron's energy ladder are closer together near the top. The top itself represents the energy of ionization—if the electron gains this much energy, 13.6 eV above the ground state in the case of hydrogen, the electron breaks free from the atom. (Any excess energy beyond that needed for ionization becomes kinetic energy of the free-moving electron.)

Because energy is always conserved, an electron cannot jump to a higher energy level unless its atom gains the energy from somewhere else. Generally, the atom gains this energy either from the kinetic energy of another particle colliding with it or from the absorption of energy carried by light. Similarly, when an electron falls to a *lower* energy level, it either transfers its energy to another particle through a collision or emits light that carries the energy away. The key point is this: *Electron jumps can occur only with the particular amounts of energy representing differences between possible energy levels.*

The result is that electrons in atoms can absorb or emit only particular amounts of energy and not other amounts in between. For example, if you attempt to provide a hydrogen atom in the ground state

Common Misconceptions: Orbiting Electrons?

Most people have been taught to think of electrons in atoms as "orbiting" the nucleus like tiny planets orbiting a tiny sun. But this representation is simply not true. Electrons and other subatomic particles do not behave at all like baseballs, rocks, or planets. In fact, the behavior of subatomic particles is so strange that human minds may be incapable of visualizing it. Nevertheless, the *effects* of electrons are easy to observe, and one of the most important effects is that electrons give atoms their size. Thus, we say that electrons in atoms are "smeared out" into an "electron cloud" around the nucleus. This description is rather vague, but it is far more accurate than the misleading picture of electrons circling like tiny planets.

with 11.1 eV of energy, the atom won't accept it because it is too high to boost the electron to level 2 but not high enough to boost it to level 3.

TIME OUT TO THINK *We will see in Chapter 6 that light comes in "pieces" called photons that carry specific amounts of energy. Can a hydrogen atom absorb a photon with 11.1 eV of energy? Why or why not? Can it absorb a photon with 10.2 eV of energy? Explain.*

If you think about it, the idea that electrons in atoms can jump only between particular energy levels is quite bizarre. It is as if you had a car that could go only particular speeds and not other speeds in between. How strange it would seem if your car suddenly jumped from 5 miles per hour to 20 miles per hour without gradually passing through a speed of 10 miles per hour! In scientific terminology, the electron's energy levels are said to be *quantized,* and the study of the energy levels of electrons (and other particles) is called *quantum mechanics.*

Electrons have quantized energy levels in all atoms, not just in hydrogen. Moreover, the allowed energy levels differ from element to element and even from one ion of an element to another ion of the same element. This fact holds the key to the study of distant objects in the universe. As we will see in Chapter 6 when we study light, the different energy levels of different elements allow light to carry "fingerprints" that can tell us the chemical composition of distant objects.

THE BIG PICTURE

In this chapter, we discussed the concepts of energy and matter in some detail. Key "big picture" ideas to draw from this chapter include the following:

■ Energy and matter are the two basic ingredients of the universe. Therefore, understanding energy and matter, and the interplay between them, is crucial to understanding the universe.

■ Energy is always conserved, and we can understand many processes in the universe by following how energy changes from one form to another in its interactions with matter. Don't forget that mass itself is a form of potential energy, called mass-energy.

■ The strange laws of quantum mechanics govern the interactions of matter and energy on the atomic level. Electrons in atoms can have only particular energies and not energies in between, and each element has a different set of allowed energy levels.

Review Questions

1. Briefly describe and differentiate between *kinetic energy, potential energy,* and *radiative energy.*

2. What is the formula for the kinetic energy of an object? Based on this formula, explain why (a) a 4-ton truck moving at 100 km/hr has 4 times as much kinetic energy as a 1-ton car moving at 100 km/hr; (b) a 1-ton car moving at 100 km/hr has the *same* kinetic energy as a 4-ton truck moving at 50 km/hr.

3. What does *temperature* measure? How is it related to kinetic energy? What is *thermal energy?*

4. What is *gravitational potential energy?* How does an object's gravitational potential energy depend on its mass, the distance it has to fall, and the strength of gravity?

5. What do we mean by *mass-energy?* Explain the meaning of the formula $E = mc^2$ and what it has to do with the Sun, nuclear bombs, and particle accelerators.

6. What is the *law of conservation of energy?* Your body is using energy right now to keep you alive. Where does this energy come from? Where does it go?

7. Briefly define *atom, element,* and *molecule.*

8. What is *electrical charge?* What type of electrical charge is carried by *protons, neutrons,* and *electrons?* Under what circumstances do electrical charges attract? Under what circumstances do they repel?

9. Describe the structure and size of an atom. How big is the *nucleus* in comparison to the entire atom?

10. Define and distinguish between *atomic number* and *atomic mass.* Under what conditions are two atoms different *isotopes* of the same element?

11. What do we mean by a *phase* of matter? Briefly describe how the idea of *bonds* between atoms (or molecules) explains the difference between the *solid, liquid,* and *gas* phases.

12. Briefly explain why a few atoms (or molecules) are always in gas phase around any solid or liquid. Then explain how *sublimation* and *evaporation* are similar and how they are different.

13. Explain why, at sufficiently high temperatures, molecules undergo *molecular dissociation* and atoms undergo *ionization.*

14. What is a *plasma?* Explain why, at very high temperatures, all matter is in this "fourth phase of matter."

15. Describe the three basic ways in which atoms can contain energy. In what way does an atom in an *excited state* contain more energy than an atom in the *ground state?*

16. How are the possible energy levels of electrons in atoms similar to the possible gravitational potential energies of a person on a ladder? How are they different?

Discussion Questions

1. *Knowledge of Mass-Energy.* Einstein's discovery that energy and mass are equivalent has led to technological developments both beneficial and dangerous. Discuss some of these developments. Overall, do you think the human race would be better or worse off if we had never discovered that mass is a form of energy? Defend your opinion.

2. *Perpetual Motion Machines.* Every so often, someone claims to have built a machine that can generate energy perpetually from nothing. Why isn't this possible according to the known laws of nature? Why do you think claims of perpetual motion machines sometimes receive substantial media attention?

3. *Indoor Pollution.* Given that sublimation and evaporation are very similar processes, why is sublimation generally much more difficult to notice? Discuss how sublimation, particularly from plastics and other human-made materials, can cause "indoor pollution."

Problems

Sensible Statements? For **problems 1–4,** decide whether the statement is sensible and explain why it is or is not.

1. The sugar in my soda will provide my body with about a million joules of energy.

2. When I drive my car at 30 miles per hour, it has three times as much kinetic energy as it does at 10 miles per hour.

3. If you put an ice cube outside the Space Station, it would take a very long time to melt, even though the temperature in Earth orbit is several thousand degrees (Celsius).

4. Someday soon, scientists are likely to build an engine that produces more energy than it consumes.

5. Two isotopes of the element rubidium differ not only in their number of neutrons, but also in their number of protons.

6. According to the laws of quantum mechanics, an electron's energy in a hydrogen atom can jump suddenly from 10.2 eV to 12.1 eV, without ever having any in-between energy such as 10.9 eV.

7. Two ions, each carrying a positive charge of $+1$, will attract each other electrically.

8. In particle accelerators, scientists can create particles where none existed previously by converting energy into mass.

9. *Gravitational Potential Energy.*

    a. Why does a bowling ball perched on a cliff ledge have more gravitational potential energy than a baseball perched on the same ledge?

    b. Why does a diver on a 10-meter platform have more gravitational potential energy than a diver on a 3-meter diving board?

    c. Why does a 100-kg satellite orbiting Jupiter have more gravitational potential energy than a 100-kg satellite orbiting the Earth, assuming both satellites orbit at the same distance from the planet centers?

10. *Einstein's Famous Formula.*

    a. What is the meaning of the formula $E = mc^2$? Be sure to define each variable.

    b. How does this formula explain the generation of energy by the Sun?

    c. How does this formula explain the destructive power of nuclear bombs?

11. *Atomic Terminology Practice.*

    a. The most common form of iron has 26 protons and 30 neutrons in its nucleus. State its atomic number, atomic mass, and number of electrons if it is electrically neutral.

    b. Consider the following three atoms: Atom 1 has 7 protons and 8 neutrons; atom 2 has 8 protons and 7 neutrons; atom 3 has 8 protons and 8 neutrons. Which two are *isotopes* of the same element?

    c. Oxygen has atomic number 8. How many times must an oxygen atom be ionized to create an O^{+5} ion? How many electrons are in an O^{+5} ion?

    d. Consider fluorine atoms with 9 protons and 10 neutrons. What are the atomic number and atomic mass of this fluorine? Suppose we could add a proton to this fluorine nucleus. Would the result still be fluorine? Explain. What if we added a neutron to the fluorine nucleus?

    e. The most common isotope of gold has atomic number 79 and atomic mass 197. How many protons and neutrons does the gold nucleus contain? If it is electrically neutral, how many elec-

trons does it have? If it is triply ionized, how many electrons does it have?

    f. The most common isotope of uranium is ^{238}U, but the form used in nuclear bombs and nuclear power plants is ^{235}U. Given that uranium has atomic number 92, how many neutrons are in each of these two isotopes of uranium?

12. *The Fourth Phase of Matter.*

    a. Explain why nearly all the matter in the Sun is in the plasma phase.

    b. Based on your answer to part (a), explain why plasma is the most common phase of matter in the universe.

    c. Given that plasma is the most common phase of matter in the universe, why is it so rare on Earth?

13. *Energy Level Transitions.* The labeled transitions below represent an electron moving between energy levels in hydrogen. Answer each of the following questions and explain your answers.

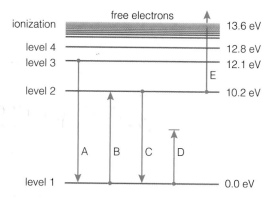

    a. Which transition could represent an electron that *gains* 10.2 eV of energy?

    b. Which transition represents an electron that *loses* 10.2 eV of energy?

    c. Which transition represents an electron that is breaking free of the atom?

    d. Which transition, as shown, is *not* possible?

    e. Describe the process taking place in transition A.

Web Projects

Find useful links for Web projects on the text Web site.

1. *Energy Comparisons.* Using information from the Energy Information Administration Web site, choose some aspect of U.S. or world energy use that interests you. Write a short report on this issue.

2. *Nuclear Power.* There are two basic ways to generate energy from atomic nuclei: through nuclear fission (splitting nuclei) and through nuclear fusion (combining nuclei). All current nuclear reactors are based on fission, but fusion would have many advantages if we could develop the technology. Research some of the advantages of fusion and some of the obstacles to developing fusion power. Do you think fusion power will be a reality in your lifetime? Explain.

If I have seen farther than others, it is because I have stood on the shoulders of giants.

ISAAC NEWTON

CHAPTER 5

Universal Motion

Everything in the universe is in constant motion, from the random meandering of molecules in the air to the large-scale drifting of galaxies in superclusters. Remarkably, just a few physical laws describe all this motion. The elucidation of these laws over the past several centuries is surely one of the greatest scientific triumphs of all time.

Much of the impetus for the discovery of the laws of motion came from the Copernican revolution (which we discussed in Chapter 3), and some of the laws were discovered through experiments by Galileo. But the task of putting all the pieces together and discovering the precise mechanics of gravity fell to Sir Isaac Newton, surely one of the most influential human beings of all time. In this chapter, we'll learn about Newton's discoveries and also discuss how they apply to modern astronomy.

As in Chapter 4, much of the subject matter of this chapter may already be familiar to you. Again, don't worry if this is not the case. The material is not difficult, and studying it carefully will greatly enhance your understanding of the astronomy in the rest of the book as well as of many everyday phenomena.

5.1 Describing Motion: Examples from Daily Life

Because everything moves in the universe, understanding motion is important to understanding the universe. We all have a great deal of experience with motion and natural intuition about how motion works, so we begin our discussion of motion with some familiar examples. Indeed, you probably are familiar with all the terms defined in this section, although their scientific definitions may differ subtly from those you use in casual conversation.

Speed, Velocity, and Acceleration

The concepts we use to determine the trajectory of a ball, a rocket, or a planet are familiar to you from driving a car. The speedometer indicates your **speed**, usually in units of both miles per hour (mi/hr) and kilometers per hour (km/hr); for example, 100 km/hr is a speed. Your **velocity** is your speed in a certain direction; "100 km/hr going due north" describes a velocity. It is possible to change your velocity without changing your speed, for example, by maintaining a steady 60 km/hr as you drive around a curve. Because your direction is changing as you round the curve, your *velocity* is also changing—even though your *speed* is constant.

Whenever your velocity is changing, you are experiencing **acceleration**. You are undoubtedly familiar with the term *acceleration* as it applies to increasing speed, such as when you accelerate away from a stop sign while driving your car. In science, we also say that you are accelerating when you slow down or turn. Slowing occurs when your acceleration is in a direction opposite to your motion; we say that your acceleration is negative, causing your velocity to decrease. Turning changes your direction, which means a change in velocity and thus involves acceleration even if your speed remains constant.

Note that you don't feel anything when you are traveling at *constant velocity*, which is why you don't feel any sensation of motion when you're traveling in an airplane on a smooth flight. In contrast, you can often feel acceleration: As you speed up in a car you feel yourself being pushed back into your seat, as you slow down you feel yourself being pulled forward from the seat, and as you drive around a curve you lean outward because of your acceleration.

The Acceleration of Gravity

One of the most important types of acceleration is that caused by gravity, which makes objects accelerate as they fall. In a famous (though probably apocryphal) experiment that involved dropping weights from the Leaning Tower of Pisa, Galileo demonstrated that gravity accelerates all objects by the same amount, regardless of their mass. This fact may be surprising because it seems to contradict everyday experience: A feather floats gently to the ground, while a rock plummets. However, this difference is caused by air resistance. If you dropped a feather and a rock on the Moon, where there is no air, both would fall at exactly the same rate.

TIME OUT TO THINK *Find a piece of paper and a small rock. Hold both at the same height, one in each hand, and let them go at the same instant. The rock, of course, hits the ground first. Next crumple the paper into a small ball and repeat the experiment. What happens? Explain how this experiment suggests that gravity accelerates all objects by the same amount.*

The acceleration of a falling object is called the **acceleration of gravity**, abbreviated g. On Earth, the acceleration of gravity causes falling objects to fall faster by 9.8 meters per second (m/s), or about 10 m/s, with each passing second. For example, suppose you drop a rock from a tall building. At the moment you let it go, its speed is 0 m/s. After 1 second, the rock will be falling downward at about 10 m/s. After 2 seconds, it will be falling at about 20 m/s. In the absence of air resistance, its speed will continue to increase by about 10 m/s each second until it hits the ground (Figure 5.1). We therefore say that the acceleration of gravity is about 10 *meters per second per second,* or 10 *meters per second squared,* which we write as 10 m/s². (More precisely, $g = 9.8$ m/s².)

FIGURE 5.1 On Earth, gravity causes falling objects to accelerate downward at about 10 m/s². That is, a falling object's downward velocity increases by about 10 m/s with each passing second. (Gravity does not affect horizontal velocity.) More precisely, $g = 9.8$ m/s².

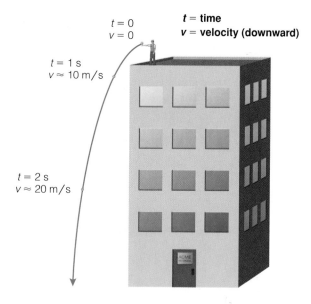

t = 0
v = 0

t = time
v = velocity (downward)

t = 1 s
$v \approx$ 10 m/s

t = 2 s
$v \approx$ 20 m/s

Momentum and Force

Imagine that you're innocently stopped in your car at a red light when a bug flying at a velocity of 30 km/hr due south slams into your windshield. What will happen to your car? Not much, except perhaps a bit of a mess on your windshield. Next imagine that a 2-ton truck runs the red light and hits you head-on with the same velocity as the bug. Clearly, the truck will cause far more damage.

Scientifically, we say that the truck imparts a much larger jolt than the bug because it transfers more **momentum** to you. Momentum describes a combination of mass and velocity. We can describe the momentum of the truck before the collision as "2 tons moving due south at 30 km/hr," while the momentum of the bug is perhaps "1 gram moving due south at 30 km/hr." Mathematically, momentum is defined as mass × velocity.

In transferring some of its momentum to your car, the truck (or bug) exerts a **force** on your car. More generally, a force is anything that can cause a change in momentum. You are familiar with many types of force besides collisional force. For example, if you shift into neutral while driving along a flat stretch of road, the forces of air resistance and road friction will continually sap your car's momentum (transferring it to molecules in the air and the pavement), slowing your velocity until you come to a stop. You

probably are also familiar with forces arising from gravity, electricity, and magnetism.

The mere presence of a force does not always cause a change in momentum. For example, if the engine works hard enough, a car can maintain constant velocity—and hence constant momentum—despite air resistance and road friction. In this case, the force generated by the engine to turn the wheels precisely offsets the forces of air resistance and road friction that act to slow the car, and we say that no **net force** is acting on the car.

As long as an object is not shedding (or gaining) mass, a change in momentum means a change in velocity (because momentum = mass × velocity), which means an acceleration. Hence, any net force will cause acceleration, and all accelerations must be caused by a force. That is why you feel forces when you accelerate in your car.

Mass and Weight

Up until now, we've been glossing over one key term: *mass*. Your **mass** refers to the amount of matter in your body, which is different from your *weight*. Imagine standing on a scale in an elevator (Figure 5.2). When the elevator is stationary or moving at constant velocity, the scale reads your "normal" weight. When the elevator is accelerating upward, the floor exerts

FIGURE 5.2 Mass is not the same as weight. The man's mass never changes, but his apparent weight does.

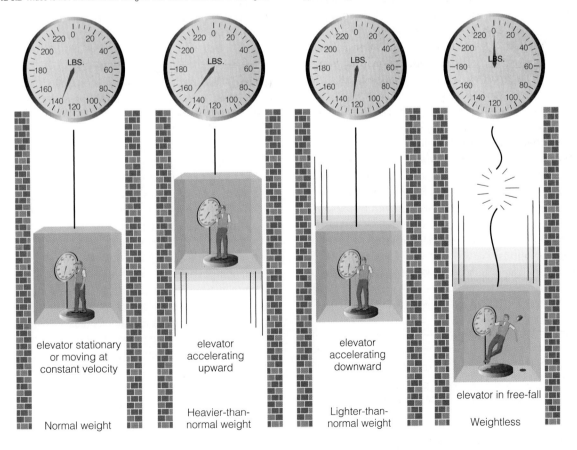

elevator stationary or moving at constant velocity

Normal weight

elevator accelerating upward

Heavier-than-normal weight

elevator accelerating downward

Lighter-than-normal weight

elevator in free-fall

Weightless

an additional force that makes you feel heavier, and the scale verifies your greater apparent weight.[1] Note that your weight is greater than its "normal" value only while the elevator is *accelerating*, not while it is moving at constant velocity. When the elevator accelerates downward, the floor and the scale are dropping away, so your weight is reduced. Thus, your weight varies with the elevator's motion, but your mass remains the same. (You can verify these facts by taking a small bathroom scale with you on an elevator.)

More precisely, your (apparent) **weight** describes the *force* that acts on your mass, and it depends on the strength of gravity and other forces acting on you (such as the force due to the elevator's acceleration). Thus, while your mass is the same anywhere, your weight can vary. For example, your mass would be the same on the Earth and on the Moon, but you would weigh less on the Moon because of its weaker gravity.

Free-Fall, Weightlessness, and Orbit

If the cable breaks so that the elevator is in **free-fall**, the floor drops away at the same rate that you fall. You lose contact with the scale, so your apparent weight is zero and you feel **weightless**. In fact, you are in free-fall whenever nothing is *preventing* you from falling. For example, you are in free-fall when you jump off a chair or spring from a diving board or trampoline. Surprising as it may seem, you have therefore experienced weightlessness many times in your life and can experience it right now simply by jumping off your chair. Of course, your weightlessness lasts for only the very short time until you hit the ground.

Astronauts are weightless for much longer periods because orbiting spacecraft are in a constant state of free-fall. To understand why, imagine a cannon that shoots a ball horizontally from a tall mountain. Once launched, the ball falls solely because of gravity and hence is in free-fall. Thus, the faster the cannon shoots the ball, the farther it goes (Figure 5.3). If the cannonball could go fast enough, its motion would keep it constantly "falling around" the Earth and it would never hit the ground (as long as we neglect air resistance). This is precisely what spacecraft such as the Space Shuttle and the Space Station do as they orbit the Earth. Their constant state of free-fall makes the spacecraft and everything in them weightless.

If an orbiting object travels at the **escape velocity** or faster, it can completely escape from the Earth, never to return. (It will still be weightless as long as

it does not fire any engines, because it is still in free-fall—only now "falling" in response to the gravity of the Sun, planets, and other objects.) The escape velocity from the Earth's surface is about 40,000 km/hr (25,000 mi/hr). We'll discuss escape velocity in more detail in Section 5.5.

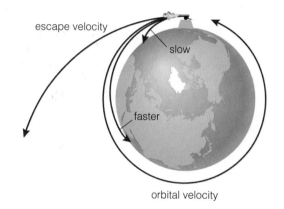

FIGURE 5.3 The faster the cannonball is shot, the farther it goes before hitting the ground. If it goes fast enough, it will continually "fall around," or orbit, the Earth. With an even faster speed, it may escape the Earth's gravity altogether.

Common Misconceptions: No Gravity in Space?

Most people are familiar with pictures of astronauts floating weightlessly in Earth orbit. Unfortunately, because we usually associate weight with gravity, many people assume that the astronauts' weightlessness implies a lack of gravity in space. Actually, there's plenty of gravity in space; even at the distance of the Moon, the Earth's gravity is strong enough to hold the Moon in orbit. In fact, in the low-Earth orbit of the Space Station, the acceleration of gravity is scarcely less than it is on the Earth's surface. Why, then, are the astronauts weightless? Because the Space Station and all other orbiting objects are in a constant state of *free-fall*, and any time you are in free-fall you are weightless. Imagine being an astronaut. You'd have the sensation of free-fall—just as when falling from a diving board—the entire time you were in orbit. This constantly falling sensation makes most astronauts sick to their stomach when they first experience weightlessness. Fortunately, they quickly get used to the sensation, which allows them to work hard and enjoy the view.

---

[1] Many physics texts distinguish between *true weight* that is due only to the effects of gravity on mass and the *apparent weight* that a scale reads when other forces (such as in an accelerating elevator) also act. In this book, you may assume that "weight" refers to apparent weight, except when stated otherwise.

5.2 Newton's Laws of Motion

The complexity of motion in daily life might lead you to guess that the laws governing how forces affect motion also would be quite complex. For example, if you watch a falling piece of paper waft lazily to the ground, you'll see it rocks irregularly back and forth in a seemingly unpredictable pattern. However, the complexity of this motion arises because the paper is affected by a variety of forces, including gravity and the changing forces caused by air currents. If you could analyze the forces individually, you'd find that each force affects the paper's motion in a simple, predictable way. The remarkable simplicity of motion was first described by Sir Isaac Newton (1642–1727).

Newton was born prematurely in Lincolnshire, England, on Christmas Day, 1642. His father, a farmer who never learned to read or write, died three months before he was born. Newton had a difficult childhood and showed few signs of unusual talent. He attended Trinity College at Cambridge, where he earned his keep by performing menial labor, such as cleaning the boots and bathrooms of wealthier students and waiting on their tables.

Shortly after he graduated, the plague hit Cambridge, and Newton returned home. By his own account, he experienced a moment of inspiration in 1666 when he saw an apple fall to the ground. In that moment, Newton shattered the remaining vestiges of the Aristotelian view of the world, which for centuries had been taken in Europe as near-gospel truth.

Aristotle's beliefs included many ideas about the physics of motion, which he had used to support the idea of an Earth-centered cosmos. Aristotle had also maintained that the heavens were totally distinct from the Earth, so that physical laws on Earth did not apply to heavenly motion. By the time Newton saw the apple fall, the Copernican revolution had already displaced the Earth from a central position, and Galileo's experiments had shown that the laws of physics were not what Aristotle had believed. Newton's sudden insight delivered the final blow to Aristotle's physics, as he realized that the force that brought the apple to the ground and the force that held the Moon in orbit were the same. With that insight, Newton eliminated the distinction between the Earth and the heavens, bringing both together in one *universe*. Newton's insight also heralded the birth of the modern science of *astrophysics* (although the term wasn't coined until much later), in which physical laws discovered on Earth are applied to phenomena throughout the cosmos.

Over the next 20 years, Newton's work completely revolutionized mathematics and science. He quantified the laws of motion and gravity, he conducted

Newton (1642–1727)

crucial experiments regarding the nature of light, he built the first reflecting telescopes, and he invented the branch of mathematics called calculus. In this section, we will discuss **Newton's three laws of motion**, which describe how forces affect motion. In the next section, we'll discuss Newton's discoveries about gravity. Newton published his laws of motion and gravity in 1687, in a book usually called *Principia*. (Its full title was *Philosophiae Naturalis Principia Mathematica,* which means "Mathematical Principles of Natural Philosophy.")

Newton's First Law of Motion

Newton's first law of motion deals with situations in which no net force is acting on an object. It states:

In the absence of a net force, an object moves with constant velocity.

Thus, objects at rest (velocity = 0) tend to remain at rest, and objects in motion tend to remain in motion with no change in either their speed or their direction.

The idea that an object at rest should remain at rest is rather obvious; after all, if a car is parked on a flat street, it won't suddenly start moving for no reason. But what if the car is traveling along a flat, straight road? Newton's first law says that the car should keep going forever *unless* a force acts on it. You know that the car eventually will come to a stop if you take your foot off the gas pedal, so we must conclude that some forces are stopping the car—in this case forces arising from friction and air resistance. If the car were in space, and therefore unaffected by friction or air, it would keep moving forever (though gravity would eventually alter its speed and direction). That is why interplanetary spacecraft, once launched into space, need no fuel to keep going.

Newton's first law also explains why you don't feel any sensation of motion when you're traveling in an airplane on a smooth flight. As long as the plane is traveling at constant velocity, no net force is acting on it or on you. Therefore, you feel no differently than you would feel at rest. You can walk around the cabin, play catch with a person a few rows forward, or relax and go to sleep just as though you were "at rest" on the ground.

Newton's Second Law of Motion

Newton's second law of motion tells us what happens to an object when a net force *is* present. We have already said that a net force changes an object's momentum, accelerating it in the direction of the force. Newton's second law quantifies this relationship, which can be stated in two equivalent ways:

force = rate of change in momentum

force = mass × acceleration (or $F = ma$)

This law explains why you can throw a baseball farther than you can throw a shot-put. For both the baseball and the shot-put, the force delivered by your arm equals the product of mass and acceleration. Because the mass of the shot-put is greater than that of the baseball, the same force from your arm gives the shot-put a smaller acceleration. Due to its smaller acceleration, the shot-put leaves your hand with less speed than the baseball and thus travels a shorter distance before hitting the ground.

Newton's Third Law of Motion

Think for a moment about standing still on the ground. The force of gravity acts downward on you, so if this force were acting alone Newton's second law would demand that you be accelerating downward. The fact that you are not falling means that the ground must be pushing back up on you with exactly the right amount of force to offset gravity. This fact is embodied in Newton's third law of motion:

For any force, there always is an equal and opposite reaction force.

According to this law, your body exerts a gravitational force on the Earth identical to the one the Earth exerts on you, except that it acts in the opposite direction. In this mutual pull, the ground just happens to be caught in the middle. Because other forces (between atoms and molecules in the Earth) keep the ground stationary with respect to the center of the Earth, the opposite, upward force is transmitted to you by the ground, holding you in place. Newton's third law also explains how rockets work: Engines generate an explosive force driving hot gas out the back, which creates an equal and opposite force propelling the rocket forward.

Conservation of Momentum and Angular Momentum

If you look more closely at Newton's laws, you will see that they all reflect aspects of a deeper principle: the *conservation of momentum*. Like the amount of energy, the total amount of momentum in the universe is conserved—that is, it does not change. Newton's first law says that an individual object's momentum will not change at all if the object is left alone. Newton's second law says that a force can change the object's momentum, but Newton's third law says that another equal and opposite force simultaneously changes some other object's momentum by a precisely opposite amount. Total momentum always remains unchanged.

In astronomy, a special kind of momentum is particularly important. Consider an ice skater spinning in place. She certainly has some kind of momentum, but it is a little different from the momentum we've discussed previously because she's not actually going anywhere. Technically, her "spinning momentum" is called **angular momentum**. (The term *angular* arises because each spin involves turning through an *angle* of 360°.) In fact, any object that is spinning or orbiting has angular momentum. We can write a simple formula for the angular momentum of an object moving in a circle:

$$\text{angular momentum} = m \times v \times r$$

where m is the object's mass, v is its speed around the circle, and r is the radius of the circle (Figure 5.4).

Just as momentum can be changed only by a force, the angular momentum of any object can be

FIGURE 5.4 The angular momentum of an object moving in a circle is $m \times v \times r$.

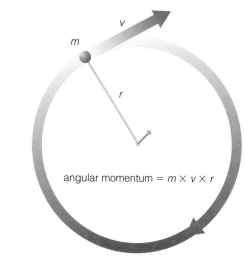

angular momentum = $m \times v \times r$

changed only by a "twisting force," or *torque*. Thus, when no net torque is present, a law very similar to Newton's first law applies—the **law of conservation of angular momentum**:

In the absence of net torque (twisting force), the total angular momentum of a system remains constant.

A spinning ice skater illustrates this law. Because there is so little friction on ice, the ice skater essentially keeps a constant angular momentum. When she pulls in her extended arms, she effectively decreases her radius. Therefore, for the product $m \times v \times r$ to remain unchanged, her velocity of rotation must increase (Figure 5.5).

5.3 The Universal Law of Gravitation

Newton's three laws of motion apply generally to all types of motion. They tell us how motion depends on any force, regardless of the source of the force. Scientists have also discovered how many forces work, and Newton himself discovered how gravity works. He described the force of gravity mathematically in what we now call the **universal law of gravitation**.

The Force of Gravity

Three simple statements summarize the universal law of gravitation:

- Every mass attracts every other mass through the force called *gravity*.

- The force of attraction between any two objects is *directly proportional* to the product of their masses. For example, doubling the mass of *one* object doubles the force of gravity between the two objects.

- The force of attraction between two objects decreases with the *square* of the distance between their centers. That is, the force follows an **inverse square law** with distance. For example, doubling the distance between two objects weakens the force of gravity by a factor of 2^2, or 4.

TIME OUT TO THINK *How does the gravitational force between two objects change if the distance between them triples? If the distance between them drops in half?*

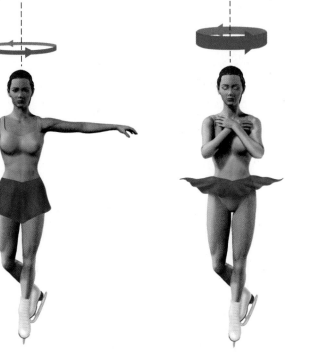

In the product $m \times v \times r$, extended arms mean larger radius and smaller velocity of rotation.

Bringing in her arms decreases her radius and therefore increases her rotational velocity.

FIGURE 5.5 A spinning skater conserves angular momentum.

Common Misconceptions: What Makes a Rocket Launch?

If you've ever watched a rocket launch, it's easy to see why many people believe that the rocket "pushes off" the ground. In fact, the ground has nothing to do with the rocket launch. The rocket takes off because of momentum conservation. Rocket engines are designed to expel hot gas with an enormous amount of momentum. To balance the explosive force driving gas out the back of the rocket, an equal and opposite force must propel the rocket forward, keeping the total momentum—gas plus rocket—unchanged. Thus, rockets can be launched horizontally as well as vertically, and a rocket can be "launched" in space (e.g., from a space station) with no need for any nearby solid ground.

Mathematically, Newton's law of universal gravitation is written:

$$F_g = G \frac{M_1 M_2}{d^2}$$

where F_g is the force of gravitational attraction, M_1 and M_2 are the masses of the two objects, and d is the distance between their *centers* (Figure 5.6). The symbol G is a constant called the **gravitational constant**. Its numerical value was not known to Newton but has since been measured by experiments to be $G = 6.67 \times 10^{-11}$ m³/(kg × s²).

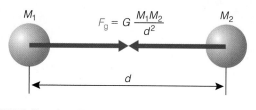

FIGURE 5.6 The universal law of gravitation.

The "Why" of Kepler's Laws, and More

For some 70 years after Kepler published his laws of planetary motion [Section 3.4], scientists debated *why* these laws hold true. Newton settled the debate. By doing a bit of mathematics with the law of universal gravitation and his laws of motion, Newton was able to derive all three of Kepler's laws. In doing so, he also extended Kepler's laws in several important ways:

- Newton found that Kepler's first two laws apply not only to planets, but to *any* object going around another object under the force of gravity. The orbits of a satellite around the Earth, of a moon around a planet, of an asteroid around the Sun, and of binary stars around each other all are ellipses (with the center of mass of the two objects at one focus) in which orbital speeds are faster when the two objects are closer together and slower when they are farther apart.

- He found that orbits do not have to be **bound orbits** as they are with ellipses (Figure 5.7). **Unbound orbits** are also possible (in the shape of parabolas or hyperbolas). A comet with an unbound orbit comes in toward the Sun just once, looping around the Sun and never returning.

- Perhaps most important, he found that Kepler's third law could be generalized in a way that often allows us to calculate the mass of one or both orbiting objects. This generalization, which we call **Newton's version of Kepler's third law**, provides the primary means by which we determine masses throughout the universe. It also

shows that the orbital period of a small object orbiting a much more massive object depends only on its average distance, not on its mass. That is why an astronaut does not need a tether to stay close to the Space Shuttle or Station (Figure 5.8).

5.4 Tides

If you've spent time near an ocean, you're probably aware of the rising and falling of the tide twice each day. What causes the tides, and why are there two each day?

We can understand the basic idea by examining the gravitational attraction between the Earth and the Moon. Gravity attracts the Earth and the Moon toward each other (with the Moon staying in orbit as it "falls around" the Earth), but it affects different parts of the Earth slightly differently: Because the strength of gravity declines with distance, the side of the Earth facing the Moon feels a slightly stronger gravitational attraction than the side facing away from the Moon. As shown in Figure 5.9, this creates two tidal bulges, one facing the Moon and one opposite the Moon. The Earth's rotation carries your location through each of the two bulges each day, creating two high tides; low tides occur when your location is at the points halfway between the two tidal bulges. (The details of tidal timing and height vary significantly from place to place depending on such factors as latitude and topography.)

In a simple sense, the reason there are *two* daily high tides is that the oceans facing the Moon bulge

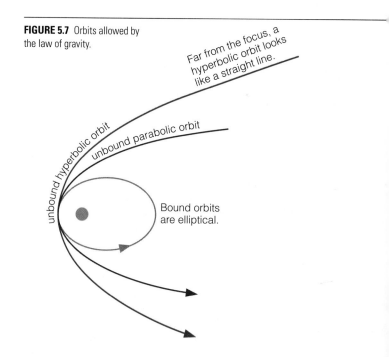

FIGURE 5.7 Orbits allowed by the law of gravity.

Far from the focus, a hyperbolic orbit looks like a straight line.

unbound hyperbolic orbit

unbound parabolic orbit

Bound orbits are elliptical.

FIGURE 5.8 Newton's version of Kepler's third law shows that, when one object orbits a much more massive object, the orbital period depends only on its average orbital distance. Thus, the astronaut and the Space Shuttle share the same orbit despite their different masses—even as both orbit the Earth at a speed of some 25,000 km/hr.

FIGURE 5.9 Tidal bulges face toward and away from the Moon because of the difference in the strength of the gravitational attraction in parts of the Earth at different distances from the Moon. (Arrows represent the strength and direction of the gravitational attraction toward the Moon.) There are two daily high tides as any location on Earth rotates through the two tidal bulges.

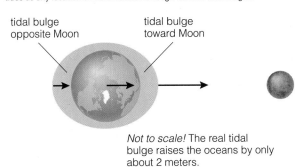

tidal bulge opposite Moon

tidal bulge toward Moon

Not to scale! The real tidal bulge raises the oceans by only about 2 meters.

FIGURE 5.10 Tides also depend on tidal force from the Sun, which is about one-third as strong as that from the Moon. In these diagrams, yellow arrows represent tidal force due to the Sun, which causes the Earth to stretch along the Sun–Earth line, and black arrows represent tidal force due to the Moon, which causes the Earth to stretch along the Earth–Moon line.

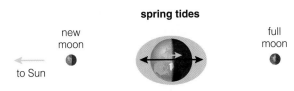

spring tides

new moon

full moon

to Sun

a At new moon and full moon, the tidal forces from both the Moon and the Sun stretch the Earth along the same line, leading to enhanced *spring tides*.

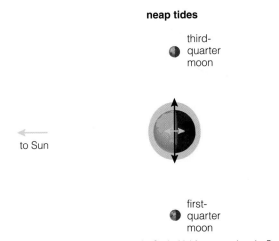

neap tides

third-quarter moon

to Sun

first-quarter moon

b At first- and third-quarter moon, the Sun's tidal force stretches the Earth along a line perpendicular to the Moon's tidal force. The tides still follow the Moon, because the Moon's tidal force is greater, but they are reduced in size, making *neap tides*.

because they are being pulled out from the Earth, while the oceans opposite the Moon bulge because the Earth is being pulled out from under them. However, a better way to look at tides is to recognize that the attraction toward the Moon gets progressively weaker with distance throughout the Earth. This varying attraction creates a "stretching force," or **tidal force**, that stretches the *entire Earth*—land and ocean—along the Earth–Moon line. The tidal bulges are more noticeable for the oceans than for the land only because liquids flow more readily than solids.

Spring and Neap Tides

The Sun also exerts a tidal force on the Earth, causing the Earth to stretch along the Sun–Earth line. You might at first guess that the Sun's tidal force would be substantial, since the Sun's mass is more than a million times the mass of the Moon. Indeed, the *gravitational* force between the Earth and the Sun is much greater than that between the Earth and the Moon, which is why the Earth orbits the Sun. However, the much greater distance to the Sun (than to the Moon) means that the *difference* in the Sun's pull on the near and far sides of the Earth is relatively small, and the overall tidal force caused by the Sun is only about one-third that caused by the Moon (Figure 5.10). When the tidal forces of the Sun and the Moon work together, as is the case at both new moon and full moon, we get the especially pronounced *spring tides* (so named because the water tends to "spring up" from the Earth). When the tidal forces of the Sun and the Moon oppose each other, as is the case at first- and third-quarter moon, we get the relatively small tides known as *neap tides*.

Tidal Friction and Synchronous Rotation

So far, we have talked as if the Earth slides smoothly through the tidal bulges as it rotates. But because it is the Earth itself that stretches, this process involves some friction, called **tidal friction**. In essence, the friction arises because the tidal bulges try to stay on the Earth–Moon line, while the Earth's rotation tries to pull the bulges around with it. The resulting "compromise" puts the bulges just ahead of the Earth–Moon line at all times (Figure 5.11), ensuring that the Earth feels continuous friction as it rotates through the tidal bulges.

This tidal friction has two important effects. First, it causes the Earth's rotation to slow gradually, so that the length of a day gradually gets longer. Second, it makes the Moon move gradually farther from the Earth, because the tidal bulge is always slightly ahead of the Earth–Moon line and thus exerts a gravitational attraction that tends to pull the Moon slightly ahead in its orbit. This makes it harder for the Earth's overall gravity to hold on to the Moon, and as a result the Moon moves slightly farther from the Earth.

Tidal friction has had even more dramatic effects on the Moon. Recall that the Moon always shows (nearly) the same face to the Earth [Section 2.3]. This trait is called **synchronous rotation**, because it means that the Moon's rotation period and orbital period are the same (see Figure 2.19). Synchronous rotation may seem like an extraordinary coincidence, but it is a natural consequence of tidal friction.

The Moon probably once rotated much faster than it does today. But just as the Moon exerts a tidal force on the Earth, the Earth exerts a tidal force on

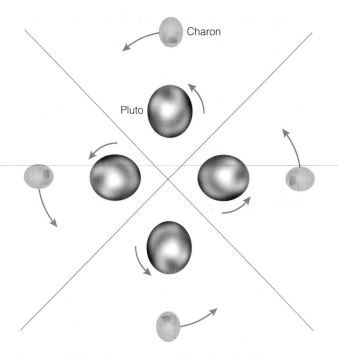

FIGURE 5.12 Pluto and Charon rotate synchronously with each other, so that each always shows the same face to the other. If you stood on Pluto, Charon would remain stationary in your sky, always showing the same face (but going through phases like the phases of our Moon). Similarly, if you stood on Charon, Pluto would remain stationary in your sky, always showing the same face (and going through phases). Tidal bulges are exaggerated in this figure.

the Moon. In fact, because of its larger mass, the Earth exerts a greater tidal force on the Moon than vice versa. This tidal force stretches the Moon along the Earth–Moon line, creating two tidal bulges similar to those on Earth. As long as the Moon rotated through these bulges, the motion created tidal friction that slowed the Moon's rotation. But once the Moon's rotation slowed to the point at which the Moon and its bulges rotated at the same rate—that is, synchronously with the orbital period—there was no further source for tidal friction.

Tidal friction has led to synchronous rotation in many other cases. Most of the moons of the jovian planets rotate synchronously; for example, Jupiter's four large moons (Io, Europa, Ganymede, and Callisto) keep nearly the same face toward Jupiter at all times. Pluto and Charon *both* rotate synchronously: Like two dancers, they always keep the same face toward each other (Figure 5.12). Some moons and planets exhibit variations on synchronous rotation. For example, Mercury rotates exactly three times for every two orbits of the Sun. This pattern ensures that Mercury's tidal bulge always aligns with the Sun at perihelion, where the Sun exerts its strongest tidal force. As you study astronomy, you will encounter many more cases where tides and tidal friction play important roles.

FIGURE 5.11 If the Earth always kept the same face to the Moon, the tidal bulges would stay fixed on the Earth–Moon line. But the Earth's rotation tries to pull them along with it. The result is that the bulges stay nearly fixed relative to the Moon but in a position slightly ahead of the Earth–Moon line. (The effect is exaggerated here for clarity.)

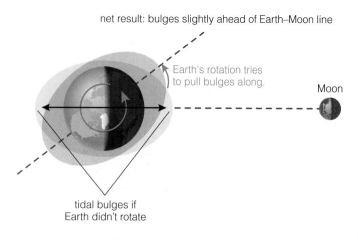

net result: bulges slightly ahead of Earth–Moon line

Earth's rotation tries to pull bulges along.

Moon

tidal bulges if Earth didn't rotate

Many people believe that tides arise because the Moon pulls the Earth's oceans toward it. But if that were the whole story, there would be a bulge only on the side of Earth facing the Moon, and hence only one high tide each day. The correct explanation for tides must account for why the Earth has *two* tidal bulges. If you think about it, you'll realize there's only one simple way to explain this fact: The Earth must be stretching from its center in both directions (toward and away from the Moon). Once you see this, it becomes clear that tides must come from the difference between the force of gravity on one side of the Earth and that on the other, since a difference makes the Earth stretch. In fact, stretching due to tides affects many objects, not just the Earth. Many moons are stretched into oblong shapes by tidal forces caused by their parent planets, and mutual tidal forces stretch close binary stars into teardrop shapes. In regions where gravity is extremely strong, such as near a black hole, tides could even stretch spaceships or people [Section 15.4].

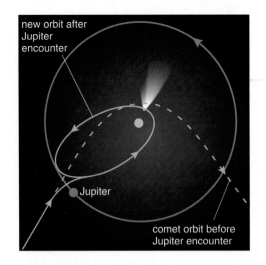

FIGURE 5.13 Depiction of a comet in an unbound orbit of the Sun that happens to pass near Jupiter. The comet loses orbital energy to Jupiter, thereby changing to a bound orbit around the sun.

5.5 Orbital Energy and Escape Velocity

Consider a satellite in an elliptical orbit around the Earth. Its gravitational potential energy is greatest when it is farthest from the Earth, and smallest when it is nearest the Earth. Conversely, its kinetic energy is greatest when it is nearest the Earth and moving fastest in its orbit, and smallest when it is farthest from the Earth and moving slowest in its orbit. Throughout its orbit, its total *orbital energy*—the sum of its kinetic and gravitational potential energies—must be conserved.

Because any change in its orbit would mean a change in its total orbital energy, a satellite's orbit around the Earth cannot change if it is left completely undisturbed. If the satellite's orbit *does* change, it must somehow have gained or lost energy. For a satellite in low-Earth orbit, the Earth's thin upper atmosphere exerts a bit of drag that can cause the satellite to lose energy and eventually plummet back to Earth. The satellite's lost orbital energy is converted to thermal energy in the atmosphere, which is why a falling satellite usually burns up. Raising a satellite to a higher orbit requires that it gain energy by firing one of its rockets. The chemical potential energy of the rocket fuel is converted to gravitational potential energy as the satellite moves to a higher orbit.

Generalizing from the satellite example shows that conservation of energy has a very important implication for motion throughout the cosmos: *Orbits cannot change spontaneously.* For example, an asteroid or a comet passing near a planet cannot spontaneously be "sucked in" to crash on the planet. It can hit the planet only if its current orbit already intersects the planet's surface or if it somehow gains or loses orbital energy so that its new orbit intersects the planet's surface. Of course, if the asteroid or comet *gains* energy, something else must *lose* exactly the same amount of energy, and vice versa.

One way two objects can exchange orbital energy is through a **gravitational encounter**, in which they pass near enough so that each can feel the effects of the other's gravity. For example, Figure 5.13 shows a gravitational encounter between Jupiter and a comet headed toward the Sun on an unbound orbit. The comet's close passage by Jupiter allows the comet and Jupiter to exchange energy: The comet loses orbital energy and changes to a bound, elliptical orbit; Jupiter must gain the energy the comet loses. However, because Jupiter is so much more massive than the comet, the effect on Jupiter is unnoticeable.

More generally, when two objects exchange orbital energy, we expect one to lose energy and fall to a lower orbit while the other gains energy and is thrown to a higher orbit. If an object gains enough energy, it may end up on an unbound orbit that allows it to *escape* from the gravitational influence of the object it is orbiting. For example, if we want to send a space probe to Mars, we must use a large rocket that gives the probe enough energy to achieve an unbound orbit (relative to Earth) and ultimately escape the Earth's gravitational influence. Although it would probably make more sense to say that the probe achieves "escape energy," we instead say that it achieves *escape velocity* (see Figure 5.3). For example, the escape velocity from the Earth's surface is about 40,000 km/hr, or 11 km/s, meaning that this is the *minimum* velocity required to escape the Earth's gravity if you start near the surface. The escape velocity does not depend on the mass of the escaping object; *any* object must travel at a velocity of 11 km/s to escape from the Earth, whether it is an individual atom or molecule escaping from the atmosphere, a spacecraft being launched into deep space, or a rock blasted into the sky by a large impact. But escape velocity does depend on whether you start from the surface or from someplace high above the surface. Because gravity weakens with distance, it takes less energy—and hence a lower escape velocity—to escape from a point high above the Earth than from the Earth's surface.

THE BIG PICTURE

We've covered a lot of ground in this chapter, from the scientific terminology of motion to the story of how universal motion was understood by Newton. Be sure you understand the following "big picture" ideas:

- Understanding the universe requires understanding motion. Although the terminology of the laws of motion may be new to you, the *concepts* are familiar from everyday experience. Think about these experiences so that you'll better understand the less familiar astronomical applications of the laws of motion.

- Motion may seem very complex, but it can be understood simply through Newton's three laws of motion. By combining these laws with his law of universal gravitation, Newton was able to explain how gravity holds planets in their orbits, and much more—including how satellites can reach and stay in orbit, the nature of tides, and why the Moon rotates synchronously with the Earth.

- Perhaps even more important, Newton's discoveries showed that the same physical laws we observe on Earth apply throughout the universe. This universality of physics opens up the entire cosmos as a possible realm of human study.

Review Questions

1. How does *speed* differ from *velocity*? Give an example in which you can be traveling at constant speed but not at constant velocity.

2. What do we mean by *acceleration*? Explain the units of acceleration.

3. What is the *acceleration of gravity* on Earth? If you drop a rock from very high, how fast will it be falling after 4 seconds (neglecting air resistance)?

4. What is *momentum*? How can momentum be affected by a *force*? What do we mean when we say that momentum will be changed only by a *net force*?

5. What is the difference between *mass* and *weight*?

6. What is *free-fall*, and why does it make you *weightless*? Briefly describe why astronauts are weightless in the Space Station.

7. Why does a spaceship require a high speed to achieve orbit? What would happen if it were launched with a speed greater than the Earth's *escape velocity*?

8. State each of Newton's three laws of motion. For each law, give an example of its application.

9. What is the *law of conservation of angular momentum*? How can a skater use this principle to vary his or her rate of spin?

10. What is the *universal law of gravitation*? Summarize what this law says in words, and then state the law mathematically. Define each variable in the formula.

11. Briefly describe how Newton explained and extended Kepler's laws of planetary motion.

12. What do we mean by *bound* and *unbound* orbits?

13. Explain how the Moon creates tides on the Earth. How do tides vary with the phase of the Moon? Why?

14. What is *tidal friction*? Briefly describe how tidal friction has affected the Earth and led to the Moon's synchronous rotation.

15. Explain why orbits cannot change spontaneously. How can atmospheric drag cause an orbit to change? How can a *gravitational encounter* cause an orbit to change?

Discussion Questions

1. *Aristotle and Modern English.* Aristotle believed that the Earth was made from the four elements fire, water, earth, and air, while the heavens were made from *ether* (literally, "upper air"). The literal meaning of *quintessence* is "fifth element," and the literal meaning of *ethereal* is "made of ether." Look up these words in the dictionary. Discuss how their modern meanings are related to Aristotle's ancient beliefs.

2. *Tidal Complications.* The ocean tides on Earth are much more complicated than they might at first seem from the simple physics that underlies tides.

Discuss some of the factors that make the real tides so complicated and how these factors affect the tides. Some factors to consider: the distribution of land and oceans; the Moon's varying distance from Earth in its orbit; the fact that the Moon's orbital plane is not perfectly aligned with the ecliptic and neither the Moon's orbit nor the ecliptic is aligned with the Earth's equator.

Problems

Sensible Statements? For **problems 1–4**, decide whether the statement is sensible and explain why it is or is not.

1. If you could go shopping on the Moon to buy a pound of chocolate, you'd get a lot more chocolate than if you bought a pound on Earth.

2. Suppose you could enter a vacuum chamber (on Earth), that is, a chamber with no air in it. Inside this chamber, if you dropped a hammer and a feather from the same height at the same time, both would hit the bottom at the same time.

3. When an astronaut goes on a space walk outside the Space Station, she will quickly float away from the station unless she has a tether holding her to the station or constantly fires thrusters on her space suit.

4. If an asteroid passed by Earth at just the right distance, it would be captured by the Earth's gravity and become our second moon.

5. *Understanding Acceleration.*

   a. Some schools have an annual ritual that involves dropping a watermelon from a tall building. Suppose it takes 6 seconds for the watermelon to fall to the ground (which would mean it's dropped from about a 60-story building). If there were no air resistance, so that the watermelon would fall with the acceleration of gravity, how fast would it be going when it hit the ground? Give your answer in m/s.

   b. As you sled down a steep, slick street, you accelerate at a rate of 4 m/s². How fast will you be going after 5 seconds? Give your answer in m/s.

   c. You are driving along the highway at a speed of 70 miles per hour when you slam on the brakes. If your acceleration is at an average rate of −20 miles per hour per second, how long will it take to come to a stop?

6. *Spinning Skater.* Suppose an ice skater wants to *start* spinning. Explain why she won't start spinning if she simply stomps her foot straight down on the ice. How should she push off the ice to start spinning? Why? What should she do when she wants to stop spinning?

7. *The Gravitational Law.* Use the universal law of gravitation to answer each of the following questions.

   a. How does tripling the distance between two objects affect the gravitational force between them?

   b. Compare the gravitational force between the Earth and the Sun to that between Jupiter and the Sun. The mass of Jupiter is about 318 times the mass of the Earth.

   c. Suppose the Sun were magically replaced by a star with twice as much mass. What would happen to the gravitational force between the Earth and the Sun?

8. *Head-to-Foot Tides.* You and the Earth attract each other gravitationally, so you should also be subject to a tidal force resulting from the difference between the gravitational attraction felt by your feet and that felt by your head (at least when you are standing). Explain why you can't feel this tidal force.

9. *Measuring Masses.* Mathematically, Newton's version of Kepler's third law reads:

$$p^2 = \frac{4\pi^2}{G(M_1 + M_2)} a^3$$

where p is the orbital period in *seconds*, a is the average distance in *meters*, M_1 and M_2 are the masses of the two objects in kilograms, and the gravitational constant is

$$G = 6.67 \times 10^{-11} \frac{m^3}{kg \times s^2}$$

   a. The Moon orbits the Earth in an average of 27.3 days at an average distance of 384,000 kilometers. Use these facts to determine the mass of the Earth. You may neglect the mass of the Moon and assume $M_{Earth} + M_{Moon} \approx M_{Earth}$.

   b. Jupiter's moon Io orbits Jupiter every 42.5 hours at an average distance of 422,000 kilometers from the center of Jupiter. Calculate the mass of Jupiter. (*Hint:* Assume $M_{Jupiter} + M_{Io} \approx M_{Jupiter}$.)

   c. Calculate the orbital period of the Space Shuttle in an orbit 300 kilometers above the Earth's surface. (*Hint:* Assume $M_{Earth} + M_{satellite} \approx M_{Earth}$.)

   d. Pluto's moon Charon orbits Pluto every 6.4 days with a semimajor axis of 19,700 kilometers. Calculate the *combined* mass of Pluto and Charon. Compare this combined mass to the mass of the Earth.

Web Projects

Find useful links for Web projects on the text Web site.

1. *Tide Tables.* Find a tide table or tide chart for a beach town that you'd like to visit. Briefly explain how to read the table or chart, and discuss any differences between the actual tidal pattern and the idealized tidal pattern described in this chapter.

2. *Space Station.* Visit a NASA site with pictures from the Space Station. Choose two photos that illustrate some facet of Newton's laws of rotation or gravity. Explain what is going on that is related to Newton's laws.

CHAPTER 6

Light, Telescopes, and Spacecraft

Ancient observers could discern only the most basic features of the light that they saw—such as color and brightness. Over the past several hundred years, we have discovered that light carries far more information. Remarkably, analysis of light with special instruments can reveal the chemical composition of distant objects, their temperature, how fast they rotate, and much more. Light, the cosmic messenger, brings the stories of distant objects to our home here on Earth.

In this chapter, we'll explore the basic properties of light. We'll learn why the light visible to our eyes is only the tip of the iceberg when it comes to the full spectrum of light, and we'll see how we can extract a treasury of information by examining light spectra in detail. We'll also discuss the basic features of telescopes, the instruments with which we collect light from the far reaches of the universe, and of spacecraft, which play an increasingly important role in astronomy.

FIGURE 6.1 A prism reveals that white light contains a spectrum of colors from red to violet.

6.1 Light in Everyday Life

Even without opening your eyes, it's clear that light is a form of energy, called *radiative energy* [Section 4.1]. Outside on a hot, sunny day, you can feel the radiative energy of sunlight being converted to thermal energy as it strikes your skin. On an economic level, electronic companies charge you for the energy needed by light bulbs (and other appliances). The rate at which a light bulb uses energy (converting electrical energy to light and heat) is usually printed on it—for example, "100 watts." A watt is a unit of **power**, which describes the *rate* of energy use: 1 watt of power means that 1 joule of energy is used each second. Thus, for every second that you leave a 100-watt light bulb turned on, you will have to pay the utility company for 100 joules of energy. Interestingly, the power requirement of an average human— about 10 million joules per day—is about the same as that of a 100-watt light bulb.

Another basic property of light is what our eyes perceive as *color*. You've probably seen a prism split light into a **spectrum** (plural, spectra), or rainbow of colors (Figure 6.1). You can also produce a spectrum with a **diffraction grating**—a piece of plastic or glass etched with many closely spaced lines. The colors in a spectrum are pure forms of the basic colors red, orange, yellow, green, blue, and violet. The wide variety of all possible colors comes from mixtures of these basic colors in varying proportions; *white* is simply what we see when the basic colors are mixed in roughly equal proportions. Your television takes advantage of this fact to simulate a huge range of colors by combining only three specific colors of red, green, and blue light.

TIME OUT TO THINK *If you have a magnifying glass handy, hold it close to your TV set to see the individual red, blue, and green dots. If you don't have a magnifying glass, try splashing a few droplets of water onto your TV screen (carefully!). What do you see? What are the drops of water doing?*

Energy carried by light can interact with matter in four general ways:

- **Emission**: When you turn on a lamp, electricity flowing through the filament of the light bulb heats it to a point at which it *emits* visible light.

- **Absorption**: If you place your hand near a lit light bulb, your hand *absorbs* some of the light, and this absorbed energy makes your hand warmer.

- **Transmission**: Some forms of matter, such as glass or air, *transmit* light; that is, they allow light to pass through them.

- **Reflection**: A mirror *reflects* light in a very specific way, similar to the way a rubber ball bounces off a hard surface, so that the direction of a reflected beam of light depends on the direction of the incident (incoming) beam of light (Figure 6.2a). Sometimes, reflection is more random, so that an incident beam of light is **scattered** in many different directions. The screen in a movie

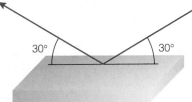

angle of incidence = angle of reflection

30°   30°

a A mirror reflects light along a path determined by the angle at which the light strikes the mirror.

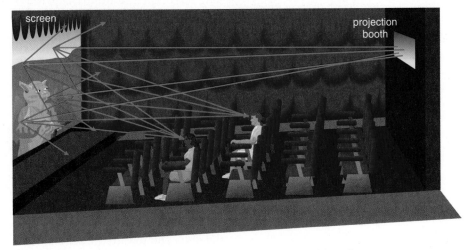

screen

projection booth

b A movie screen scatters light into an array of beams that reach every member of the audience.

FIGURE 6.2 Light reflection and scattering.

theater scatters a narrow beam of light from the projector into an array of beams that reach every member of the audience (Figure 6.2b).

Particular materials affect different colors of light differently. For example, red glass transmits red light but absorbs other colors. A green lawn reflects green light but absorbs all other colors.

Now let's put all these ideas together and think about what happens when you walk into a dark room and turn on the light switch. The light bulb begins to emit white light, which is a mix of all the colors in the spectrum. Some of this light exits the room, transmitted through the windows. The rest of the light strikes the surfaces of objects inside the room, and each object's material properties determine the colors absorbed or reflected. The light coming from each object therefore carries an enormous amount of information about the object's location, shape and structure, and material makeup. You acquire this information when light enters your eyes, where it is absorbed by special cells (called *cones* and *rods*) that use the energy of the absorbed light to send signals to your brain. Your brain interprets the messages carried by the light, recognizing materials and objects in the process we call *vision*.

6.2 Properties of Light

In fact, light carries even more information than your ordinary vision can recognize. Just as a microscope can reveal structure that is invisible to the naked eye, modern instruments can reveal otherwise invisible details in the spectrum of light. Learning to interpret these details is the key to unlocking the vast amount of information carried by light. Let's begin by examining fundamental properties of light.

Particle or Wave?

In our everyday lives, waves and particles appear to be very different. After all, no one would confuse the ripples on a pond with a baseball. However, light behaves as *both* a particle and a wave. Like particles, light comes in individual "pieces," called **photons**, that can hit a wall one at a time. Like ripples on a pond, light can make particles bob up and down.

First, let's ignore the particle properties of light and look at its wave nature. Light is an **electromagnetic wave**—a wave in which electric and magnetic fields vibrate. (The concept of a *field* is a bit abstract, but you can think of electric or magnetic fields as

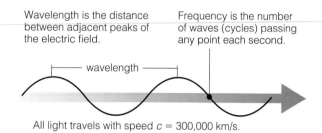

a A row of electrons would wriggle up and down as light passes by, showing that light carries a vibrating electric field. It also carries a magnetic field (not shown) that vibrates perpendicular to the direction of the electric field vibrations.

Wavelength is the distance between adjacent peaks of the electric field.

Frequency is the number of waves (cycles) passing any point each second.

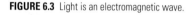

wavelength

All light travels with speed $c = 300{,}000$ km/s.

b Characteristics of light waves. Because all light travels at the same speed, light of longer wavelength must have lower frequency.

FIGURE 6.3 Light is an electromagnetic wave.

places where charged particles would feel electric or magnetic forces.) Like a leaf on a rippling pond, an electron will bob up and down when an electromagnetic wave passes by; if you could set up a row of electrons, they would wriggle like a snake (Figure 6.3a). The **wavelength** is the distance between adjacent peaks of the electric or magnetic field, and the **frequency** is the number of peaks that pass by any point each second (Figure 6.3b). All light travels at the same speed—about 300,000 kilometers per second, or 3×10^8 m/s—regardless of its wavelength or frequency. Therefore, light with a shorter wavelength must have a higher frequency, and vice versa.

Measuring the particle properties of light requires thinking of each photon as a distinct entity. Just as a baseball carries a specific amount of kinetic energy, each photon of light carries a specific amount of radiative energy. The shorter the wavelength of the light (or, equivalently, the higher its frequency), the higher the energy of the photons. For example, a photon with a wavelength of 100 nanometers (nm) has more energy than a photon with a 120-nm wavelength. (A nanometer is a billionth of a meter: 1 nm = 10^{-9} m.)

The Many Forms of Light

Because light consists of electromagnetic waves, the spectrum of light is called the **electromagnetic spectrum**. Photons of light can have *any* wavelength or frequency, so in principle the complete electromagnetic spectrum extends from a wavelength of zero to infinity. For convenience, we usually refer to

different portions of the electromagnetic spectrum by different names (Figure 6.4).

The **visible light** that we see with our eyes has wavelengths ranging from about 400 nm at the blue end of the rainbow to about 700 nm at the red end. Light with wavelengths somewhat longer than red light is called **infrared**, because it lies beyond the red end of the rainbow. Light with very long wavelengths is called **radio** (or *radio waves*)—thus, radio is a form of light, *not* a form of sound.

On the other end of the spectrum, light with wavelengths somewhat shorter than blue light is called **ultraviolet**, because it lies beyond the blue (or violet) end of the rainbow. Light with even shorter wavelengths is called **X rays**, and the shortest-wavelength light is called **gamma rays**. Note that visible light is an extremely small part of the entire electromagnetic spectrum: The reddest red that our eyes can see has only about twice the wavelength of the bluest blue, but the radio waves from your favorite radio station are a billion times longer than the X rays used in a doctor's office.

Because wavelengths decrease as we move from the radio end toward the gamma-ray end of the spectrum, the frequencies and energies must increase. Visible photons happen to have enough energy to activate the molecular receptors in our eyes. Ultraviolet photons, with a shorter wavelength than visible light, carry more energy—enough to harm our skin

Common Misconceptions: You Can't Hear Radio Waves and You Can't See X Rays

Most people associate the term *radio* with sound, but radio waves are a form of *light* with long wavelengths—too long for our eyes to see. Radio stations encode sounds (e.g., voices, music) as electrical signals, which they broadcast as radio waves. What we call "a radio" in daily life is an electronic device that receives these radio waves and decodes them to re-create the sounds played at the radio station. Television is also broadcast by encoding information (both sound and pictures) in the form of light called radio waves.

X rays are also a form of light, with wavelengths far too short for our eyes to see. In a doctor's or dentist's office, a special machine works rather like the flash on an ordinary camera but emits X rays instead of visible light. This machine flashes the X rays at you, and a piece of photographic film records the X rays that are transmitted through your body. Note that you never *see* the X rays; you see only an image left on film by the transmitted X rays.

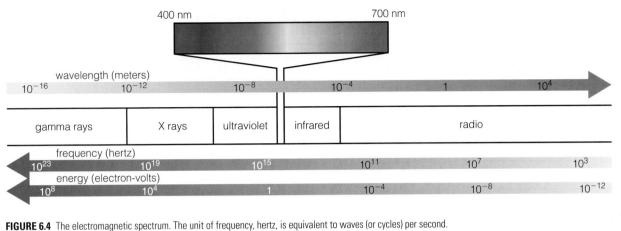

FIGURE 6.4 The electromagnetic spectrum. The unit of frequency, hertz, is equivalent to waves (or cycles) per second. For example, 10^3 hertz means that $10^3 = 1,000$ wave peaks pass by a point each second.

cells, causing sunburn or skin cancer. X-ray photons have enough energy to transmit easily through skin and muscle but not so easily through bones or teeth. That is why photographs taken with X-ray light allow doctors and dentists to see our underlying bone structures.

6.3 Light and Matter

Whenever matter and light interact, matter leaves its fingerprints. Examining the color of an object is a crude way of studying the clues left by the matter it contains. If we take light and disperse it into a spectrum, we can see the spectral fingerprints in more detail.

Figure 6.5 shows a schematic spectrum of light from a celestial body such as a planet. The spectrum is a graph that shows the amount of radiation, or **intensity**, at different wavelengths. At wavelengths where a lot of light is coming from the celestial body, the intensity is high; at wavelengths where there is little light, the intensity is low. Our goal is to see how

the bumps and wiggles in this graph convey a wealth of information about the celestial body in question. Let's begin by going through a short list of ways that matter interacts with light and showing the spectra that result from these interactions.

Absorption and Emission by Thin Gases

You know that gases can absorb light; for example, ozone in the Earth's atmosphere absorbs ultraviolet light from space, preventing it from reaching the ground. Gases can also emit light, which is what makes interstellar clouds glow so beautifully. But *how* do gases absorb or emit light?

Recall that the electrons in atoms can have only specific energies, somewhat like the specific heights of the rungs on a ladder [Section 4.4]. If an electron in an atom is bumped from a lower energy level to a higher one—by a collision with another atom, for example—it will eventually fall back to the lower level. The energy that the atom loses when the electron falls back down must go somewhere, and often it goes to *emitting* a photon of light. The emitted photon must have

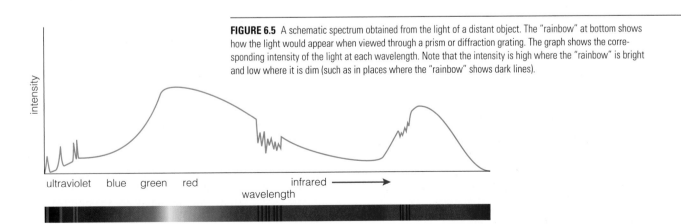

FIGURE 6.5 A schematic spectrum obtained from the light of a distant object. The "rainbow" at bottom shows how the light would appear when viewed through a prism or diffraction grating. The graph shows the corresponding intensity of the light at each wavelength. Note that the intensity is high where the "rainbow" is bright and low where it is dim (such as in places where the "rainbow" shows dark lines).

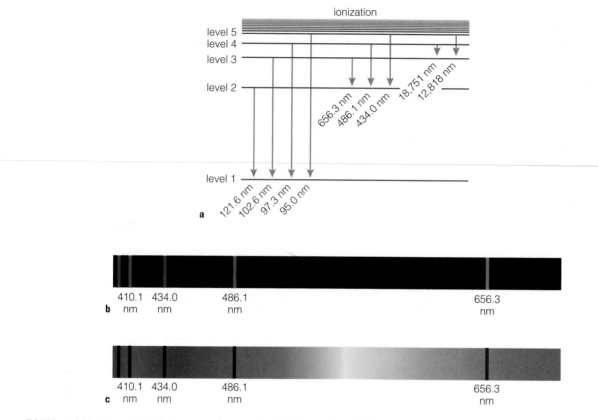

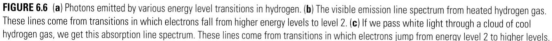

FIGURE 6.6 (**a**) Photons emitted by various energy level transitions in hydrogen. (**b**) The visible emission line spectrum from heated hydrogen gas. These lines come from transitions in which electrons fall from higher energy levels to level 2. (**c**) If we pass white light through a cloud of cool hydrogen gas, we get this absorption line spectrum. These lines come from transitions in which electrons jump from energy level 2 to higher levels.

exactly the same amount of energy that the electron loses, which means that it has a specific wavelength (and frequency).

Figure 6.6a shows the allowed energy levels in hydrogen, along with the wavelengths of the photons emitted by various downward *transitions* of an electron from a higher energy level to a lower one. For example, transitions from level 2 to level 1 emit an ultraviolet photon of wavelength 121.6 nm, and transitions from level 3 to level 2 emit a red visible-light photon of wavelength 656.3 nm. If you heat some hydrogen gas so that collisions are continually bumping electrons to higher energy levels, you'll get an **emission line spectrum** consisting of the photons emitted as each electron falls back to lower levels (Figure 6.6b).

TIME OUT TO THINK *If nothing continues to heat the hydrogen gas, all the electrons eventually will end up in the lowest energy level (the ground state, or level 1). Use this fact to explain why we should not expect to see an emission line spectrum from a very cold cloud of hydrogen gas.*

Photons of light can also be absorbed, causing electrons to jump *up* in energy—but only if an incoming photon happens to have precisely the right amount of energy. For example, just as an electron moving downward from level 2 to level 1 in hydrogen emits a photon of wavelength 121.6 nm, absorbing a photon with this wavelength will cause an electron in level 1 to jump up to level 2.

Suppose a lamp emitting white light illuminates a cloud of hydrogen gas from behind. The cloud will absorb photons with the precise energies needed to bump electrons from a low energy level to a higher one, while all other photons pass right through the cloud. The result is an **absorption line spectrum** that looks like a rainbow with light missing at particular wavelengths (Figure 6.6c).

If you compare the bright emission lines in Figure 6.6b to the dark absorption lines in Figure 6.6c, you will see that the lines occur at the same wavelengths regardless of whether the hydrogen is absorbing or emitting light. Absorption lines simply correspond to upward jumps of the electron between energy levels, while emission lines correspond to downward jumps.

helium

sodium

neon

FIGURE 6.7 Emission line spectra for helium, sodium, and neon. The patterns and wavelengths of lines are different for each element, giving each a unique spectral fingerprint.

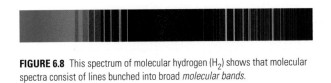

FIGURE 6.8 This spectrum of molecular hydrogen (H_2) shows that molecular spectra consist of lines bunched into broad *molecular bands*.

The energy levels of electrons in each chemical element are unique [Section 4.4]. As a result, each element produces its own distinct set of spectral lines, giving it a unique "spectral fingerprint." For example, Figure 6.7 shows emission line spectra for helium, sodium, and neon. Each different type of molecule (made from two or more atoms) also produces its own unique set of spectral lines; these lines tend to come in bunches called **molecular bands** (Figure 6.8).

Thermal Radiation: Every Body Does It

While thin gases tend to produce simple emission line or absorption line spectra, more "complex" objects—such as planets, stars, and you—produce what we call **thermal radiation** (sometimes called *blackbody* radiation). This term arises because, if an object emitted *only* thermal radiation, its spectrum would depend only on its *temperature*. No real object is a perfect "thermal emitter," but many objects make close approximations. Therefore, it is useful to understand the two rules that describe how a thermal radiation spectrum depends on temperature:

1. *Hotter objects emit more total radiation per unit surface area.* The radiated energy is proportional to the *fourth* power of the temperature expressed in Kelvin (*not* in Celsius or Fahrenheit). For example, a 600 K object has twice the temperature of a 300 K object and therefore radiates $2^4 = 16$ times as much total energy per unit surface area.

2. *Hotter objects emit photons with a higher average energy,* which means a shorter average wavelength.

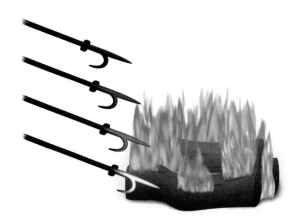

FIGURE 6.9 A fireplace poker gets brighter as it is heated, demonstrating rule 1 for thermal radiation (hotter objects emit more total radiation per unit surface area). In addition, its "color" moves from infrared to red to white as it is heated, demonstrating rule 2 (hotter objects emit photons with higher average energy).

You can see the first rule in action by playing with a light that has a dimmer switch: When you turn the switch up, the filament in the light bulb gets hotter and the light brightens; when you turn it down, the filament gets cooler and the light dims. (You can verify the changing temperature by placing your hand near the bulb.) You can see the second rule in action by observing a fireplace poker (Figure 6.9). When the poker is relatively cool, it emits only infrared radiation, which we cannot see. As it gets hot, it begins to glow red ("red hot"). If the poker continues to heat up, the average wavelength of the emitted photons gets shorter, moving toward the blue end of the visible spectrum. By the time it gets very hot, the mix of colors emitted by the poker looks white ("white hot").

Figure 6.10 shows how these two rules affect idealized thermal radiation spectra. The spectra of hotter objects show bigger "humps" because they emit more total radiation per unit area (rule 1). Hotter objects also have the peaks of their humps at

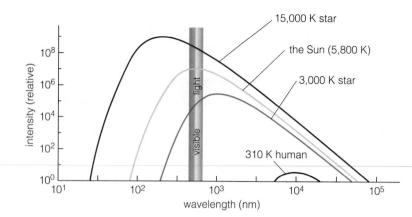

FIGURE 6.10 Graphs of idealized thermal radiation spectra. Note that, per unit surface area, hotter objects emit more radiation at every wavelength, demonstrating rule 1 for thermal radiation. The peaks of the spectra occur at shorter wavelengths (higher energies) for hotter objects, demonstrating rule 2 for thermal radiation. *Technical Note:* The curves appear to fall sharply to zero on both sides because of the logarithmic (power-of-10) scale used on the vertical axis. On a finer scale, we would see that the curves continue at low values on the long-wavelength (falling asymptotically to zero).

shorter wavelengths because of the higher average energy of their photons (rule 2). Note that hotter objects emit more light at *all* wavelengths but the biggest difference appears at the shortest wavelengths. An object with a temperature of 310 K, which is about human body temperature, emits mostly in the infrared and emits no visible light at all—that explains why we don't glow in the dark! A relatively cool star, with a 3,000 K surface temperature, emits mostly red light; that is why some bright stars in our sky appear red, such as Betelgeuse (in Orion) and Antares (in Scorpius). The Sun's 5,800 K surface emits most strongly in green light (around 500 nm), but the Sun looks yellow or white to our eyes because it also emits other colors throughout the visible spectrum. Hotter stars emit mostly in the ultraviolet, but because our eyes cannot see ultraviolet they appear blue or blue-white in color. If an object were heated to a temperature of millions of degrees, it would radiate mostly X rays. Some astronomical objects are indeed hot enough to emit X rays, such as disks of gas encircling exotic objects like neutron stars and black holes (see Chapter 15).

Summary of Spectral Formation

We can now summarize the circumstances under which objects produce thermal, absorption line, or emission line spectra. (These rules are often called *Kirchhoff's laws.*)

■ Any "complex" object produces thermal radiation over a broad range of wavelengths. If the object is hot enough to produce visible light, as is the case with the filament of a light bulb, we see a smooth, *continuous* rainbow when we disperse the light through a prism or a diffraction grating (Figure 6.11a). On a graph of intensity versus wavelength, the rainbow becomes the characteristic hump of a thermal radiation spectrum.

■ When thermal radiation passes through a thin cloud of gas, the cloud leaves fingerprints that may be either absorption lines or emission lines, depending on its temperature. If the background source of thermal radiation is hotter than the cloud, the balance between emission and absorption in the cloud's spectral lines tips toward absorption. We then see absorption lines cutting into the thermal spectrum. This is the case when the light from the hot light bulb passes through a cool gas cloud (Figure 6.11b). On the graph of intensity versus wavelength, these absorption lines create dips in the thermal radiation spectrum; the width and depth of each dip depend on how much light is absorbed by the chemical responsible for the line.

■ If the background source is colder than the cloud, or if there is no background source at all, the spectrum is dominated by bright emission lines produced by the cloud's atoms and molecules (Figure 6.11c). These lines create narrow peaks on a graph of intensity versus wavelength.

Reflected Light

We've now covered enough material to understand spectra emitted by objects generating their own light, such as stars and clouds of interstellar gas. But most of our daily experience involves *reflected* (or *scattered*) light. The source of the light is thermal radiation from

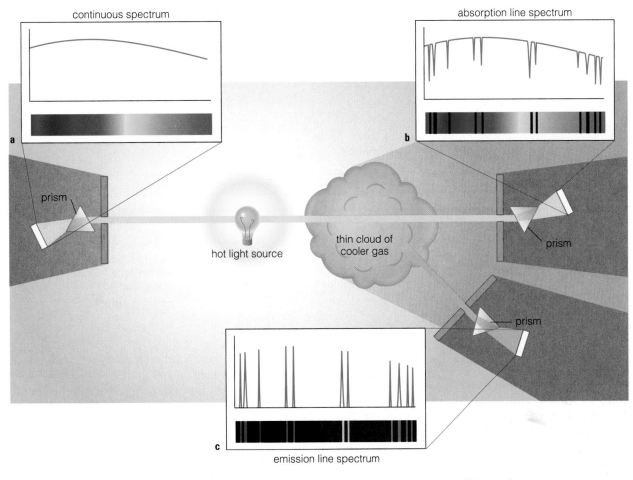

continuous spectrum

absorption line spectrum

prism

hot light source

thin cloud of cooler gas

prism

prism

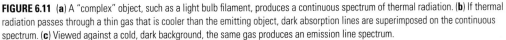

emission line spectrum

FIGURE 6.11 (**a**) A "complex" object, such as a light bulb filament, produces a continuous spectrum of thermal radiation. (**b**) If thermal radiation passes through a thin gas that is cooler than the emitting object, dark absorption lines are superimposed on the continuous spectrum. (**c**) Viewed against a cold, dark background, the same gas produces an emission line spectrum.

the Sun or a lamp. After this light strikes the ground, clouds, people, or other objects, we see only the wavelengths of light that are reflected. For example, a red sweatshirt absorbs blue light and reflects red light, so its visible spectrum looks like the thermal radiation spectrum of its light source—the Sun—but with blue light missing.

In the same way that we distinguish lemons from limes, we can use color information in reflected light to learn about celestial objects. Different fruits, different rocks, and even different atmospheric gases reflect and absorb light at different wavelengths. Although the absorption features that show up in spectra of reflected light are not as distinct as the emission and absorption lines for thin gases, they still provide useful information. For example, the surface materials of a planet determine how much light of different colors is reflected or absorbed. The reflected light gives the planet its color, while the absorbed light heats the surface and helps determine its temperature.

Putting It All Together

Figure 6.12 again shows the complicated spectrum we began with in Figure 6.5, but this time with labels indicating the processes responsible for its various features. What can we say about this object from its spectrum? The hump of thermal emission shows that this object has a surface temperature of about 225 K, well below the freezing point of water. The absorption bands in the infrared come mainly from carbon dioxide, which tells us that the object has a carbon dioxide atmosphere. The emission lines in the ultraviolet come from hot gas in a high, thin layer of the object's atmosphere. The reflected light looks like the Sun's 5,800 K thermal radiation except that the blue light is missing, so the object must be reflecting sunlight and must look red in color. Perhaps by now you have guessed that this figure represents the spectrum of the planet Mars.

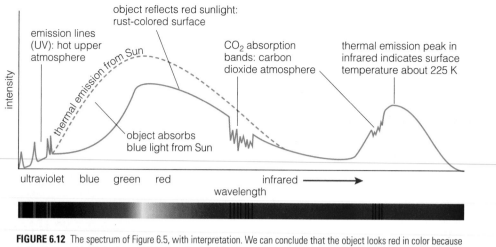

FIGURE 6.12 The spectrum of Figure 6.5, with interpretation. We can conclude that the object looks red in color because it absorbs more blue light than red light from the Sun. The absorption lines tell us that the object has a carbon dioxide atmosphere, and the emission lines tell us that its upper atmosphere is hot. The hump in the infrared (near the right of the diagram) tells us that the object has a surface temperature of about 225 K. It is a spectrum of the planet Mars.

6.4 The Doppler Shift

The volume of information about temperature and composition that light contains is truly amazing, but we can learn even more once we identify the various lines in a spectrum. Among the most important pieces of information contained in light is information about motion. In particular, we can determine the radial motion (toward or away from us) of a distant object from changes in its spectrum caused by the **Doppler effect**.

You've probably noticed the Doppler effect on *sound;* it is especially easy to notice when you stand near train tracks and listen to the whistle of a train (Figure 6.13). As the train approaches, its whistle is relatively high pitched; as it recedes, the sound is relatively low pitched. Just as the train passes by, you hear the dramatic change from high to low pitch—a sort of "weeeeeeee-ooooooooooh" sound. You can visualize the Doppler effect by imagining that the train's sound waves are bunched up ahead of it, resulting in shorter wavelengths and thus the high pitch you hear as the train approaches. Behind the train, the sound waves are stretched out to longer wavelengths, resulting in the low pitch you hear as the train recedes.

The Doppler effect causes similar shifts in the wavelengths of light. If an object is moving toward us, then its entire spectrum is shifted to shorter wavelengths. Because shorter wavelengths are bluer when we are dealing with visible light, the Doppler shift of an object coming toward us is called a **blueshift**. If an object is moving away from us, its light is shifted to longer wavelengths; we call this a **redshift** because longer wavelengths are redder when we are dealing with visible light. Note that the terms *blue-*

shift and *redshift* are used even when we are not dealing with visible light.

Spectral lines provide the reference points we use to identify and measure Doppler shifts (Figure 6.14). For example, suppose we recognize the pattern of hydrogen lines in the spectrum of a distant object. We know the **rest wavelengths** of the hydrogen lines—that is, their wavelengths in stationary clouds of hydrogen gas—from laboratory experiments in which a tube of hydrogen gas is heated so the wavelengths of the spectral lines can be measured. If the hydrogen lines from the object appear at longer wavelengths, then we know that they are redshifted and the object is moving away from us; the larger the shift, the faster the object is moving. If the lines appear at shorter wavelengths, then we know that they are blueshifted and the object is moving toward us.

6.5 Collecting Light with Telescopes

Now that we understand what we can learn from light, it's time to turn our attention to how we collect light. Telescopes are essentially giant eyes that allow us to collect far more light than we could with our naked eyes alone. Telescopes are characterized by two key properties:

- **Light-collecting area**, which depends on the telescope's size, determines how much light the telescope can focus at a given instant. The light-collecting area of the world's largest visible-light telescopes is more than a million times that of the human eye.

true wavelength; normal pitch for sound
true wavelength; normal pitch for sound
a
longer wavelength; sound lower pitched
shorter wavelength; sound higher pitched
b
c

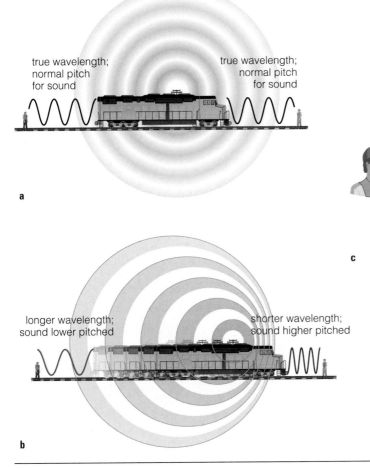

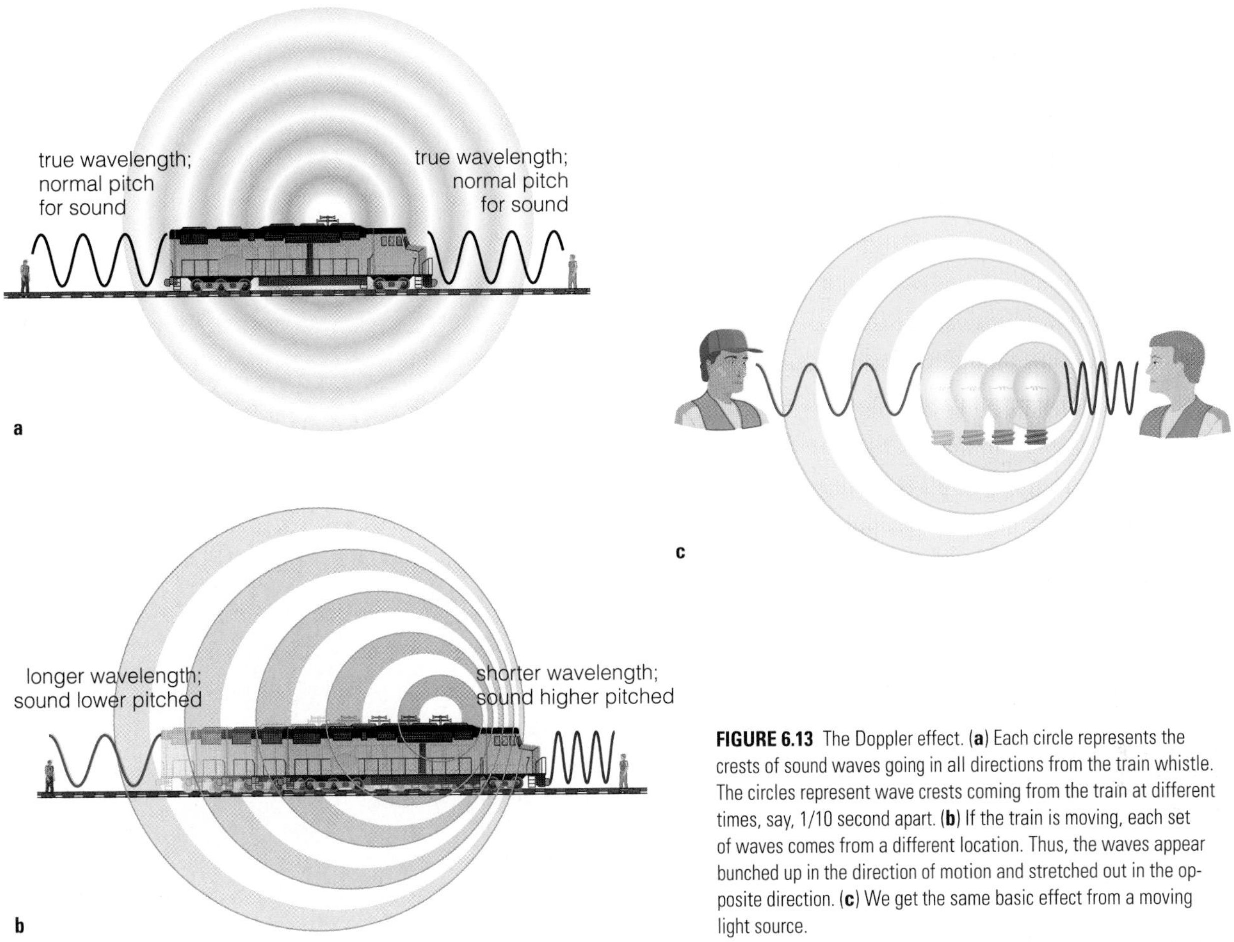

FIGURE 6.13 The Doppler effect. (**a**) Each circle represents the crests of sound waves going in all directions from the train whistle. The circles represent wave crests coming from the train at different times, say, 1/10 second apart. (**b**) If the train is moving, each set of waves comes from a different location. Thus, the waves appear bunched up in the direction of motion and stretched out in the opposite direction. (**c**) We get the same basic effect from a moving light source.

- **Angular resolution** tells us how much detail we can see in the telescope's images. For example, an angular resolution of 1 arcminute (about that of the human eye) means that two stars will appear distinct only if they appear greater than 1 arcminute ($\frac{1}{60}°$) apart in the sky. (Otherwise, they will look like a single star.) The smaller the angular resolution, the more powerful the telescope. The Hubble Space Telescope has an angular resolution of 0.05 arcsecond, which would allow you to read this book from a distance of about 800 meters ($\frac{1}{2}$ mile).

In principle, larger telescopes have both a greater light-collecting area and a better angular resolution. However, the angular resolution may be degraded if a telescope is poorly made, and the Earth's atmosphere also can limit the angular resolution of a telescope (unless it is in space). Therefore, astronomers seek ways to maximize both light-collecting area and angular resolution when designing new telescopes.

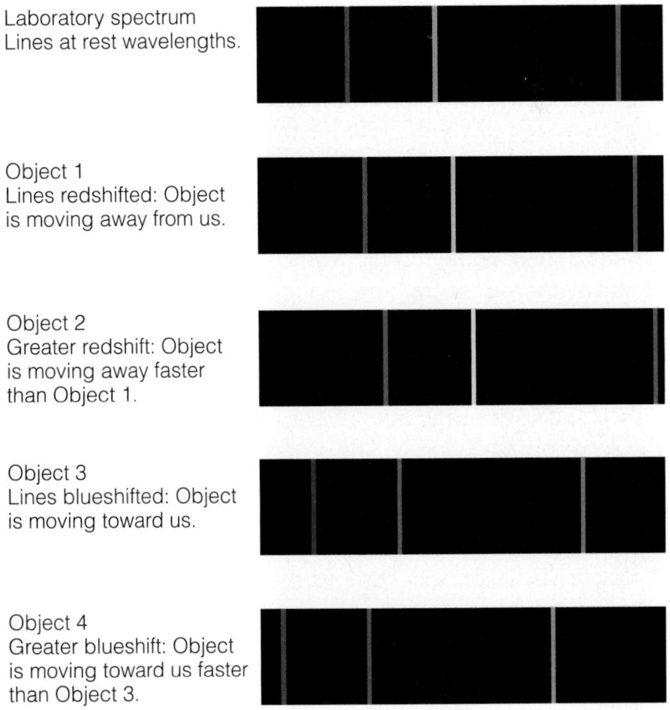

Laboratory spectrum
Lines at rest wavelengths.
Object 1
Lines redshifted: Object is moving away from us.
Object 2
Greater redshift: Object is moving away faster than Object 1.
Object 3
Lines blueshifted: Object is moving toward us.
Object 4
Greater blueshift: Object is moving toward us faster than Object 3.

FIGURE 6.14 Spectral lines provide the crucial reference points for measuring Doppler shifts.

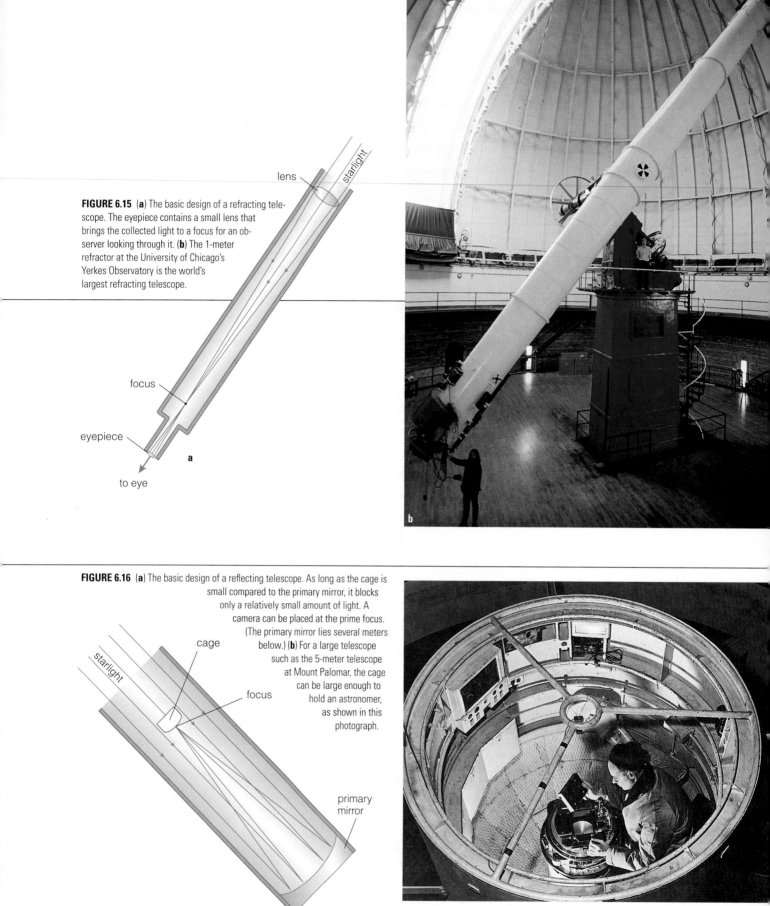

FIGURE 6.15 (**a**) The basic design of a refracting telescope. The eyepiece contains a small lens that brings the collected light to a focus for an observer looking through it. (**b**) The 1-meter refractor at the University of Chicago's Yerkes Observatory is the world's largest refracting telescope.

lens

starlight

focus

eyepiece

a

to eye

b

FIGURE 6.16 (**a**) The basic design of a reflecting telescope. As long as the cage is small compared to the primary mirror, it blocks only a relatively small amount of light. A camera can be placed at the prime focus. (The primary mirror lies several meters below.) (**b**) For a large telescope such as the 5-meter telescope at Mount Palomar, the cage can be large enough to hold an astronomer, as shown in this photograph.

cage

starlight

focus

primary mirror

a

b

Basic Telescope Design

Telescopes come in two basic designs: *refracting* and *reflecting*. A **refracting telescope** operates much like an eye, using transparent glass lenses to focus the light from distant objects (Figure 6.15a). The earliest telescopes, including those built by Galileo, were refracting telescopes. The world's largest refracting telescope, completed in 1897, has a lens that is 1 meter (40 inches) in diameter and a telescope tube that is 19.5 meters (64 feet) long (Figure 6.15b).

A **reflecting telescope** uses a precisely curved **primary mirror** to gather and focus light (Figure 6.16a). Today, most astronomical research uses reflecting telescopes. Note that the *prime focus*, the point at which rays that come straight into the telescope converge, lies *in front of* the mirror. Thus, if we place a camera at the prime focus, it will block some of the light entering the telescope. This does not create a serious problem as long as the camera (and its mount) is small compared to the primary mirror. If the telescope is large enough, a "cage" in which an astronomer can work can be suspended over the prime focus (Figure 6.16b).

An alternative and more common arrangement for reflecting telescopes is to place a **secondary mirror** just below the prime focus. The secondary mirror reflects the light to a location that is more convenient for viewing or for attaching instruments (Figure 6.17): through a hole in the primary mirror (a *Cassegrain focus*), to a hole in the side (a *Newto-*

nian focus, particularly common in smaller telescopes), or to a third mirror that deflects light toward instruments that are not attached to the telescope (a *coudé focus* or a *Nasmyth focus*).

For most of the latter half of the twentieth century, telescopes hardly changed at all. The 5-meter Hale telescope on Mount Palomar outside San Diego remained the world's most powerful for more than 40 years after its opening in 1947. (A Russian telescope built in 1976 was larger, but it suffered optical quality problems that made it far less useful.) Build-

FIGURE 6.17 Alternative designs for reflecting telescopes. Each of these designs incorporates a small, secondary mirror just below the prime focus. The secondary mirror reflects the light to a location that is more convenient for viewing or for attaching instruments.

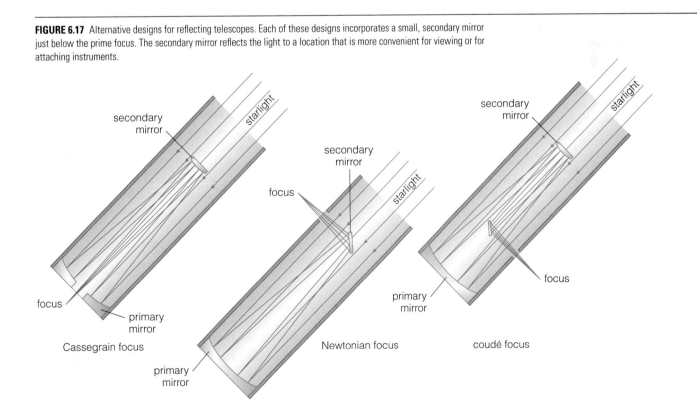

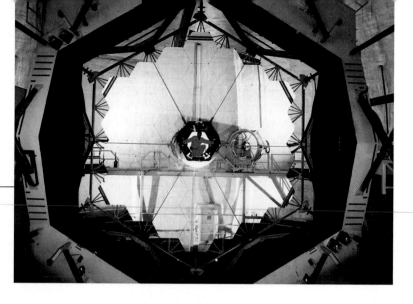

FIGURE 6.18 The Keck telescopes use 36 hexagonal mirrors to function as a single 10-meter primary mirror. These mirrors make the honeycomb pattern that surrounds the hole with the man in this photo.

ing larger telescopes posed a technological challenge, because large mirrors tend to sag under their own weight. Recent technological innovations have made it possible to build very large, low-weight mirrors, fueling a boom in large telescope construction. For example, each of the twin 10-meter Keck telescopes in Hawaii has a primary mirror consisting of 36 smaller mirrors that function together (Figure 6.18). Indeed, the Hale telescope is no longer even among the top 10 in size (including telescopes currently under construction).

Uses of Telescopes

Astronomers use many different kinds of instruments and detectors to extract the information contained in the light collected by a telescope. Every astronomical observation is unique, but most observations fall into one of three basic categories:

- **Imaging** yields pictures of astronomical objects. At its most basic, an imaging instrument is simply a camera. Astronomers sometimes place *filters* in front of the camera that allow only particular colors or wavelengths of light to pass through.

- **Spectroscopy** involves dispersing light into a spectrum. Instruments called *spectrographs* use diffraction gratings or other techniques to disperse light into spectra, which are then recorded with a detector such as a CCD.

- **Timing** tracks how the light intensity hitting the detector varies with time. For a slowly varying object, a timing experiment may be as simple as comparing a set of images or spectra obtained on different nights. For more rapidly varying sources, special instruments essentially make rapid multiple exposures, in some cases recording the arrival time of every individual photon.

Major observatories typically have several different instruments capable of each of these tasks, and some instruments can perform all three basic tasks.

Atmospheric Effects on Observations

A telescope on the ground does not look directly into space but rather looks through the Earth's sometimes murky atmosphere. In fact, the Earth's atmosphere creates several problems for astronomical observations. The most obvious problem is weather—an optical telescope is useless under cloudy skies. Another problem is that our atmosphere scatters the bright lights of cities, creating what astronomers call **light pollution**. Somewhat less obvious is the fact that the atmosphere distorts light. The ever-changing motion, or **turbulence**, of air in the atmosphere bends light in constantly shifting patterns. This turbulence causes the familiar twinkling of stars and blurs astronomical images.

TIME OUT TO THINK *If you look down a long, paved street on a hot day, you'll notice the images of distant cars and buildings rippling and distorting. How are these distortions similar to the twinkling of stars? Why do you think these distortions are more noticeable on hot days than on cooler days?*

Many of these atmospheric problems can be mitigated by choosing appropriate sites for observatories. The key criteria are that the sites be dark (limiting light pollution), dry (limiting rain and clouds), calm (limiting turbulence), and high (placing them above at least part of the atmosphere). Islands are often ideal, and the 4,300-meter (14,000-foot) summit of Mauna Kea on the big island of Hawaii is home to many of the world's best observatories (Figure 6.19).

FIGURE 6.19 Observatories on the summit of Mauna Kea in Hawaii. The twin domes at the far right house the two Keck telescopes.

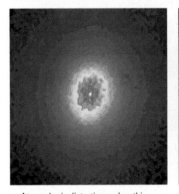

a Atmospheric distortion makes this ground-based image of a double star look like a single star.

b Using the same telescope, but with adaptive optics, the two stars are clearly distinguishable. The angular separation between the two stars is 0.38 arcsecond. (Both images were taken in near-infrared light with the Canada-France-Hawaii telescope and are shown in false color.)

FIGURE 6.20 Adaptive optics can overcome atmospheric distortion.

Modern technology also provides solutions to some of the problems caused by the atmosphere. For example, an amazing technique called **adaptive optics** can eliminate most atmospheric distortion. Atmospheric turbulence causes a stellar image to dance around in the focal plane of a telescope. Adaptive optics essentially make the telescope's mirrors (usually the secondary mirror) do an opposite dance, canceling out the atmospheric distortions (Figure 6.20).

Of course, the ultimate solution to atmospheric problems is to put telescopes in space. That is the primary reason why the Hubble Space Telescope was built and why it is so successful even though its 2.4-meter primary mirror is much smaller than that of many telescopes on the ground (Figure 6.21).

FIGURE 6.21 The Hubble Space Telescope orbits the Earth. Although it orbits at a relatively low altitude, it is high enough to be above the distorting effects of the Earth's atmosphere and to observe infrared and ultraviolet light. This photograph shows astronauts working on the telescope in the Space Shuttle cargo bay during a servicing mission in 1997.

Common Misconceptions: Closer to the Stars?

Many people mistakenly believe that space telescopes are advantageous because their locations above the Earth make them closer to the stars. You can quickly realize the error of this belief by thinking about scale. On the 1-to-10-billion scale discussed in Section 1.2, the Hubble Space Telescope is so close to the surface of the millimeter-diameter Earth that you would need a microscope to resolve its altitude. Thus, the distances to the stars are effectively the same no matter whether a telescope is on the ground or in space. The real advantages of space telescopes all arise from being above the many observational problems presented by the Earth's atmosphere.

Telescopes Across the Spectrum

If we studied only visible light, we'd be missing much of the picture. Planets are relatively cool and emit primarily infrared light. The hot upper layers of stars like the Sun emit ultraviolet and X-ray light. Many objects emit radio waves, including some of the most exotic objects in the cosmos—objects that may contain *black holes* buried in their centers [Section 15.4]. Some violent events even produce gamma rays that travel through space to the Earth. Indeed, most objects emit light over a broad range of wavelengths.

Unfortunately, the atmosphere poses a major problem that no Earth-bound technology can overcome: It prevents most forms of light from reaching the ground at all. Figure 6.22 shows the depth to which different forms of light penetrate into the Earth's atmosphere. Only radio waves, visible light,

FIGURE 6.22 Diagram showing the approximate depths to which different wavelengths of light penetrate the Earth's atmosphere. Note that most of the electromagnetic spectrum—except for visible light, a small portion of the infrared, and radio—can be observed only from very high altitudes or from space.

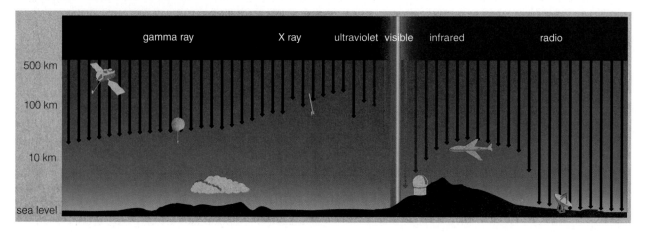

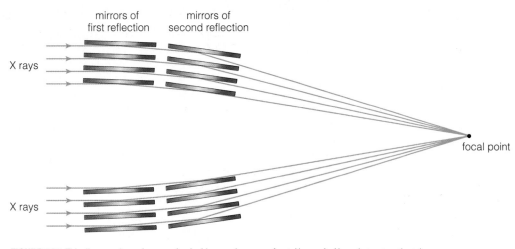

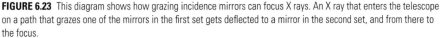

mirrors of
first reflection

mirrors of
second reflection

X rays

X rays

focal point

FIGURE 6.23 This diagram shows how grazing incidence mirrors can focus X rays. An X ray that enters the telescope on a path that grazes one of the mirrors in the first set gets deflected to a mirror in the second set, and from there to the focus.

parts of the infrared spectrum, and the longest wavelengths of ultraviolet light reach the ground. Observing other wavelengths requires placing telescopes in space (or, in a few cases, very high in the atmosphere on airplanes or balloons).

Today, astronomers study light across the entire spectrum. With the exception of telescopes for radio, some infrared, and visible-light observations, most of these telescopes are in space. Indeed, while the Hubble Space Telescope is the only major visible-light observatory in space (it is also used for infrared and ultraviolet observations), many less famous observatories are in Earth orbit for the purpose of making observations in nonvisible wavelengths.

Telescopes for nonvisible wavelengths generally require the use of somewhat different telescope or detector technologies. For example, ordinary telescopes are warm enough to emit significant amounts of infrared light, and this emission can interfere with attempts to observe infrared light from the cosmos. As a result, some infrared telescopes are equipped with sophisticated cooling systems. At the other end of the spectrum, X rays and gamma rays have sufficient energy to penetrate ordinary telescope mirrors, making them useless for these forms of light. X rays can be focused by slightly bending their paths with what astronomers call *grazing incidence* mirrors (Figure 6.23). (Other technologies are used for gamma rays.)

Radio telescopes use large metal dishes as "mirrors" to reflect radio waves. However, the long wavelengths of radio waves mean that very large telescopes are necessary to achieve reasonable angular resolution. The largest radio dish in the world, the Arecibo Telescope, stretches 305 meters (1,000 feet) across a

FIGURE 6.24 The Arecibo radio telescope stretches across a natural valley in Puerto Rico. At 305 meters across, it is the world's largest single radio dish.

natural valley in Puerto Rico (Figure 6.24). Despite its large size, Arecibo has an angular resolution of only about 1 arcminute at commonly observed radio wavelengths (e.g., 21 cm)—a few hundred times worse than the visible-light resolution of the Hubble Space Telescope.

In the 1950s, radio astronomers developed an ingenious technique for improving the angular resolution of radio telescopes. This technique, called **interferometry**, links several separate radio dishes

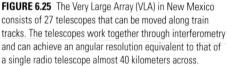
FIGURE 6.25 The Very Large Array (VLA) in New Mexico consists of 27 telescopes that can be moved along train tracks. The telescopes work together through interferometry and can achieve an angular resolution equivalent to that of a single radio telescope almost 40 kilometers across.

together in an array that resolves details as if it were one enormous radio telescope. One famous example, the Very Large Array (VLA) near Socorro, New Mexico, consists of 27 individual radio dishes that can be moved along railroad tracks laid down in the shape of a Y (Figure 6.25). The light-gathering capability of the VLA's 27 dishes is simply equal to their combined area, equivalent to a single telescope 130 meters across. But the VLA's best angular resolution, achieved by spacing the individual dishes as widely as possible, is like that of a single radio telescope with a diameter of almost 40 kilometers (25 miles). Today, astronomers can achieve even higher angular resolution by linking radio telescopes around the world.

In principle, interferometry can improve angular resolution not only for radio waves but also for any other form of light. In practice, interferometry becomes increasingly difficult for light with shorter wavelengths, and astronomers are only beginning to succeed at infrared and visible interferometry. Nevertheless, one reason why *two* Keck telescopes were built on Mauna Kea is so they can be used for infrared interferometry and someday for optical interferometry. The potential value of such interferometers is enormous. In the future, infrared interferometers may be able to obtain spectra from individual planets around other stars, allowing spectroscopy that could determine the compositions of their atmospheres and help determine whether they harbor life.

6.6 Spacecraft

Most people probably don't equate spacecraft with light and telescopes, but nearly all astronomical spacecraft use light in some way, and many carry sophisticated telescopes and detectors (such as cameras and spectrographs). Thus, we conclude this chapter with a brief discussion of spacecraft.

With a few exceptions, such as the Space Shuttle, Space Station, and the Apollo missions to the Moon, nearly all spacecraft are *robotic*, meaning that

FIGURE 6.26 The trajectory of *Voyager 2*, which made flybys of each of the four jovian planets in our solar system.

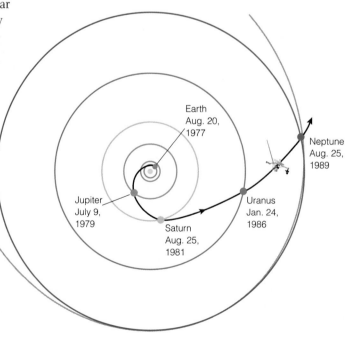

FIGURE 6.27 An artist's conception of a lunar observatory.

no humans are aboard. Their operations are partly automatic and partly controlled by people who send instructions via radio signals from Earth. Broadly speaking, most astronomical spacecraft fall into four main categories:

1. **Earth-orbiters**: Simply being in space is enough to overcome the observational problems presented by the Earth's atmosphere, so most space-based astronomical observatories orbit the Earth.

2. **Flybys**: Close-up study of other planets requires sending spacecraft to them. Flybys follow unbound orbits past their destinations—they fly past a world just once and then continue on their way. For example, *Voyager 2*'s orbit allowed it to make flybys of Jupiter, Saturn, Uranus, and Neptune (Figure 6.26).

3. **Orbiters** (of other worlds): A flyby can study a planet or moon close up—but only once. For more detailed or longer-term studies, we need a spacecraft that goes into orbit around another world. For example, NASA's *Cassini* spacecraft will orbit Saturn, beginning in 2004.

4. **Probes and landers**: The most "up close and personal" study of other worlds comes from spacecraft that send probes into their atmospheres or landers to their surfaces. For example, NASA's *Pathfinder* landed on Mars in 1997 (see Figure 8.23c).

The primary constraint on building more spacecraft is cost: Besides the costs of developing, building, and operating a spacecraft, it can cost tens to hundreds of millions of dollars just to launch it into space. Fortunately, new technologies (particularly miniaturization, which reduces launch weight) are making spacecraft cheaper, and you're likely to read about more and more new space missions as time goes on. Someday astronomers hope to build an observatory on the Moon (Figure 6.27). Because the Moon has no atmosphere, it offers all the advantages of telescopes in space while also offering the ease of operating on a solid surface.

THE BIG PICTURE

We've discussed a lot about light, telescopes, and spacecraft in this chapter. "Big picture" ideas that will help you keep your understanding in perspective include the following:

■ There is far more to light than meets the eye. By dispersing light into a spectrum with a prism or a diffraction grating, we discover a wealth of information about the object from which the light has come. Most of what we know about the universe comes from information that we receive in the form of light.

■ The visible light that our eyes can see is only a small portion of the complete electromagnetic spectrum. Different portions of the spectrum may contain different pieces of the story of a distant object, so it is important to study spectra at many wavelengths.

■ By studying the spectra of a distant object, we can determine its composition, surface temperature, motion toward or away from us, and more.

■ Telescopes work much like giant eyes, enabling us to see the universe in great detail. New technologies for making larger telescopes, along with advances in adaptive optics and interferometry, are making ground-based telescopes more powerful than ever.

■ For the ultimate in observing the universe, space is the place! Telescopes in space allow us to detect light from across the entire spectrum while also avoiding any distortion caused by Earth's atmosphere. Flybys, orbiters, and landers allow us to make detailed studies of other worlds in our solar system that are not possible from Earth.

Review Questions

1. What is the difference between *energy* and *power*, and what units do we use to measure each?

2. What do we mean when we speak of a *spectrum* produced by a prism or *diffraction grating*?

3. Describe each of the four basic ways that light can interact with matter.

4. What is a *photon*? In what way is a photon like a particle? In what way is it like a wave? Explain why we say that light is an *electromagnetic wave*.

5. Explain why light with a shorter wavelength must have a higher frequency, and vice versa.

6. Describe the *electromagnetic spectrum*.

7. Explain how spectral lines can be used to determine the composition and temperature of their source.

8. Describe the two laws of *thermal radiation* and give a few examples of the use of each law.

9. Summarize the circumstances under which objects produce thermal, emission line, or absorption line spectra as shown in Figure 6.11.

10. Study Figure 6.12 carefully, and explain each feature in the spectrum.

11. Describe the *Doppler effect* for light.

12. Briefly distinguish between a *refracting telescope* and a *reflecting telescope*.

13. What do we mean by a telescope's *light-collecting area* and *angular resolution*?

14. Define and differentiate between the three basic categories of astronomical observation: *imaging*, *spectroscopy*, and *timing*.

15. Explain how the Earth's atmosphere affects astronomical observations made from the ground.

16. Briefly describe the advantages of putting telescopes in space.

17. What is *interferometry*, and why is it useful?

18. Describe each of the four main categories of robotic spacecraft.

Discussion Questions

1. *The Changing Limitations of Science.* In 1835, French philosopher Auguste Comte stated that the composition of stars could never be known by science. Although spectral lines had been seen in the Sun's spectrum by that time, not until the mid-1800s did scientists recognize that spectral lines give clear information about chemical composition (primarily through the work of Foucault and Kirchhoff). Why might our present knowledge have seemed unattainable in 1835? Discuss how new discoveries can change the apparent limitations of science. Today, other questions seem beyond the reach of science,

such as the question of how life began on Earth. Do you think that such questions will ever be answerable by science? Defend your opinion.

2. *A Lunar Observatory.* Do the potential benefits of building an astronomical observatory on the Moon justify its costs at the present time? If it were up to you, would you recommend that Congress begin funding such an observatory? Defend your opinions.

Problems

Sensible Statements? For **problems 1–8**, decide whether the statement is sensible and explain why it is or is not.

1. If you could view a spectrum of light reflecting off a blue sweatshirt, you'd find the entire rainbow of color.

2. Because of their higher frequency, X rays must travel through space faster than radio waves.

3. If the Sun's surface became much hotter (while the Sun's size remained the same), the Sun would emit more ultraviolet light but less visible light than it currently emits.

4. If a distant galaxy has a substantial redshift (as viewed from our galaxy), then anyone living in that galaxy would see a substantial redshift in a spectrum of the Milky Way Galaxy.

5. New technologies will soon allow astronomers to use X-ray telescopes on the Earth's surface.

6. Thanks to interferometry, a properly spaced set of 10-meter radio telescopes can achieve the angular resolution of a single, 100-kilometer radio telescope.

7. Scientists are planning a flyby of Uranus in order to do long-term monitoring of the planet's weather.

8. An observatory on the Moon's surface could have telescopes monitoring light from all regions of the electromagnetic spectrum.

9. *Spectral Summary.* Clearly explain how studying an object's spectrum can allow us to determine each of the following properties of the object.
   a. The object's surface chemical composition.
   b. The object's surface temperature.
   c. Whether the object is a thin cloud of gas or something more substantial.
   d. Whether the object has a hot upper atmosphere.
   e. The speed at which the object is moving toward or away from us.

10. *Planetary Spectrum.* Suppose you take a spectrum of light coming from a planet that looks blue to the eye. Do you expect to see any visible light in the planet's

spectrum? Is the visible light emitted by the planet, reflected by the planet, or both? Which (if any) portions of the visible spectrum do you expect to find "missing" in the planet's spectrum? Explain your answers clearly.

11. *Hotter Sun.* Suppose that the surface temperature of the Sun were about 12,000 K, rather than 6,000 K.
   a. How much more thermal radiation would the Sun emit?
   b. How would the thermal radiation spectrum of the Sun be different?
   c. Do you think it would still be possible to have life on Earth? Explain.

12. *The Doppler Effect.* In hydrogen, the transition from level 2 to level 1 has a rest wavelength of 121.6 nm. Suppose you see this line at a wavelength of 120.5 nm in Star A, at 121.2 nm in Star B, at 121.9 nm in Star C, and at 122.9 nm in Star D. Which stars are coming toward us? Which are moving away? Which star is moving fastest relative to us (either toward or away from)? Explain your answers without doing any calculations.

13. *The Expanding Universe.* Recall from Chapter 1 that we know the universe is expanding because (1) all galaxies outside our Local Group are moving away from us and (2) more distant galaxies are moving faster. Explain how Doppler shift measurements allow us to know these two facts.

14. *Project: Twinkling Stars.* Using a star chart, identify 5 to 10 bright stars that should be visible in the early evening. On a clear night, observe each of these stars for a few minutes. Note the date and time, and for each star record the following information: approximate altitude and direction in your sky, brightness compared to other stars, color, how much the star twinkles compared to other stars. Study your record. Can you draw any conclusions about how brightness and position in your sky affect twinkling? Explain.

Web Projects

Find useful links for Web projects on the text Web site.

1. *Planetary Mission.* Choose a current mission to a planet and find as much information as you can about the mission. Write a two- to three-page summary of the mission, including discussion of its purpose, design, and cost.

2. *Major Observatories.* Take a virtual tour of one of the world's major astronomical observatories. Write a short report on why the observatory is useful to astronomy.

PART III
Learning from Other Worlds

The evolution of the world may be compared to a display of fireworks that has just ended: some few red wisps, ashes and smoke. Standing on a cooled cinder, we see the slow fading of the suns, and we try to recall the vanished brilliance of the origin of the worlds.

G. LEMAÎTRE (1894–1966),
ASTRONOMER AND CATHOLIC PRIEST

CHAPTER 7

Formation of the Solar System

How old is the Earth, and how did it come to be? Is it unique? Our ancestors could do little more than guess, but today we can answer with reasonable confidence: The Earth and the rest of our solar system formed from a great cloud of gas and dust about 4.6 billion years ago. Whereas ancient astronomers believed the Earth to be fundamentally different from objects in the heavens, we now see the Earth as just one of many worlds.

As we examine the evidence that helped shape the modern scientific view of the formation of our solar system, we'll highlight one common technique of scientific research: sifting through seemingly unrelated facts to find the most important trends, patterns, and characteristics—and then attempting to explain them. The explanation must be based on known physical laws, and the success or failure of a theory depends on how well it matches the facts. As we discover more and more planets around other stars, we hope to test our theories further and to modify them as necessary.

7.1 Comparative Planetology

Galileo's telescopic observations began a new era in astronomy in which the Sun, the Moon, and the planets could be studied for the first time as *worlds*, rather than merely as lights in the sky. During the early part of this new era, astronomers studied each world independently, but in recent decades we discovered that the similarities among worlds are often more profound than the differences.

Today, we know that the Sun, planets, moons, and other bodies in our solar system all formed at about the same time from the same cloud of interstellar gas and dust (*interstellar* means "between the stars") and in accord with the same physical laws. Thus, the differences between these worlds must be attributable to physical processes that we can study and understand by comparing the worlds to one another. This approach is called **comparative planetology**; the term *planetology* is used broadly to include moons, asteroids, and comets as well as planets. The idea is that we can learn more about the Earth, or any other world, by studying it in the context of other objects in our solar system—rather like learning about a person by studying his or her family, friends, and culture.

Before beginning the comparative study of worlds, we must be sure that the comparison will provide valid lessons. The basic premise of comparative planetology—that the similarities and differences among worlds in our solar system can be traced to common physical processes—assumes that the planets share a common origin. Thus, we begin our study of comparative planetology by seeking clues to our origins from the current state of our solar system; then we will examine the modern theory of solar system formation that explains much of what we see.

7.2 The Origin of the Solar System: Four Challenges

Learning about the origin of the solar system is easiest if we begin by looking at our solar system with a "big picture" view, rather than focusing on individual worlds. Imagine that we have the perspective of an alien spacecraft making its first scientific survey of our solar system. What would we see?

First, as we learned from our tour of a scale model of our solar system in Chapter 1, we'd find that our solar system is mostly empty—interplanetary space is a near-vacuum containing only very low density gas and tiny, solid grains of dust. But we'd soon focus on the widely spaced planets, mapping their orbits, measuring their sizes, compositions, and densities, and taking inventory of their moons and ring systems. Figure 7.1 shows the resulting schematic map, and Table 7.1 summarizes the planetary properties. Asteroids and comets are also listed in the table because these objects orbit the Sun, though with orbital properties somewhat different from those of the planets.

Studying the map and the table reveals several general features of our solar system. Let's investigate by grouping our observations into four major challenges that must be met by any theory claiming to explain how our solar system developed.

FIGURE 7.1 Side view of the solar system. Arrows indicate the orientation of the rotation axes of the planets and their orbital motion. (Planetary tilts in this diagram are aligned in the same plane for easier comparison. Planets not to scale.) Seen from above, all orbits except those of Mercury and Pluto are nearly circular. Most moons orbit in the same direction as the planets orbit and rotate—counterclockwise when seen from above Earth's North Pole.

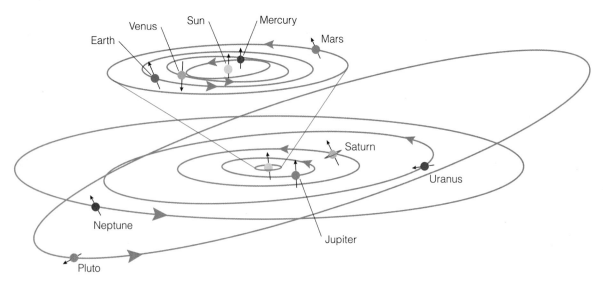

Table 7.1 Planetary Facts*

Photo	Planet	Average Distance from Sun (AU)	Temperature[†]	Relative Size	Average Equatorial Radius (km)	Average Density (g/cm³)	Composition	Known Moons	Rings?
	Mercury	0.387	700 K	·	2,440	5.43	Rocks, metals	0	No
	Venus	0.723	740 K	•	6,051	5.24	Rocks, metals	0	No
	Earth	1.00	290 K	•	6,378	5.52	Rocks, metals	1	No
	Mars	1.52	240 K	·	3,397	3.93	Rocks, metals	2 (tiny)	No
	Most asteroids	2–3	170 K	·	≤500	1.5–3	Rocks, metals	?	No
	Jupiter	5.20	125 K	●	71,492	1.33	H, He, hydrogen compounds[‡]	28	Yes
	Saturn	9.53	95 K	●	60,268	0.70	H, He, hydrogen compounds[‡]	30	Yes
	Uranus	19.2	60 K	●	25,559	1.32	H, He, hydrogen compounds[‡]	21	Yes
	Neptune	30.1	60 K	●	24,764	1.64	H, He, hydrogen compounds[‡]	8	Yes
	Pluto	39.5	40 K	·	1,160	2.0	Ices, rock	1	No
	Most comets	10–50,000	A few K[§]	·	A few km?	<1?	Ices, dust	?	No

*Appendix E gives a more complete list of planetary properties.

†Surface temperatures for all objects except Jupiter, Saturn, Uranus, and Neptune, for which cloud-top temperatures are listed.

‡Includes water (H_2O), methane (CH_4), and ammonia (NH_3).

§Comets passing close to the Sun warm considerably, especially their outer layers.

Challenge 1: Patterns of Motion

If we look closely at the map of solar system orbits in Figure 7.1, we'll see several striking patterns:

- All planets orbit the Sun in the same direction—counterclockwise when seen from high above the Earth's North Pole.

- All planetary orbits lie nearly in the same plane.

- Almost all planets travel on nearly circular orbits, and the spacing between planetary orbits increases with distance from the Sun according to a fairly regular trend. The most notable exception to this trend is an extra-wide gap between Mars and Jupiter that is populated with asteroids.

- Most planets rotate in the same direction in which they orbit—counterclockwise when seen from above the Earth's North Pole—with fairly small axis tilts (i.e., less than about 25°).

- Almost all moons orbit their planet in the same direction as the planet's rotation and near the planet's equatorial plane.

- The Sun rotates in the same direction in which the planets orbit.

These patterns show that motion in the solar system is generally quite organized. If each planet had come into existence independently, we might expect planetary motions to be much more random. (The motions of asteroids are slightly more random, and those of comets even more so, as we will discuss shortly.)

The first challenge for any theory of solar system formation is to explain why motions in the solar system are generally so orderly.

Challenge 2: Categorizing Planets

Before reading any further, study the properties of the planets in Table 7.1. Can you categorize the planets into groups? How many categories do you need? What characteristics do planets within a category share?

Astronomers classify most of the planets in two distinct groups: the rocky **terrestrial planets** and the gas-rich **jovian planets**. The word *terrestrial* means "Earth-like" (*terra* is the Greek word for Earth), and the terrestrial planets are the four planets of the inner solar system—Mercury, Venus, Earth, and Mars. These are the worlds most like our own. They are relatively small, close to the Sun, and close together. They have solid, rocky sur-

faces and an abundance of metals deep in their interiors. They have few moons, if any, and none have rings. Note that many scientists count our Moon as a fifth terrestrial world because it shares these general characteristics, although it's not technically a planet.

The word *jovian* means "Jupiter-like," and the jovian planets are the four outer solar system planets—Jupiter, Saturn, Uranus, and Neptune. The jovian planets are much larger than the terrestrial planets, farther from the Sun, and widely separated from each other. These huge planets have little in common with Earth. They are made mostly of hydrogen, helium, and **hydrogen compounds** such as water, ammonia, and methane. They contain relatively small amounts of rocky material only deep in their cores. Jovian planets do not have solid surfaces; if you plunged into the atmosphere of a jovian planet, you would sink deeper and deeper until you were crushed by overwhelming pressure. Each jovian planet has rings and an extensive system of moons. These solid satellites are made mostly of low-density ices and rocks. Table 7.2 contrasts the general traits of the terrestrial and jovian planets.

The second challenge for any theory of solar system formation is to explain why the inner and outer solar system planets divide so neatly into two classes.

TIME OUT TO THINK *Are the distinctions in Table 7.2 clear-cut? Examine Table 7.1 to compare, for example, the radii of the largest terrestrial planet and the smallest jovian planet.*

Note that Pluto is left out in the cold, both literally and figuratively. On one hand, it is small and solid like the terrestrial planets. On the other hand, Pluto is far from the Sun, cold, and made of low-density ices. Pluto also has an unusual orbit, far more eccentric than the orbit of any other planet and substantially inclined to the plane in which the other planets orbit the Sun. For a long time, scientists considered Pluto to be a lone misfit. However, some scientists now classify Pluto with other icy bodies in the outer solar system: comets.

Table 7.2 Comparison of Terrestrial and Jovian Planets

Terrestrial Planets	Jovian Planets
Smaller size and mass	Larger size and mass
Higher density (rocks, metals)	Lower density (light gases, hydrogen compounds)
Solid surface	No solid surface
Closer to the Sun (and closer together)	Farther from the Sun (and farther apart)
Warmer	Cooler
Few (if any) moons and no rings	Rings and many moons

Challenge 3: Asteroids and Comets

No formation theory is complete without an explanation of the most numerous objects in the solar system: asteroids and comets. **Asteroids** are small, rocky bodies that orbit the Sun primarily in the **asteroid belt** between the orbits of Mars and Jupiter. Asteroids orbit the Sun in the same direction as the planets. Their orbits generally lie close to the plane of planetary orbits, although they are usually a bit tilted. Some asteroids have elliptical orbits that are quite eccentric compared to the nearly circular orbits of the planets. Almost 9,000 asteroids have been identified and cataloged, but these are probably only the largest among a much greater number of small asteroids. The largest asteroids are a few hundred kilometers in radius—much less than half the Moon's radius.

Comets are small, icy bodies that spend most of their lives well beyond the orbit of Pluto; we generally recognize them only on the rare occasions when one dives into the inner solar system and grows a spectacular tail. Based on the orbits of the relatively few comets that reach the inner solar system, astronomers have determined that many billions of comets must be orbiting the Sun, primarily in two broad regions. The first region, called the **Kuiper belt** (pronounced koy-per), begins somewhere in the vicinity of the orbit of Neptune (about 30 AU from the Sun) and extends to perhaps 100 AU from the Sun. Comets in the Kuiper belt have orbits that lie fairly close to the plane of planetary orbits, and they travel around the Sun in the same direction as the planets. The second region populated by comets, called the **Oort cloud**, is a huge, spherical region centered on the Sun and extending perhaps halfway to the nearest stars. The orbits of comets in the Oort cloud are completely random, with no pattern to their inclinations, orbital directions, or eccentricities.

The third challenge for any theory of solar system formation is to explain the existence and general properties of the large numbers of asteroids and comets.

Challenge 4: Exceptions to the Rules

The first three challenges involve explaining general patterns in our solar system. But some objects don't fit these patterns. For example, Mercury and Pluto have larger orbital eccentricities and inclinations than the other planets. The rotational axes of Uranus and Pluto are substantially tilted, and Venus rotates backward—clockwise, rather than counterclockwise, as viewed from high above Earth's North Pole. Unlike the other terrestrial planets, Earth has a large moon. Pluto has a moon almost as big as itself. While most moons of the jovian planets orbit with the same orientation as their planet's rotation, a few orbit in the opposite direction.

Table 7.3 Four Major Characteristics of the Solar System

Large bodies in the solar system have orderly motions. All planets and most satellites have nearly circular orbits going in the same direction in nearly the same plane. The Sun and most of the planets rotate in this same direction as well.

Planets fall into two main categories: small, rocky terrestrial planets near the Sun and large, hydrogen-rich jovian planets farther out. The jovian planets have many moons and rings made of rock and ice.

Swarms of asteroids and comets populate the solar system. Asteroids are concentrated in the asteroid belt, and comets populate the regions known as the Kuiper belt and the Oort cloud.

Several notable exceptions to these general trends stand out, such as planets with unusual axis tilts or surprisingly large moons, and moons with unusual orbits.

Allowing for these and other exceptions to the general patterns is the fourth challenge for any theory of solar system formation.

Summarizing the Challenges

Table 7.3 summarizes the major characteristics of our solar system embodied in the four challenges to be explained by our solar system formation theory. In the remainder of this chapter, we will examine the leading theory of solar system formation and put it to the test to see whether it can meet the four challenges. In this sense, we will be following the idealized scientific method by starting with the facts, putting forward a theory, and putting the theory to the test. In doing so, we have the benefit of hindsight: The historical path to our modern theory of the solar system was actually quite complex, with many competing theories being proposed and eventually discarded because they were unable to meet the challenges we have laid out.

7.3 The Nebular Theory of Solar System Formation

Over the past three centuries, several models have been put forth in attempts to meet the four challenges for a solar system formation theory. In the past few decades, a tremendous amount of evidence has accumulated in support of one model. This model, which we will call the **nebular theory**, holds that our solar system formed from a giant, swirling **interstellar cloud** of gas and dust (Figure 7.2). Such a cloud is also called a **nebula** (the Latin word for "cloud"). More generally, we now have evidence that all **star systems**—systems like our solar system

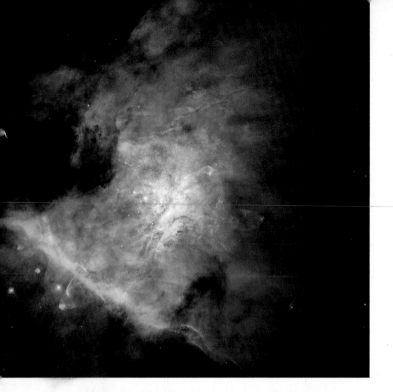

FIGURE 7.2 This photograph shows the central region of the Orion Nebula, an interstellar cloud in which star systems—possibly including planets—are forming. The photo is actually a composite of more than a dozen separate images taken with the Hubble Space Telescope. (See Figure 16.12 for a complete view of the Orion Nebula.)

that contain at least one star and, in some cases, planets or other orbiting material—form similarly from interstellar clouds.

Our galaxy, like the entire universe, originally contained only hydrogen and helium (and trace amounts of lithium). By the time our solar system formed, 4.6 billion years ago, about 2% of the original hydrogen and helium in the galaxy had been converted into heavier elements by fusion inside stars. This material was mixed into the interstellar clouds by **stellar winds** and stellar explosions. Although 2% sounds like a small amount, it was more than enough to form the rocky terrestrial planets—and us. Nevertheless, hydrogen and helium remain the most abundant elements by far and make up the bulk of the Sun and the jovian planets.

TIME OUT TO THINK *Could a solar system like ours have formed with the first generation of stars after the Big Bang? Explain.*

Collapse of the Solar Nebula

An individual star system forms from just a small part of a giant interstellar cloud. We refer to the collapsed piece of cloud that formed our own solar system as the **solar nebula**. The solar nebula collapsed under its own gravity; the collapse may have been triggered by a cataclysmic event such as the impact of a shock wave from a nearby exploding star. Before the collapse, the low-density gas cloud may have spanned a few light-years. As it collapsed to a diameter of about 200 AU—roughly twice the present-day diameter of Pluto's orbit—three important processes gave form to our solar system (Figure 7.3).

First, the temperature of the solar nebula increased as it collapsed. Such heating represents energy conservation in action [Section 4.2]. As the cloud shrank, its gravitational potential energy was converted to the kinetic energy of individual gas particles falling inward. These particles crashed into one another, converting the kinetic energy of their inward fall into the random motions of thermal energy. Some of this energy was radiated away as thermal radiation. The solar nebula became hottest near its center, where much of the mass collected to form the **protosun** (the prefix *proto* comes from a Greek word meaning "earliest form of"). The protosun eventually became so hot that nuclear fusion ignited in its core—at which point our Sun became a full-fledged star.

Second, like an ice skater pulling in her arms as she spins, the solar nebula rotated faster and faster as it shrank in radius. This increase in rotation rate represents the conservation of angular momentum in action [Section 5.2]. The rotation helped ensure that not all of the material in the solar nebula collapsed onto the protosun: The greater the angular momentum of a rotating cloud, the more spread out it will be.

Third, the solar nebula flattened into a disk—the **protoplanetary disk** from which the planets eventually formed. This flattening is a natural consequence of collisions between particles, which explains why flat disks are so common in the universe (e.g., the disks of spiral galaxies like the Milky Way, ring systems around planets, and *accretion disks* around neutron stars or black holes [Section 15.3]). A cloud may start with any size or shape, and different clumps of gas within the cloud may be moving in random directions at random speeds. When the cloud collapses, these different clumps collide and merge, giving the new clumps the average of their differing velocities. The result is that the random motions of the clumps in the original cloud become more orderly as the cloud collapses, changing the cloud's original lumpy shape into a rotating, flattened disk. Similarly, collisions between clumps of material in highly elliptical orbits reduce their ellipticities, making their orbits more circular. You can see a similar effect if you shake some pepper into a bowl of water and quickly stir it in a random way. Because the water molecules are always colliding with one another, the motion of the pepper grains will settle down into a slow rotation representing the average of the original, random velocities.

These three processes—heating, spinning, and flattening—explain the tidy layout of our solar system. The flattening of the protoplanetary disk explains why all planets orbit in nearly the same plane. The

a The original cloud is large and diffuse, and its rotation is almost imperceptibly slow.

b The cloud heats up and spins faster as it contracts.

c The result is a spinning, flattened disk, with mass concentrated near the center.

FIGURE 7.3 This sequence of paintings shows the collapse of an interstellar cloud. In our solar nebula, the hot, dense central bulge became the Sun, and the planets formed in the disk.

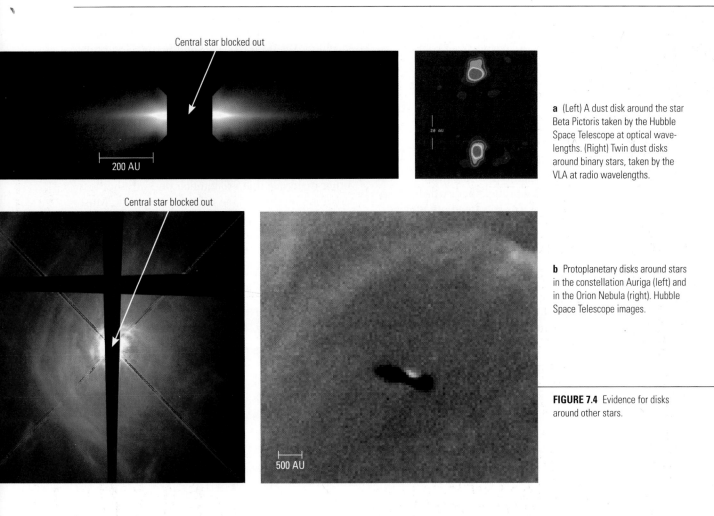

Central star blocked out

200 AU

Central star blocked out

500 AU

a (Left) A dust disk around the star Beta Pictoris taken by the Hubble Space Telescope at optical wavelengths. (Right) Twin dust disks around binary stars, taken by the VLA at radio wavelengths.

b Protoplanetary disks around stars in the constellation Auriga (left) and in the Orion Nebula (right). Hubble Space Telescope images.

FIGURE 7.4 Evidence for disks around other stars.

spinning explains why all planets orbit in the same direction and also plays a role in making most planets rotate in this same direction. The fact that collisions in the protoplanetary disk tend to make highly elliptical orbits more circular explains why most planets in our solar system have nearly circular orbits today.

Evidence Concerning Nebular Collapse

The theory that our solar system formed from interstellar gas may sound reasonable, but we cannot ac-

cept it without hard evidence. Fortunately, we have strong observational evidence that the same processes are occurring elsewhere in our galaxy. A collapsing nebula emits thermal radiation [Section 6.3], primarily in the infrared. We've detected such infrared radiation from many other nebulae where star systems appear to be forming. We've even seen several structures around other stars that look similar to protoplanetary disks (Figure 7.4). Other support for the nebular theory comes from sophisticated computer models that simulate the formation processes it describes.

All in all, the nebular theory meets the challenge of explaining most of the orderly motions of the planets and satellites. Now let's turn to our second challenge: explaining the strikingly different characteristics of the terrestrial and jovian planets.

7.4 Building the Planets

The churning and mixing of the gas in the solar nebula ensured that its composition was about the same throughout: roughly 98% hydrogen and helium, and 2% heavier elements such as carbon, nitrogen, oxygen, silicon, and iron. How did the planets and other bodies in our solar system end up with such a wide variety of compositions when they came from such uniform material? To answer this question, we must investigate how the material in the protoplanetary disk came together to form the planets.

Condensation: Sowing the Seeds of Planets

In the center of the collapsing solar nebula, gravity drew much of the material together into the protosun. In the rest of the spinning protoplanetary disk, however, the gaseous material was so spread out that, at first, gravity was not strong enough to pull it together to form planets. The formation of planets therefore required "seeds"—solid chunks of matter that gravity could eventually draw together. Understanding these seeds is the key to explaining the differing compositions of the planets.

The vast majority of the material in the solar nebula was gaseous: High temperatures kept virtually all the ingredients of the solar nebula vaporized near the protosun. Farther out the nebula was still primarily gaseous because the hydrogen and helium that made up 98% of the solar nebula remain gaseous even at extremely low temperatures. But the 2% of material consisting of heavier elements could form solid seeds where temperatures were low enough. The formation of solid or liquid particles from a cloud of gas is called **condensation**. Pressures in the solar nebula were so low that liquid droplets rarely formed, but solid particles could condense in the same way that snowflakes condense from water vapor in our atmosphere. We refer to such solid particles as **condensates**. The different kinds of planets and satellites formed out of the different kinds of condensates present at different locations in the solar nebula.

The ingredients of the solar nebula fell into four categories based on their condensation temperatures, see Figure 7.5.

- **Metals** include iron, nickel, aluminum, and other materials that are familiar on Earth but less common on the surface than ordinary rock. Most metals condense into solid form at temperatures between 1,000 K and 1,600 K. Metals made up less than 0.2% of the solar nebula's mass.

FIGURE 7.5 Characteristics of the materials in the solar nebula.

Materials in the Solar Nebula				
	Metals	**Rocks**	**Hydrogen Compounds**	**Light Gases**
Examples	iron, nickel, aluminum	silicates	water (H_2O) methane (CH_4) ammonia (NH_3)	hydrogen, helium
Typical Condensation Temperature	1,000–1,600 K	500–1,300 K	<150 K	(do not condense in nebula)
Relative Abundance (by mass)	.	■	■	■
	(0.2%)	(0.4%)	(1.4%)	(98%)

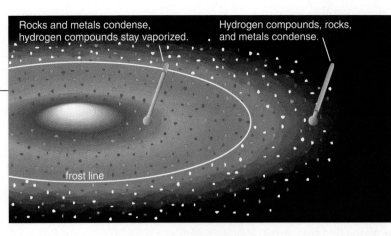

Rocks and metals condense, hydrogen compounds stay vaporized.

Hydrogen compounds, rocks, and metals condense.

frost line

FIGURE 7.6 Temperature differences in the solar nebula led to different kinds of condensed materials, sowing the seeds for two different kinds of planets.

- **Rocks** are materials common on the surface of the Earth, primarily silicon-based minerals. Rocks are solid at temperatures and pressures found on Earth but typically melt or vaporize at temperatures of 500–1,300 K depending on their type. Rocky materials made up about 0.4% of the nebula by mass.

- **Hydrogen compounds** are molecules such as methane (CH_4), ammonia (NH_3), and water (H_2O) that solidify into **ices** below about 150 K. These compounds were significantly more abundant than rocks and metals, making up 1.4% of the nebula's mass.

- **Light gases** (hydrogen and helium) never condense under solar nebula conditions. These gases made up the remaining 98% of the nebula's mass.

The order of condensation temperatures is easy to remember: It's the same as the order of densities.

Temperature differences between the hot inner regions and the cool outer regions of the nebula determined what kinds of condensates were available to form planets (Figure 7.6). Very near the protosun, where the nebula temperature was above 1,600 K, there were no condensates—everything remained gaseous. Farther out, where the temperature was slightly lower, metal flakes appeared. Near the distance of Mercury's orbit, flakes of rock joined the mix. Moving outward past the orbits of Venus and Earth, more varieties of rock minerals condensed. Near the location of the future asteroid belt, temperatures were low enough to allow minerals containing small amounts of water to condense as well. Dark, carbon-rich materials also condensed here.

Only beyond the **frost line**, which lay between the present-day orbits of Mars and Jupiter, were temperatures low enough (150 K ≈ −123°C) for hydrogen compounds to condense into ices. (Water ice did not form at the familiar 0°C because the pressures in the nebula were 10,000 times lower than on Earth.) Thus, the outer solar system contained condensates of all kinds: rocks, metals, and ices. However, ice flakes were nearly three times more abundant than flakes of rock and metal because of the greater abundance of hydrogen compounds in the solar nebula.

Common Misconceptions: Solar Gravity and the Density of Planets

You might think that the dense rocky and metallic materials were simply pulled to the inner part of the solar nebula by the Sun's gravity or that light gases simply escaped from the inner nebula because gravity couldn't hold them. But this is not the case; all the ingredients were orbiting the Sun together under the influence of the Sun's gravity. The orbit of a particle or a planet does *not* depend on its size or density, so the Sun's gravity is *not* the cause of the different kinds of planets. Rather, the different temperatures in the solar nebula are the cause.

Accretion: Assembling the Planetesimals

The first solid flakes that condensed from the solar nebula were microscopic in size. They orbited the protosun in the same orderly, circular paths as the gas from which they condensed. Individual flakes therefore moved with nearly the same speed as their neighboring flakes, allowing them to collide very gently. At this point, the flakes were far too small to attract one another by gravity, but they were able to stick together through electrostatic forces—the same "static electricity" that makes hair stick to a comb. Thus, the flakes grew slowly into larger particles. As the particles grew in mass, gravity began to aid the process of their sticking together, accelerating their growth. This process of growing by colliding and sticking is called **accretion** (Figure 7.7). The growing objects formed by accretion are called **planetesimals**, which essentially means "pieces of planets." Small planetesimals probably came in a variety of shapes, still reflected in many small asteroids today. Larger planetesimals (i.e., those several hundred kilometers across) became spherical due to the force of gravity pulling everything toward the center.

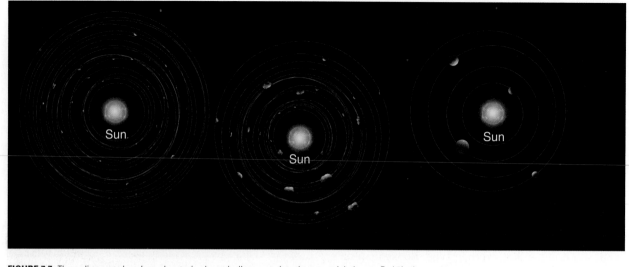

FIGURE 7.7 These diagrams show how planetesimals gradually accrete into the terrestrial planets. Early in the accretion process, there are many Moon-size planetesimals on crisscrossing orbits (left). As time passes, a few planetesimals grow larger by accreting smaller ones, while others shatter in collisions (center). Ultimately, only the largest planetesimals avoid shattering and grow into full-fledged planets (right). Diagram not to scale.

The sizes and compositions of the planetesimals depended on the temperature of the surrounding solar nebula. In the inner solar system, where only rocky and metallic flakes condensed, planetesimals were made of rock and metal. That is why the terrestrial planets ended up being composed of rocks and metals. Moreover, because rocky and metallic elements made up only about 0.6% of the material in the solar nebula, the planetesimals in the inner solar system could not grow very large, which explains why the terrestrial planets are relatively small.

Beyond the frost line between the orbits of Mars and Jupiter, where temperatures were cold enough for ices to condense, planetesimals were built from ice flakes in addition to flakes of rock and metal. Because ice flakes were much more abundant, these planetesimals were made mostly of ices and could grow to much larger sizes than could planetesimals in the inner solar system. The largest icy planetesimals of the outer solar system became the cores of the jovian planets.

The densities of the various objects in the solar system support this theory. Objects near the Sun are composed of higher-density materials like metals and rocks. In the outer solar system, solid objects such as moons and comets are made mostly of low-density ices with a smaller proportion of rocks and metals.

Nebular Capture: Making the Jovian Planets

Accretion proceeded rapidly in the outer solar system, since the presence of ices meant much more material in solid form. Some of the icy planetesimals of the outer solar system quickly grew to sizes many times larger than the Earth. At these large sizes, their gravity was strong enough to capture the far more abundant hydrogen and helium gas from the surrounding nebula. As they accumulated substantial amounts of gas, the gravity of these growing planets grew larger still—allowing them to capture even more gas. The process by which icy planetesimals act as seeds for capturing far larger amounts of hydrogen and helium gas, called **nebular capture**, led directly to the formation of the jovian worlds (Figure 7.8). It explains their huge sizes and the large abundance of hydrogen and helium reflected in their low average densities.

Nebular capture also explains the formation of the diverse satellite systems of the jovian planets. As the early jovian planets captured large amounts of gas from the solar nebula, the same processes that formed the protoplanetary disk—heating, spinning, and flattening—formed similar but smaller disks of material around these planets. Condensation (of metals, rocks, and a lot of ice) and accretion took place within these **jovian nebulae**, essentially creating a miniature solar system around each jovian planet. The spinning disks of the jovian nebulae explain why most of the jovian planet satellites orbit in nearly circular paths lying close to the equatorial plane of their parent planet and also why they orbit in the same direction in which their planet rotates. The composition of the jovian nebulae explains why jovian planets possess systems of large, icy satellites. Their densities of 1–3 g/cm^3 reflect their mixtures of icy and rocky condensates. Temperature differences within the jovian nebulae may have led to density

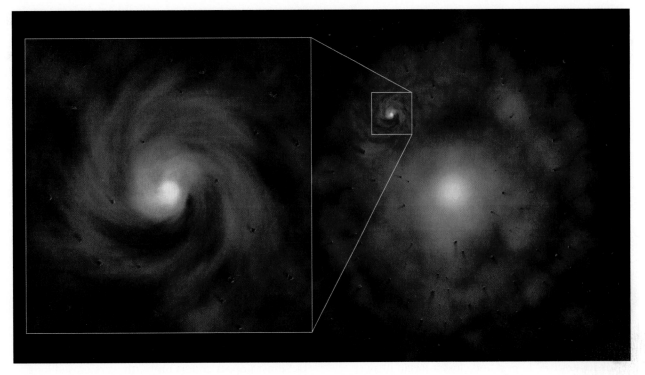

FIGURE 7.8 Large icy planetesimals in the cold outer regions of the solar nebula captured significant amounts of hydrogen and helium gas. This process of nebular capture created jovian nebulae, resembling the solar nebula in miniature, in which the jovian planets and satellites formed. This painting shows a jovian nebula (enlarged at left) located within the entire solar nebula.

differences between satellites analogous to the density differences between planets.

Our theory of condensation, accretion, and nebular capture meets the second challenge for our solar system formation theory: It explains the general differences between terrestrial and jovian planets. (Our theory can also explain the presence of planetary rings, but we postpone this interesting discussion until Chapter 9.) However, our theory does not yet explain why the spacing between the planets increases with distance from the Sun, although we can come up with some reasonable hypotheses. For example, the rapid accretion in the outer solar system may have allowed a few protoplanets to gobble up all their neighbors, leaving the jovian planets more widely spaced.

The Solar Wind: Clearing Away the Nebula

What happened to the remaining gas of the solar nebula? Apparently, this gas was blown into interstellar space by the **solar wind**, a flow of charged particles [Section 4.3] ejected by the Sun in all directions. Although the solar wind is fairly weak today, we have evidence that it was much stronger when the Sun was young—strong enough to have swept huge quantities of gas out of our solar system. The clearing of the gas interrupted the cooling process in the nebula.

Had the cooling continued longer, ices might have condensed in the inner solar system. Instead, the solar wind swept the still-vaporized hydrogen compounds away from the inner solar system, along with the remaining hydrogen and helium throughout the solar system. When the gas cleared, the compositions of objects in the early solar system were essentially set.

7.5 Leftover Planetesimals

We are now ready to turn to our third and fourth challenges: explaining the existence of asteroids and comets and explaining the various exceptions to general trends. Both challenges involve the planetesimals that remained between the forming planets in the early solar system.

Origin of Asteroids and Comets

The strong wind from the young Sun cleared excess gas from the solar nebula, but many planetesimals remained scattered between the newly formed planets. These "leftovers" became comets and asteroids. Like the planetesimals that formed the planets, they formed from condensation and accretion in the solar nebula. Thus, their compositions followed the pattern determined by condensation: planetesimals of rock and metal in the inner solar system, and icy planetesimals in the outer solar system.

Leftover planetesimals originally must have had nearly circular orbits in the same plane as the orbits of the planets. But the strong gravity of the jovian planets tugged and nudged the orbits of planetesimals even at great distances. The result was that the remaining planetesimals ended up with more highly elliptical orbits than the planets, and sometimes with large tilts relative to the plane of planetary orbits. Most of these asteroids and comets eventually either crashed into one of the planets or were flung out of the solar system, but many others still survive today.

Asteroids are the rocky leftover planetesimals of the inner solar system. Some asteroids are scattered throughout the inner solar system, but most are concentrated in the "extra-wide" gap between Mars and Jupiter that contains the *asteroid belt*. This region probably once contained enough rocky planetesimals to form another terrestrial planet. However, the gravity of Jupiter (the largest and closest jovian planet) tended to nudge the orbits of these planetesimals, sending most of them on collision courses with the planets or with one another. The present-day asteroid belt contains the remaining planetesimals from this "frustrated planet formation." Although thousands of asteroids remain in the asteroid belt, their *combined* mass is less than 1/1,000 of Earth's mass. Jupiter's gravity continues to nudge these asteroids, changing their orbits and sometimes leading to violent, shattering collisions. Debris from these collisions often crashes to Earth in the form of meteorites.

Comets are the icy leftover planetesimals of the outer solar system. The icy planetesimals that cruised the space between Jupiter and Neptune couldn't grow to more than a few kilometers in size before suffering either a collision or a gravitational encounter with one of the jovian planets. Many were flung into distant, randomly oriented orbits, becoming the comets of the *Oort cloud*. Beyond the orbit of Neptune, the icy planetesimals were much less likely to be destroyed by collisions or cast off by gravitational encounters. Instead, they remained in orbits going in the same direction as planetary orbits and concentrated near the plane of planetary orbits (but with somewhat more randomness than the orbits of the planets). They were also able to continue their accretion, and many may have grown to hundreds or even thousands of kilometers in diameter. These are the comets of the *Kuiper belt*. Pluto is probably the largest member of this class.

The nebular theory has thus met our third challenge: explaining the existence of asteroids and comets. Moreover, it has suggested an explanation for the seemingly anomalous planet Pluto. Now we turn to our fourth and final challenge: explaining the exceptions to the general trends in our solar system.

The Early Bombardment: A Rain of Rock and Ice

The collision of a leftover planetesimal with a planet is called an **impact**, and the responsible planetesimal is called the **impactor**. On planets with solid surfaces, impacts leave the scars we call **impact craters**. Impacts were extremely common in the young solar system; in fact, impacts were part of the accretion process in the late stages of planetary formation. The vast majority of impacts occurred in the first few hundred million years of our solar system's history.

The heavy bombardment of planetary surfaces in the early solar system would have resembled a rain of rock and ice from space (Figure 7.9). The surface of the Earth was once scarred like the Moon's surface, but most impact craters on Earth were erased long ago by erosion and other geological processes. Only the outlines of a few large craters are recognizable on Earth. But vast numbers of craters remain on worlds that experience less erosion or other geological activity, such as the Moon and Mercury. Indeed, one way of estimating the age of a planetary surface (the time since the surface last changed in a substantial way) is to count the number of craters: If there are many craters, the surface must still look much as it did at the time of the early bombardment, about 4 billion years ago. Although most impacts occurred long ago, asteroids and comets still occasionally crash into planets.

The early rain of rock and ice did more than just scar planetary surfaces. It also brought the materials from which atmospheres, oceans, and polar caps eventually formed. Remember that the terrestrial planets were built from planetesimals of metal and rock. But Earth's oceans, the polar caps of Earth and Mars, and the atmospheres of Venus, Earth, and Mars are all made from hydrogen compounds that remained gaseous in the inner solar nebula. These materials must have arrived on the terrestrial planets after their initial formation, most likely brought by impacts of planetesimals formed farther out in the solar system. We don't yet know whether the impactors came from the asteroid belt, where rocky planetesimals contained small amounts of water and other hydrogen compounds that had condensed as ices, or whether they were comets containing huge amounts of ice. Either way, the water we drink and the air we breathe probably once were part of planetesimals floating beyond the orbit of Mars.

FIGURE 7.9 Around 4 billion years ago, Earth, its Moon, and the other planets were heavily bombarded by leftover planetesimals. This painting shows the young Earth and Moon glowing with the heat of accretion, and with an impact in progress on the Earth.

Captured Moons

We can easily explain the orbits of most jovian planet satellites by their formation in a jovian nebula that swirled around the forming planet. But some moons have unusual orbits—orbits in the "wrong" direction (opposite the rotation of their planet) or with large inclinations to the planet's equator. These unusual moons are probably leftover planetesimals that were *captured* into orbit around a planet. Because of the random nature of the capture process, the captured moons would not necessarily orbit in the same direction as their planet or in its equatorial plane.

The two small moons of Mars—Phobos and Deimos—probably were asteroids captured by this process (Figure 7.10). They resemble asteroids seen in the asteroid belt and are much darker and lower in density than Mars. Jupiter probably also captured several of its moons. Other jovian planets may have captured moons, including one that is particularly large—Triton, the largest moon of Neptune. Triton is considerably larger than Pluto and orbits Neptune in a direction opposite to Neptune's rotation. It may be a captured object from the Kuiper belt [Section 9.5]. The one unusual moon that cannot be explained by this process is our own: Our Moon is much too large to have been captured by Earth.

Giant Impacts and the Formation of Our Moon

The largest planetesimals remaining as the planets formed may have been huge—some may have been the size of Mars. When one of these planet-size planetesimals collided with a planet, the spectacle would have been awesome. Such a **giant impact** could have significantly altered the planet's fate.

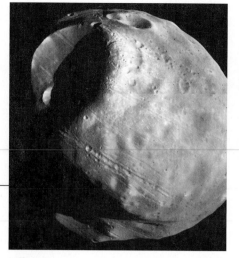
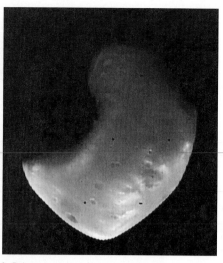

FIGURE 7.10 The two moons of Mars, shown here in photos taken by the Viking spacecraft, are probably captured asteroids. Phobos is only about 13 km across and Deimos is only about 8 km across—making each of these two moons small enough to fit within the boundaries of a typical large city.

a Phobos   **b** Deimos

What if an object the size of Mars had collided with Earth? Computer simulations address this interesting question. The Earth would have shattered from the impact, and material from the outer layers would have ended up in orbit around the Earth (Figure 7.11). There this material could have reaccreted to form a large satellite. Depending on exactly where and how fast the giant impactor struck the Earth, the blow might also have tilted the Earth's axis, changed its rotation rate, or completely torn it apart. Today, such a giant impact is the leading hypothesis for ex-plaining the origin of our Moon. The Moon's composition is similar to that of the Earth's outer layers, exactly what we would predict for the kind of collision shown in Figure 7.11. Moreover, we can rule out the idea of the Moon forming simultaneously with Earth. In that case, Earth and the Moon would have formed from the same material and therefore should have the same density, but the Moon's density is considerably lower than the Earth's.

In fact, many of the unusual properties of specific planets—properties that defy the general trends ex-

FIGURE 7.11 Artist's conception of the impact of a Mars-size object with Earth, as may have occurred soon after Earth's formation. The ejected material comes mostly from the outer rocky layers and accretes to form the Moon, which is poor in metal.

pected by the nebular theory—may be the results of giant impacts. Mercury may have lost much of its outer, rocky layer in a giant impact, leaving it with a huge metallic core. A giant impact might even have contributed to the slow, backward rotation of Venus, which may have had a "normal" rotation until the arrival of the giant impactor. Giant impacts probably were also responsible for the axis tilts of many planets (including Earth) and for tipping Uranus on its side. Pluto's moon Charon may have formed in a giant-impact process similar to the one that formed our Moon.

Unfortunately, we can do little to test whether a particular giant impact really occurred billions of years ago. But no other idea so effectively explains the formation of our Moon and other "oddities" that we've discussed. Moreover, giant impacts certainly should have occurred, given the number of large leftover planetesimals predicted by the nebular theory. Random giant impacts are the most promising explanation for the many exceptional circumstances noted in the fourth challenge.

Summary: Meeting the Challenges

The nebular theory explains the great majority of important facts contained in our four challenges. But you should not be left with the impression that competing theories were never put forth or that solar system formation is now a "solved problem." Theories have evolved hand-in-hand with the discovery of the nature of our solar system, and what we have presented here is the culmination of that great endeavor to date. Planetary scientists are still struggling with more quantitative aspects, seeking reasons for the exact sizes, locations, and compositions of the planets. Perhaps as we study other planetary systems and examine their properties, we will be able to improve our understanding of solar system formation.

Assuming that the nebular theory is correct, it is interesting to ask whether our solar nebula was "destined" to form the solar system we see today. The first stages of planet formation were orderly and inevitable according to the nebular theory. Nebular collapse, condensation, and the first stages of accretion were relatively gradual processes that probably would happen all over again if we turned back the clock. But the final stages of accretion, and giant impacts in particular, are inherently random in nature and probably would not happen again in just the same way. A larger or smaller planet might form at Earth's location and might suffer from a larger giant impact or from none at all. We don't yet know whether these differences would fundamentally alter the solar system or simply change a few "minor" details—such as the possibility of life on Earth.

As we conclude the discussion of *how* our solar system formed, it's natural to ask *when*. We have a surprisingly good answer: The solar system began forming about 4.6 billion years ago. The answer comes from a meticulous examination of meteorites that are leftovers from that period. All rocks, including meteorites, contain minute fractions of various radioactive elements that decay or break apart and become new elements. The rates of decay for the various elements can be measured to very high accuracy in the laboratory, so scientists can determine how long those elements have been decaying within the rock. Virtually all meteorites, and all the various elements within the meteorites, give the same answer of about 4.6 billion years as the age of our solar system.

7.6 Other Planetary Systems

How common are planetary systems? What are the odds of another Earth-like planet? Are habitable planets as common in the real universe as they are in science fiction? We cannot yet answer these questions definitively, but we've made great progress in our understanding of planetary systems.

As we've discussed, observations have confirmed that protoplanetary disks are common, as predicted by our theory of solar system formation. Even binary and multiple star systems, which many astronomers once thought unlikely to harbor planets, now seem to be reasonable candidates for planets. Observations show protoplanetary disks around at least some binary star systems (see Figure 7.4a), and recent calculations show that planets can safely orbit in such systems either by being close to one star or by being in a large orbit around both stars.

More importantly, rapid advances in observational technology now allow us to search for actual planets, not just for protoplanetary disks. At the beginning of the 1990s, we had no conclusive proof that planets existed around any star besides our Sun. By 2001, dozens of planetlike objects had been detected around other stars. We cannot always determine precise masses for these objects, and some may turn out to be more like small stars than the planets of our solar system. In addition, our current technology is not yet capable of detecting planets as small as Earth, so we still have no observational evidence concerning the existence of Earth-like planets. Nevertheless, we can now be quite sure that our solar system is not the only planetary system, and we know of many more planets outside our solar system than within it.

Detecting Extrasolar Planets

You might think that the easiest way to discover **extrasolar planets**, or planets around other stars, would be simply to photograph them through powerful telescopes. Unfortunately, current observational technology cannot produce such images. The primary problem arises from the fact that any light from an orbiting planet would be overwhelmed by light from the star it orbits. For example, a Sun-like star would be a *billion times* brighter than the reflected light from an Earth-like planet. Because even the best telescopes blur the light from stars at least a little, finding the small blip of planetary light amid the glare of scattered starlight would be very difficult. Astronomers are working on technologies that may overcome this problem, but for now we rely primarily on techniques that observe the star itself to find indirect evidence of planets.

To date, the most commonly used strategy involves searching for a planet by watching for the small gravitational tug it exerts on its star. In most cases, the easiest way to find this tug is to identify small Doppler shifts in the star's spectrum [Section 6.4]. An orbiting planet causes its star to alternately move slightly toward and away from us, which makes the star's spectral lines alternately shift toward the blue and toward the red (Figure 7.12). Remarkably, current techniques can measure a star's velocity to better than 3 m/s—jogging speed—which is good enough to find gravitational tugs caused by planets the size of Jupiter, and in some cases even smaller. For example, the data in Figure 7.12b reveal a planet with roughly 60% of the mass of Jupiter orbiting the star called 51 Pegasi. This planet lies so close to its star that its "year" lasts only four of our days and its temperature is probably over 1,000 K.

By 2001, the Doppler-shift method had been used to identify more than 50 planets around other stars (Figure 7.13), and discoveries are coming so rapidly that the number may be far greater by the time you read this. Highlights of recent discoveries include a bona fide "planetary system" of three planets orbiting the star Upsilon Andromedae, a handful of planets as small as Saturn around various stars (see Figure 7.13), and a dust disk plus a planet orbiting the star Epsilon Eridani.

The Nature of Extrasolar Planets

Although the Doppler-shift technique allows us to discover orbiting objects, by itself it cannot tell us the nature of the orbiting bodies. The first problem is that the amount of a Doppler shift depends both on the mass of the orbiting object and on the tilt of its orbit: Orbits seen edge-on (from our vantage point on Earth) produce the largest Doppler shifts, while face-on orbits produce no Doppler shift at all. Therefore, an observed Doppler shift could be produced either by a low-mass planet in an edge-on orbit or by a more massive planet (or even a *brown dwarf* [Section 14.2] or small star) in a more tilted orbit. The masses listed in Figure 7.13 are therefore underestimates in some cases, but probably by no more than a factor of 2. The second problem is that, even with precise mass estimates, we still would not know the size of the planet or whether the planet is more terrestrial or jovian in nature.

For the planet orbiting HD209548, the uncertainties in mass and size vanished in fall 1999 when

---

FIGURE 7.12 Detecting planets with Doppler shifts.

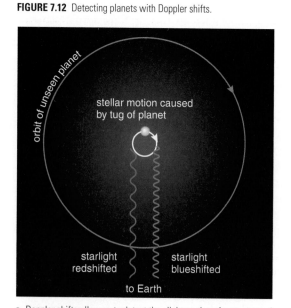

a Doppler shifts allow us to detect the slight motion of a star caused by an orbiting planet.

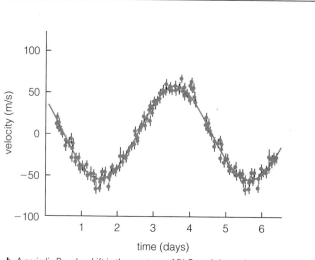

b A periodic Doppler shift in the spectrum of 51 Pegasi shows the presence of a large planet with an orbital period of about 4 days. Dots are actual data points; bars through dots represent measurement uncertainty.

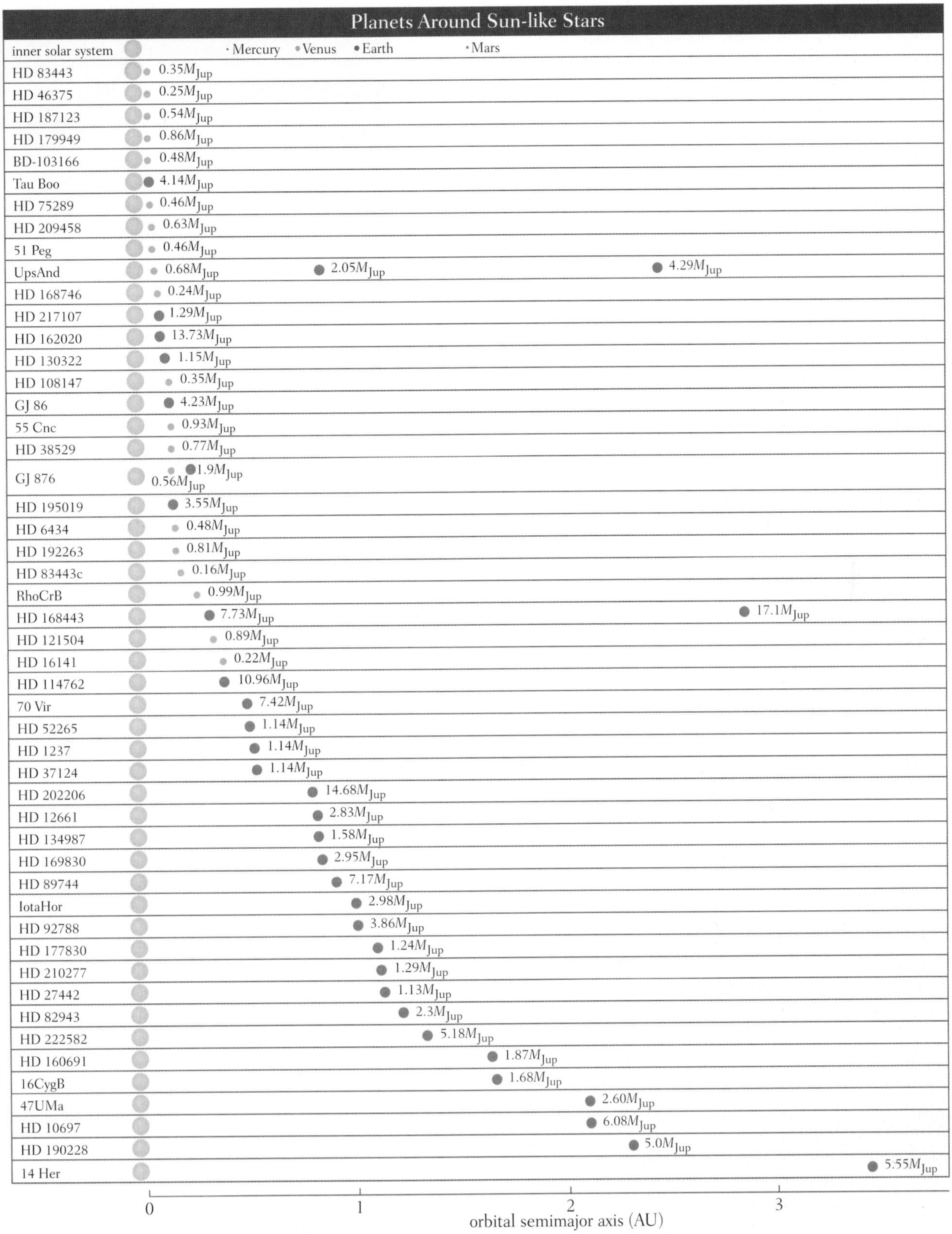

Planets Around Sun-like Stars

Star	Planets (mass)
inner solar system	·Mercury ·Venus ·Earth ·Mars
HD 83443	0.35M_{Jup}
HD 46375	0.25M_{Jup}
HD 187123	0.54M_{Jup}
HD 179949	0.86M_{Jup}
BD-103166	0.48M_{Jup}
Tau Boo	4.14M_{Jup}
HD 75289	0.46M_{Jup}
HD 209458	0.63M_{Jup}
51 Peg	0.46M_{Jup}
UpsAnd	0.68M_{Jup} 2.05M_{Jup} 4.29M_{Jup}
HD 168746	0.24M_{Jup}
HD 217107	1.29M_{Jup}
HD 162020	13.73M_{Jup}
HD 130322	1.15M_{Jup}
HD 108147	0.35M_{Jup}
GJ 86	4.23M_{Jup}
55 Cnc	0.93M_{Jup}
HD 38529	0.77M_{Jup}
GJ 876	0.56M_{Jup} 1.9M_{Jup}
HD 195019	3.55M_{Jup}
HD 6434	0.48M_{Jup}
HD 192263	0.81M_{Jup}
HD 83443c	0.16M_{Jup}
RhoCrB	0.99M_{Jup}
HD 168443	7.73M_{Jup} 17.1M_{Jup}
HD 121504	0.89M_{Jup}
HD 16141	0.22M_{Jup}
HD 114762	10.96M_{Jup}
70 Vir	7.42M_{Jup}
HD 52265	1.14M_{Jup}
HD 1237	1.14M_{Jup}
HD 37124	1.14M_{Jup}
HD 202206	14.68M_{Jup}
HD 12661	2.83M_{Jup}
HD 134987	1.58M_{Jup}
HD 169830	2.95M_{Jup}
HD 89744	7.17M_{Jup}
IotaHor	2.98M_{Jup}
HD 92788	3.86M_{Jup}
HD 177830	1.24M_{Jup}
HD 210277	1.29M_{Jup}
HD 27442	1.13M_{Jup}
HD 82943	2.3M_{Jup}
HD 222582	5.18M_{Jup}
HD 160691	1.87M_{Jup}
16CygB	1.68M_{Jup}
47UMa	2.60M_{Jup}
HD 10697	6.08M_{Jup}
HD 190228	5.0M_{Jup}
14 Her	5.55M_{Jup}

orbital semimajor axis (AU)

0   1   2   3

FIGURE 7.13 This diagram shows the orbital distances and approximate masses of the first 55 planets discovered around other stars. Most of the planets found so far are closer to their stars and more massive than the planets in our solar system. (Planet sizes are not to scale.)

meticulous observations by two teams of astronomers revealed a 1.7% drop in the brightness of the star (Figure 7.14). The drop occurred at exactly the time that a newly discovered planet was expected to pass on the side of the orbit nearest Earth. The planet must have passed directly on the line of sight between the Earth and the star, temporarily blocking some of the star's light from reaching the Earth. At that moment, we learned that the orbit was truly edge-on and therefore that the mass estimate was correct. We also learned that the planet's disk covers about 1.7% of the star's disk, which allowed us to estimate the planet's radius and confirm that it is a jovian planet.

Lessons for Solar System Formation

The discovery of extrasolar planets presents us with opportunities to test our theory of solar system formation. Can our existing theory explain other planetary systems, or will we have to go back to the drawing board? So far, the discoveries have presented at least one significant challenge.

The new challenge arises from the fact that most of the recently discovered planets are quite different from those of our solar system. Many have highly elliptical orbits rather than the nearly circular orbits of planets in our solar system, and most of the planets are more massive than Jupiter. Most intriguing is the fact that many of these large planets lie quite close to their stars—a very different situation from that in our solar system, where the large planets are found in the outer solar system. Moreover, as we discussed in Section 7.4, our theory has a good explanation for why large planets should form only in the outer solar system: That is the only place where the solar nebula was cool enough for ices to condense and grow into the large seeds needed for jovian planets. Thus, the existence of dozens of systems with large, close-in planets seems to contradict our theory.

Fortunately, there are at least two reasonable ways out of this dilemma. First, recent theoretical work has shown that the gas in a forming solar system can exert drag on young jovian planets, tending to make them migrate closer to their stars. Thus, if a star's wind does not sweep the remaining nebular gas into space soon enough, the jovian planets may end up close to their stars—which could explain the large, close-in planets we've detected. In some cases, the planets may even end up being consumed by their parent stars! If this idea proves correct, the layout of our solar system suggests that the solar wind blew out the remaining gas of the solar nebula relatively early compared to other star systems.

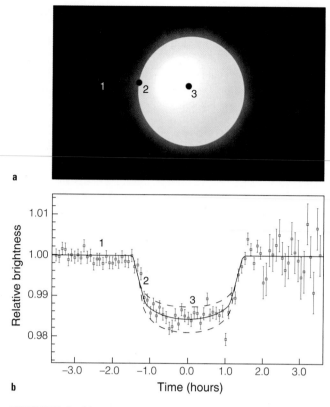

FIGURE 7.14 Careful measurements of the brightness of the star called HD209548 revealed that an orbiting planet passes directly in front of it as seen from Earth, which means that the planet's orbit must be edge-on as seen from Earth. (**a**) Artist's conception of the planet as it passes directly in front of its star as seen from Earth. (**b**) These data show the 1.7% drop in the star's brightness that proved the planet is passing in front of the star as seen from Earth.

Second, the extrasolar planets discovered to date may turn out to be relatively rare exceptions to general rules. Astronomers have succeeded in detecting planets around only a few percent of the stars they've studied so far. It's possible that planets also circle most of the other stars but have eluded detection because they are more like the planets in our own solar system. The Doppler technique makes it much easier to find large planets with close-in orbits, because such planets exert greater gravitational tugs on their stars than smaller or more distant planets. Thus, our early discoveries of extrasolar planets may be like glancing at animals in the rain forest: The jungle appears full of brightly colored parrots and frogs, but far more animals fail to catch our eye. In the same way, we may have discovered the rare exceptions but so far missed the more common cases that are like our own solar system. Some of the most recent discoveries, in fact, are jovian planets almost as far from their star as ours are from the Sun.

As we discover more planets, we should gain a better understanding of how well our theory of solar system formation holds up. The discoveries should come rapidly, thanks to new technologies and planned space missions. In particular, NASA has placed a high priority on building ambitious orbital telescopes for the detection and study of extrasolar planets. Within a couple of decades, NASA hopes to launch high-resolution interferometers [Section 6.5] that will enable us to obtain images and spectra of planets in other solar systems. Then, at last, we will know whether solar systems like ours—and planets like Earth—are rare or common.

THE BIG PICTURE

We've seen that the nebular theory accounts for the major characteristics of our solar system. As you continue your study of the solar system, keep in mind the following "big picture" ideas.

- Close examination of the solar system unveils a wealth of patterns, trends, and groupings, leading us to the conclusion that all the planets formed from the same cloud of gas at about the same time.

- Chance events may have played a large role in determining how individual planets turned out. No one knows how different the solar system might be if it started over.

- Planet-forming processes are apparently universal. The discovery of protoplanetary disks and full-fledged planets around other stars brings planetary science to the brink of an exciting new era.

Review Questions

1. What is *comparative planetology?* What is its basic premise? What are its primary goals?

2. Briefly summarize the observed patterns of motion in our solar system.

3. Summarize the differences between terrestrial planets and jovian planets. Why is the Moon grouped with the terrestrial planets? Where does Pluto fit in?

4. What are *asteroids?* Where are they found? What are *comets,* and how do those of the *Oort cloud* and the *Kuiper belt* differ in terms of their orbits?

5. Summarize the four challenges that any theory of the solar system must explain.

6. What is the *nebular theory?* How does it get its name? What do we mean by the *solar nebula?*

7. Describe the processes that led the solar nebula to collapse into a spinning disk. Also describe the evidence supporting the idea that our solar system had a *protosun* and a *protoplanetary disk* early in its history.

8. Distinguish between *metals, rocks, hydrogen compounds,* and *light gases* in terms of condensation temperatures and relative abundances in the solar nebula.

9. Explain how temperature differences in the solar nebula led to the condensation of different materials at different distances from the protosun. What do we mean by the *frost line* in the solar nebula?

10. What is *accretion?* What are *planetesimals?*

11. Describe the process of *nebular capture* that allowed the jovian planets to grow to very large sizes. How does this process explain why jovian planets have extensive satellite systems?

12. What is the *solar wind?* How did its strength in the past differ from its strength today? What role did it play in ending the growth of the planets?

13. Briefly describe and explain the present distribution of asteroids and comets in our solar system.

14. What was the early bombardment? How can we use *impact craters* to estimate the age of a planetary surface?

15. What clues suggest that a planetary satellite was captured instead of forming with its planet?

16. Describe the process by which a giant impact may have led to the formation of the Moon. What other oddities of our solar system might be explained by giant impacts?

17. Summarize how the nebular theory meets the four challenges laid out in this chapter. Describe a few of the remaining unanswered questions.

18. Describe how we can detect extrasolar planets today, and summarize the current evidence concerning planets in other solar systems.

19. What have we learned about solar system formation from discoveries of extrasolar planets?

Discussion Question

1. *Lucky to Be Here?* In considering the overall process of solar system formation, do you think it was very likely for a planet like Earth to have formed? Could random events in the early history of the solar system have prevented us from even being here today? Defend your opinion. Do your opinions on these questions have any implications for your belief in the possibility of Earth-like planets around other stars?

Problems

Surprising Discoveries? For **problems 1–8**, suppose we found a solar system with the property described. (These are *not* real discoveries.) In light of our theory of solar system formation, decide whether the discovery should be considered reasonable or surprising. Explain.

1. A solar system has five terrestrial planets in its inner solar system and three jovian planets in its outer solar system.

2. A solar system has four large jovian planets in its inner solar system and seven small planets made of rock and metal in its outer solar system.

3. A solar system has ten planets that all orbit the star in approximately the same plane. However, five planets orbit in one direction (e.g., counterclockwise), while the other five orbit in the opposite direction (e.g., clockwise).

4. A solar system has 12 planets that all orbit the star in the same direction and in nearly the same plane. The 15 largest moons in this solar system orbit their planets in nearly the same direction and plane as well. However, several smaller moons have highly inclined orbits around their planets.

5. A solar system has six terrestrial planets and four jovian planets. Each of the six terrestrial planets has at least five moons, while the jovian planets have no moons at all.

6. A solar system has four Earth-size terrestrial planets. Each of the four planets has a single moon nearly identical in size to Earth's Moon.

7. A solar system has many rocky asteroids and many icy comets. However, most of the comets orbit the star in a belt much like the asteroid belt of our solar system, while the asteroids inhabit regions much like the Kuiper belt and Oort cloud of our solar system.

8. A solar system has several planets similar in composition to our jovian planets but similar in mass to our terrestrial planets.

9. *Two Classes of Planets.* Explain in terms a friend or roommate would understand why the jovian planets are lower in density than the terrestrial planets even though they all formed from the same cloud.

10. *A Cold Solar Nebula.* Suppose the entire solar nebula had cooled to 50 K before the solar wind cleared it away. How would the composition and sizes of the terrestrial planets be different from what we see today? Explain your answer in a few sentences.

11. *No Nebular Capture.* Suppose the solar wind had cleared away the solar nebula before the process of nebular capture was completed in the outer solar system. How would the jovian planets be different? Would they still have satellites? Explain your answer in a few sentences.

12. *Angular Momentum.* Suppose our solar nebula had begun with much more angular momentum than it did. Do you think planets could still have formed? Why or why not? What if the solar nebula had started with zero angular momentum? Explain your answers in one or two paragraphs.

13. *Jupiter's Action.* Suppose that, for some reason, the planet Jupiter had never formed. How do you think the distribution of asteroids and comets in our solar system would be different? How would these differences have affected Earth? Explain your answer in a few sentences.

14. *51 Pegasi.* The star 51 Pegasi has about the same mass as our Sun, and the planet discovered around it has an orbital period of 4.23 days. The mass of the planet is estimated to be 0.6 times the mass of Jupiter.

    a. Use Kepler's third law to find the planet's average distance (semimajor axis) from its star. (*Hint:* Because the mass of 51 Pegasi is about the same as the mass of our Sun, you can use Kepler's third law in its original form, $p^2 = a^3$ [see Chapter 3]; be sure to convert the period into years.)

    b. Briefly explain why, according to our theory of solar system formation, it is surprising to find a planet the size of the 51 Pegasi planet orbiting at this distance.

    c. Hypothesize as to how the 51 Pegasi planet might have come to exist. Explain your hypothesis in a few sentences.

15. *Transiting Planets.* The star HD209548 is about the same size as our Sun, and the planet discovered around it blocks 1.7% of the star's area when it passes in front of the star. (Appendix E contains useful data for this problem.)

    a. How large a planet is required to block the observed fraction of the star's area? Give your answer in kilometers. (*Hint:* Remember that the brightness drop tells us that the planet blocked 1.7% of the star's visible *area*.)

    b. The mass of the planet is estimated to be 0.6 times the mass of Jupiter. What is the density of the planet?

    c. Compare the density of this planet to that of planets in our solar system, and comment on whether this density makes the planet terrestrial or jovian in nature.

Web Projects

Find useful links for Web projects on the text Web site.

1. *New Planets.* Find up-to-date information about discoveries of planets in other solar systems. Create a personal "Web journal," complete with pictures from the Web, describing at least three recent discoveries of new planets. For each case, write one or two paragraphs in your journal describing the method used in the discovery, comparing the discovered planet to the planets of our own solar system, and summarizing whether the discovery poses any new challenges to our theory of solar system formation.

2. *Missions to Search for Planets.* Learn about one proposed space mission to search for planets around other stars, such as Kepler, SIM, or Terrestrial Planet Finder. Write a short report about the plans for the mission and its current status.

CHAPTER 8
The Terrestrial Worlds

As we live our daily lives, we can't help but take Earth's pleasant conditions for granted: a temperature neither boiling nor freezing, abundant water, a protective atmosphere, and a relatively stable environment thanks to a moderate level of geological activity. We need look only as far as our neighboring terrestrial worlds to see how fortunate we are.

Since the beginning of the space age, we've documented the widely different surfaces and atmospheres of the planets, and we can now make detailed comparisons among them. We've learned that, even though all the terrestrial worlds are similar in composition and formed at about the same time, their subsequent histories diverged dramatically because of a few basic properties of each world. Thus, by comparing the geological and atmospheric evolution of the terrestrial worlds, we can gain a better appreciation for how the Earth works and why it has turned out to be advantageous for us.

Mercury

Venus

FIGURE 8.1 Global views of the terrestrial planets to scale and representative surface close-ups, each a few hundred kilometers across. The global view of Venus shows its surface without its atmosphere, based on radar data from the Magellan spacecraft; all other images are photos (or composite photos) taken from spacecraft.

8.1 Principles of Comparative Planetology

The five terrestrial worlds—Mercury, Venus, Earth, the Moon, and Mars—share a common ancestry in their birth from the solar nebula, but their present-day surfaces and atmospheres show vast differences (Figure 8.1). Mercury and the Moon are ancient, battered worlds densely covered by craters, with some areas of volcanic plains. Venus is shrouded in a thick, broiling atmosphere, and its surface has been flooded with lava and twisted by internal stresses. Mars, despite its middling size, has the solar system's largest volcanoes and is the only planet other than Earth where running water played a major role in shaping the surface. Earth has surface features similar to all those on other terrestrial worlds and more—including a unique layer of living organisms that covers almost the entire surface of the planet. Our purpose here is to understand how the profound differences among the terrestrial worlds came to be.

Geology and atmospheric science began as studies just of the Earth, but the space age has broadened these endeavors to include all the planets and their satellites. *Comparative planetology* is the study of the solar system through examining and understanding the similarities and differences among the planets. It is based on the idea that the planets have a common origin, are subject to the same physical laws, and undergo the same physical processes to varying degrees. Differences among the planets are rarely accidents but rather can be traced to a planet's fundamental properties, such as size, location in the solar system, composition, or other important factors. Comparative planetology allows a far better appreciation of our world—and our place in the universe—than could possibly be achieved by even the most advanced study of the Earth alone.

In this chapter, we will begin by examining the governing principles of comparative planetology, focusing first on geological processes and then on atmospheres. We'll then examine our neighboring terrestrial worlds in detail, using the tools developed in the beginning of the chapter.

8.2 Planetary Geology

The study of surface features and the processes that create them is called **geology**. The root *geo* means "Earth," and *geology* originally referred only to the study of the Earth. Today we speak of *planetary geology*, the extension of geology to include all the solid bodies in the solar system, whether rocky or icy.

Geology is relatively easy to study on Earth, where we can examine the surface in great detail, but is clearly much more challenging on the other planets. The Moon is the only alien world from which we've collected rocks, some gathered by the Apollo

Earth

Earth's Moon

Mars

astronauts and others by Russian robotic landers in the 1970s. We can also study the dozen or so meteorites from Mars that have landed on Earth. Aside from these few rocks, we have little more than images taken from spacecraft from which to decode the history of the other planets over the past 4.6 billion years. It's like studying people's facial expressions to understand what they're feeling inside—and what their childhood was like! Fortunately, this decoding is considerably easier for planets than for people.

Viewing the Terrestrial Worlds

Spacecraft have visited and photographed all the terrestrial worlds. Besides ordinary, visible-light photographs, we also have spacecraft images taken with infrared and ultraviolet cameras, spectroscopic data, and in some cases three-dimensional data compiled with the aid of radar. Finally, we have close-up photographs of selected locations on all the terrestrial worlds but Mercury, taken by spacecraft that have landed on their surfaces (Figure 8.2). The result is that we now understand the geology of the terrestrial worlds well enough to make detailed and meaningful comparisons among them.

Comparative planetary geology (one part of comparative planetology) hinges on the principle that a planet's surface features can be traced back to its fundamental properties. For example, we saw in Chapter 7 that a planet's composition depends pri-

marily on how far from the Sun it formed. In the rest of this section, we look for similar relationships between surface features and the fundamental properties of the planets. But because surface geology depends largely on a planet's interior, we must first look inside the terrestrial worlds.

Inside the Terrestrial Worlds

Planets are all approximately spherical and quite smooth relative to their size. For example, compared to the Earth's radius of 6,378 km, the tallest mountains could be represented by grains of sand on a typical globe. It's easy to understand why the gaseous jovian planets ended up nearly spherical: Gravity always acts to pull material together, and a sphere is the most compact shape possible. It's a bit subtler when we deal with the rocky terrestrial worlds, because rock can resist the pull of gravity. Indeed, many small moons and asteroids are "potato-shaped" precisely because their weak gravity is unable to overcome the rigidity of their rocky material. We need to take a closer look at the behavior of rock.

We often think of rocks as the very definition of strength ("solid as a rock"), but rocks are not always as solid as they may seem. When subjected to sustained stress over millions or billions of years, rocky material slowly deforms and flows. In fact, rock acts more like Silly Putty™, which stretches when you

a Mercury

FIGURE 8.2 Surface views of the terrestrial worlds. No spacecraft have landed on Mercury, so an artist's conception is shown; all other images are photos. The Venus photo shows only a small patch of surface at the base of the Russian Venera lander, visible in the lower right.

b Venus

c Earth

d Moon

e Mars

pull it slowly but breaks if you pull it sharply (Figure 8.3). And just as Silly Putty becomes more pliable if you heat it, warm rocks are weaker and more deformable than cooler rocks.

The rocky terrestrial worlds became spherical because of rock's ability to flow. When objects exceed about 500 km in diameter, gravity can overcome the strength of solid rock and make a world spherical in less than a billion years. If the object is molten, as was the case with the terrestrial worlds in their early histories, it can become spherical much more quickly. (For similar reasons, Earth's ocean surface is smoother and more spherical than its rocky surface.)

Gravity also gives the terrestrial worlds similar internal structures. You know that in a mixture of oil and water the less dense oil rises to the top while the denser water sinks to the bottom. The process by which gravity separates materials according to density is called **differentiation** (because it results in layers made of *different* materials). The terrestrial worlds underwent differentiation early in their histories, at the time when they were molten throughout their interiors. The result was the formation of three layers of differing composition within each terrestrial world (Figure 8.4). Dense metals such as iron and nickel sank through molten rocky material to form a **core**. Rocky material came to rest above the core, forming a thick **mantle**. A "scum" of the lowest-density rocks rose to form a **crust**.

The terms *core, mantle,* and *crust* are defined by the composition within each layer. But characterizing the layers by *rock strength* rather than composition turns out to be more useful for understanding geological activity. From this view, we speak of an outer layer of relatively rigid rock called the **lithosphere** (*lithos* means "stone" in Greek). The lithosphere generally encompasses the crust and the uppermost portion of the mantle. Beneath the lithosphere, the higher temperatures allow rock to deform and flow much more easily even though it is not actually molten. Thus, the lithosphere is essentially a layer of rigid rock that "floats" on the softer rock below. (On the Earth, the

FIGURE 8.3 Silly Putty stretches when pulled slowly but breaks cleanly when pulled rapidly. Rock behaves just the same, but on a longer time scale.

lithosphere is broken into *plates* that move as the underlying rock flows gradually, creating the phenomenon called *continental drift* [Section 11.2].)

The strength of the lithosphere determines the level of geological activity on the planet's surface. Interior heat and motion will lead to volcanoes and other geological activity on a planet with a relatively thin lithosphere. In contrast, geological activity is

FIGURE 8.4 Interior structures of the terrestrial worlds, in order of size. The thickness of outer layers is exaggerated for clarity.

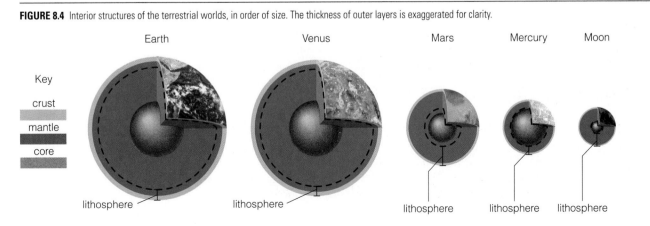

inhibited on a planet with a thick, strong lithosphere. The most important factor determining lithospheric thickness is internal temperature: Higher internal temperature makes rocks softer, leading to a thinner rigid lithosphere. Figure 8.4 compares the interior structures of the terrestrial worlds. The greatest difference is that smaller worlds have thicker lithospheres, indicating that they have cooler interiors.

The secret to understanding the differences between planetary lithospheres lies in understanding how planets are heated and how they cool off. Planetary *surfaces* are all warmed by sunlight, but the high temperatures *inside* the planets today are due to radioactive heating. Rocks contain small amounts of radioactive elements such as uranium, thorium, and potassium. When the nuclei of radioactive elements decay, subatomic particles fly off at high speeds, colliding with neighboring atoms and heating them up. In essence, this transfers some of the mass-energy of the radioactive elements ($E = mc^2$ [Section 4.2]) to the heat of the planetary interior.

Some of a planet's internal heat is always escaping the interior and being radiated away to space. Heat flows outward from the hot interior toward the cooler surface through *conduction* and *convection*. **Conduction** transfers heat between solid objects in contact with each other; it makes heat flow from your hand to a glass full of ice water. Conduction is the main process by which heat flows upward through the lithosphere. Planets with warm interiors also lose heat through **convection**, in which hot material expands and rises while cooler material contracts and falls. Convection can occur any time a substance is strongly heated from underneath; you can see convection whenever you heat a pot of soup on the stove, and you're probably also familiar with it in the context of weather, in which warm air rises while cool air falls in our atmosphere. Each small region of rising and falling material is called a **convection cell**.

The interiors of the terrestrial planets are slowly cooling as their heat escapes. By now, 4.6 billion years after their formation, much of the planets' original heat from accretion and differentiation (Figure 7.9) has leaked away. Even the rate of radioactive decay declines as the planets age (because a particular nucleus can decay only once). The interior cooling gradually changes a planet's structure, making the lithosphere thicker and leaving any regions of molten rock lying deeper inside.

Ultimately, the single most important factor in determining how long a planet stays hot is its size: Larger planets stay hot longer, just as larger baked potatoes stay hot longer than small ones. You can see why size matters most by picturing a large planet as a smaller planet wrapped in extra layers of rock. The extra rock acts as an insulating blanket, so heat from

the center takes longer to reach the surface on the larger planet than it would on the smaller one. The same principle explains many everyday phenomena. For example, crushing a cube of ice into smaller pieces increases the total amount of surface area from which heat is lost, while the total amount of ice remains the same. Thus, the crushed ice will cool a drink more quickly than would the original ice cube.

TIME OUT TO THINK *Can you think of other examples in everyday life of a small object cooling off faster than a large one? What about a smaller object warming up more quickly than a large one? How do these examples relate to the issue of geological activity on the terrestrial worlds?*

The overall result is that the large terrestrial worlds—Earth and Venus—have cooled the least over the life of the solar system. These worlds therefore have thin lithospheres and substantial geological activity. The smaller worlds—Mercury and the Moon—have mostly cooled off and therefore have very thick lithospheres that essentially make them geologically "dead." Mars, intermediate in size, has cooled significantly, resulting in an intermediate interior temperature.

A planet's interior is also where its magnetic field, if any, is generated. The metallic cores of all the terrestrial worlds are electrically conducting; if the core materials are also in motion, they can create a magnetic field that surrounds the planet. Although our understanding of planetary magnetic fields is still rudimentary, we know that the combined motions of planetary rotation and convection (in a molten

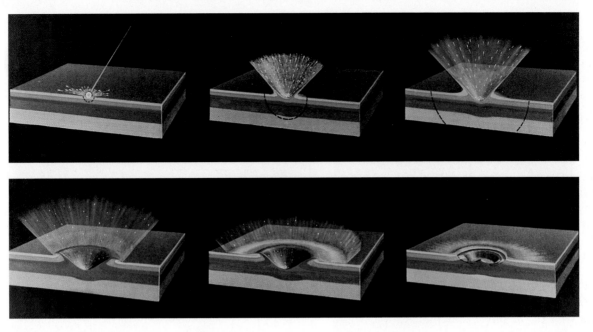

FIGURE 8.5 Artist's conception of the impact process. The last frame shows how, in a larger crater, the center can rebound just as water does after you drop a pebble into it.

core) are necessary to create a magnetic field. Earth, with its convecting core and rapid rotation, has the strongest magnetic field, while the Moon, with its small, frozen core and slow rotation, has none. Mercury lies in between. Mars lacks a significant magnetic field today, but the latest measurements from the Mars Global Surveyor spacecraft reveal the imprint of a stronger field in Mars's first billion years. Apparently, as the core cooled and solidified, the magnetic field faded away. Venus lacks any magnetic field, due to either its slow rotation or an as-yet-unexplained lack of core convection.

The presence or absence of a magnetic field may not seem important to the massive rocky worlds, but it actually matters a great deal to a planet's atmosphere. Magnetic fields are responsible for forming a protective bubble around a planet called a *magnetosphere*; as we'll see, this can have significant effects on the evolution of a planet's atmosphere.

8.3 Shaping Planetary Surfaces

When we look around the Earth, we find an apparently endless variety of geological surface features. The diversity increases when we survey the other planets. But on closer examination, geologists have found that almost all the features observed result from just four major **geological processes** that affect planetary surfaces:

- **Impact cratering**: the excavation of bowl-shaped depressions (*impact craters*) by asteroids or comets striking a planet's surface.

- **Volcanism**: the eruption of molten rock, or *lava*, from a planet's interior onto its surface.

- **Tectonics**: the disruption of a planet's surface by internal stresses.

- **Erosion**: the wearing down or building up of geological features by wind, water, ice, and other phenomena of planetary weather.

Planetary geology once consisted largely of cataloging the number and kinds of geological features found on the planets. The field has advanced remarkably in recent decades, however, and it is now possible to describe how planets work in general and what kinds of geological features we expect to find on different planets. This progress has been made possible by an examination of the geological processes in detail and the determination of what factors affect and control them.

Impact Cratering

Impact cratering occurs when a leftover planetesimal (such as a comet or an asteroid) crashes into the surface of a terrestrial world. Impacts can have devastating effects on planetary surfaces (Figure 8.5). Impactors typically hit planets at speeds of between 30,000 and 250,000 km/hr (10–70 km/s) and thus pack enough energy to vaporize solid rock and excavate a **crater**. Craters are generally circular because the impact blasts out material in all directions, no matter which direction the impactor came from.

Lunar maria are huge impact basins that were flooded by lava. Only a few small craters appear on the maria.

Lunar highlands are ancient and heavily cratered.

a. The Moon's surface, as seen in this Apollo photograph, shows both heavily and lightly cratered areas.

b This map of the entire lunar surface shows the differences between the lightly cratered maria (dark areas) and the heavily cratered highlands. At the lower left is a multi-ring basin on the far side of the moon known as Mare Orientale. The mare in the center and right portions of the map make the "man-in-the-moon" face we see from Earth.

FIGURE 8.6 Geology of the Moon.

Debris from the blast shoots high into the atmosphere and then rains down over an area much larger than that of the impact crater itself. In large impacts, some of the atmosphere may be blasted away into space, and some of the rocky ejecta may completely escape from the planet.

Craters come in all sizes and a few different shapes, but most are small and bowl-shaped. Small craters far outnumber large ones because there are far more small objects orbiting the Sun than large ones. When the largest (and rarest) impactors strike a planet, they form **impact basins**. The *lunar maria*, easily visible through binoculars, are filled impact basins up to 1,100 kilometers across (Figure 8.6). These large impacts violently fractured the Moon's lithosphere, making cracks through which lava escaped to flood the impact basins. When the lava solidified, it left a smooth surface in the region of the lava flood.

At the small extreme, the most common impactors are sand-size particles called *micrometeorites.* Such tiny particles burn up as meteors in the atmospheres of Venus, Earth, and Mars. But on worlds that lack a significant atmosphere, such as Mercury and the Moon, the countless impacts of micrometeorites gradually pulverize the surface rock to create a layer of powdery "soil." On the Moon, the Apollo astronauts and their rovers left marks in this powdery surface (Figure 8.7). Because of the lack of

wind and rain, these footprints and tire tracks will last millions of years, but they will eventually be erased by micrometeorite impacts.

Of the four geological processes, impact cratering is the most universal. Craters formed on all the terrestrial worlds, regardless of their size or location in the solar system. The "rain of rock and ice" that ended around 3.8 billion years ago saturated (completely covered) all solid surfaces with impact craters. We can therefore use the present-day abundance of craters on a planet's surface to learn about its subsequent geological history. A surface region that is still saturated with craters, such as the *lunar highlands,* must have remained essentially undisturbed. (See Figure 8.6.) In contrast, in regions like the *lunar maria* that now have few craters, the original craters must have been somehow "erased." The flood of lava that formed the lunar maria covered any craters that had formed within the impact basin, and the few craters that exist today on the maria must have formed from impacts occurring after the lava flows solidified. These craters tell us that the impact rate since the end of heavy bombardment has been quite small: The lunar maria have only 3% as many craters as the lunar highlands, but radioactive dating of moon rocks shows that the maria are still 3–3.5 billion years old.

FIGURE 8.7 Reminders of the Apollo missions to the Moon, including this footprint, will last for millions of years.

FIGURE 8.8 Eruption of an active volcano on the flanks of Kilauea on the Big Island in Hawaii.

TIME OUT TO THINK *Earth must also have been saturated with impact craters early in its history, but we see relatively few impact craters on Earth today. What processes erase impact craters on Earth?*

Volcanism

The second major geological process is volcanism. We find evidence of volcanoes and lava flows on all the terrestrial worlds, as well as on a few moons of the outer solar system. Volcanoes erupt when underground lava finds a path through the lithosphere to the surface (Figure 8.8). Molten rock is generally less dense than solid rock, so it has a natural tendency to rise. The lava source may be squeezed by tectonic forces that drive the magma upward under pressure. Any trapped gases expand as the molten rock rises, further pushing lava upward.

The appearance of volcanic features depends on the consistency of the lava that erupts onto the surface ("runny" vs. "sticky"). The common rock type known as **basalt** makes relatively runny lava. The runniest basalt lavas flow far and flatten out before solidifying, creating vast volcanic plains such as the lunar maria (Figure 8.9a). Somewhat less runny basalt lavas solidify before they can completely spread out, resulting in **shield volcanoes** (so named because they are shield-shaped). Shield volcanoes can be very tall, but they are not very steep; most have slopes of only 5°–10° (Figure 8.9b). The mountains of the Hawaiian Islands are shield volcanoes; measured from the ocean floor to their summits, the Hawaiian mountains are the tallest (and widest) on Earth. Tall, steep volcanoes such as Mount St. Hel-

ens are made from very "sticky" lavas that can't flow very far before solidifying (Figure 8.9c).

Volcanism occurs on planets with high internal temperatures and thin lithospheres, where molten rock lies relatively close to the surface. While all the terrestrial worlds had relatively thin lithospheres early in their histories, the smaller worlds have cooled off and their lithospheres have thickened. Only the largest terrestrial planets (Earth and Venus) have warm enough interiors and thin enough lithospheres for ongoing volcanic activity. On these worlds, volcanoes have obliterated most of the ancient craters that once covered the surface.

Volcanism plays an additional key role in planetary evolution through **outgassing**, the release of gases from a planet's interior. Most of the ingredients that make up today's planetary atmospheres were locked in the interior after formation but are released

FIGURE 8.9 Volcanoes produce different types of features depending primarily on the consistency of the lava erupted.

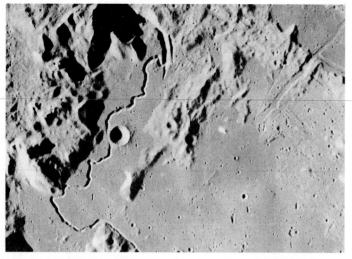

a "Runny" lava makes flat lava plains. (Lava plains [maria] on the Moon)

b Intermediate lava makes shallow-sloped shield volcanoes. (Olympus Mons [Mars])

c "Sticky" lava makes steep-sloped stratovolcanoes. (Mount St. Helens.)

when lava reaches the surface. The gases that help drive volcanic eruptions accumulate to form the atmosphere (and even oceans), as we'll discuss further later in this chapter.

Tectonics

Tectonics, our third geological process, refers to the processes by which the internal forces and stresses acting on the lithosphere create surface features. Tectonic features take a wide variety of forms (Figure 8.10). Mountains may rise where the crust is compressed, as in the case of the Appalachian Mountains of the eastern United States. Huge valleys and cliffs may form where the crust is pulled apart; examples include the Guinevere Plains on Venus and New Mexico's Rio Grande Valley. Other forms of internal stress come from plumes of hot mantle material that push up on the lithosphere or from the expansion or contraction that comes with the warming or cooling of a planet.

Like volcanoes, tectonic features are more common on larger worlds that remain hot inside, and the two kinds of features are often found together. The thick lithospheres of smaller worlds are generally too strong to allow widespread deformation.

Erosion

The last of the four major geological processes is erosion, which encompasses a variety of processes connected by a single theme: the breakdown and transport of rocks by an atmosphere. Wind, rain, rivers, flash floods, and glaciers are just a few examples of processes that contribute to erosion on Earth. Erosion not only breaks down existing geological features (wearing down mountains and forming gullies, riverbeds, and deep valleys), but also builds new ones (such as sand dunes, river deltas, and lake bed deposits). If enough material is deposited over time, the layers can compact to form **sedimentary rocks**.

Erosion is significant only on planets with substantial atmospheres, which means the larger planets. Larger planets are better able to generate atmospheres by outgassing; their stronger gravity also tends to prevent their atmospheres from escaping to space. In general, a thick atmosphere causes more erosion than a thin atmosphere, but a planet's rotation rate is also important. Slowly rotating planets have correspondingly slow winds and therefore weak or nonexistent wind erosion. Surface temperature also matters: Under the right conditions, some atmospheric ingredients can form rain and snow, which are very effective at eroding rock. We will expand on all these issues in the next section.

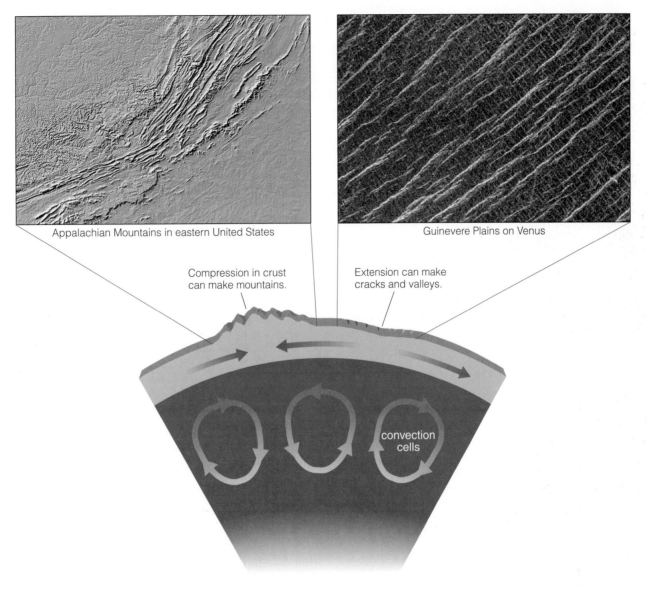

Appalachian Mountains in eastern United States

Guinevere Plains on Venus

Compression in crust can make mountains.

Extension can make cracks and valleys.

convection cells

FIGURE 8.10 Tectonic forces can produce a wide variety of features. Mountains and fractured plains are among the most common.

8.4 Planetary Atmospheres

Planetary atmospheres play a controlling role in climate and planetary habitability. We have also seen how a planet's geology and atmosphere can be closely intertwined, as in the cases of volcanism and erosion. In the pages that follow, we'll examine what atmospheres are like, how they behave, and how they change over time.

All the planets of our solar system have atmospheres to varying degrees, as do several of the solar system's larger moons. The jovian planets are essentially atmosphere throughout—they are largely made of gaseous material. In contrast, the atmospheres of solid bodies—terrestrial worlds, jovian moons with atmospheres, and Pluto—are relatively thin layers of gas that contain only a minuscule fraction of a world's mass. Although we will concentrate on understanding the atmospheres of the terrestrial worlds, similar principles apply to the atmospheres of all the planets and satellites.

Comparing Planetary Atmospheres

The terrestrial atmospheres are even more varied than the terrestrial geologies (Table 8.1). The Moon and Mercury have very little atmosphere at all—you would seem to be facing the blackness of space even when standing on their surfaces. If you could condense the entire atmosphere of the Moon into solid form, you would have so little material that you could

Table 8.1 Atmospheres of the Terrestrial Worlds

World	Composition	Pressure Relative to Earth	Winds, Weather Patterns	Clouds, Hazes
Mercury	helium, sodium, oxygen	10^{-14}	None: too little atmosphere	None
Venus	96% CO_2 3.5% N_2	90	Slow winds, no violent storms, acid rain	Sulfuric acid clouds
Earth	77% N_2 21% O_2 1% argon H_2O (variable)	1	Winds, hurricanes	H_2O clouds, pollution
Moon	helium, sodium, argon	10^{-14}	None: too little atmosphere	None
Mars	95% CO_2 2.7% N_2 1.6% argon	0.007	Winds, dust storms	H_2O and CO_2 clouds, dust

store it in a basement. At the other extreme, Venus is shrouded by a thick, cloudy atmosphere made of carbon dioxide. The pressure is 90 times greater than that on Earth, with a searing temperature hotter than your oven set on high. The atmosphere of Mars is also made mostly of carbon dioxide, but much less of it than on Venus. The result is very thin air with a pressure so low that your body tissues would bulge painfully if you stood on the surface without wearing a full space suit. In fact, the pressure and temperature are both so low that liquid water would rapidly disappear, with some evaporating and some freezing into ice. Thus, the fact that we see dried-up riverbeds on Mars tells us that its atmosphere must have been very different in the past. And Earth's unusual atmosphere of nitrogen and oxygen is the only one under which human life is possible, at least without the creation of an artificial environment.

Why did the atmospheres of the planets turn out to be so different, and what makes Earth's atmosphere so special? The best way to answer these questions is to examine and compare planetary atmospheres. Each atmosphere is unique, but, as we saw with geology in the previous sections, similar properties and processes determine the characteristics of all atmospheres. First we will explore the general behavior of atmospheres, and then we will learn how and why the terrestrial atmospheres came to differ so profoundly.

Atmospheric Structure

If you've ever driven up a mountain, you know that the atmosphere changes with altitude. Your popping ears and shortness of breath are proof that the pressure is lower as you go higher, and the chill tells you that temperature also decreases with altitude. Let's investigate the origin of atmospheric structure—that is, the variation in pressure and temperature with altitude.

Planetary atmospheres exist in a perpetual balance between the downward weight of their gases and the upward push of their gas pressure. The higher you go in an atmosphere, the less the weight of the gas above you. Thus, the pressure must also become less as you go upward, which explains why the pressure decreases as you climb a mountain or ascend in an airplane. You can visualize what happens by imagining the atmosphere as a very big stack of pillows. The pillows at the bottom are very compressed because of the weight of all the pillows above. As you go upward, the pillows are less and less compressed because less weight lies on top of them. But atmospheres do not have clear upper boundaries; instead, they gradually fade away with increasing altitude (Figure 8.11).

The temperature variations in an atmosphere arise from the way different wavelengths of sunlight are absorbed in different layers of the atmosphere. Figure 8.12 shows the structure of a "generic" planetary atmosphere; it shows how and why the temperature varies with altitude. We will now investigate the processes that create these basic atmospheric layers in detail, from the ground up.

Common Misconceptions: Higher Altitudes Are Always Colder

Many people think that the low temperatures in the mountains are just the result of lower pressures, but Figure 8.12 shows that it's not that simple. The higher temperatures near sea level on Earth are a result of the greenhouse effect trapping more heat at lower altitudes. If Earth had no greenhouse gases, mountaintops wouldn't be so cold—or, more accurately, sea level wouldn't be so warm.

a Most of Earth's atmosphere is confined to a layer less than 100 km thick.

FIGURE 8.11 Two perspectives on Earth's atmosphere.

b The atmosphere does not end suddenly; some gas is present even at altitudes of hundreds of kilometers, causing the glow visible around the tail of the Space Shuttle as it orbits.

FIGURE 8.12 The structure of a generic planetary atmosphere: Solar X rays are absorbed in the thermosphere, ultraviolet light is absorbed in the stratosphere, and visible light reaches the ground. Planets that lack ultraviolet-absorbing molecules will lack a stratosphere, and planets with very little gas will have only an exosphere.

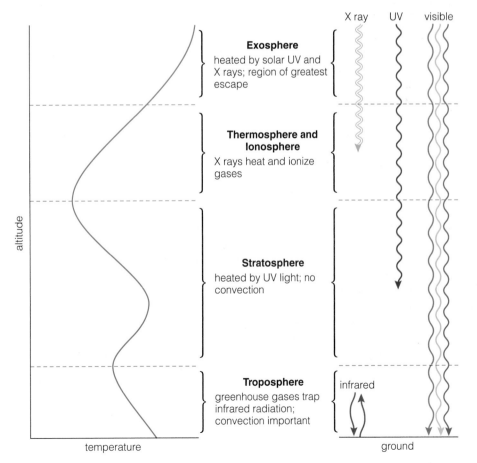

Visible Light: Warming the Surface and Coloring the Sky

The Sun emits most of its energy in the form of visible light, and atmospheric gases are generally transparent in this wavelength range. Thus, it's primarily visible light that reaches a planet's surface and warms it. In the absence of an atmosphere (or at least in the absence of any greenhouse effect), the surface temperature of any planet is determined by three basic properties: its distance from the Sun (which determines the intensity of sunlight), the fraction of sunlight absorbed by the surface (i.e., not reflected away), and the planet's rotation rate (which determines whether the planet is warmed evenly or bakes on one side while freezing on the other). The planet absorbs a certain amount of sunlight and reaches the temperature at which it emits exactly the same amount of energy in the form of thermal radiation. Most planetary surfaces are in the temperature range at which they emit their thermal radiation in the infrared [Section 6.3].

Although most visible light passes directly through the atmosphere, some is *scattered* (redirected) by molecules in the atmosphere. This scattering of sunlight makes the sky bright and prevents us from seeing stars in the daytime. If our atmosphere did not scatter light, you'd be able to block out the light of the Sun simply by holding your hand in front of it, and then you'd have no trouble seeing other stars.

The scattering of light by the atmosphere also explains why Earth's sky is blue. It turns out that molecules scatter blue light much more effectively than red light. Thus, although the Sun illuminates the atmosphere with all colors of light, effectively only the blue light gets scattered. When the Sun is overhead, this scattered blue light reaches your eyes from all directions, which explains why the sky appears blue (Figure 8.13). At sunset (or sunrise), the

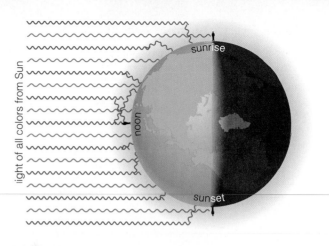

FIGURE 8.13 Atmospheric gases scatter blue light more than they scatter red light. During most of the day, you therefore see blue photons coming from most directions in the sky, making the sky look blue. But only the red photons reach your eyes at sunrise or sunset, when the light must travel a longer path through the atmosphere to reach you.

sunlight must pass through a greater amount of atmosphere on its way to you. Thus, most of the blue light is scattered away from you altogether (making someone else's sky blue), leaving only red light to color your sunset.

Infrared Light, the Greenhouse Effect, and the Troposphere

Figure 8.14 shows how the **greenhouse effect** works. Water vapor, carbon dioxide, and other *greenhouse gases* in the atmosphere absorb some of the infrared radiation emitted upward from the planet's surface. These gases therefore warm up and emit infrared thermal radiation themselves—but in all directions. Some of this radiation is directed back down toward the surface, making the surface warmer than it would be from absorbing visible sunlight alone. The more greenhouse gases present, the greater the degree of greenhouse warming. (The name *greenhouse effect* comes from botanical greenhouses, usually built mostly of glass, which keep the air inside them warmer than it would be otherwise. But it turns out that greenhouses trap heat simply by not letting hot air rise—a very different way of trapping heat than that taking place in planetary atmospheres.)

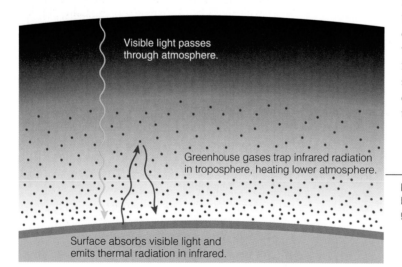

Visible light passes through atmosphere.

Greenhouse gases trap infrared radiation in troposphere, heating lower atmosphere.

Surface absorbs visible light and emits thermal radiation in infrared.

FIGURE 8.14 The greenhouse effect: The troposphere becomes warmer than it would be if it had no greenhouse gases.

Table 8.2 Temperatures of the Terrestrial Worlds

World	Distance to Sun (AU)	Reflectivity[†] (0 = black, 1 = white)	Length of Day (Earth Days)	"No Greenhouse" Temperature*	Observed Average Surface Temperature
Mercury	0.39	0.11	176	440 K	700 K (day), 100 K (night)
Venus	0.72	0.72	117	230 K	740 K
Earth	1.00	0.36	1	250 K	288 K
Moon	1.00	0.07	28	273 K	400 K (day), 100 K (night)
Mars	1.52	0.25	≈1	218 K	223 K

*Assumes rapid rotation, in which case day and night temperatures would be the same.
[†]Fraction of light reflected by planets.

FIGURE 8.15 Earth's winds and storms are driven in part by the planet's relatively rapid rotation.

TIME OUT TO THINK *Clouds on Earth are made of water, and H_2O is a very effective greenhouse gas—even when the water vapor condenses into droplets in clouds. Use this fact to explain why clear nights tend to be colder than cloudy nights. (Hint: Sunlight warms the surface only in the daytime, but the surface radiates its heat away both day and night.)*

Table 8.2 lists the temperatures of the terrestrial worlds with and without the greenhouse effect. The greenhouse effect is responsible for warming Mars (where it adds 5 K), Earth (adding 40 K), and Venus (adding nearly 500 K!). Ironically, the greenhouse effect on Earth is almost entirely due to gases that make up less than 2% of our atmosphere: mainly water vapor, carbon dioxide, and methane. The major constituents of Earth's atmosphere, N_2 and O_2, have no effect on infrared light and do not contribute to the greenhouse effect. If they did, Earth might be too warm for life.

Most of what we call weather occurs in the troposphere. The fact that the greenhouse effect warms air near the ground drives *convection* in the troposphere. The warm air near the ground expands and rises, while cooler air near the top of the troposphere contracts and falls. Convection continually churns the air in the troposphere and raises air laden with water vapor to cooler altitudes, where it condenses to form clouds and rain. Winds in the troposphere are driven in part by the planet's rotation; fast winds and hurricane-like storms are found only on planets with rapid enough rotation (Figure 8.15).

Common Misconceptions:
The Greenhouse Effect Is Bad

The greenhouse effect is often in the news, usually in discussions about environmental problems. But the greenhouse effect itself is not a bad thing. In fact, we could not exist without it: Without the greenhouse effect, Earth's average temperature would be 250 K ($-23°C$, $-9°F$). Thus, the greenhouse effect is the only reason why our planet is not frozen over. Why, then, is the greenhouse effect discussed as an environmental problem? It is because human activity is adding more greenhouse gases to the atmosphere—which might change the Earth's climate [Section 11.5]. After all, while the greenhouse effect makes the Earth livable, it is also responsible for the searing 740 K temperature of Venus—proving that it's possible to have too much of a good thing.

Ultraviolet Light and the Stratosphere

As we move upward through the stratosphere, the primary source of atmospheric heating is absorption of ultraviolet light from the Sun. This heating is stronger at higher altitudes because the ultraviolet light gets absorbed before it reaches lower altitudes. As a result, the stratosphere gets warmer with increasing altitude—the opposite of the situation in the troposphere. Convection therefore cannot occur in the stratosphere: Heat cannot rise if the air is even hotter higher up. The stratosphere gets its name because the lack of convection makes its air relatively stagnant and *stratified* (layered), rather like a sitting jar of oil and water. The lack of convection also means that the stratosphere essentially has no weather and no rain.

Note that a planet can have a stratosphere *only* if its atmosphere contains molecules that are particularly good at absorbing ultraviolet photons. In fact, Earth is the only terrestrial planet with such an ultraviolet-absorbing layer and hence the only terrestrial planet with a stratosphere—at least in our solar system. We'll discuss why Earth is unique in this way in Chapter 11.

X Rays and the Thermosphere and Exosphere

A planet's upper atmosphere is strongly heated by solar X rays in the daytime, usually to much higher temperatures than the gases at the surface. That is why the upper region of the atmosphere is called the *thermo*sphere. Despite its high temperatures, the gas in the thermosphere would not feel hot to your skin because its density and pressure are so low.

If a planet has very little atmosphere, it will consist of just an exosphere. The exosphere marks the fuzzy boundary between the atmosphere and space. (The prefix *exo* means "outermost" or "outside.") The high temperatures in the exosphere give the individual atoms or molecules high speeds, allowing some of them to escape from the planet. The Space Shuttle and many satellites actually orbit the Earth within its exosphere, and the small amount of gas in the exosphere can exert a bit of atmospheric drag. (See Figure 8.11b.) That is why satellites in low-Earth orbit eventually spiral deeper into the atmosphere and burn up.

Magnetospheres and the Solar Wind

The Sun emits one more type of energy that we haven't yet considered: the low-density breeze of charged particles that we call the solar wind [Section 7.4]. Although the solar wind does not significantly affect atmospheric structure, it can have other important influences. Among the terrestrial planets, only Earth has a strong enough magnetic field to divert the charged particles of the solar wind. Earth's strong magnetic field creates a **magnetosphere** that acts like a protective bubble surrounding the planet (Figure 8.16a). Charged particles cannot easily pass through a magnetic field, so a magnetosphere diverts most of the solar wind around a planet. In contrast, solar wind particles can impact the exospheres of Venus and Mars—and the surfaces of Mercury and the Moon—because they lack strong magnetic fields and magnetospheres.

Some solar wind particles manage to infiltrate Earth's magnetosphere at its most vulnerable points, near the magnetic poles. Once inside, they have a difficult time escaping. The ions and electrons accumulate to make **charged particle belts** encircling the planet. (The charged particle belts around the Earth are called the *Van Allen belts* after their discoverer.) The high energies of the particles in charged particle belts can be very hazardous to spacecraft and astronauts passing through them.

Particles trapped in magnetospheres also cause the beautiful spectacle of light called *auroras*. If a trapped charged particle has enough energy, it can follow the magnetic field all the way down to a planet's atmosphere in the north and south polar regions. There the charged particles collide with atmospheric atoms and molecules, causing them to radiate and produce auroras (Figure 8.16b). On Earth, auroras are best viewed from Canada, Alaska, and Russia in the Northern Hemisphere (the *aurora borealis,* or northern lights) and from Australia, Chile, and Argentina in the Southern Hemisphere (the *aurora australis,* or southern lights).

8.5 Atmospheric Origins and Evolution

Today's headlines are full of stories about the human effects on Earth's atmosphere, such as ozone depletion, global warming, and air pollution. But the atmospheres of Earth and the other terrestrial planets have seen much larger changes in their billions of years of history. What causes such large changes? To answer this question, we must examine both how a planet gets its atmospheric gases and how it can lose them. As we will see, atmospheric changes have been just as important as surface changes, and the two are usually connected.

Where did the terrestrial worlds get their atmospheres? They were too small to capture significant amounts of gas from the solar nebula before the solar wind cleared it away [Section 7.4]. Any gas that the terrestrial worlds did capture from the solar nebula consisted primarily of hydrogen and helium (just like

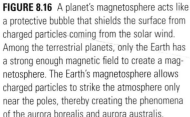

a This diagram shows how Earth's magnetosphere interacts with the solar wind.

Most solar wind particles are deflected around planets with strong magnetic fields.

N

charged particle belts

Earth

aurora

S

Some solar wind particles infiltrate magnetosphere near poles.

b Earth's aurora as seen from the Space Shuttle.

FIGURE 8.16 A planet's magnetosphere acts like a protective bubble that shields the surface from charged particles coming from the solar wind. Among the terrestrial planets, only the Earth has a strong enough magnetic field to create a magnetosphere. The Earth's magnetosphere allows charged particles to strike the atmosphere only near the poles, thereby creating the phenomena of the aurora borealis and aurora australis.

the solar nebula as a whole). But these light gases escape easily from the terrestrial worlds and by now are long gone. The atmospheres of the terrestrial worlds therefore must have formed after the worlds themselves.

The most important source of atmospheric gases is **outgassing**, the release of gases by volcanic eruptions. Recall that these gases were originally brought to the terrestrial planets by comets and asteroids during accretion and the early bombardment that ended 3.8 billion years ago [Section 7.5]. The high pressures inside the planets trapped these gases in rocks in much the same way that a pressurized bottle traps the bubbles of carbonated beverages. When the rocks melt and erupt onto the surface as lava, the release of pressure expels the gases. The gases may be released either with the lava itself or from nearby vents (Figure 8.17). The most common gases ex-

pelled are water (H_2O), carbon dioxide (CO_2), nitrogen (N_2), and sulfur-bearing gases (H_2S or SO_2). Because outgassing depends on volcanism, it is most important on larger planets which retain their internal heat longer. Outgassing supplied most of the gas that became the atmospheres of Venus, Earth, and Mars. The Moon and Mercury, however, are not completely airless: Bombardment by the solar wind, micrometeorites, and even larger impactors supplies a small amount of material to their atmospheres.

An atmosphere can lose gas in several important ways: *thermal escape, bombardment, atmospheric cratering, condensation,* and *chemical reactions* (Figure 8.18). The terrestrial planets lost any original hydrogen and helium they once captured from the solar nebula through the process of **thermal escape**, in which an atom or a molecule in the exosphere

FIGURE 8.17 Volcanoes give off H_2O, CO_2, N_2, and sulfur-bearing gases. Some gas is given off from the lava itself, and some leaks out from nearby gas vents. (Volcanoes National Park, Hawaii.)

moves fast enough to escape the pull of gravity. Three factors determine whether an atmospheric gas can be lost to space by thermal escape:

1. The planet's *escape velocity,* or the speed at which a particle must travel to escape the pull of gravity—assuming it is traveling in the right direction and doesn't collide with anything else. The more massive the planet, the higher its escape velocity [Section 5.5].

2. The temperature. Individual particles in a gas move with a wide range of random speeds [Sec-

tion 4.2], but the higher the temperature, the faster their *average* speed.

3. The mass of the gas particles. At a particular temperature, lighter particles (such as H or He) move around at faster average speeds than heavier ones (such as O_2 or CO_2) and therefore are more likely to achieve escape velocity.

Escape can be driven from the outside by impacts, both large and small. *Bombardment* by the particles in the solar wind can lead to slow and steady atmospheric loss, ejecting even atoms that are too heavy for thermal escape. In general, planets with protective magnetic fields—such as Earth—are less susceptible to this type of atmospheric loss. Larger impacts can also cause major atmospheric escape. The same impacts that create craters on the surface may blast away a significant amount of the atmosphere as well in a process sometimes called *atmospheric cratering.* For the most part, atmospheric cratering only happened early in the solar system's history. We expect it to have been most important on smaller worlds, where gravity's hold on the atmosphere is weaker. Mars may have lost a significant amount of atmospheric gas through atmospheric cratering.

Two other atmospheric loss processes allow atmospheric gas to be absorbed back into the solid planet. Gases may simply *condense* into liquid or solid form if a planet cools down, especially in the cold polar regions. Finally, gases may become locked up in the surface through *chemical reactions* with rocks or liquids—another example of the strong links between planetary atmospheres and surfaces. We encounter examples of these five atmospheric loss processes as we tour the terrestrial worlds in the next section.

FIGURE 8.18 The five major processes by which atmospheres lose gas.

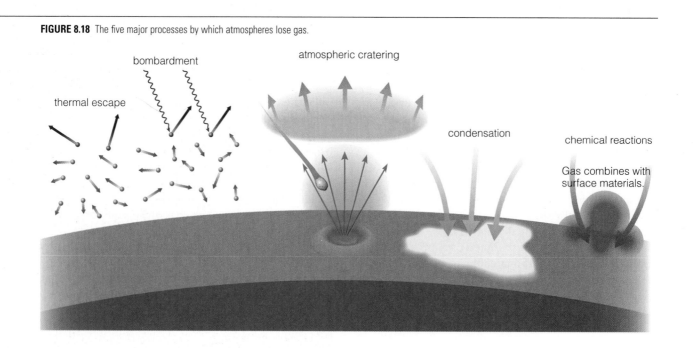

8.6 A Tour of the Terrestrial Worlds

Now that we've examined the processes that shape the geology and atmospheres of the terrestrial worlds, we're ready to return to the issue that lies at the heart of this chapter: why the terrestrial worlds ended up so different from one another. We'll take a brief "tour" of the terrestrial worlds. We could organize the tour by distance from the Sun, by density, or even by alphabetical order. But we'll choose the planetary property that has the strongest effect on geology: size, which controls a planet's internal heat. We'll start with our Moon, the smallest terrestrial world.

The Moon (1,738-km radius, 1.0 AU from Sun)

On a clear night, you can see much of the Moon's global geological history with your naked eye. The Moon is unique in this respect; other planets are too far away for us to see any surface details, and Earth is so close that we see only the local geological history. Twelve Apollo astronauts visited the lunar surface between 1969 and 1972, taking photographs, making measurements, and collecting rocks. Thanks to these visits and more recent observations of the Moon from other spacecraft, we know more about the Moon's geology than that of any other object, with the possible exception of the Earth.

The Moon's composition differs somewhat from the Earth's. According to the leading theory, our Moon formed early in the history of our solar system when a giant impact blasted away some of Earth's rocky outer layers [Section 7.5]. The impact did not dredge up metallic core material, which explains why the Moon has a lower metal content than Earth and a smaller metal core. In addition, lunar rocks collected by astronauts contain a much smaller proportion of volatiles (such as water) than Earth rocks, presumably because heat from the giant impact evaporated the volatiles and allowed them to escape into space before the Moon formed from the debris.

As the Moon accreted following the giant impact, the heat of accretion melted its outer layers. The lowest-density molten rock rose upward through the process of differentiation, forming an ocean of lava. This ocean cooled and solidified during the heavy bombardment that took place early in the history of the solar system, leaving the Moon's surface crowded with craters. We still see this ancient, heavily cratered landscape in the lunar highlands (Figure 8.19a). Lunar samples confirm that the highlands are composed of low-density rocks. As the bombardment trailed off, around 3.8 billion years ago, a few larger impactors struck the surface and formed impact basins.

Even as the heat of accretion leaked away, radioactive decay kept the Moon's interior molten long enough for volcanism to reshape parts of its surface. Between about 3 and 4 billion years ago, molten rock welled up through cracks in the deepest impact basins, forming the lunar maria (Figure 8.6). Long, winding channels (Figure 8.19b, a close-up of Figure 8.9a) must once have been rivers of molten rock

FIGURE 8.19 Impact cratering and volcanism are the most important geological processes on the Moon.

a Astronaut explores a small crater.

b Site of an ancient lava river (as seen in Figure 8.9a).

that helped fill the lunar maria. Because the lava that filled the maria rose up from the Moon's mantle, the maria contain dense, iron-rich rock that is darker in color than the rock of the lunar highlands. The contrasts between the light-colored rock of the highlands and the dark rock of the maria make the "man-in-the-Moon" pattern that some people imagine when they look at the full moon.

Almost all the geological features of the Moon were formed by impacts and volcanism. Tectonic activity was inhibited by the Moon's thick lithosphere, and erosion never occurred due to the lack of a significant atmosphere. Over time, the Moon's small size allowed its interior to cool, thickening the lithosphere. Today, the Moon's lithosphere probably extends to a depth of 1,000 km, making it far too thick to allow further volcanic or tectonic activity. The Moon has probably been in this geologically "dead" state for 3 billion years.

Although geological activity and outgassing ceased billions of years ago, the Moon still holds some surprises. The *Lunar Prospector* orbiter recently discovered evidence of ice locked up in the Moon's polar regions. The water probably came from comet impacts that momentarily created a thin lunar atmosphere. The water molecules from these impacts bounced around the surface at random until they condensed into ice in the polar craters. The bottoms of these craters lie in nearly perpetual shadow, keeping them so cold that the water remains perpetually frozen. A thin atmosphere supplied by bombardment also surrounds the Moon, but it is continually lost by thermal escape and the impact of the solar wind.

The Moon is of prime interest to those hoping to build colonies in space. The *Lunar Prospector* arrived at the Moon in 1998 and began mapping the surface in search of promising locations for a permanent human base. Although no concrete plans are in place, several nations are exploring the possibility of building a human outpost on the Moon within the next couple of decades. The possibility of a water supply in polar craters may become an important issue.

Mercury (2,440-km radius, 0.39 AU from Sun)

Mercury (Figure 8.20) is the least studied of the terrestrial worlds. Its proximity to the Sun makes it difficult to study through telescopes, and it has been visited by only one spacecraft: *Mariner 10,* which collected data during three rapid flybys of Mercury in 1974–75. *Mariner 10* obtained images of only one hemisphere of Mercury (Figure 8.20b). The influence of Mercury's gravity on *Mariner 10*'s orbit allowed planetary scientists to determine Mercury's mass and high average density. Based on its high density, Mercury must be about 61% iron by mass, with a core

that extends to perhaps 75% of its radius. Mercury's high metal content is due to its formation close to the Sun, possibly enhanced by a giant impact that blasted away its outer, rocky layers [Section 7.5].

Mercury has craters almost everywhere, indicating an ancient surface (Figure 8.20c). A huge impact basin called the *Caloris Basin* (Figure 8.20a) covers a large portion of one hemisphere. Smooth patches between many of the craters tell us that volcanic lava once flowed and covered up many of the smallest craters. Mercury has no vast volcanic maria, but smaller lava plains are found almost everywhere, suggesting that Mercury went through at least as much volcanic activity as the Moon. Much like the Moon, Mercury passed through phases of accretion, differentiation, and heavy cratering. Radioactive heat accumulated enough to melt the interior again, leading to the volcanism that created the lava plains.

Tectonic processes played a more significant role on Mercury than on the Moon, as evidenced by tremendous cliffs. The cliff shown in Figure 8.20c is several hundred kilometers long and up to 3 kilometers high in places. The crater in the center of the image apparently crumpled during the formation of the cliff. This cliff, and many others on Mercury, probably formed when tectonic forces compressed the crust. However, nowhere on Mercury do we find evidence of extension (stretching) to match this crustal compression. Can it be that the whole planet simply shrank? Apparently so. Early in its history, Mercury gained much more internal heat from accretion and differentiation than did the Moon, owing to its larger proportion of iron and its stronger gravity. Then, as the core cooled, it contracted by perhaps as much as 20 kilometers in radius. This contraction crumpled and compressed the crust, forming the many compressional cliffs. The contraction probably also closed off volcanic vents, ending Mercury's period of volcanism.

Mercury lost its internal heat relatively quickly because of its small size. Its days of volcanism and tectonic shrinking probably ended within its first billion years. Mercury, like the Moon, has been geologically dead for most of its existence. But recent radar observations suggest that Mercury, like the Moon, may have water ice in its polar craters, probably for the same reason as on the Moon. Mercury also has a tenuous atmosphere created by bombardment, although its exact origin remains unclear.

Mercury may soon lose its unenviable status of being least explored. Another of NASA's "better, cheaper, faster" Discovery missions may be launched to Mercury in 2004 to map its entire surface, explore its unexpected magnetic fields, and study its atmosphere.

a Edge of Caloris Basin

c Close-up view shows small lava plains and cliffs (indicated by arrow) hundreds of kilometers long.

b Mercury

FIGURE 8.20 Mercury, like the Moon, is a heavily cratered object with evidence of volcanism.

Mars (3,397-km radius, 1.52 AU from Sun)

Mars has captured the human imagination, from the first astronomers to science fiction writers to today's scientists and aerospace engineers. Much of the appeal comes from the fact that it has long seemed the most promising place to look for life beyond Earth. Early telescopic observations of Mars revealed several uncanny resemblances to Earth. A Martian day is just over 24 hours, and the Martian rotation axis is tilted about the same amount as Earth's. Mars also has polar caps, which we now know to be composed primarily of frozen carbon dioxide, with smaller amounts of water ice. Telescopic observations also showed seasonal variations in surface coloration over the course of the Martian year (about 1.9 Earth years). All these discoveries led to the perception that Mars and Earth were at least cousins, if not twins. By the early 1900s, many astronomers—as well as the public—envisioned Mars as nearly Earth-like, possessing water, vegetation that changed with the seasons, and possibly intelligent life.

In 1877, Italian astronomer Giovanni Schiaparelli reported seeing linear features across the surface of Mars through his telescope. He named these features *canali,* the Italian word for "channels." This report inspired the American astronomer Percival Lowell to build an observatory in Flagstaff, Arizona, for the study of Mars. In 1895, Lowell claimed that canali were *canals* being used by intelligent Martians to route water around their planet. He suggested that Mars was falling victim to unfavorable climate changes and that water was a precious resource that had to be managed carefully. Lowell's studies drove rampant speculation about the nature of the Martians, fueling science fiction fantasies. The public mania drowned out the skepticism of astronomers who saw no canals through their telescopes or in their photographs.

The debate about Martian canals and cities was finally put to rest in 1965 with photos of the barren, cratered surface taken during the *Mariner 4* flyby and by many other spacecraft since. Schiaparelli's *canali*

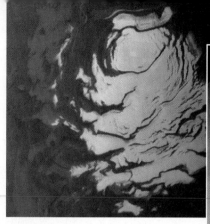

a This photo, taken by the *Viking Orbiter,* shows the ice cap at a Martian pole during the Martian winter.

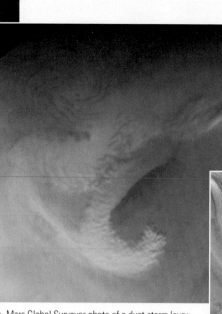

b Mars Global Surveyor photo of a dust storm (curving clouds) brewing over Mars's northern polar cap.

FIGURE 8.21 Polar ice caps and dust storms on Mars.

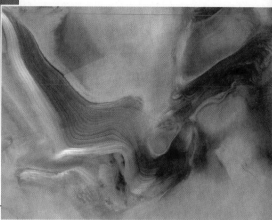

c Close-up of the edge of a polar cap, showing alternating layers of ice and dust. Each layer is a few dozen meters thick and may take thousands of years to form.

turned out to be nothing more than dark dust deposits redistributed by seasonal dust storms (Figure 8.21b). Dust also settles onto the polar caps, making alternating layers of darker dust and brighter frost (Figure 8.21a and c). The dust storms leave Mars with a perpetually dusty sky, giving it a pale pink color. Several other spacecraft studied Mars in the 1960s and 1970s, culminating with the impressive Viking missions in 1976. Each of the two Viking spacecraft deployed an orbiter that mapped the Martian surface from above and landers that returned surface images and regular weather reports for almost 5 years. More than 20 years passed before the next successful missions to Mars, the Mars Pathfinder and Mars Global Surveyor missions in 1997. More missions to Mars are planned for every favorable launch window (roughly every 2 years) in the next decade and beyond.

Mars Geology The Viking and Mars Global Surveyor images show a much wider variety of geological features than seen on the Moon or Mercury. Some areas of the southern hemisphere are heavily cratered (Figure 8.22a), while the northern hemisphere is covered by huge volcanoes surrounded by extensive volcanic plains. Volcanism was expected on Mars—40% larger than Mercury, it should have retained a hot interior much longer.

Mars has a long, deep system of valleys called *Valles Marineris* running along its equator (Figure 8.22b). Named for the Mariner 9 spacecraft that first imaged it, Valles Marineris is as long as the United States is wide and almost four times deeper than the Grand Canyon. No one knows exactly how it formed; in fact, parts of the canyon are completely enclosed by high cliffs on all sides, so neither flowing lava nor water could have been responsible. Nevertheless, the linear "stretch marks" around the valley are evidence of tectonic stresses on a very large scale.

More cracks are visible on the nearby *Tharsis Bulge,* a continent-size region that rises well above the surrounding Martian surface (Figure 8.22c). Tharsis was probably created by a long-lived plume of rising mantle material that bulged the surface upward while stretching and cracking the crust. The plume was also responsible for releasing vast amounts of basaltic lava that built up several gigantic shield volcanoes, including *Olympus Mons,* the largest shield volcano in the solar system (Figure 8.22d). Olympus Mons has roughly the same shallow slope as the volcanic island of Hawaii, but it is three times larger in every dimension. Its base is some 600 kilometers wide, and it stands 26 kilometers high—as wide as Arizona and three times higher than Mount Everest. Planetary geologists estimate that the volcanoes went dormant hundreds of millions of years ago. Geologically speaking, this wasn't so long ago, and it remains possible that the Martian volcanoes will someday come back to life.

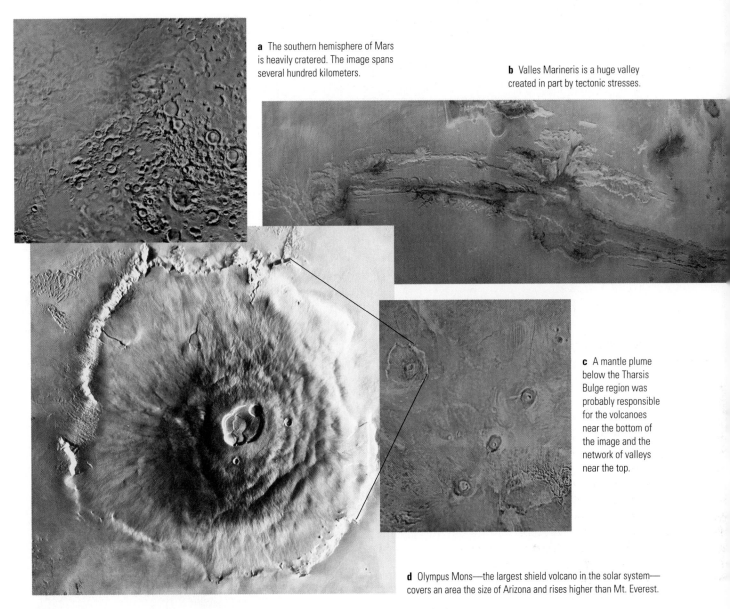

a The southern hemisphere of Mars is heavily cratered. The image spans several hundred kilometers.

b Valles Marineris is a huge valley created in part by tectonic stresses.

c A mantle plume below the Tharsis Bulge region was probably responsible for the volcanoes near the bottom of the image and the network of valleys near the top.

d Olympus Mons—the largest shield volcano in the solar system—covers an area the size of Arizona and rises higher than Mt. Everest.

FIGURE 8.22 These photos of the Martian surface, taken from orbiting spacecraft, show that impact cratering has been an important process on Mars, but volcanism and tectonics have been even more important in shaping the current Martian surface.

Martian Water The Martian robotic explorers have also revealed geological features unlike any seen on the Moon or Mercury—features caused by erosion. Figure 8.23a no doubt reminds you of a dry riverbed on Earth seen from above. The eroded craters visible on some of the oldest terrain suggest that rain fell on the Martian surface billions of years ago. In some places, erosion has even acted below the surface, where underground water and ice have broken up or dissolved the rocks. When the water and ice disappeared, a wide variety of odd pits and troughs were left behind. This process may have contributed to the formation of parts of Valles Marineris.

Underground ice may be the key to one of the biggest floods in the history of the solar system. Figure 8.23b shows a landscape hundreds of kilometers across that was apparently scoured by rushing water. Geologists theorize that heat from a volcano melted great quantities of underground ice, unleashing a catastrophic flood that created winding channels and large sandbars along its way.

The *Mars Pathfinder* spacecraft went to this interesting region to see the flood damage close up. The lander released the *Sojourner* rover, named after Sojourner Truth, an African-American heroine of the Civil War era who traveled the nation advocating

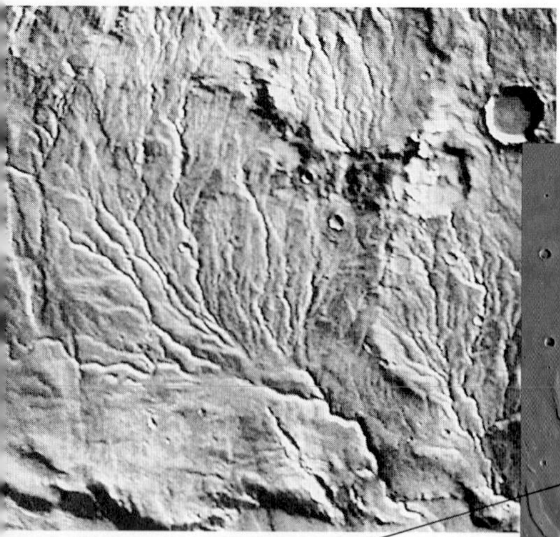

a This Viking photo (from orbit) shows ancient riverbeds that were probably created billions of years ago.

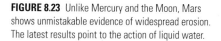

FIGURE 8.23 Unlike Mercury and the Moon, Mars shows unmistakable evidence of widespread erosion. The latest results point to the action of liquid water.

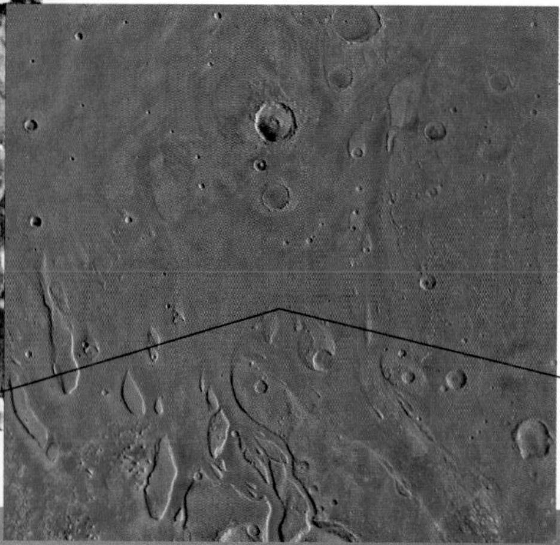

b This photo shows winding channels and sandbars that were probably created by catastrophic floods.

c View from the floodplain (see **b** above) from the *Mars Pathfinder;* the *Pathfinder* landing site is now known as Carl Sagan Memorial Station. The *Sojourner* rover is visible near the large rock.

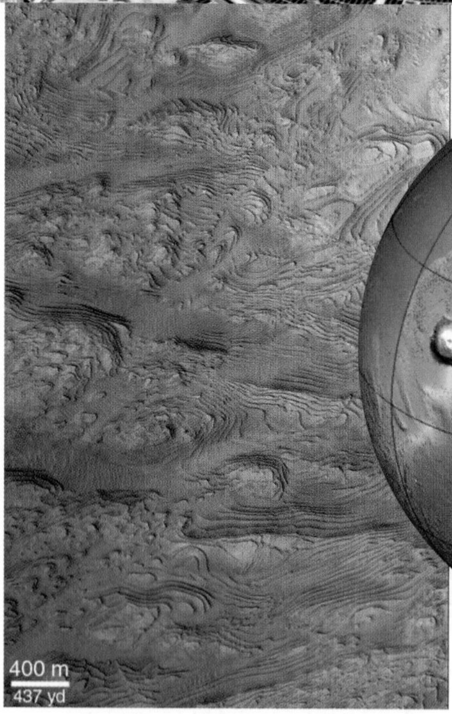

400 m
437 yd

d Close-up view of the floor of an eroded crater. Billions of years ago, the crater was apparently a pond, and layer upon layer of sediment was deposited on the bottom. The water is long gone, and winds have since sculpted the layers into the astonishing patterns captured in this photograph from the *Mars Global Surveyor*.

e Topographic map of Mars with low-lying regions in blue and higher elevations in red, brown, and white. The white patches are tall volcanoes; Valles Marineris cuts through the terrain to their right. Some planetary geologists believe that an ocean once filled the smooth, low-lying northern regions shown in dark blue.

400 m
437 yd

f This photograph from *Mars Global Surveyor* shows gullies on a crater wall; scientists suspect they were formed by water seeping out from the ground during episodic flash floods. The gullies are geologically young, but no one yet knows whether similar gullies may still be forming today.

equal rights for women and blacks. The six-wheeled rover, no larger than a microwave oven, carried cameras and instruments to measure the chemical composition of nearby rocks. Together the lander and rover confirmed the flood hypothesis: Gentle, dry, winding channels were visible in stereo images; rocks of many different types had been jumbled together in the flood; and the departing waters had left rocks stacked against each other in just the same manner that floods do here on Earth (Figure 8.23c).

In the decades since we first learned of past water on Mars, scientists have debated when water last flowed. Recent images from the *Mars Global Surveyor* have only deepened the mystery. Figure 8.23d shows a close-up view of the floor of a crater. The sculpted patterns in this picture were probably created as erosion exposed layer upon layer of sedimentary rock, much like the sedimentary layers visible in Earth's Grand Canyon. This suggests that ancient rains once filled the crater like a pond, with the sedimentary layers built up by material that settled to the bottom. Figure 8.23e, made from the spacecraft's precise topographic measurements, shows evidence of even more widespread water, suggesting that much of Mars's northern hemisphere was once covered by an ocean. Perhaps the most surprising finding is the presence of small gullies on the walls of many craters and valleys (Figure 8.23f). The gullies probably formed when underground water broke out in episodic flash floods, carrying boulders and soil into a broad fan at the bottom of the slope. Moreover, the gullies must be geologically young, because they have not been covered over by ongoing dust storms, nor have small craters formed on their surfaces. Unfortunately, "young" can be quite ambiguous in a geological sense: No one knows if these gullies formed "just" millions of years ago or if some might still be forming today.

Even if water still flows on occasion, it seems clear that Mars was much wetter in the past than it is today. Ironically, Percival Lowell's supposition that Mars was drying up has turned out to be basically correct, even though he did not have real evidence to support it.

The fact that liquid water once flowed on Mars means that its past must have been very different from its current "freeze-dried" state. The surface temperature must have been much warmer, and therefore Mars must have had a much stronger greenhouse effect and much higher atmospheric pressure. Computer simulations show that these requirements can be met by a carbon dioxide atmosphere about 400 times denser (i.e., having greater pressure) than the current Martian atmosphere—or about twice as thick as Earth's atmosphere.

The idea that Mars had a much denser atmosphere in the past is reasonable. Mars has plenty of ancient volcanoes, and calculations show that these volcanoes could have supplied all of the carbon dioxide needed to warm the planet. Even CO_2 ice clouds could have helped hold heat in the Martian troposphere. Moreover, if Martian volcanoes outgas carbon dioxide and water in the same proportions as do volcanoes on Earth, Mars would have had enough water to fill oceans tens or even hundreds of meters deep.

The bigger question is not whether Mars once had a denser atmosphere but where those atmospheric gases are today. Mars must somehow have lost most of its carbon dioxide gas. This loss reduced the strength of the greenhouse effect until the planet essentially froze over. The fate of the lost carbon dioxide is not completely clear. Some is locked up in the polar caps, and some in carbonate minerals like those identified in the Martian meteorites. Some of the gas in the early Martian atmosphere may have been blasted away by large impacts (atmospheric cratering). But, according to magnetic field measurements from the *Mars Global Surveyor,* the final blow to Mars's atmosphere was probably bombardment by solar wind particles. Early in Mars's history, the warm convecting interior produced both volcanic outgassing and a magnetospheric bubble to protect the atmosphere. As the small planet cooled, the magnetic field weakened, and eventually bombardment by the solar wind stripped away the atmosphere. Had Mars kept its magnetic field, it might have retained much of its atmosphere and kept its moderate climate.

The vast amounts of water once present on Mars are also gone. Some is tied up in the polar caps and underground ice, but most was probably lost forever. Mars lacks an ultraviolet-absorbing stratosphere, and water molecules are easily broken apart by ultraviolet photons. Once water molecules were broken apart, the hydrogen atoms would have been rapidly lost to space through thermal escape. Some of the remaining oxygen was probably lost by solar wind bombardment, and the rest was probably drawn out of the atmosphere through chemical reactions with surface rock. This oxygen literally rusted the Martian rocks, giving the "red planet" its distinctive tint.

The history of the Martian atmosphere holds important lessons for us on Earth. Mars was apparently once a world with pleasant temperatures and streams, rain, glaciers, lakes, and possibly oceans. It had all the necessities for "life as we know it." But the once hospitable planet turned into a frozen and barren desert at least 3 billion years ago, and it is unlikely that Mars will ever be warm enough for its frozen water to flow again (without human intervention). If life once existed on Mars, it is either extinct or hidden away in a few choice locations, such as hot springs around not-quite-dormant volcanoes. As we think about the possibility of future climate change on Earth, Mars presents us with an ominous example of how much things can change.

The Mars Global Surveyor and Pathfinder missions are paving the way for more adventurous missions. The Mars Odyssey spacecraft will map out surface composition in fine detail, searching for clues to the disappearance of Mars's water. *Mars Global Surveyor* is searching for optimal landing sites for future rovers—one will search for evidence of life past and present, and another will gather rock samples for return to Earth. For the latest news on Mars missions, check the book Web site and Web sites listed in Appendix H.

Venus (6,051-km radius, 0.72 AU from Sun)

Venus is our closest neighbor in distance and in size, but it could hardly be more different from Earth. Its thick carbon dioxide atmosphere creates an extreme greenhouse effect that bakes the surface to unbearable temperatures. Venus's larger size allowed it to retain more interior heat than Mars, leading to greater volcanic activity. The associated outgassing produced the vast quantities of carbon dioxide in Venus's atmosphere, as well as the much smaller amounts of nitrogen and sulfur dioxide.

Venus is perpetually shrouded by some of the thickest and most unusual clouds in the solar system (Figure 8.24). The clouds are made of sulfuric acid dissolved in water droplets. These droplets form high in Venus's troposphere, where temperatures are 400 K cooler than on the surface. The droplets grow in the clouds until they form a sulfuric acid rain that falls toward the surface. But the droplets evaporate completely long before they reach the ground. The sulfuric acid is formed by chemical reactions involving sulfur dioxide and water, both of which come from volcanic eruptions. The atmosphere continuously loses sulfur dioxide through chemical reactions with surface rocks, so the presence of sulfuric acid clouds suggests that volcanoes on Venus must still be active (at least occasionally) in order to replenish the cloud-making ingredients.

Venus's thick cloud cover prevents us from seeing through to its surface, but we can study its geological features with radar. *Radar mapping* involves bouncing radio waves off the surface from a spacecraft and using the reflections to create three-dimensional images of the surface. From 1990 to 1993, the Magellan spacecraft used radar to map the surface of Venus, discerning features as small as 100 meters across. During its 4 years of observations, *Magellan* returned more data than all previous planetary missions combined. Like the planet itself, almost all the geological features are named for female goddesses and famous women in history.

Venus is a big step up in size from Mars—it is almost as big as Earth. Thus, we would expect it to have a history of early impact cratering, followed by even more volcanism and tectonics than we have seen on Mars. *Magellan's* observations have borne out this expectation. Most of Venus is covered by relatively smooth, rolling plains with few mountain ranges. The surface has some impact craters (Figure 8.25a), but very few compared to the Moon or even Mars. Venus lacks very small craters, because small impactors burn up in its dense atmosphere. But even large craters are rare, indicating that they must have been erased by other geological processes. Volcanism is clearly at work on Venus, as evidenced by an abundance of volcanoes and lava flows. Many are shield volcanoes, suggesting familiar basaltic eruptions (Figure 8.25b).

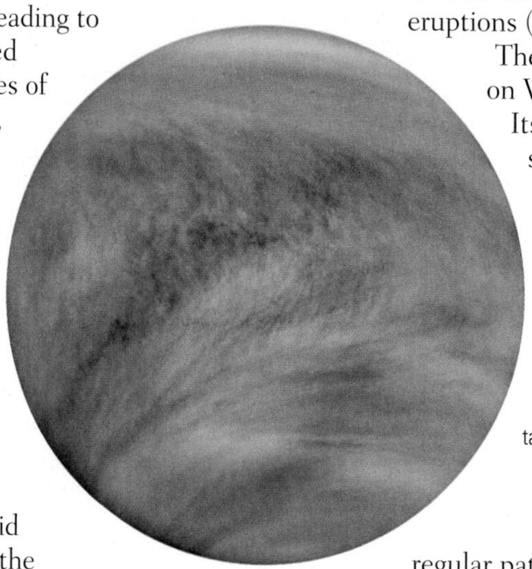

The most remarkable features on Venus are tectonic in origin. Its crust is quite contorted; in some regions, the surface appears to be fractured in a

FIGURE 8.24 Clouds on Venus are all that can be seen in this ultraviolet image taken by the *Pioneer Venus Orbiter.*

regular pattern (Figure 8.25c; see also Figure 8.10). Many of the tectonic features are associated with volcanic features, suggesting a strong link between volcanism and tectonics on Venus. One striking example of this link is a type of roughly circular feature called a *corona* (Latin for "crown") that probably resulted from a hot, rising plume in the mantle (Figure 8.25d). The plume pushed up on the crust, forming rings of tectonic stretch marks on the surface. The plume also forced lava to the surface, dotting the area with volcanoes.

Does Venus, like Earth, have plate tectonics that moves pieces of its lithosphere around? Convincing evidence for plate tectonics might include deep trenches and linear mountain ranges like those we see on Earth, but *Magellan* found no such evidence on Venus. The apparent lack of plate tectonics suggests that Venus's lithosphere is quite different from

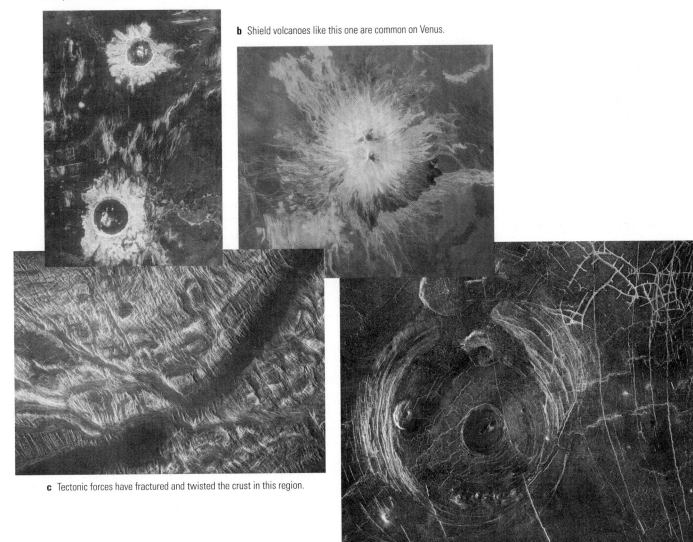

a Impact craters like these are rare on Venus.

b Shield volcanoes like this one are common on Venus.

c Tectonic forces have fractured and twisted the crust in this region.

d A mantle plume probably created the round corona, which is surrounded by tectonic stress marks.

FIGURE 8.25 The surface of Venus is covered with abundant lava flows and tectonic features, along with a few large impact craters. Because these images were taken by the Magellan spacecraft radar, dark and light areas correspond to how well radio waves are reflected, not visible light. Nonetheless, geological features stand out well. All images are several hundred kilometers across.

Earth's. Some geologists theorize that Venus's lithosphere is thicker and stronger, preventing its surface from fracturing into plates. Another possibility is that plate tectonics happens only occasionally on Venus. The uniform but low abundance of craters suggests that most of the surface formed around a billion years ago. Plate tectonics is one candidate for the process that wiped out older surface features and created a fresh surface at that time. For the last billion years, only volcanism, a bit of impact cratering, and small-scale tectonics have occurred. Changing levels of volcanism may have affected the atmosphere and even the strength of the greenhouse effect.

One might expect Venus's thick atmosphere to drive strong erosional processes, but the view both from orbit and from the surface suggests otherwise. The Soviet Union landed two probes on the surface of Venus in 1975. Before being destroyed by the 700 K heat, the probes returned images of a bleak, volcanic landscape with little evidence of erosion (see Figure 8.2b). Venus's slow rotation is apparently insufficient to drive significant surface winds, and its high surface temperature evaporates the corrosive rain before it reaches the surface. The lack of strong erosion on Venus leaves the tectonic contortions of its surface exposed to view, even though many features probably approach a billion years in age. Earth's terrain might look equally stark and rugged from space if not for the softening touches of wind, rain, and life.

As is the case for Mars, our understanding of Venus is incomplete until we consider its atmospheric evolution and its history of water. Venus is incredibly dry, with only minuscule amounts of water vapor in its atmosphere and not a drop on the surface, even though outgassing should have supplied a substantial amount of water. Water vapor is the most abundant gas from volcanic outgassing on Earth, and we expect similar gases to come out of volcanoes on Venus since both planets have similar overall compositions. Did Venus once have huge amounts of water? If so, what happened to it?

The leading theory suggests that Venus did indeed outgas plenty of water. But Venus's proximity to the Sun kept surface temperatures too high for oceans to form as they did on Earth. Outgassed water vapor accumulated in the atmosphere along with carbon dioxide and other gases. Because both water and carbon dioxide are strong greenhouse gases, Venus rapidly grew warmer still. Before the water disappeared, the surface temperature may have been double its current high value. Solar ultraviolet photons broke apart water molecules, and the high temperature allowed rapid escape of hydrogen (just as on Mars). The remaining oxygen was lost to bombardment by the solar wind (made easier by Venus's lack of a protective magnetosphere) and by chemical reactions with surface rocks. Thus, the water that once was on Venus is gone forever.

It's difficult to prove that Venus lost an ocean's worth of water billions of years ago, but some evidence comes from the gases that didn't escape. Recall that most hydrogen nuclei contain just a single proton. But a tiny fraction of all hydrogen atoms in the universe contain a neutron in addition to the proton. This isotope of hydrogen, called *deuterium*, can be present in water molecules in place of one of the hydrogen atoms of "heavy water." One deuterium atom is therefore released into the atmosphere when an ultraviolet photon breaks apart such a water molecule. Both hydrogen and deuterium escape from the atmosphere, but less deuterium escapes due to its higher mass. If Venus once had huge amounts of water molecules that broke apart, the more rapid escape of ordinary hydrogen should have left its atmosphere enriched in deuterium. Indeed, measurements show that deuterium is at least a hundred

times more abundant (relative to ordinary hydrogen) on Venus than anywhere else in the solar system.

The next time you see Venus shining brightly as the morning or evening "star," consider the radically different path it has taken from that of Earth—and thank your lucky star. If Earth had formed a bit closer to the Sun or if the Sun had been a shade hotter, our planet might have suffered the same greenhouse-baked fate as Venus.

Earth (6,378-km radius, 1.0 AU from Sun)

We've toured our neighboring terrestrial worlds and found a clear relationship between size and volcanic and tectonic activity. Thus, it should come as no surprise that Earth, the largest of the terrestrial worlds, has a high level of ongoing volcanism and tectonics.

Figure 8.26 summarizes the trends among the terrestrial worlds for volcanism and tectonics. It shows one of the key rules in planetary geology: Larger planets stay volcanically and tectonically active much longer than smaller planets. The situation for erosion is somewhat more complex. Earth has far more erosion than any other terrestrial world, but the most similar erosion is found on Mars, whose surface is sculpted by flowing water despite its small size. Yet even this analysis doesn't come close to explaining Earth's unique geology, its unusual atmosphere, or even its suitability for life.

It's ironic that astronomers began their studies of Mars and Venus expecting these planets to be variants of Earth. Instead, we've discovered that they are fundamentally different from Earth, particularly in the areas of atmospheric composition and climate. Rather than being a "role model" for understanding other planets, Earth looks more like the odd one out. The lesson of planetary exploration is that in order to understand Earth we must understand the solar system as a whole: its formation, its sister terrestrial worlds, and even the jovian planets and the smallest bodies orbiting the Sun. We are halfway through this list already, and after two more chapters we will come full circle back to Earth, ready to explore and explain our unique circumstances in Chapter 11.

TIME OUT TO THINK *Suppose another star system has a rocky terrestrial planet that is much larger than Earth. What kind of geology would you expect it to have: a surface saturated with craters, or widespread volcanic and tectonic activity? What about erosion? Explain.*

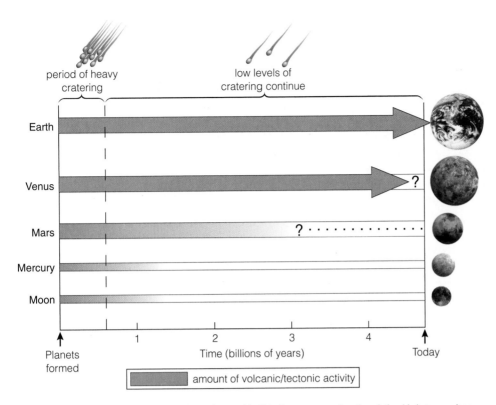

FIGURE 8.26 Volcanic/tectonic histories of the ancient worlds. This diagram summarizes the relationship between planetary size and volcanic/tectonic activity for the terrestrial worlds. The length of the red arrow shows the duration of activity: Earth is the only terrestrial world that we know to be active today, but we expect that Venus is still active as well. Most ancient craters have been erased on both. Mars may still have low levels of volcanism, and some ancient craters still remain. The narrower width of the arrows for Mercury and the Moon indicates that their volcanic and tectonic activity was never strong enough to wipe out all their heavily cratered surfaces. The diagram also shows the period of heavy cratering in the early history of the solar system. Erosion is not shown; it is negligible everywhere except Earth (where it is significant) and Mars (where it was once significant and endures to the present day at low levels).

THE BIG PICTURE

With what we have learned in this chapter, we now have a complete "big picture" view of how the terrestrial planets started out so similar yet ended up so different. As you continue your study of the solar system, keep the following important ideas in mind:

- Much of planetary geology can be distilled down to a few basic geological processes that depend on a handful of basic planetary properties.

- Every terrestrial world was once as heavily cratered as the Moon is today, but craters have been erased on the other worlds to varying degrees—depending mainly on each planet's size.

- Planetary atmospheres are not static. Complete atmospheric transformation over the age of the solar system appears to be the rule for large planets, not the exception.

- Volcanism and tectonics depend primarily on a world's size, but erosion depends on characteristics of planetary atmospheres.

- Our understanding of planetary geology and atmospheres explains much of what we see on terrestrial worlds in our solar system and might help us learn what happens on planets in other solar systems.

1. Briefly explain how the terrestrial planets ended up spherical in shape. Would you also expect small asteroids to be spherical? Why or why not?

2. How does *differentiation* occur, and under what circumstances? Explain how differentiation led to the *core–mantle–crust* structure of the terrestrial worlds.

3. What is the *lithosphere* of a planet, and what determines its thickness? What primary factor determines how long a planetary interior remains hot? Explain.

4. What conditions appear to give a planet a *magnetic field?* Contrast the magnetic fields of the terrestrial planets. What is a *magnetosphere?* What are *charged particle belts?* Describe how the solar wind affects magnetospheres and how auroras are produced.

5. Define each of the four major geological processes: *impact cratering, volcanism, tectonics,* and *erosion.*

6. Explain how we can estimate the geological age of a planetary surface from its number of impact craters. How do we know that the *lunar maria* are younger than the *lunar highlands?*

7. Give two reasons why lava may be forced to a planetary surface. What kinds of volcanic features are made by "runny" versus "sticky" lava?

8. Describe several types of internal stresses that can drive tectonic activity on a planet. Briefly explain how interior temperature affects tectonics.

9. What factors lead to erosional activity on a planet? Which terrestrial planets have erosional features?

10. In a few words, summarize the basic characteristics of the atmospheres of each of the five terrestrial worlds.

11. Explain how to interpret the generic atmospheric structure profile shown in Figure 8.12. Define the *exosphere, thermosphere, stratosphere,* and *troposphere.* Briefly describe how the *greenhouse effect* works and on what planets it is most important.

12. Briefly explain why our sky is blue in the daytime and why sunrises and sunsets are red.

13. What is *convection?* Why is it important in a planet's interior? Why is it important in a planet's atmosphere?

14. What is *outgassing?* Which planets undergo a lot of outgassing, and why? Briefly describe the five processes by which a planetary atmosphere can lose gas.

15. What factors determine whether thermal escape will be important for an atmospheric gas on a planet?

Use this idea to explain why Mercury and the Moon retain no substantial atmospheres at all and why the other terrestrial planets retain most gases but not hydrogen or helium.

16. Briefly summarize the geological history of the Moon. How and when did the maria form? Why is the Moon geologically "dead" today?

17. Briefly summarize the geological history of Mercury. What is the *Caloris Basin?* How did Mercury get its many huge cliffs? How do we know that Mercury has a huge iron core (relative to its size)?

18. Describe the *Tharsis Bulge, Valles Marineris,* and *Olympus Mons.* What geological process or processes were responsible for making them?

19. What geological processes are most important on Venus's surface? Does Venus have plate tectonics?

20. Briefly describe the evidence for climate change on Mars. Where did all the atmospheric gas go? Is it likely that Mars will ever again have a much thicker and warmer atmosphere? Why or why not?

21. Briefly explain how Earth's geology fits the general patterns we should expect based on study of other terrestrial worlds.

22. Why should we expect Venus and Earth to have had very similar early atmospheres? How did the two planetary atmospheres become so different? In particular, what happened to the water on Venus? Explain the evidence supporting the idea that Venus has lost enormous amounts of water.

Discussion Questions

1. *Geological Connections.* Consider the four geological processes: impact cratering, volcanism, tectonics, and erosion. Which two do you think are most closely connected to each other? Explain your answer, giving several ways in which the processes are connected.

2. *Mars: Past and Future.* Summarize the evolution of the Martian atmosphere since planetary formation, emphasizing the important source and loss processes. What is your prediction for the future of the Martian atmosphere?

Problems

Surprising Discoveries? For **problems 1–7,** suppose a spacecraft were to make the observations described. (These are *not* real discoveries.) In light of your understanding of planetary geology, decide whether the discovery should be considered reasonable or surprising. Explain your answer, tracing your logic back to the planet's formation properties. Explain.

1. The next mission to Mercury photographs part of the surface never seen before and detects vast fields of sand dunes.

2. A radar mapper in orbit around Venus detects a new volcano larger than any other on Venus but not seen by the Magellan spacecraft.

3. A Venus radar mapper discovers extensive regions of layered sedimentary rocks similar to those found on Earth and Mars.

4. Clear-cutting in the Amazon rain forest exposes vast regions of ancient terrain that is as heavily cratered as the lunar highlands.

5. If the Earth's atmosphere did not contain molecular nitrogen, X rays from the Sun would reach the surface.

6. The Earth's oceans probably formed at a time when no greenhouse effect operated on Earth.

7. Mars once may have been warmer, but because it is farther from the Sun than is Earth, it's not possible that it could ever have been as warm as the Earth.

8. *Earth vs. Moon.* Why is the Moon heavily cratered, but not Earth? Explain in a paragraph or two, relating your answer to their planetary formation properties.

9. *Miniature Mars.* Suppose Mars had turned out to be significantly smaller than its current size—say, the size of our Moon. How would this have affected the number of geological features due to each of the four major geological processes (impact cratering, volcanism, tectonics, and erosion)? Do you think Mars would still be a good candidate for harboring extraterrestrial life? Why or why not? Summarize your answers in two or three paragraphs.

10. *Mystery Planet.* It's the year 2098, and you are designing a robotic mission to a newly discovered planet around a star that is nearly identical to our Sun. The planet is as large in radius as Venus, rotates with the same daily period as Mars, and lies 1.2 AU from its star. Your spacecraft will orbit but not land.

    a. Some of your colleagues believe that the planet has no metallic core. How could you support or refute their hypothesis? (*Hint:* Remember that metal is dense and electrically conducting.)

    b. Other colleagues believe that the planet formed with few if any radioactive elements. How could images of the surface support or refute their hypothesis?

    c. Others suspect it has no atmosphere, but the instruments designed to study the planet's atmosphere fail due to a software error. How could you use the spacecraft's photos of geological features to determine whether a significant atmosphere is (or was) present on this planet?

11. *Atmospheric Structure.* Study Figure 8.12, which shows the atmospheric structure for a generic terrestrial planet. For each of the following cases, make a sketch similar to Figure 8.12 showing how the atmospheric structure would be different and explain the differences in words.

    a. Suppose the planet had no greenhouse gases.

    b. Suppose the Sun emitted no solar ultraviolet light.

    c. Suppose the Sun had a higher output of X rays.

12. *Project: Geological Properties of Silly Putty.*

    a. Roll room-temperature Silly Putty into a ball and measure its diameter. Place the ball on a table and gently place one end of a heavy book on it. After 5 seconds, measure the height of the squashed ball.

    b. Warm the Silly Putty in hot water and repeat the experiment in part (a). Then chill the Silly Putty in ice water and repeat. How did the different temperatures affect the rate of "squashing"?

    c. Did heating and cooling have a small or large effect on the rate of "squashing"? How does this experiment relate to planetary geology?

13. *Project: Planetary Cooling in a Freezer.* To simulate the cooling of planetary bodies of different sizes, use a freezer and two small *plastic* containers of similar shape but different size. Fill them with cold water and place them in the freezer. Checking every hour or so, record the time and your estimate of the thickness of the "lithosphere" (the frozen layer) in the two tubs.

    a. On a graph, plot the lithospheric thickness versus time for each tub. In which tub does the lithosphere thicken faster?

    b. How long does it take each tub to freeze completely? What is the ratio of the two freezing times?

    c. Describe in a few sentences the relevance of your experiment to planetary geology.

14. *Amateur Astronomy: Observing the Moon.* Any amateur telescope is adequate to identify geological features on the Moon. The light highlands and dark maria should be evident, and the shading of topography will be seen in the region near the terminator (the line between night and day). Try to observe near first- or third-quarter phase. Sketch the Moon at low magnification, and then zoom in on a region of interest, noting its location on your sketch. Sketch your field of view, label its features, and identify the geological process that created them. Look for craters, volcanic plains, and tectonic features. Estimate the horizontal size of the features by comparing them to the size of the whole Moon (radius = 1,738 km).

Web Projects

Find useful links for Web projects on the text Web site.

1. *Future Planetary Missions.* Learn about at least one upcoming mission designed to study a terrestrial world. Write a short summary of the mission and what it is designed to accomplish.

2. *Martian Weather.* Find the latest weather report for Mars from spacecraft and other satellites. What season is it in the northern hemisphere? When was the most recent dust storm? What surface temperature was most recently reported from Mars's surface, and where was the lander located? Summarize your findings by writing a 1-minute script for a television news update on Martian weather.

3. *Terraforming Mars.* Some people have proposed that we might someday *terraform* Mars, making it more Earth-like so that we might live there more easily. Research some proposals for terraforming Mars. Do you think the proposals could work? Do you think it would be a good idea to terraform Mars? Write a one- to two-page essay summarizing your findings and defending your opinions.

Do there exist many worlds, or is there but a single world? This is one of the most noble and exalted questions in the study of Nature.

St. Albertus Magnus (1206–1280)

CHAPTER 9
Jovian Planet Systems

In Roman mythology, the namesakes of the jovian planets are rulers among gods: Jupiter is the king of the gods, Saturn is Jupiter's father, Uranus is the lord of the sky, and Neptune rules the sea. But even the most imaginative of our ancestors did not foresee the true majesty of the four jovian planets. The smallest, Neptune, is still large enough to contain the volume of more than 50 Earths. The largest, Jupiter, has a volume some 1,400 times that of Earth. These worlds are totally unlike the terrestrial planets. They are essentially giant balls of gas, with no solid surface on which to stand.

Why should we care about a set of worlds so different from our own? Apart from satisfying natural curiosity about the diversity of the solar system, the jovian planet systems serve as a testing ground for our general theories of comparative planetology. Are our theories good enough to predict the nature of these planets and their elaborate satellite systems, or are different processes at work? We'll find out in this chapter as we investigate the nature of the jovian planets, along with their intriguing moons and beautifully complex rings.

FIGURE 9.1 Jupiter, Saturn, Uranus, and Neptune, shown to scale with Earth for comparison.

9.1 The Jovian Worlds: A Different Kind of Planet

The jovian planets have long held an important place in many human cultures. Jupiter is the third-brightest object that appears regularly in our night skies, after the Moon and Venus, and ancient astronomers carefully charted its 13-year trek through the constellations of the zodiac. Saturn is fainter and slower-moving, but people of many cultures also mapped its 27-year circuit of the sky. Ancient skywatchers did not know of the planets Uranus and Neptune, but astronomers charted their orbits soon after their discoveries in 1781 and 1846, respectively.

Scientific study of the jovian worlds dates back to Galileo in the early 1600s; however, scientists did not fully appreciate the differences between terrestrial and jovian planets until they established the absolute scale of the solar system (that is, in units such as kilometers instead of AU). Once they had established distances, they could calculate the truly immense sizes of the jovian planets from their *angular* sizes (as measured through telescopes) [Section 6.5]. Knowing the distance scale also allowed them to measure the masses of the jovian planets; recall that Newton's version of Kepler's third law allows the calculation of a planet's mass from the orbital period and average distance of one of its moons [Section 5.3]. Together, the measurements of size and mass revealed

the low densities of the jovian planets, proving that these worlds are very different from the Earth.

The slow pace of discovery about the jovian planets gave way to revolutionary advances with the era of spacecraft exploration. The first spacecraft to the outer planets, *Pioneer 10* and *Pioneer 11*, flew past Jupiter and Saturn in the early 1970s. The Voyager missions followed less than a decade later. *Voyager 2* continued on a "grand tour" of the jovian planets, flying past Uranus in 1986 and Neptune in 1989. The results were spectacular. Figure 9.1 shows a montage of the jovian planets compiled by the Voyager spacecraft. Exploration of the jovian planet systems continued with the Galileo spacecraft, which went into orbit around Jupiter in 1995. Next is the Cassini mission, scheduled for arrival at Saturn in 2004.

Jupiter, Saturn, Uranus, and Neptune are so different from the terrestrial planets that we must create an entirely new mental image of the term *planet*. Table 9.1 summarizes the bulk properties of these giant worlds. The most obvious difference between the jovian and terrestrial planets is size. Earth's mass in comparison to that of Jupiter is like the mass of a squirrel in comparison to that of a full-grown adult. By volume, Earth in comparison to Jupiter is like a pea in a cup of soup.

The overall composition of the jovian planets—particularly Jupiter and Saturn—is more similar to that of the Sun than to any of the terrestrial worlds.

Table 9.1. Comparison of Bulk Properties of the Jovian Planets

Planet		Average Distance from Sun (AU)	Mass (Earth masses)	Radius (Earth radii)	Average Density (g/cm³)	Bulk Composition
Jupiter		5.20	317	11.2	1.33	Mostly H, He
Saturn		9.53	90	9.4	0.70	Mostly H, He
Uranus		19.2	14	4.11	1.32	Hydrogen compounds and rocks, H and He
Neptune		30.1	17	3.92	1.64	Hydrogen compounds and rocks, H and He

Their primary chemical constituents are the light gases hydrogen and helium, although their atmospheres also contain many hydrogen compounds and their cores consist of a mixture of rocks, metals, and hydrogen compounds. Although these cores are small in proportion to the jovian planet volumes, each is still more massive than any terrestrial planet. Moreover, the pressures and temperatures near the cores are so extreme that the "surfaces" of the cores do not resemble terrestrial surfaces at all. In a sense, the general structures of the jovian planets are opposite those of the terrestrial planets: Whereas the terrestrial planets have thin atmospheres around rocky bodies, the jovian planets have proportionally small, rocky cores surrounded by massive layers of gas (Figure 9.2). A hypothetical spacecraft descending into Jupiter, for example, would have to travel *tens of thousands* of kilometers before reaching the core.

The jovian planets are Sun-like in composition, but their masses are far too low to provide the interior

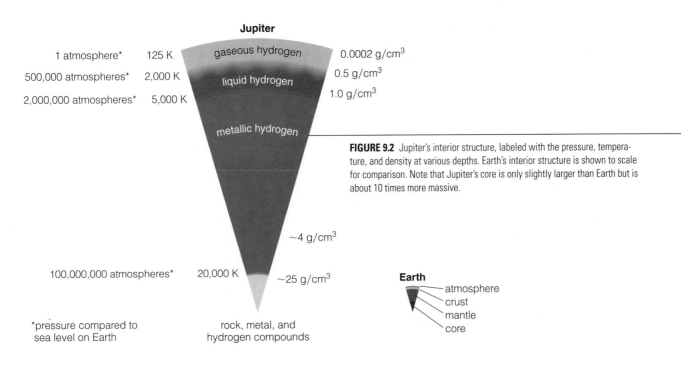

Jupiter

1 atmosphere* — 125 K — gaseous hydrogen — 0.0002 g/cm³

500,000 atmospheres* — 2,000 K — liquid hydrogen — 0.5 g/cm³

2,000,000 atmospheres* — 5,000 K — 1.0 g/cm³

metallic hydrogen

~4 g/cm³

FIGURE 9.2 Jupiter's interior structure, labeled with the pressure, temperature, and density at various depths. Earth's interior structure is shown to scale for comparison. Note that Jupiter's core is only slightly larger than Earth but is about 10 times more massive.

100,000,000 atmospheres* — 20,000 K — ~25 g/cm³

Earth
- atmosphere
- crust
- mantle
- core

*pressure compared to sea level on Earth

rock, metal, and hydrogen compounds

temperatures and densities needed for nuclear fusion. Calculations show that nuclear fusion is possible only with a mass at least 80 times that of Jupiter.

The jovian planets rotate much more rapidly than any of the terrestrial worlds, but precisely defining their rotation rates can be difficult because they are not solid. In fact, different latitude ranges on the jovian planets appear to rotate at different rates. Despite these difficulties, measurements show that the jovian "days" range from only about 10 hours on Jupiter and Saturn to 16–17 hours on Uranus and Neptune.

Even the shapes of the jovian planets are somewhat different from the shapes of the terrestrial planets. Gravity makes the jovian planets approximately spherical, but their rapid rotation rates make them slightly squashed. Material near the equator, where speeds around the rotation axis are highest, is flung outward in the same way that you feel yourself flung outward when you ride a merry-go-round. The result is that jovian planets bulge around the equator.

We can trace almost all the major characteristics of the jovian planets back to their formation [Section 7.4]. The jovian planets formed far from the Sun where icy grains condensed in the solar nebula, yielding far more solid matter to accrete into planetesimals. As the massive ice/rock cores of the future jovian planets accreted, their strong gravity captured hydrogen and helium gas from the surrounding nebula. Some of this gas formed flattened "miniature solar nebulae" (or *jovian nebulae*) around the jovian planets. Just as solid grains condensed and combined into planets in the full solar nebula, solid grains in the nebulae surrounding the jovian planets condensed to form satellite systems. The deviations from the general patterns, such as the large 98° axis tilt of Uranus, probably arose from giant impacts as the planets were forming. In the rest of this chapter, we will explore the features of the jovian planets and their satellites in greater depth, putting our theories of planetary development to the test.

9.2 Jovian Planet Interiors

As was the case with the terrestrial planets, developing a true understanding of the jovian planets requires looking deep inside them. Probing jovian planet interiors is even more difficult than probing terrestrial interiors, but several techniques allow us to learn what lies below the clouds. We've already discussed how Earth-based observations yielded the sizes and average densities of the jovian planets and how spacecraft measurements of magnetic and gravitational fields provide additional clues to their interior structure.

Spectroscopy from Earth and from spacecraft reveals the chemical compositions of the jovian upper atmospheres. Their deeper atmospheres are generally more difficult to probe, but two recent events provided rare opportunities. In July 1994, fragments of comet Shoemaker–Levy 9 slammed into Jupiter, blasting material from deeper in the atmosphere out into the open, where astronomers could study it through telescopes [Section 10.6]. Then, in December 1995, NASA's Galileo spacecraft dropped a scientific probe into Jupiter's atmosphere. The Galileo probe survived to a depth of about 300 kilometers—a significant distance, but still much less than 1% of Jupiter's 70,000-kilometer radius—providing our first direct data from within a jovian world.

We'll begin our discussion of the jovian interiors by using Jupiter as a prototype; then we'll apply the principles of comparative planetology to understand the other jovian planet interiors.

Inside Jupiter

Imagine plunging head-on into Jupiter in a futuristic space suit that allows you to survive the incredible interior conditions. Near the cloud tops, you'll find the temperature to be a brisk 125 K ($-148°C$), the density to be a low 0.0002 g/cm^3, and the atmospheric pressure to be about 1 "atmosphere"—the same as the pressure at sea level on Earth. The deeper you go, the higher the temperature, density, and pressure become (see Figure 9.2). By a depth of 7,000 km—about 10% of the planet's radius—you'll find that the temperature has increased to a scorching 2,000 K, the density has reached 0.5 g/cm^3 (half that of water), and the pressure is about 500,000 atmospheres. Under these conditions, hydrogen acts more like a liquid than a gas. At a depth of 14,000 km, you'll find that the density is about the same as that of water, the temperature is near 5,000 K, and the pressure is 2 million atmospheres. This extreme pressure is so great that it forces hydrogen into an even more compact form: *metallic hydrogen,* which is so compact that all the molecules share electrons, just as happens in everyday metals. The metallic and hence electrically conducting nature of this interior hydrogen is important in generating Jupiter's strong magnetic field.

Continuing your descent, you'll reach Jupiter's core at a depth of 60,000 km, about 10,000 km from the center. The core is a mix of hydrogen compounds, rocks, and metals, but these materials bear little resemblance to familiar solids or liquids because of the extreme 20,000-K temperature and 100-million-atmosphere pressure. The core materials probably remain mixed together, with a density of about 25 g/cm^3, rather than separating into layers of different composition as in the terrestrial planets. The total mass of Jupiter's core is about 10 times that of Earth, so the core alone would make an impressive planet.

Our study of the terrestrial planets taught us that internal heat sources can have a strong effect on the behavior of planets, so it is natural to ask if Jupiter and the other jovian planets are also affected by internal heat sources. Jupiter has a tremendous amount of internal heat—so much that it emits almost twice as much energy as it receives from the Sun. (For comparison, Earth's internal heat adds only 0.005% as much energy to the surface as does sunlight.) This internal heat source is much larger than can be explained by radioactive decay, so we must search for other explanations. The most probable explanation is that Jupiter is still slowly contracting, as if it has not quite finished forming from a jovian nebula. This contraction is so gradual that we cannot measure it directly, but it converts gravitational potential energy to thermal energy, keeping Jupiter hot inside.

Comparing Jovian Planet Interiors

Now that we've discussed the interior of Jupiter, what can we say about the other jovian planets? Look back at the summary of the bulk properties of the jovian planets in Table 9.1. These properties may seem difficult to explain at first, because no clear correlation exists between size, density, and composition. But further examination of the behavior of gaseous materials removes the mystery.

Building a planet of hydrogen and helium is a bit like making one out of very fluffy pillows (Figure 9.3a). Imagine assembling a planet pillow by pillow. As each new pillow is added, those on the bottom are more compressed by those above. As the lower layers are forced closer together, their mutual gravitational attraction increases, compressing them even further. At first the stack grows substantially with each additional pillow, but eventually the growth slows until adding pillows hardly increases the height of the stack. This analogy explains why Jupiter is only slightly larger than Saturn in radius even though it is more than three times more massive. The extra mass of Jupiter compresses its interior to a much greater extent, making Jupiter's average density almost twice that of Saturn. In fact, Saturn's average density of 0.7 g/cm³ is less than that of water. More precise calculations show that Jupiter's radius is almost the maximum possible for a jovian planet: If more hydrogen and helium were added to Jupiter, they would actually compress the interior enough to make the planet *smaller* instead of larger (Figure 9.3b). In fact, the smallest stars are significantly smaller in radius than Jupiter, even though they are 80 times more massive.

The pillow analogy suggests that a hydrogen/helium planet less massive than Saturn should have an even lower density. Because Uranus and Neptune have higher densities, we conclude that they cannot have the same composition as Jupiter and Saturn. Instead, Uranus and Neptune must have a much larger fraction of higher-density material, such as

---

FIGURE 9.3 Understanding jovian planet sizes and densities.

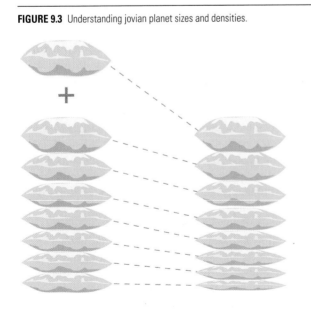

a Adding pillows to a stack may increase its height at first, but eventually it just compresses all the pillows in the stack. In a similar way, adding mass to a jovian planet would eventually just compress the planet to higher density without increasing the planet's radius.

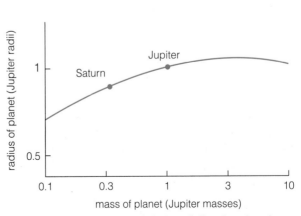

b This graph shows how the radius of a hydrogen/helium planet depends on the planet's mass. Jupiter's radius is only slightly larger than Saturn's, although it is three times more massive. For a planet much more massive than Jupiter, gravitational compression would actually make it smaller in size.

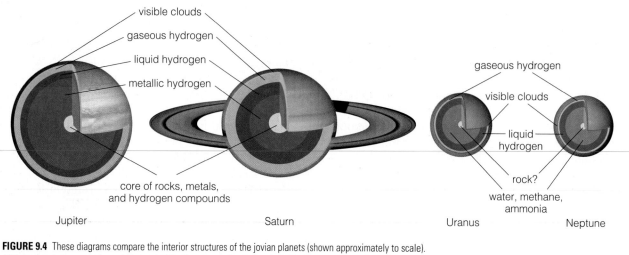

visible clouds
gaseous hydrogen
liquid hydrogen
metallic hydrogen

gaseous hydrogen
visible clouds
liquid hydrogen
rock?
water, methane, ammonia

core of rocks, metals, and hydrogen compounds

Jupiter   Saturn   Uranus   Neptune

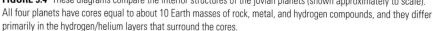

FIGURE 9.4 These diagrams compare the interior structures of the jovian planets (shown approximately to scale). All four planets have cores equal to about 10 Earth masses of rock, metal, and hydrogen compounds, and they differ primarily in the hydrogen/helium layers that surround the cores.

hydrogen compounds and rocks, and a smaller fraction of "plain" hydrogen and helium (Figure 9.4).

Saturn is large enough for interior densities, temperatures, and pressures to eventually become high enough to force gaseous hydrogen into liquid and then metallic form, just as in Jupiter. But these conditions occur relatively deeper in Saturn than in Jupiter because of its smaller size. Uranus and Neptune have even smaller layers of hydrogen and helium, so pressures are not high enough to form liquid or metallic hydrogen. However, their cores of rock, metal, and hydrogen compounds may be liquid, making for very odd "oceans" buried deep inside them.

Like Jupiter, Saturn and Neptune emit nearly twice as much energy as they receive from the Sun. But Uranus has no detectable excess internal heat. The reasons for these differences are not yet clear, but they probably have an effect on the weather occurring on these planets.

9.3 Jovian Planet Atmospheres

The jovian planets lack the geology of the terrestrial planets because they do not have solid surfaces. But whatever they lack in geology they more than make up for with their atmospheres. Fortunately, their atmospheric processes are quite similar to those we've already discussed for the terrestrial atmospheres, despite their much greater extent and very different compositions. Let's again begin by studying Jupiter as a prototype.

Jupiter's Atmosphere

Jupiter's atmosphere is almost entirely hydrogen and helium, but it also contains trace amounts of an in-

teresting and important assortment of other hydrogen compounds, including methane (CH_4), ammonia (NH_3), and water (H_2O). Although these hydrogen compounds make up only a minuscule fraction of Jupiter's atmosphere, they are responsible for virtually all aspects of its appearance. Some of these compounds condense to form the clouds so prominent in telescope and spacecraft images, and others are responsible for Jupiter's great variety of colors. Without these compounds, Jupiter would be a uniform, colorless ball of gas.

TIME OUT TO THINK *Jupiter possesses substantial amounts of gases like methane, propane, and acetylene, highly flammable fuels used here on Earth. There is plenty of lightning on Jupiter to provide sparks, so why aren't we concerned about Jupiter exploding?* (Hint: *What's missing from Jupiter's atmosphere that's necessary for ordinary fire?*)

Jupiter is the only jovian planet that we've directly sampled. On December 7, 1995, following a 6-year trip from Earth, the Galileo spacecraft released a scientific probe (the size of a large suitcase) into Jupiter's atmosphere (Figure 9.5). The probe collected temperature, pressure, composition, and radiation measurements for about an hour as it descended, until it was destroyed by the ever-increasing pressures and temperatures. The probe sent its data via radio signals back to the Galileo spacecraft in orbit around Jupiter, which relayed the data back to Earth. When combined with observations by other spacecraft and telescopes, the probe data give astronomers a good understanding of Jupiter's atmosphere.

As on the terrestrial planets, Jupiter's weather occurs mostly in its troposphere, where higher temperatures at lower altitudes drive vigorous convection. This convection is responsible for the thick clouds that enshroud Jupiter. As a parcel of gas rises upward through the troposphere, it encounters gradually lower temperatures. Soon the surrounding temperature is low enough for water vapor to condense into liquid droplets or flakes of ice, forming water clouds similar to those on Earth. As the gas continues to rise, temperatures become low enough for another minor atmospheric ingredient, ammonium hydrosulfide (NH_4SH), to condense and form clouds. Still higher, the temperatures fall to the point at which ammonia (NH_3) condenses to form clouds of ammonia crystals—the prominent white clouds visible on Jupiter. Thus, Jupiter has several layers of clouds of different compositions. The relentless motion of Jupiter's atmosphere continuously regenerates the extensive clouds.

As on Earth, Jupiter's wind patterns are driven by the planet's rotation. But Jupiter's rotation is so much more rapid than Earth's that the atmosphere breaks up into many swirling bands encircling the entire planet (Figure 9.6a). The bands of rising air are called **zones**, and they appear white in color because ammonia clouds form as the air rises to high, cool altitudes. The adjacent **belts** of falling air are depleted in the cloud-forming ingredients and do not contain any white ammonia clouds. Instead, they appear dark in color because we can see down to the red or tan ammonium-hydrosulfide clouds that form at lower altitudes. Infrared images confirm the temperature differences between Jupiter's warm belts and cool zones (Figure 9.6b).

Jupiter's **Great Red Spot** is perhaps the most dramatic weather pattern in the solar system (Figure 9.7). We know that it has been prominent in

FIGURE 9.5 Artist's conception of the Galileo probe entering Jupiter's atmosphere. This view, looking outward from beneath the clouds, shows the suitcase-size probe falling with its parachute extended above it.

Jupiter's southern hemisphere for at least three centuries, because it has been seen ever since telescopes became powerful enough to detect it. The Great Red Spot is huge—more than twice as wide as the Earth. Could it be a tremendous storm like a hurricane on Earth? Not quite. The Great Red Spot actually spins in the opposite direction than a hurricane would.

FIGURE 9.6 Jupiter's belts and zones.

Belts are warm, red, low-altitude clouds.

Zones are cool, white, high-altitude clouds.

a The color difference between belts and zones is evident in this Hubble Space Telescope image.

b In this infrared image taken nearly simultaneously with (**a**), brightness indicates high temperatures.

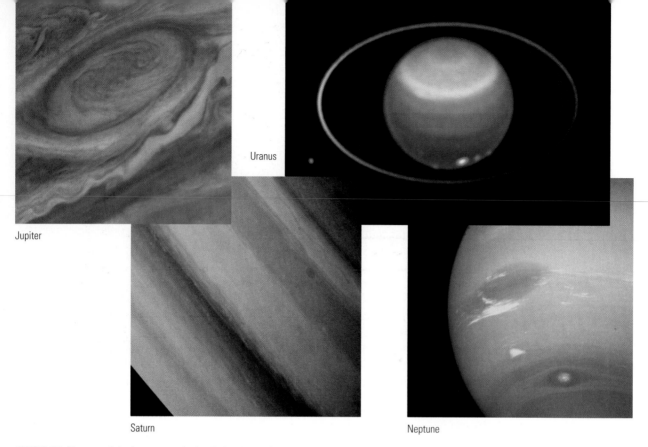

Jupiter

Uranus

Saturn

Neptune

FIGURE 9.7 Close-ups of cloud patterns on the four jovian planets. Scientists combine series of images like these to make weather movies similar to those for Earth shown on the evening news. Jupiter's Great Red Spot is a huge high-pressure storm that is large enough to swallow two or three Earths. The Great Dark Spot is visible in the Neptune image. Jupiter, Saturn, and Neptune images by the Voyager spacecraft at visible wavelengths, Uranus image by the Hubble Space Telescope at infrared wavelengths.

Astronomers speculate that the storm is kept spinning by the influence of nearby belts and zones.

Jupiter's vibrant colors remain perplexing despite decades of intense study. Observations tell us that the high clouds of ammonia in the zones are usually white and that the lower clouds of ammonium hydrosulfide in the belts are brown or red. But pure ammonium-hydrosulfide crystals are as white as ammonia crystals, so the colors of the belt clouds must come from ingredients besides the ammonium-hydrosulfide crystals themselves. Sulfur compounds are a plausible suspect, because they form a variety of reds and tans (also seen on the surface of Jupiter's moon Io). Phosphorus compounds (including phosphine, PH_3) are another possibility. Whatever their origin, the red and tan compounds are found primarily at lower (and therefore warmer) altitudes. The chemical reactions that form them evidently require the extra energy of Jupiter's internal heat, or possibly lightning deep below the clouds.

As far as we know, Jupiter's climate is steady and unchanging. Jupiter has no appreciable axis tilt and therefore has no seasons. In fact, Jupiter's polar temperatures are quite similar to its equatorial temperatures. Solar heating alone might leave the poles relatively cool, but Jupiter's internal heat source keeps the planet uniformly warm.

Comparing Jovian Planet Atmospheres

The most striking difference among the jovian atmospheres is their colors. Starting with Jupiter's distinct red colors, the jovian planets turn to Saturn's more subdued yellows, to Uranus's faint blue-green tinge, and finally to Neptune's pronounced blue hue (see Figure 9.1). What's responsible for this regular progression of colors? The primary constituents in all the jovian atmospheres (hydrogen and helium) are colorless, so the planetary colors must come from trace gases or chemical reactions that create colored compounds.

Saturn's reds and tans almost certainly come from the same compounds that produce these colors on Jupiter—whatever those compounds may be. The blue colors of Uranus and Neptune have a completely different explanation: abundant methane gas, which absorbs red light but transmits blue light. The clouds then reflect (scatter) the blue light upward, where it is again transmitted through the gas above.

The Voyager, Galileo, Cassini, and Hubble Space Telescope cameras have monitored the meteorology

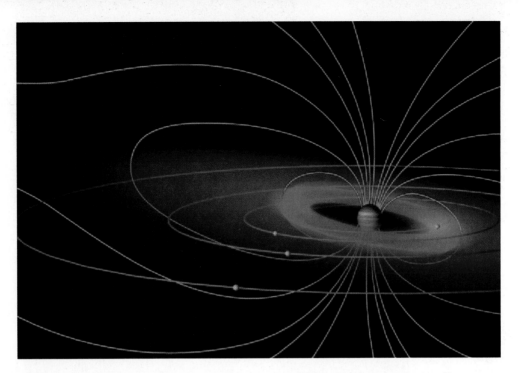

FIGURE 9.8 Artist's conception of Jupiter's magnetosphere, with Io embedded deep within. An orange glow around Io represents its escaping atmosphere, and the red donut represents the charged particle belt formed by Io. The blue lines show Jupiter's strong magnetic field.

of the four jovian planets with excellent resolution (Figure 9.7). Jupiter's weather phenomena (clouds, belts, zones, and other storms) are by far the strongest and most active. Saturn also possesses belts and zones (in more subdued colors), along with some small storms and an occasional large storm, but it lacks any weather feature as prominent as Jupiter's Great Red Spot. (For unknown reasons, Saturn's winds are even stronger than Jupiter's.) Neptune's atmosphere is also banded and has a high-pressure storm, called the Great Dark Spot, that acts much like Jupiter's Great Red Spot. The greatest surprise in the weather patterns on the jovian planets is the weather on Uranus, the planet tipped on its side. When the Voyager spacecraft flew past in 1986, it was midsummer in Uranus's northern hemisphere, with the summer pole pointed almost directly at the Sun. Photographs revealed virtually no clouds, belts, or zones, which scientists attributed to the lack of a significant internal heat source on Uranus. But recently the Hubble Space Telescope has detected storms raging on the planet (bright spots in Figure 9.7). The seasons are changing, and the winter hemisphere is seeing sunlight for the first time in decades.

9.4 Jovian Planet Magnetospheres

Each jovian planet is surrounded by a bubble-like magnetosphere consisting of the planet's magnetic field and the particles trapped within it. The jovian planets themselves dwarf the terrestrial planets, and their magnetospheres are even more impressive.

Jupiter's magnetic field is awesome—about 20,000 times stronger than Earth's. This strong magnetic field deflects the solar wind some 3 million km (about 40 Jupiter radii) before it even reaches Jupiter. If our eyes could see this part of Jupiter's magnetosphere, it would be larger than the full moon in our sky.

Jupiter's magnetosphere contains a great number of charged particles, some captured from the solar wind but most originating from Jupiter's volcanically active moon, Io. The charged particles bombard the surfaces of Jupiter's icy moons and can generate thin bombardment atmospheres—just as bombardment creates the thin atmospheres of Mercury and the Moon [Section 8.5]. On Io, the bombardment leads to the continuous escape of the atmospheric gases released by volcanic outgassing. As a result, Io loses atmospheric gases faster than any other object in the solar system. The escaping gases (sulfur, oxygen, and a hint of sodium) are ionized and feed a donut-shaped charged particle belt, called the *Io torus,* that approximately traces Io's orbit (Figure 9.8).

None of the other jovian planets has anything like the Io torus (because none has satellites quite like Io), but their magnetospheres do hold some big surprises. Both Uranus's and Neptune's magnetic fields are tipped more than 45°. Planetary scientists are still at a loss to explain these large tilts.

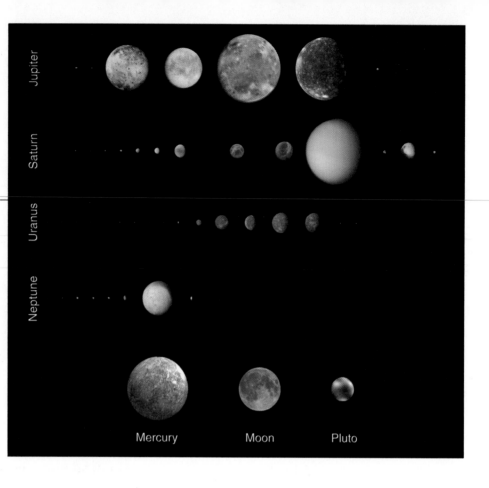

FIGURE 9.9 The larger satellites of the jovian planets, with sizes (but not distances) shown to scale. Mercury, the Moon, and Pluto are included for comparison.

Labels in figure: Jupiter, Saturn, Uranus, Neptune, Mercury, Moon, Pluto

9.5 A Wealth of Worlds: Satellites of Ice and Rock

Nearly 90 known moons (*satellites*) orbit the jovian planets. Figure 9.9 shows the larger ones, and Appendix C lists fundamental data for all the known moons. It's worth taking a few minutes to study the figure and appendix, looking for patterns, before you read on.

Broadly speaking, the jovian moons can be divided into three groups: small moons less than about 300 km across, medium-size moons ranging from about 300 to 1,500 km in diameter, and large moons more than 1,500 km in diameter. Most of the known moons fall into the small category, and additional small moons may yet be discovered. The small moons generally look more like potatoes than spheres, because their gravities are too weak to force their rigid material into spheres.

Nearly all the medium-size and large moons follow the orbital patterns we expect from formation in "miniature solar nebulae": They have approximately circular orbits that lie close to the equatorial plane of their parent planet, and they orbit in the same direction in which their planet rotates. These moons are planetlike in almost all ways. They are approximately spherical, each has a solid surface with its own unique geology, and some possess atmospheres, hot interiors,

and even magnetic fields. A few are planetlike in size as well. The two largest moons—Jupiter's moon Ganymede and Saturn's moon Titan—are larger than the planet Mercury. Four others are larger than Pluto: Jupiter's moons Io, Europa, and Callisto and Neptune's moon Triton.

Nearly all jovian moons share an uncanny trait: They always keep the same face turned toward their planet, just as our Moon always shows the same face to Earth (see Figure 2.19). This *synchronous rotation* arose from the strong tidal forces [Section 5.4] exerted by the jovian planets, which caused each moon to end up with equal rotational and orbital periods regardless of how fast the moon rotated when it formed.

Ice Geology

Prior to the Voyager missions, most scientists expected the jovian moons to be cold and geologically dead. After all, only two of the moons are even as large as Mercury, and Mercury has been geologically dead for a long time. Instead, Voyager provided many surprises, revealing worlds with spectacular past and present geological activity. We've since learned even more—about Jupiter's moons in particular—thanks primarily to data from the Galileo spacecraft.

How can moons have so much more geological activity than similar-size (or larger) planets? In a few cases, the answer involves unforeseen heat sources that we'll discuss shortly. But another important factor is the icy composition of most jovian moons, because ice melts and deforms at much lower temperatures than rock.

All in all, the *ice geology* of the jovian moons bears many similarities to the rock geology of the terrestrial planets [Section 8.3]. *Impact cratering,* occurring mostly during the time of the early bombardment more than 4 billion years ago, probably left all the satellites with battered surfaces. *Volcanism* is certainly present on some jovian moons, and *tectonics* seems likely as well. Both of these processes are driven by internal heat and hence take place more easily on icy moons than on rocky planets. We know very little about *erosion* in the outer solar system, although we expect it to be relatively rare. Most satellites possess either no atmosphere or a very thin one, so in most cases wind or rain erosion is unlikely. Nevertheless, many of the jovian moons have interesting features that go beyond simple generalities. There are far too many moons to study all of them in depth here, but as we undertake our brief tour of the jovian moon systems we will stop to investigate a few of the more intriguing moons in detail.

The Galilean Satellites of Jupiter: Io, Europa, Ganymede, and Callisto

The first stops on our tour are the Galilean satellites of Jupiter—the four moons that Galileo saw through his telescope and that disproved the idea that all celestial bodies circle the Earth [Section 3.4]. These four moons are large enough that they would count as planets if they orbited the Sun. They bear the names of four mythological lovers of the Roman god Jupiter: Io, Europa, Ganymede, and Callisto.

Io Just a few months before the Voyager spacecraft reached Jupiter, a group of scientists made the astonishing prediction that we would find active volcanism on Io. This prediction went against the common belief that only much larger bodies could have substantial geological activity. But the prediction proved correct. Direct proof came from the discovery of towering volcanic plumes reaching hundreds of kilometers above the surface and spreading fallout over almost a million square kilometers of Io's surface (Figure 9.10a). The indirect proof was just as mind-boggling: the complete lack of impact craters on the surface—not a single one, to the resolution of Voyager's cameras. Io's eruptions have buried *all* its impact craters.

Despite Io's unusual colors and frost-covered surface, planetary scientists found volcanic eruptions remarkably similar to those of basalt volcanoes on Earth (Figure 9.10b). When the flowing lava comes in contact with frost, the resulting "steam" jets off at high speed to make the plume. The process is similar to lava flowing into the ocean on Earth, but Io's relative lack of atmosphere and low gravity let the plumes reach great altitude (Figure 9.10c). Not all of Io's volcanoes erupt so violently. Lava flows out of some vents, and sulfur gas expelled from a volcano coats the surface in red patches. Some of the higher mountains were probably built by basaltic lava flows similar to those of terrestrial volcanoes. Many of Io's volcanic vents glow red-hot. Tectonic processes are probably also active on Io, but evidence for them is covered by the frequent lava flows and plume deposits everywhere on Io's surface. Galileo and Voyager images showed that fallout from volcanic plumes can cover an area the size of Arizona in a matter of months.

Why is Io so geologically active? After all, it is about the size of our own geologically dead Moon, and it is not as icy in composition as most other jovian satellites. Io must have an additional internal heat source besides the usual combination of accretion, differentiation, and radioactivity that heats the interiors of the terrestrial worlds [Section 8.2]. This fourth internal heat source is **tidal heating**, and its theory was the basis for the pre-Voyager prediction of volcanic activity on Io.

How does tidal heating work? Tidal forces lock Io in synchronous rotation so that it keeps the same face toward Jupiter. But Io's orbit is slightly elliptical, so its orbital speed and distance from Jupiter vary. As a result, the strength and even the direction of the tidal force change very slightly over its orbit: Io is continuously flexed by Jupiter (Figure 9.11a). The flexing of Io's rocky crust and interior releases heat, just as flexing warms Silly Putty. In fact, Io's tidal heating releases more than 200 times more heat (per gram of mass) than the radioactive heat driving Earth's geology, which explains why Io is the most volcanically active body in the solar system.

But why is Io's orbit slightly elliptical, when almost all other large satellites' orbits are virtually circular? Io's neighbors are responsible: Gravitational nudges between Io, Europa, and Ganymede maintain **orbital resonances** between them (Figure 9.11b). During the time in which Ganymede completes one orbit of Jupiter, Europa completes exactly two orbits and Io completes exactly four orbits. The satellites therefore line up periodically, and the gravitational tugs these satellites exert on one another add up over time. The satellites are always tugged in exactly the same direction, and the satellite orbits become very slightly elliptical as a result. (If the satellites didn't line up periodically, the tugs would be exerted in random directions at random times—as ineffective as pushing a child on a swing at random times.) The

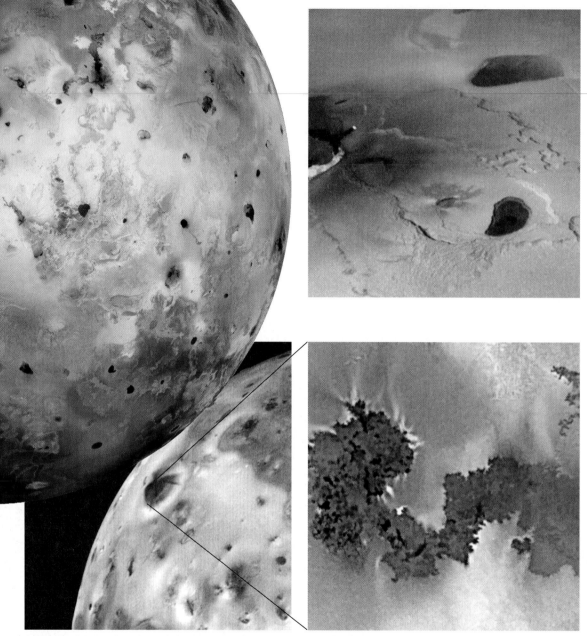

a Most of the black, brown, and red spots are recently active volcanic features. The white and yellow areas are sulfur and sulfur dioxide deposits from volcanic gases. Colors in this Galileo image are slightly enhanced.

b Galileo close-up of eruptions reveals intensely hot lava, probably similar in composition to basalt volcanoes on Earth.

c When basaltic lava flows over sulfur dioxide ice, the explosive sublimation creates huge plumes. This plume rises 80 km high.

FIGURE 9.10 Io is the most volcanically active body in the solar system.

resulting tidal forces are strongest on Io because it is closest to Jupiter, but they are also important on Europa and Ganymede.

Europa Europa's bizarre, fractured crust is proof enough that tidal heating has taken hold there (Figure 9.12a). The icy surface is nearly devoid of impact craters and may be only a few million years old. Jumbled icebergs (Figure 9.12b) suggest that an ocean of liquid water lies just a few kilometers below the surface, extending down to perhaps 100 km deep. The icy shell overlying the ocean is repeatedly flexed by tidal forces, resulting in a global network of cracks (Figure 9.12c).

FIGURE 9.11 Tidal heating explained.

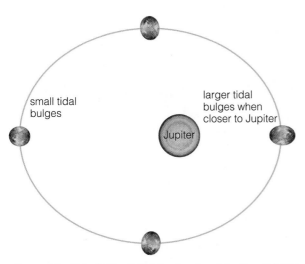

a Because Io's orbit is slightly elliptical, the strength and direction of Io's tidal bulges change. The bulges and orbital eccentricity are exaggerated.

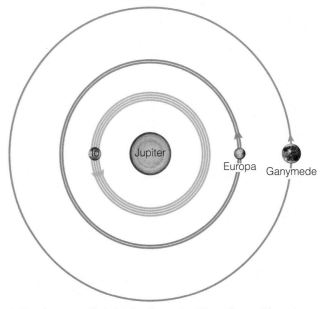

b About every seven Earth days (one Ganymede orbit, two Europa orbits, and four Io orbits), the three moons line up as shown. The small gravitational tugs repeat and make all three orbits slightly elliptical.

FIGURE 9.12 Europa is one of the most intriguing moons in the solar system.

b Some regions show jumbled crust with icebergs, apparently frozen in slush. This figure combines low-resolution images and high-resolution close-ups from the Galileo spacecraft.

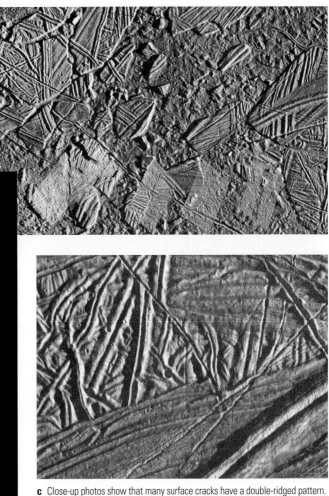

a Europa's icy crust is criss-crossed with cracks.

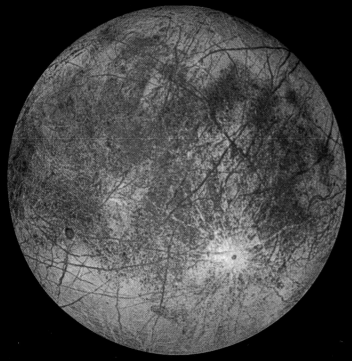

c Close-up photos show that many surface cracks have a double-ridged pattern.

FIGURE 9.13 Tidal heating may give Europa a subsurface ocean beneath its icy crust. This artist's conception imagines a region where the crust has been disrupted by an undersea volcano.

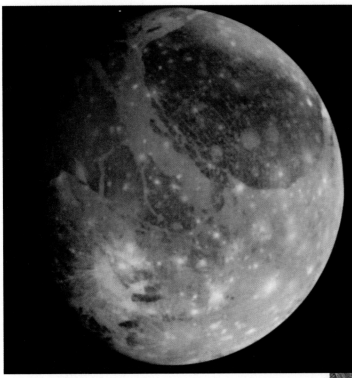

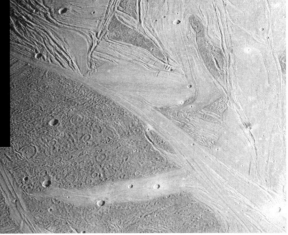

b The brighter, ridged regions of Ganymede's surface, called grooved terrain, have few craters and indicate relatively recent geological activity.

a Ganymede's numerous craters (bright spots) show that its surface is older than Europa's.

FIGURE 9.14 Ganymede, the largest moon in the solar system.

In some ways, Europa resembles a rocky terrestrial planet wrapped in an icy crust. The rocky object that makes up most of Europa is undoubtedly geologically active below the layer of ocean and ice. Perhaps basaltic lavas erupt on Europa's seafloors, sometimes violently enough to jumble up the icy crust above (Figure 9.13). And, just possibly, tidal heating makes Europa's oceans as hospitable for life as our own oceans [Section 11.6].

Ganymede Photographs of Ganymede tell the story of an interesting geological history. Parts of the surface have many impact craters, indicating that these regions are billions of years old (Figure 9.14a). But other areas show unusual "grooved terrain" unlike anything we've seen in terrestrial geology (Figure 9.14b). The grooves may have formed as Ganymede's crust expanded and fractured early in its history. Ganymede also shows some evidence of tectonics similar to plate tectonics on Earth—some chunks of the surface appear to have shifted relative to one another. Surprisingly, Ganymede has its own magnetic field—perhaps indicating a molten, convecting core. It may even harbor a liquid water layer

below the ice, but evidence for such an ocean is still quite tentative.

Callisto The outermost Galilean satellite, Callisto, looks most like what scientists first expected for outer solar system satellites: a heavily cratered iceball (Figure 9.15a). The bright patches on its surface are caused by cratering: Large impactors dig up cleaner ice from deep down and spread it over the surface. The odd rings in the upper left of Callisto's image also come from impacts: The largest impactors shatter the icy sphere in a way that produces concentric rings around the impact site.

Despite its relatively large size (the third-largest moon in the solar system), Callisto lacks volcanic and tectonic features. Internal heat sources must be very small. Callisto has no tidal heating because it shares no orbital resonances with other satellites. Nonetheless, Callisto appears to have a magnetic field, which could be generated in a salty ocean below the ice. The surface also holds some surprises. Close-up images show the surface to be covered by a dark, powdery substance concentrated in low-lying areas, leaving ridges and crests bright white (Figure 9.15b). No one knows the nature of this material or how it got there.

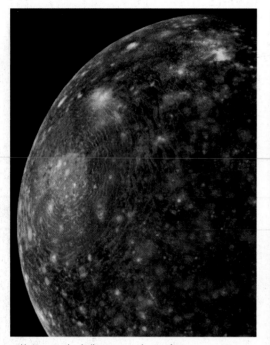

a Heavy cratering indicates an ancient surface.

b Close-up photos show a dark powder overlying the low areas of the surface.

FIGURE 9.15 Callisto shows no evidence of volcanic or tectonic activity.

Titan and the Medium-Size Moons of Saturn

Leaving Jupiter's moons behind, our satellite tour takes us next to Saturn. Here we find an amazing moon shrouded in a thick atmosphere: Titan. It is Saturn's only large moon and is the second-largest moon in the solar system after Ganymede (see Figure 9.9).

Titan's hazy and cloudy atmosphere hides its surface (Figure 9.16a). The atmosphere is about 90% nitrogen—making it the only world besides Earth where nitrogen is the dominant atmospheric constituent. The rest of Titan's atmosphere consists of argon, methane, and other hydrogen compounds, including ethane (C_2H_6). The methane and ethane actually give Titan an appreciable greenhouse effect, but the surface temperature is still a frigid 93 K ($-180°C$). The surface pressure on Titan is only somewhat higher than on Earth—about 1.5 bars at the surface, which would be fairly comfortable if not for the lack of oxygen and the cold temperatures.

How did Titan end up with such an unusual atmosphere? Titan is composed mostly of ices, including methane and ammonia ice. Some of this ice sublimed long ago to form an atmosphere on Titan. Over billions of years, solar ultraviolet light broke apart the ammonia molecules (NH_3) into hydrogen and nitrogen. The light hydrogen escaped from the atmosphere (through thermal escape [Section 8.5]), but the heavier nitrogen molecules remained and accumulated. Similarly, methane (CH_4) may have been converted to ethane, which could make clouds, rain, and even an ocean on Titan (Figure 9.16b). Recent infrared pictures (Figure 9.16c) have probed through the clouds to the surface, but we cannot yet determine whether such oceans really exist.

Because we cannot see its surface, we do not know whether Titan is geologically active. However, given its relatively large size and icy composition, it seems likely that Titan has a rich geological history. Our understanding of Titan should improve dramatically when the Cassini spacecraft reaches Saturn in 2004 [Section 6.6]. In addition to cameras and spectrographs, *Cassini* carries a radar mapper similar to the *Magellan* radar that mapped the surface of Venus [Section 8.6]. Thus, we will see the surface of Titan at high resolution and at last will learn about its geology—and whether the surface is covered by an ethane ocean. *Cassini* will also drop a probe (named *Huygens,* for a seventeenth-century pioneer in astronomy and optics) into Titan's atmosphere that will take pictures and measure atmospheric properties on the way down. Just in case, the probe is designed to float in liquid ethane.

a Titan is enshrouded by a hazy, cloudy atmosphere.

b Artist's conception of the surface of Titan, showing the possible ethane oceans.

c A recent image from the Keck Telescope taken at infrared wavelengths can see through Titan's clouds to the surface. The dark areas might be oceans.

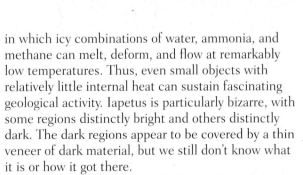

FIGURE 9.16 Saturn's moon Titan.

TIME OUT TO THINK *Based on your understanding of the four geological processes on icy satellites, what kind of surface features do you think might be present on Titan?*

Saturn's medium-size moons (Rhea, Iapetus, Dione, Tethys, Enceladus, and Mimas) show evidence of substantial geological activity (Figure 9.17). While all show some cratered surfaces, Enceladus has grooved terrain similar to Ganymede's, and the others show evidence of flows of "ice lava." But these satellites are considerably smaller than Ganymede: Enceladus is barely 500 kilometers across—small enough to fit inside the boundaries of Colorado. How can such small moons support so much geological activity? The answer probably lies with "ice geology," in which icy combinations of water, ammonia, and methane can melt, deform, and flow at remarkably low temperatures. Thus, even small objects with relatively little internal heat can sustain fascinating geological activity. Iapetus is particularly bizarre, with some regions distinctly bright and others distinctly dark. The dark regions appear to be covered by a thin veneer of dark material, but we still don't know what it is or how it got there.

Mimas (radius = 195 km), the smallest of the "medium-size" moons of Saturn, is essentially a heavily cratered iceball (Figure 9.17). The impact that created its largest crater probably came close to breaking Mimas apart.

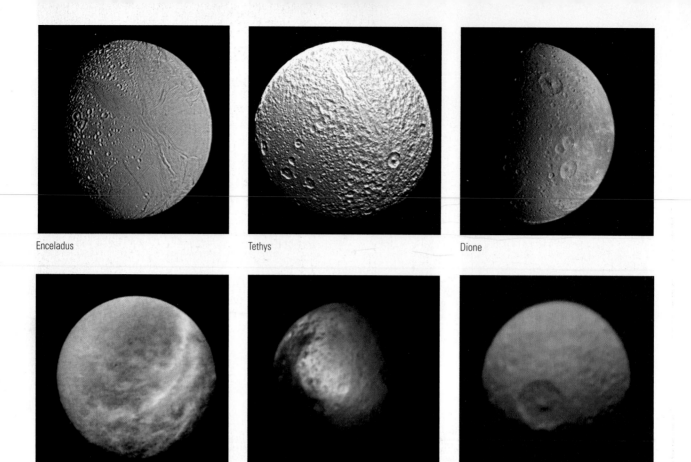

FIGURE 9.17 Saturn's six medium-size moons, which range in diameter from 390 km (Mimas) to 1,530 km (Rhea). All but Mimas show some evidence of volcanism and/or tectonics.

Enceladus Tethys Dione Rhea Iapetus Mimas

The Medium-Size Moons of Uranus

Uranus is orbited by 5 medium-size moons (see Figure 9.9) and 10 smaller moons discovered by Voyager. Astronomers have recently discovered 3 more moons, including 2 that orbit the planet backward. Miranda, the smallest of Uranus's medium-size moons (radius = 235 km), is the most surprising (Figure 9.18). Prior to the Voyager flyby, scientists expected Miranda to be a cratered iceball like Saturn's similar-size moon Mimas (see Figure 9.17). Instead, Voyager images of Miranda show tremendous tectonic features and relatively few craters. Why should Miranda be so much more geologically active than Mimas? Our best guess is an episode of tidal heating from a temporary orbital resonance with another satellite of Uranus billions of years ago. Evidently, Mimas never underwent a similar period of tidal heating.

Triton, the Backward Moon of Neptune

Last stop on our satellite tour is Neptune, so distant that only two of its moons (Triton and Nereid) were known to exist before the Voyager visit. Triton is the only large moon in the Neptune system, and only one other moon (Proteus) even makes it into our "medium-size" category.

Triton may appear to be a typical satellite, but it is not. Its orbit is retrograde (it travels in a direction opposite to Neptune's rotation) and highly inclined to Neptune's equator. These are telltale signs of a captured satellite. But Triton is not small and potato-shaped like most captured asteroids. Instead it is large, spherical, and icy. In fact, it is larger than the planet Pluto. Nonetheless, it is almost certain that Triton once orbited the Sun instead of orbiting Neptune.

A quick study of the Triton photos in Figure 9.19 reveals a surface unlike any other in the solar system, and one that poses many unanswered questions. Are the flat surfaces like the lunar maria? Are the wrinkly ridges (nicknamed "cantaloupe terrain") tectonic in nature? Are the bright polar caps similar to those on Mars? Clearly, Triton is more than a cratered iceball. It appears to have undergone some sort of icy volcanism in the past. Such geological activity is astonishing in light of Triton's extremely low (40 K) surface

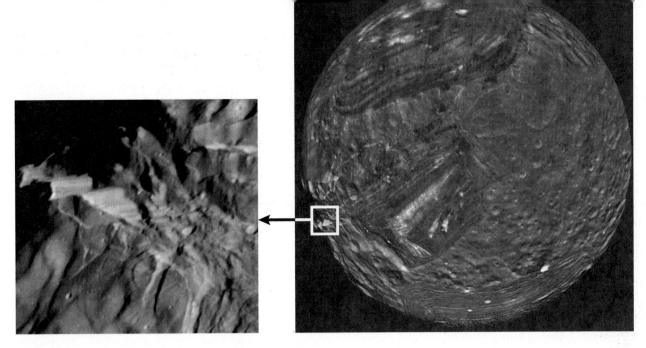

FIGURE 9.18 The surface of Miranda shows astonishing tectonic activity despite its small size. The cliff walls seen in the inset are higher than those of the Grand Canyon on Earth.

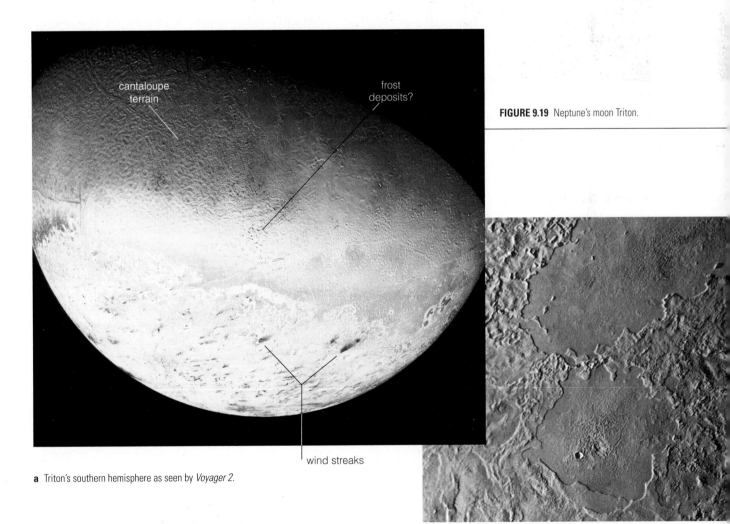

cantaloupe terrain

frost deposits?

FIGURE 9.19 Neptune's moon Triton.

wind streaks

a Triton's southern hemisphere as seen by *Voyager 2*.

b The flat regions in this close-up photo of Triton's icy surface may be "lava"-filled impact basins similar to the lunar maria.

temperature, but it probably involved low-melting-point mixtures of water, methane, and ammonia ices. The internal heating source for Triton's geological activity is not known, but it may have involved tidal heating during its capture. To cap off Triton's odd geology, the sublimation of surface ices creates a thin atmosphere. Thin as it is, the atmosphere creates wind streaks on the surface, and unknown processes pump unusual plumes of gas and particles into the atmosphere.

If Triton really is a captured satellite, where did it come from? Could other objects like it still be orbiting the Sun? Intriguingly, we know of at least one object that appears to be quite similar to Triton: the planet Pluto. But that's a story for the next chapter.

Satellite Summary: The Active Outer Solar System

Our brief tour has taken us to most of the medium-size and large moons in the solar system. The major lesson of the tour is that icy-satellite geology is harder to predict than terrestrial-planet geology. Size isn't everything when it comes to geological activity. Ices form lavas more easily than rocks, so icy worlds can have more geological activity than a rocky world of the same size. Satellites made of ices that include methane and ammonia may have even greater potential geological activity. In addition, we saw many cases that demonstrate how tidal heating can supply added heat even when other internal heat sources are no longer important. The jovian planet satellites have turned out to be far more interesting and instructive than anyone could have imagined before the Voyager expeditions.

9.6 Jovian Planet Rings

We have completed our comparative study of the jovian planets themselves and of their major moons, but we have one more topic left to cover: their amazing rings. Saturn's rings have dazzled and puzzled astronomers since Galileo first saw them through his small telescope and suggested that they resembled "ears" on Saturn. For a long time, Saturn's rings were thought to be unique in the solar system, but we now know that all four jovian planets have rings. As we did for the planets themselves, we'll look at one ring system in detail and then examine the others for important similarities and differences. Saturn's rings are the clear choice as the standard for comparison.

Saturn's Rings

You can see Saturn's rings through a backyard telescope, but learning about their nature requires higher resolution. Even through large telescopes on Earth, the rings appear to be continuous, concentric sheets of material separated by gaps (Figure 9.20a). Spacecraft images reveal these "sheets" to be made of many more individual rings, each broken up into even smaller concentric *ringlets,* and different regions of a ring to range from transparent to opaque (Figure 9.20b). But even these appearances are somewhat deceiving. If we could wander into the rings, we'd see that they are made of countless individual particles orbiting Saturn together, each obeying Kepler's laws and occasionally colliding with nearby particles (Figure 9.20c). The particles range in size from large boulders to dust grains—far too small to be photographed, even when spacecraft like *Voyager* or *Cassini* pass nearby.

Spectroscopy reveals that Saturn's ring particles are made of relatively reflective (high-albedo) water ice. The rings look bright where there are enough particles to intercept most of the Sun's light and scatter it back toward us and they appear more transparent where there are fewer particles. In regions where light cannot easily pass through, neither can another ring particle. In the densest parts of the rings, each particle collides with another particle every few hours.

TIME OUT TO THINK *Which ring particles travel faster: those at the inner edge of Saturn's rings, or those at the outer edge? (Hint: Think about Kepler's third law.) Can you think of a way to confirm your answer with telescopic observations?*

Saturn's rings lie in a thin plane directly above the planet's bulging equator, so they share Saturn's 27° axis tilt. Over the course of Saturn's year, the rings present a changing appearance to the Earth and the Sun, appearing wide open at Saturn's solstices and edge-on at its equinoxes. The rings are perhaps the thinnest known astronomical structure: They are over 270,000 kilometers across but only a few tens of *meters* thick.

Where do the rings come from? An important clue is that the rings lie close to the planet in a region where tidal forces are very influential. Within two to three planetary radii (of any planet), the tidal forces tugging an object apart become comparable to the gravitational forces holding it together. This region is called the **Roche tidal zone**. Only relatively small objects held together by nongravitational forces (such as the electrostatic forces that hold solid rock—or spacecraft or human beings—together) can survive within the Roche tidal zone. Thus, there are two basic scenarios for the origin of the rings. One possibility is that a wandering moon strayed too close to Saturn and was torn apart. A more likely scenario is that tidal forces prevented the material in Saturn's rings from accreting into a single large moon, instead forming many smaller moons.

Rings and Gaps

By the time the Voyager spacecraft approached Saturn, astronomers knew of at least six distinct rings around Saturn. Voyager scientists had hoped to find reasons for the six rings, but instead they found that the total number of individual rings and gaps may be as high as 100,000. Theorists are still struggling to explain all the rings and gaps, but some general ideas are now clear.

Rings and gaps are caused by particles bunching up at some orbital distances and being forced out at others. This bunching happens when gravity nudges the orbits of ring particles in some particular way. One source of nudging is tiny moons—called *gap moons*—located within the rings themselves. The gravity of a gap moon tugs gently on nearby particles, effectively nudging them farther away to other orbits. This clears a gap in the rings around the moon's orbit. In some cases, Voyager images show two gap moons forcing particles trapped between them into line. The gap moons are said to act as *shepherd moons* in such cases, because they shepherd the ring particles.

In some cases, ring particles orbiting Saturn may also be nudged by the gravity from Saturn's larger and more distant moons. For example, a ring particle orbiting about 120,000 km from Saturn's center will circle the planet in exactly half the time it takes the moon Mimas to orbit. Every time Mimas returns to a certain location, the ring particle will also return to its original location and will experience the same gravitational nudge from Mimas. The nudges reinforce one another and eventually clear a gap in the rings (Figure 9.21a). This "Mimas 2:1 resonance" is essentially the same type of *orbital resonance* that affects the orbits of Io, Europa, and Ganymede. It is responsible for the large gap in Saturn's rings easily visible from Earth (called the *Cassini division*). Many similar resonances are responsible for other gaps, and resonances can even create beautiful ripples within the rings (Figure 9.21b).

Although we now know three causes of rings and gaps—gap moons, shepherd moons, and orbital resonances—we have not yet identified specific causes for the vast majority of the features within Saturn's rings. Perhaps the Cassini spacecraft will identify more gap moons and shepherd moons, or perhaps it will reveal even more puzzling structures.

Comparing Planetary Rings

The ring systems of Jupiter, Uranus, and Neptune are so much fainter than Saturn's ring system that it took almost four centuries longer to discover them. Ring particles in these three systems are far less numerous, generally smaller, and much darker. Saturn's rings were the only ones known until 1977,

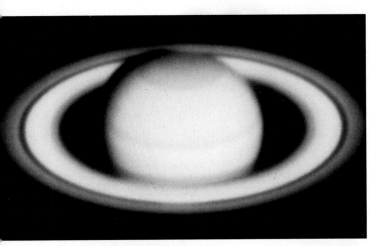

a Earth-based telescopic view of Saturn.

b Voyager image of Saturn's rings against the disk.

c Artist's conception of particles in a ring system. All the particles are moving slowly relative to one another and occasionally collide.

FIGURE 9.20 Looking closer at Saturn's rings.

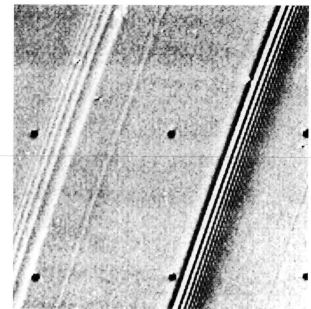

a The largest gap in Saturn's rings, called the Cassini division, is caused by an orbital resonance with the moon Mimas.

Cassini division

b Another Mimas resonance creates remarkable ripples in Saturn's rings. The dark spots in the image are calibration marks for the camera.

FIGURE 9.21 Gaps in Saturn's rings.

when the rings of Uranus were discovered during observations of a *stellar occultation*—a star passing behind Uranus as seen from the Earth. During the occultation, the star "blinked" on and off nine times before it disappeared behind Uranus, and nine more times as it emerged. Scientists concluded that these nine "blinks" were caused by nine thin rings encircling Uranus. Similar observations of stars passing behind Neptune gave more confounding results: Rings appeared to be present some of the time, but not at other times. Could Neptune's rings be incomplete or transient?

The Voyager spacecraft provided some definitive answers. Voyager cameras first discovered thin rings around Jupiter in 1979. Then, after providing incredible images of Saturn's rings, *Voyager 2* confirmed the existence of rings around Uranus as it flew past in 1986. In 1989, *Voyager 2* passed by Neptune and found that it does, in fact, have partial rings—at least when seen from Earth. The space between the ring segments is filled with dust not detectable from Earth.

A family portrait of the jovian ring systems shows many similarities (Figure 9.22). All rings lie in their planet's equatorial plane within the Roche tidal zone. Particle orbits are fairly circular, with small orbital tilts relative to the equator. Gaps and ringlets are probably due to gap moons, shepherd moons, and orbital resonances. The differences also offer important insights: The larger size, higher reflectivity, and much greater number of particles in Saturn's rings

compel us to wonder if different processes are at work there.

Origin of the Rings

Where do rings come from? How old are they? Did they form with their planet out of the nebula, or are they a more recent phenomenon? These questions long puzzled astronomers when they had only one case (Saturn) to study, but the discovery that all jovian planets have rings has made the answers clearer.

Part of the answer lies in determining how long rings will last. Unlike planets and large satellites, ring particles may not last very long. Ring particles orbiting Saturn bump into one another every few hours, and even with low-velocity collisions the particles are chipped away over time. Basketball-size ring particles around Saturn are ground away to dust in a few million years. Other processes dismantle rings around the other jovian planets. In summary, particles in the four ring systems cannot last as long as the 4.6-billion-year age of our solar system.

If ring particles are rapidly disappearing, some source of new particles must keep the rings supplied. The most likely source is collisions. As ring particles are ground away by collisions, small moons (perhaps the gap moons) occasionally collide and create more ring particles. Meteorites may also strike these moons, with the impacts releasing even more ring particles. We now believe that the "miniature solar nebulae" that surrounded the jovian planets formed

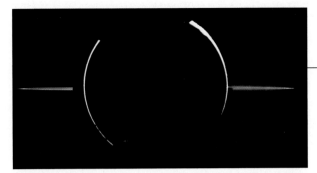

Jupiter

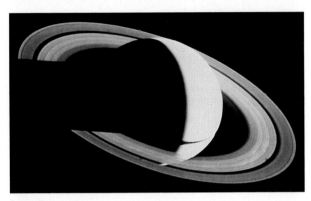

Saturn

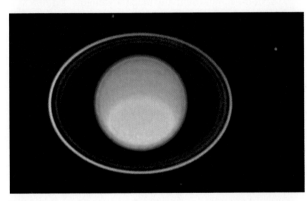

Uranus

Neptune

FIGURE 9.22 Four ring systems. The planets are not shown to scale. Uranus's rings were photographed by the Hubble Space Telescope, the others by Voyager. The Neptune frame is made of two images taken on either side of the bright planet.

many small satellites and that the gradual dismantling of these satellites creates the dramatic ring systems we see today. The appearance of any particular ring system probably changes dramatically over millions and billions of years. The spectacle of Saturn's brilliant rings may be a special treat of our epoch, one that could not have been seen a billion years ago and that may not last long on the time scale of our solar system.

THE BIG PICTURE

In this chapter, we saw that the jovian planets really are a "different kind of planet"—and, indeed, a different kind of planetary system. The jovian planets dwarf the terrestrial planets, and some of their moons are as large as terrestrial worlds. As you continue your study of the solar system, keep the following "big picture" ideas in mind:

- The jovian planets may lack solid surfaces on which geology can work, but they are interesting and dynamic worlds with rapid winds, huge storms, strong magnetic fields, and interiors in which common materials behave in unfamiliar ways.

- Despite their relatively small sizes and frigid temperatures, many jovian satellites are geologically active by virtue of their icy compositions. Ironically, it was cold temperatures in the solar nebula that led to icy compositions and hence geological activity.

- Ring systems owe their existence to small satellites formed from the "miniature solar nebulae" that produced the jovian planets billions of years ago. The rings we see today are composed of particles liberated from those satellites surprisingly recently.

- Understanding the jovian planets forced us to modify many of our earlier ideas about the solar system, in particular by adding the concepts of ice geology, tidal heating, and orbital resonances. Each new set of circumstances offers further opportunities to learn how the universe works.

Review Questions

1. Briefly describe the past and present space exploration of the jovian worlds. Summarize what we've learned about the jovian planets in terms of composition, rotation rate, and shape. How did the jovian planets form?

2. Describe the interior structure of Jupiter. Contrast Jupiter's interior structure with the structures of the other jovian planets. Why is Jupiter denser than Saturn? Why is Neptune denser than Saturn?

3. How do we think Jupiter generates its internal heat? Does a similar process operate on any of the terrestrial planets?

4. What is the *Great Red Spot?* Is it the same type of storm as a hurricane on Earth? Explain.

5. Describe the possible origins of Jupiter's vibrant colors. Contrast these with the origins of the colors of the other jovian planets.

6. Does Jupiter have seasons? Why or why not?

7. What is the *Io torus?* Why don't Saturn, Uranus, and Neptune have a similar "torus"? What is unusual about the magnetospheres of the other jovian planets?

8. Briefly describe some of the general characteristics of the jovian moons.

9. How is the *ice geology* of the jovian moons similar to the rock geology of the terrestrial planets? How is it different?

10. Describe the volcanoes of Io and the mechanism of *tidal heating* that generates Io's internal heat.

11. What are *orbital resonances?* Explain how the resonance between Io, Europa, and Ganymede makes their orbits slightly elliptical. How do orbital resonances relate to tidal heating?

12. Describe the evidence suggesting that Europa may have a subsurface ocean of liquid water.

13. Describe the general surface characteristics of Ganymede and Callisto. What do these characteristics tell us about internal heating on these worlds?

14. Describe the atmosphere of Titan. What evidence suggests that Titan may have oceans of liquid ethane?

15. Describe a few general features of Saturn's medium-size moons. Why is it surprising to find evidence of geological activity on some of these moons, and how do we explain it? How do we explain the odd appearance of Uranus's moon Miranda?

16. How do we know that Triton is *not* a typical large jovian satellite? Describe Triton's general characteristics and give possible ideas of where Triton came from.

17. Describe the appearance and structure of Saturn's rings. Where do the rings lie in relation to Saturn's *Roche tidal zone?*

18. Briefly describe how *gap moons* and orbital resonances can lead to gaps within ring systems.

19. Describe two possible scenarios for the origin of planetary rings. What makes us think that ring systems must be continually replenished?

Discussion Questions

1. *Jovian Mission.* We can study terrestrial planets up close by landing on them, but jovian planets have no surfaces to land on. Suppose you are in charge of planning a long-term mission to "float" in the atmosphere of a jovian planet. Describe the technology you will use and how you will ensure survival for any people assigned to this mission.

2. *Extrasolar Planets.* Many of the newly discovered planets orbiting other stars are more massive than Jupiter and much closer to their stars. Assuming that they are in fact jovian (as opposed to terrestrial or something else), how would you expect these new planets to differ from the jovian planets of our solar system? How would any magnetospheres, satellites, or rings be different? Explain.

Problems

Surprising Discoveries? For **problems 1–7**, suppose someone claimed to make the discovery described. (These are *not* real discoveries.) Decide whether the discovery should be considered reasonable or surprising. More than one right answer may be possible, so explain your answer clearly.

1. Saturn's core is pockmarked with impact craters and dotted with volcanoes erupting basaltic lava.

2. A jovian planet in another star system has a moon as big as Mars.

3. An extrasolar planet is discovered that is made primarily of hydrogen and helium; it has approximately the same mass as Jupiter but is 10 times larger than Jupiter in radius.

4. A new small moon is discovered to be orbiting Jupiter. It is smaller than any other of Jupiter's moons and orbits slightly farther from Jupiter than any of the other moons, but it has several large, active volcanoes.

5. A new moon is discovered to be orbiting Neptune. The moon orbits in Neptune's equatorial plane and in the same direction in which Neptune rotates, but it is made almost entirely of metals such as iron and nickel.

6. An icy, medium-size moon is discovered to be orbiting a jovian planet in a star system that is only a few hundred million years old. The moon shows evidence of active tectonics.

7. A jovian planet is discovered in a star system that is much older than our solar system. The planet has no moons at all, but it has a system of rings as spectacular as the rings of Saturn.

8. *The Importance of Rotation.* Suppose the material that formed Jupiter came together without any rotation, so that no "jovian nebula" formed and the planet today wasn't spinning. How else would the jovian system be different? Think of as many effects as you can, and explain each in a sentence.

9. *Minor Ingredients Matter.* Suppose the jovian planet atmospheres were composed 100% of hydrogen and helium rather than 98% of hydrogen and helium. How would the atmospheres be different in terms of color and weather? Explain.

10. *Disappearing Satellite.* Io loses about a ton (~1,000 kg) of sulfur dioxide per second to Jupiter's magnetosphere.

    a. At this rate, what fraction of its mass would Io lose in 5 billion years?

    b. Suppose sulfur dioxide currently makes up 1% of Io's mass. When will Io run out of this gas at the current loss rate?

11. *Project: Jupiter's Moons.* Using binoculars or a small telescope, view the moons of Jupiter. Make a sketch of what you see, or take a photograph. Repeat your observations several times (nightly, if possible) over a period of a couple of weeks. Can you determine which moon is which? Can you measure their orbital periods? Can you determine their approximate distances from Jupiter? Explain.

12. *Project: Saturn and Its Rings.* Using binoculars or a small telescope, view the rings of Saturn. Make a sketch of what you see, or take a photograph. What season is it in Saturn's northern hemisphere? How far above Saturn's atmosphere do the rings extend? Can you identify any gaps in the rings? Describe any other features you notice.

Web Projects

Find useful links for Web projects on the text Web site.

1. *Galileo and Cassini Missions.* Find the latest news on the Galileo and Cassini missions.

   a. Print out (or describe) an image from the Galileo mission of interest to you, and interpret it using the techniques developed in this chapter.

   b. In a few sentences, describe the main scientific goals of the Cassini mission. What is the current health of the spacecraft, and how is the mission progressing?

2. *Oceans of Europa.* The possibility of subsurface oceans on Europa holds great scientific interest and may even mean that life could exist on Europa. Find some of the latest research concerning Europa and learn about proposed missions to Europa. Write a one- to two-page summary of what you learn.

As we look out into the Universe and identify the many accidents of physics and astronomy that have worked to our benefit, it almost seems as if the Universe must in some sense have known that we were coming.

FREEMAN DYSON

CHAPTER 10

Remnants of Rock and Ice
Asteroids, Comets, and Pluto

Asteroids and comets might seem insignificant in comparison to the planets and satellites we've been discussing. But there is strength in numbers, and the trillions of small bodies orbiting our Sun are far more important than their small size might suggest.

Few topics in astronomy have the potential to touch our lives more directly than asteroids and comets. The appearance of comets more than once altered the course of human history, when our ancestors acted upon superstitions related to comet sightings. More profoundly, asteroids or comets falling to the Earth have scarred our planet with impact craters and altered the course of biological evolution.

Asteroids and comets are also important for a very different reason: They are remnants from the birth of our solar system and therefore provide an important testing ground for our theories of how our solar system came to be. In this chapter, we will explore asteroids, comets, and those bodies that fall to Earth as meteorites. We will also examine the smallest planet, Pluto, and see that, while it may be a misfit among planets, it may be right at home among the smaller objects of our solar system.

10.1 Remnants from Birth

The objects we've studied so far—terrestrial planets, jovian planets, and large moons—have changed dramatically since their formation. Most planets have been transformed in almost every way: Their interiors have differentiated, their surfaces have been reshaped by geological processes, and their atmospheres have changed as some gases escape while others are added. But many small bodies remain virtually unchanged since their formation some 4.5 billion years ago. Thus, comets, asteroids, and meteorites carry the history of our solar system encoded in their compositions, locations, and numbers.

Using these small bodies to understand planetary formation is a bit like picking through the trash in a carpenter's shop to see how furniture is made. By studying the scraps, sawdust, paint chips, and glue, we can develop a rudimentary understanding of how the carpenter builds furniture. Similarly, the "scraps" left over from the formation of our solar system allow us to understand how the planets and larger moons came to exist. Indeed, much of our modern theory of solar system formation was developed from studies of asteroids, comets, and meteorites.

In this chapter, we will investigate what we have learned from these remnants from the birth of our solar system. We'll also discuss the cosmic collisions that sometimes occur between leftover planetesimals and the existing planets. Before we continue, however, it's worth noting that the terminology used to describe these objects has a long history that can sometimes be confusing. The word *comet* comes from the Greek word for "hair," and comets get their name from the long, hairlike tails they display on the rare occasions when they come close enough to the Sun to be visible in our sky. *Asteroid* means "starlike," but asteroids are starlike only in their appearance through a telescope, not in their fundamental properties. The term *meteor* refers to a bit of interplanetary dust burning up in our atmosphere; literally, it means "a thing in the air." Meteors are also sometimes called *shooting stars* or *falling stars* because some people once believed that they really were stars falling from the sky. Larger chunks of rock that survive the plunge through the atmosphere and hit the ground are called *fallen stars* or *meteorites*, which means "associated with meteors." To keep potential confusion to a minimum, we will use only the following terms and definitions in this book:

- *Asteroid*: a rocky leftover planetesimal orbiting the Sun.

- *Comet*: an icy leftover planetesimal orbiting the Sun—regardless of its size or whether or not it has a tail.

- *Meteor*: a flash of light in the sky caused by a particle entering the atmosphere, whether the particle comes from an asteroid or a comet.

- *Meteorite*: any piece of rock that fell to the ground from the sky, whether from an asteroid, a comet, or even another planet.

10.2 Asteroids

Asteroids are virtually undetectable to the naked eye and went unnoticed for almost two centuries after the invention of the telescope. The first asteroids were discovered about 200 years ago when astronomers were searching the "extra wide" gap between the orbits of Mars and Jupiter in hopes of discovering a previously unknown planet. It took 50 years to discover the first 10 asteroids (all large and bright); today modern telescopes with advanced detectors discover that many on a typical night. Once an asteroid's orbit is calculated and verified, the asteroid is assigned a number (in order of verification), and the discoverer chooses a name (subject to the approval of the International Astronomical Union). The earliest discovered asteroids bear names of mythological figures; more recent discoveries carry names of scientists, cartoon heroes, pets, and rock stars.

The radius of the largest asteroid, Ceres, is about 500 kilometers—about half that of Pluto. About a dozen others are large enough that we would call them "medium-size" moons if they orbited a planet. Most asteroids are much smaller. There are probably more than 100,000 asteroids larger than 1 kilometer in diameter. Despite their large numbers, asteroids don't add up to much: if they could all be put together, they'd make an object less than 1,000 kilometers in radius—far smaller than any terrestrial planet.

Painstaking observations have revealed the precise orbits of thousands of asteroids. The main **asteroid belt** lies between 2.2 and 3.3 AU from the Sun, though some interesting asteroids lie outside the belt (Figure 10.1). Asteroid orbits are distinctly elliptical and can be inclined relative to the ecliptic by as much as 20°–30°, but all orbit the Sun in the same direction as the planets. Science fiction movies often show the asteroid belt as a crowded and hazardous place, but the average distance between asteroids is millions of kilometers. Spacecraft must be carefully guided to fly close enough to an asteroid to study it, and accidental collisions are extremely unlikely.

Origin and Evolution of the Asteroid Belt

Why are asteroids concentrated in the asteroid belt, and why didn't a full-fledged planet form in this re-

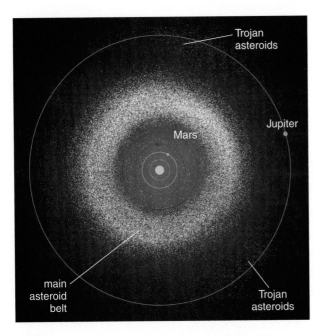

FIGURE 10.1 Positions of nearly 80,000 asteroids for midnight, 1 January 2002. The asteroids themselves are much smaller than the dots on this scale. The Trojan asteroids are found 60° ahead of and behind Jupiter in its orbit.

orbital resonances between Saturn's moons and particles in its rings: The moons are much larger than the ring particles and therefore force the ring particles out of resonant orbits, clearing *gaps* in the rings [Section 9.6]. In the asteroid belt, similar orbital resonances occur between asteroids and Jupiter, the most massive planet by far. For example, any asteroid in an orbit that takes 6 years to circle the Sun—half of Jupiter's 12-year orbital period—would receive the same gravitational nudge from Jupiter every 12 years and thus would soon be pushed out of this orbit. The same is true for asteroids with orbital periods of 4 years ($\frac{1}{3}$ of Jupiter's period) and 3 years ($\frac{1}{4}$ of Jupiter's period).

Today, we see the results of these orbital resonances when we graph the number of asteroids with various orbital periods. Just as we would expect, there are virtually no asteroids with periods exactly $\frac{1}{2}$, $\frac{1}{3}$, or $\frac{1}{4}$ of Jupiter's. Thus, there are no asteroids with average distances from the Sun (semimajor axes) that correspond to these orbital periods. Figure 10.2 shows several other gaps in the asteroid belt, corresponding to orbital resonances that make other simple ratios with Jupiter's orbital period. (By a *gap* in the asteroid belt we mean a gap in *average* distance; because asteroid orbits are elliptical, some asteroids pass through the gaps on parts of their orbits.)

The gravitational tugs from Jupiter that created the gaps in the asteroid belt probably also explain why no planet ever formed in this region. Early in the history of the solar nebula, the asteroid belt probably contained enough rocky planetesimals to form another planet as large as Earth or Mars. Just as in other parts of the solar nebula, collisions among

gion? The answers probably lie with *orbital resonances*. Recall that an orbital resonance occurs whenever one object's orbital period is a simple ratio of another object's period, such as $\frac{1}{2}$, $\frac{1}{4}$, or $\frac{5}{3}$. In such cases, the two objects periodically line up with each other, and the extra gravitational attractions at these times can affect the objects' orbits. Consider, for example, the

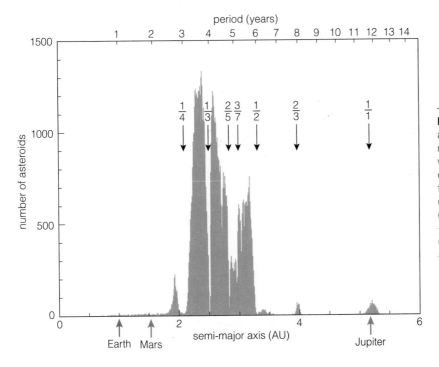

FIGURE 10.2 This graph shows the numbers of asteroids with different average distances (semimajor axes) from the Sun. Note the gaps—distances where we see few if any asteroids. The ratio labeling each gap shows how the orbital period of objects at this distance compares to the 12-year orbital period of Jupiter. For example, the label $\frac{1}{3}$ means that the orbital period of objects with this average distance is $\frac{1}{3}$ of Jupiter's orbital period, or about 4 years. Most orbital resonances result in gaps, though some (such as $\frac{1}{1}$ and $\frac{2}{3}$) happen to gather up asteroids.

these planetesimals frequently disrupted their orbits. However, the asteroid belt had an additional source of disruption: gravitational resonances with the forming planet Jupiter. The result was that planetesimals could not collect together to form a planet. Over the past 4.5 billion years, the ongoing orbital disruptions have gradually kicked pieces of this "unformed planet" out of the asteroid belt altogether. Once booted from the main asteroid belt, these objects eventually either crash into a planet or a moon or are flung out of the solar system. Thus, the asteroid belt slowly loses mass, explaining why the total mass of all its asteroids is now less than that of any terrestrial planet.

Most of the asteroids found outside the asteroid belt—including the *Earth-approaching asteroids* that pass near Earth's orbit—are probably "impacts waiting to happen." But asteroids can safely congregate in two stable zones outside the main belt. These zones are found along Jupiter's orbit 60° ahead of and behind Jupiter (see Figure 10.1). The asteroids found in these two zones are called the *Trojan asteroids*. The Trojan asteroids are stable because of a different type of orbital resonance with Jupiter. In this case, any asteroid that wanders away from one of theses zones is nudged back *into* the zone by Jupiter's gravity.

Learning About Asteroids

The first step in learning about an asteroid is determining its precise orbit. Asteroids are recognizable in telescopic images because they move noticeably relative to the stars in just a short time (Figure 10.3).

FIGURE 10.3 Because asteroids orbit the Sun, they move relative to the stars just as planets do. In this photograph, stars show up as white dots, and the motion of an asteroid makes it show up as a short streak.

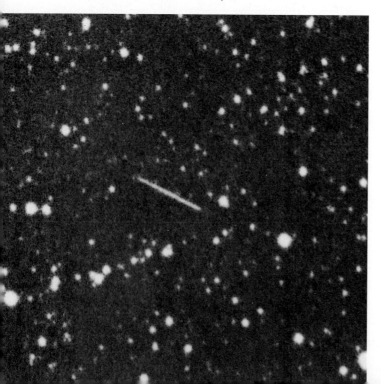

Once we detect an asteroid, we can calculate its orbit by applying Kepler's laws [Section 3.4]—even if we've observed only a small part of one complete orbit.

Most asteroids appear as little more than points of light even to our largest Earth-based telescopes, so the best way to study an asteroid in depth is to send a spacecraft out to meet it. Spacecraft have visited only a handful of asteroids close-up in their native environment of the asteroid belt and a few other presumed asteroids "in captivity" orbiting other planets (e.g., Phobos and Deimos orbiting Mars, and the backward outer satellites of Jupiter). The Galileo spacecraft on its way to Jupiter photographed the asteroid Gaspra (Figure 10.4a) and an asteroid pair known as Ida and Dactyl (Figure 10.4b). The Near-Earth Asteroid Rendezvous mission (NEAR) photographed the asteroid Mathilde (Figure 10.4c) on its way to the asteroid Eros. NEAR orbited Eros (Figure 10.4d) for a full year, studying its entire surface at close range, before setting down on its surface. These spacecraft images reveal battered worlds that are probably fragments of larger asteroids shattered by collisions.

10.3 Meteorites

No spacecraft has yet landed on an asteroid, but we have pieces of asteroids nonetheless—the rocks called *meteorites* that fall from the sky. The reality of rocks falling from the sky wasn't always accepted. In fact, Thomas Jefferson (who worked in science as well as politics), upon hearing of a meteorite fall in Connecticut, reportedly said, "It is easier to believe that Yankee professors would lie than that stones would fall from heaven." Today we know that rocks really do sometimes fall from the heavens. Some are large enough to do significant damage, blasting huge craters upon impact. Despite the forces of erosion, more than 100 impact craters can still be seen on Earth.

Rocks from space continue to rain down on Earth, and observers find a few meteorites every year by following the trajectories of very bright meteors called *fireballs*. Meteorites are often blasted apart in their fiery descent through our atmosphere, scattering fragments over an area several kilometers across. A direct hit on the head by a meteorite would be fatal, but there are no reliable accounts of human deaths from meteorites. However, meteorites have injured at least one human (by a ricochet, not a direct hit), damaged houses and cars, and killed animals. Most meteorites fall into the oceans, which cover three-quarters of the Earth's surface area.

Comparing Meteorites

Most meteorites are difficult to distinguish from terrestrial rocks without a detailed scientific analysis,

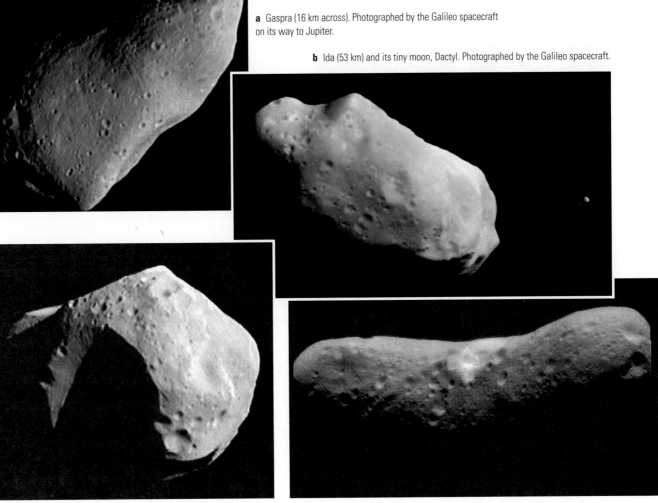

a Gaspra (16 km across). Photographed by the Galileo spacecraft on its way to Jupiter.

b Ida (53 km) and its tiny moon, Dactyl. Photographed by the Galileo spacecraft.

c Mathilde (59 km). Photographed by NEAR on its way to Eros.

d Eros (40 km). The NEAR spacecraft orbited Eros for a year before landing on it.

FIGURE 10.4 Close-up views of asteroids studied by spacecraft.

but a few clues can help. Meteorites are usually covered with a dark, pitted crust resulting from their fiery passage through the atmosphere. Some can be readily distinguished from terrestrial rocks by their high metal content. The ultimate judge of extraterrestrial origin is laboratory analysis of the rock's composition. (Most types of meteorites contain enough metal to attract a magnet hanging on a string. If you suspect that you have found a meteorite, most science museums will analyze a small chip free of charge.) The presence of certain rare elements such as iridium is one major indicator of extraterrestrial origin; nearly all of Earth's iridium sank to the core long ago and hence is absent from surface rocks.

The tens of thousands of known meteorites fall into two basic categories. The vast majority of meteorites appear to be composed of a random mix of flakes from the solar nebula, and radioactive dating shows that they formed at the same time as the solar system itself—about 4.6 billion years ago [Section 7.5]. These **primitive meteorites** are our best source

of information about conditions in the solar nebula (Figure 10.5a). Most primitive meteorites are composed of rocky minerals with an important difference from Earth rocks: A small but noticeable fraction of pure metallic flakes is mixed in. (Iron in Earth rocks is chemically bound in minerals.) Some primitive meteorites also contain substantial amounts of carbon compounds, and some even contain a small amount of water.

A much smaller group of meteorites appears to have undergone substantial change since the formation of the solar system, and radioactive dating shows some of these meteorites to be younger than the primitive meteorites. We call this group **processed meteorites** because they apparently were once part of a larger object that "processed" the original material of the solar nebula into another form (Figure 10.5b). Some processed meteorites resemble the Earth's core in composition: a dense iron/nickel mixture with trace amounts of other metals. Other processed meteorites have much lower densities and are made of rocks more similar to the Earth's crust or

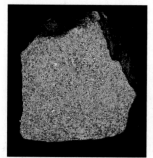

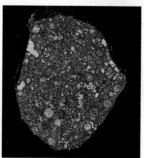

a

Stony primitive meteorite: metal flakes intermixed with rocky material.

Carbon-rich primitive meteorite: similar to meteorite at left, but with carbon compounds added in.

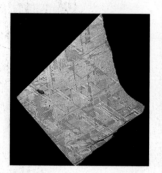

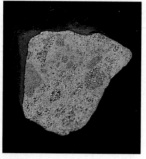

b

Differentiated iron meteorite: similar to a planetary core.

Differentiated stony meteorite: similar to volcanic rocks.

FIGURE 10.5 (a) Two examples of primitive meteorites. (b) Two examples of processed meteorites.

mantle. A few even have a composition remarkably close to that of the lava erupted from terrestrial volcanoes [Section 8.3].

The Origin of Meteorites

We've seen that primitive meteorites may be either rocky or carbon-rich. Why are there two varieties? The answer lies in where they formed in the solar nebula. Carbon compounds condense only at the relatively low temperatures that were found in the solar nebula beyond about 3 AU from the Sun. Thus, carbon-rich meteorites must come from the outer portion of the asteroid belt (beyond about 3 AU from the Sun). The remainder of the primitive meteorites lack carbon compounds and thus presumably formed in the inner warmer part of the asteroid belt.

The processed meteorites tell a more complex story. Their compositions appear similar to the cores, mantles, or crusts of the terrestrial worlds. Thus, they must be fragments of worlds that underwent *differentiation* just like the terrestrial worlds [Section 8.2]. That is, they are fragments of worlds that must have been heated to high enough temperatures to melt inside, allowing metals to sink to the center and rocks to rise to the surface. The processed meteorites with basaltic compositions must come from lava flows that occurred on the surfaces of some of the

larger asteroids during a time when they had active volcanism.

Meteorites resembling lava may simply have been chipped off the surface of a large asteroid by relatively small collisions. In contrast, the processed meteorites with corelike or mantlelike compositions must be fragments of worlds that completely shattered in collisions. These shattered worlds not only lost a chance to grow into planets but sent many fragments onto collision courses with other asteroids and other planets, including Earth. Thus, processed meteorites essentially offer us an opportunity to study a "dissected planet."

If fragments such as the basaltic meteorites can be chipped off the surface of an asteroid, is it possible that they might also be chipped off a larger object such as a moon or a planet? In fact, a few processed meteorites have been found that don't appear to match the compositions of asteroids but instead appear to match the composition either of the Moon or of Mars. We now believe that some of these meteorites in fact were chipped off the Moon in impacts and that others were chipped off Mars. The analysis of these *lunar meteorites* and *Martian meteorites* is providing new insights into the conditions on the Moon and Mars. In at least one case, a Martian meteorite may be offering us clues about whether life once existed on Mars [Section 11.6].

10.4 Comets

Humans have watched comets in awe for millennia. The occasional presence of a bright comet in the night sky was hard to miss before the advent of big cities, inexpensive illumination, and television. In some cultures, these rare intrusions into the unchanging heavens foretold bad or good luck, and in most cultures there was little attempt to interpret the event in astronomical terms. Nowadays, these leftover icy planetesimals can teach us about formation processes in the outer solar system, just as asteroids teach us about the inner solar system.

After Kepler developed his laws of planetary motion, other astronomers applied them to comets and found extremely eccentric orbits. Some comets visit the inner solar system only once before being ejected into interstellar space, others return on orbits spanning thousands of years, and a few come by more frequently. The most famous is Halley's comet, named for the English scientist Edmund Halley (1656–1742). Halley did not "discover" his comet but rather gained fame for recognizing that a comet seen in 1682 was the same one seen on a number of previous occasions. He used Newton's law of gravitation to calculate the comet's 76-year orbit and, in a book published in 1705, predicted that the comet would return in 1758. The comet

a Comet Hyakutake in 1996.

b Comet Hale–Bopp in 1997, photographed at Mono Lake in California.

FIGURE 10.6 Brilliant comets can appear at almost any time, as demonstrated by the back-to-back appearances of comet Hyakutake and comet Hale–Bopp.

c Comet SOHO-6's final blaze of glory. This "Sun grazing" comet was observed by the SOHO spacecraft a few hours before it passed just 50,000 km above the Sun's surface. The comet did not survive its passage, due to the intense solar heating and tidal forces. In this image, the Sun is blocked out behind the large, orange disk; the sun's size is indicated by the white circle.

was given his name when it reappeared as predicted, 16 years after his death. Halley's comet has returned on schedule ever since, including twice in the twentieth century—spectacularly in 1910 when it passed near the Earth, and unimpressively in 1986 when it passed by at a much greater distance. It will next visit in 2061. Several comets are discovered every year, and most have never been seen before by human eyes. In 1996 and 1997, we were treated to back-to-back brilliant comets: comet Hyakutake and comet Hale–Bopp (Figure 10.6). The surprise champion of comet discovery is the SOHO spacecraft, an orbiting solar observatory. SOHO has detected more than 200 "Sun grazing" comets, most on their last pass by the Sun (Figure 10.6c).

The Flashy Lives of Comets

Comets are icy planetesimals from the outer solar system. Their composition has been described as "dirty snowballs": ices mixed with rocky dust. Comets are completely frozen when they are far from the Sun, and most are just a few kilometers across. But a comet takes on an entirely different appearance when it comes close to the Sun (Figure 10.7). The dirty snowball is the comet's **nucleus**. The sublimation of ices in the nucleus creates a rapidly escaping dusty atmosphere called the **coma**, along with a **tail** that points away from the Sun regardless of which way the comet is moving through its orbit. In most

cases, we actually see two tails: a **plasma tail** made of ionized gas, and a **dust tail** made of small solid particles. Although much of comet science focuses on the dramatic tail and coma, we begin our closer look at comets with the humbler nucleus.

Comet nuclei are rarely observable with telescopes because they are quite small and are shrouded by the dusty coma. A spacecraft flyby of comet Halley revealed a dark, lumpy, potato-shaped nucleus about 16 kilometers long and 8 kilometers in width and depth. Despite its icy composition, comet Halley's nucleus is darker than charcoal, reflecting only 3% of

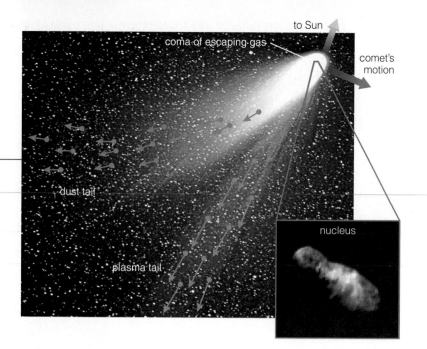

to Sun

comet's motion

coma of escaping gas

dust tail

plasma tail

nucleus

FIGURE 10.7 Anatomy of a comet. The inset photo is the nucleus of comet Borrelly taken by the Deep Space 1 spacecraft in September 2001.

the light that falls on it. It doesn't take much rocky soil or carbon-rich material to darken a comet.

How can such a small nucleus put on such a spectacular show? Comets spend most of their lives in the frigid outer limits of our solar system. As a comet accelerates toward the Sun, its surface temperature increases and ices begin to sublime into gaseous form. By the time the comet comes within about 5 AU of the Sun, sublimation begins to form a noticeable atmosphere that easily escapes the comet's weak gravity. The escaping atmosphere drags away dust particles that were mixed with the ice and begins to create the dusty coma around the nucleus. Sublimation rates increase as the comet approaches the Sun, and gases jet away from patches on the nucleus at speeds of hundreds of meters per second (Figure 10.8).

As gas and dust move away from the comet, they are influenced by the Sun in different ways. The gases are ionized by ultraviolet photons from the Sun and then carried straight outward away from the Sun with the solar wind at speeds of hundreds of kilometers per second. These ionized gases form the *plasma tail,* which may extend hundreds of millions of kilometers away from the Sun. Dust-size particles experience a much weaker push caused by the pressure of sunlight itself (called *radiation pressure* [Section 14.4]), so the *dust tail* is swept away from the Sun in a slightly different direction (see Figure 10.8). The comet wheels around the Sun, always keeping its tails of dust and plasma pointed roughly away from the Sun. Comets also eject some larger, pebble-size particles that are not pushed away from the Sun. These form another, invisible, "tail" extending along the comet's orbit and are the particles responsible for meteor showers [Section 10.6].

After the comet loops around the Sun and begins to head back outward, sublimation declines, the coma dissipates, and some of the dust settles back to the surface. Nothing happens until the comet again comes sunward—in a century, a millennium, a million years, or perhaps never.

Active comets cannot last forever. In a close pass by the Sun, a comet may shed a layer 1 meter thick. Dust that is too heavy to escape accumulates on the surface. This thick dusty deposit helps make comets dark and may eventually block the escape of interior gas that makes comets so phenomenal. It is estimated that a comet loses about 0.1% of its volatiles on every pass around the Sun. No one is certain what happens after the volatiles stop escaping. Either the remaining dust disguises the dead comet as an asteroid until it crashes into a planet or is ejected from the solar system, or the comet comes "unglued" without the binding effects of ices and simply disintegrates along its orbit.

The Origin of Comets

Comets are a rarity in the inner solar system, with only a handful coming within the orbit of Jupiter at any one time. However, because comets cannot last for very many passes by the Sun, comets must come from enormous reservoirs at much greater distances: the Kuiper belt and the Oort cloud (Figure 10.9).

Most comet orbits fit no pattern—they do not orbit the Sun in the same direction as the planets, and their elliptical orbits can be pointed in any direction. Comets Hyakutake and Hale–Bopp both fall into this class. It is believed that such comets are visitors from the Oort cloud. Based on the number of comets that fall into the inner solar system, the Oort cloud must contain about a trillion (10^{12}) comets.

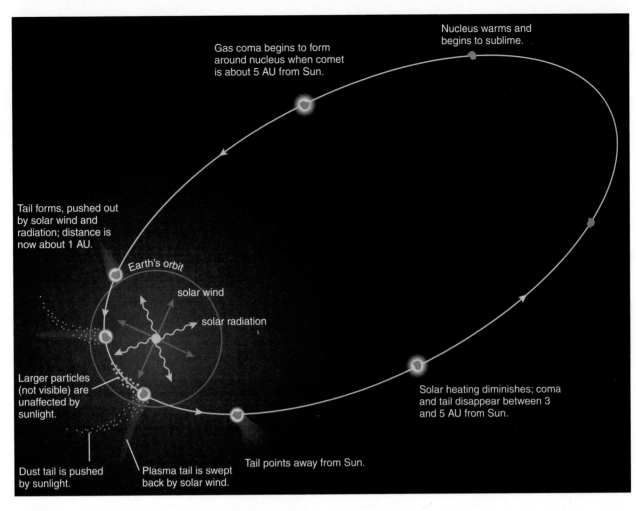

FIGURE 10.8 Comets exist as bare nuclei over most of their orbits and grow a coma and tails only when they approach the Sun. (Diagram not to scale.)

However, Oort cloud comets normally orbit the Sun at such great distances—up to about 50,000 AU—that we have so far seen only the rare visitors to the inner solar system. The comets of the Oort cloud probably formed in the vicinity of the jovian planets and were flung into their large, random orbits by gravitational encounters with these planets.

In contrast, the comets of the Kuiper belt probably still lie in the same general region in which they formed, which explains why their orbits match the pattern of the solar nebula as a whole. In the 1990s, astronomers began using the largest telescopes to search for Kuiper belt objects in their native environment rather than on

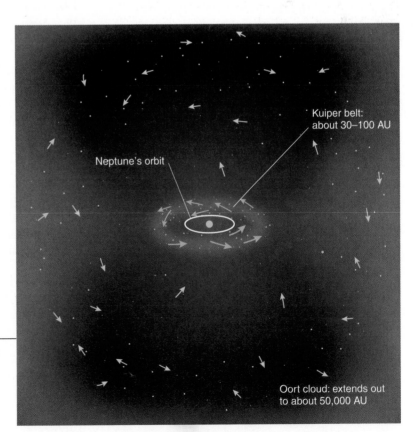

FIGURE 10.9 The Kuiper belt and the Oort cloud. Arrows indicate representative orbital motions of objects in the two regions. (Figure not to scale.)

their excursions into the inner solar system. The first discovery, given the name 1992QB1, orbits beyond Pluto with an orbital period of 296 years. It is a comet approximately 280 kilometers across. Subsequent searches have identified dozens of other objects that match the predicted orbital properties of Kuiper belt comets: fairly circular orbits in the same direction as the planets, with small orbital tilts. More than 400 Kuiper belt comets have been discovered so far, suggesting that at least 100,000 comets more than 100 km across must populate the Kuiper belt, along with a billion objects about 10 km across (the typical size of comets that enter the inner solar system). Thus, the total mass of the Kuiper belt is far greater than that of the asteroid belt.

Other solar systems may well have their own Kuiper belts. The star CW Leonis reveals vast amounts of water vapor, apparently released as the heat of the dying star encroaches on the distant realm of comets orbiting it.

How big can Kuiper belt objects be? The largest Kuiper belt comet discovered so far, designated 2001KX76, may be nearly 1,200 km across, larger than the asteroid Ceres. If a comet can be 1,200 km across, could one be larger still? It's time to consider the odd planet Pluto, only twice the diameter of 2001KX76.

10.5 Pluto: Lone Dog or Part of a Pack?

Pluto was discovered in 1930 by American astronomer Clyde Tombaugh, culminating a search that began not long after the discovery of Neptune in 1846. Neptune's existence and location were predicted before its actual discovery by mathematical analysis of irregularities in the orbit of Uranus. Further analysis of Uranus's orbit suggested the existence of a "ninth planet," and Pluto was found only 6° from the predicted position of this planet. Initial estimates suggested that Pluto was much larger than Earth, but successively more accurate measurements derived ever-smaller sizes. We now know that Pluto has a radius of only 1,195 kilometers and a mass of just 0.0025 Earth mass—making it far too small to affect the orbit of Uranus. In fact, the discovery of Pluto near the predicted position of the "ninth planet" was coincidental; Uranus's supposed orbital irregularities were apparently just errors in measurement. Pluto might have escaped detection for decades without those errors.

Pluto has long seemed to be a misfit among the planets, fitting into neither the terrestrial nor the jovian category. Its 248-year orbit is also unusually elliptical and significantly tilted relative to the ecliptic. (See Figure 7.1.) Near perihelion, it actually comes closer to the Sun than Neptune—such was

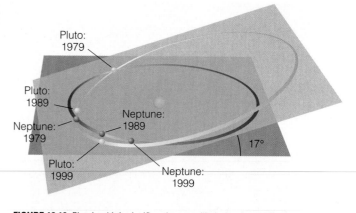

FIGURE 10.10 Pluto's orbit is significantly more elliptical and more tilted relative to the ecliptic than that of any other planet. It comes closer to the Sun than Neptune for 20 years in each 248-year orbit, as was the case between 1979 and 1999. But there's no danger of a collision between Pluto and Neptune, because they share an orbital resonance in which Neptune completes three orbits for every two orbits by Pluto.

the case between 1979 and 1999. At aphelion, it is 50 AU from the Sun, putting it far beyond the realm of the jovian planets. Despite the fact that Pluto sometimes comes closer to the Sun than Neptune, there is no danger of the two planets colliding: For every two Pluto orbits, Neptune circles the Sun three times (Figure 10.10). Because of this stable *orbital resonance*, whenever Pluto is near Neptune's orbit, Neptune is always a safe distance away. The two will probably continue their dance of avoidance until the end of the solar system.

Pluto and Its Moon

Pluto's great distance and small size make it difficult to study, but the task became much easier with the discovery of Pluto's moon Charon in 1978. The original discovery was made when astronomers noticed a "bump" on Pluto's blurry image that moved from side to side over a 6.4-day period. Today we have much clearer pictures of Pluto and Charon from the Hubble Space Telescope. The discovery of the moon enabled astronomers to accurately determine the mass of Pluto by applying Newton's version of Kepler's third law. They also learned that Pluto's rotation axis is tipped 118° relative to its orbit, making it the third planet (in addition to Venus and Uranus) that rotates backward relative to the majority of planets.

Pluto's moon was discovered just in time, because astronomers soon learned that Charon's orbit was about to go edge-on as seen from the Earth—something that happens only every 124 years. From 1985 to 1990, astronomers carefully monitored the combined brightness of Pluto and Charon as they alternately eclipsed each other every few days. Detailed analysis of brightness variations allowed calculation of accurate sizes, masses, and densities for both Pluto and Charon, as well as the compilation of rough maps

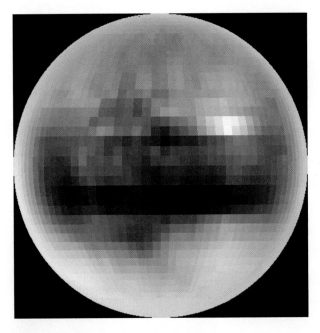

FIGURE 10.11 The surface of Pluto in approximate true color, as derived from brightness measurements made during mutual eclipses between Pluto and Charon. The brown color probably comes from methane ice, but the origin of the dark equatorial band is unknown.

thousand times fainter than it appears here on Earth, and it would be no larger in angular size than Jupiter is in our skies.

Pluto: Planet or Kuiper Belt Object?

Pluto is an object in search of a category. It is a misfit among the planets, but it seems somewhat less out of place when considered in the context of the Kuiper belt. After all, it orbits in the same vicinity as Kuiper belt comets, it has a cometlike composition of icy and rocky materials, and it has a cometlike atmosphere that grows as it comes nearer the Sun in its orbit and fades as it recedes. Could Pluto be simply the largest known member of the Kuiper belt?

A closer look at the orbits of Kuiper belt objects uncovers an uncanny resemblance to the orbit of Pluto. Like Pluto, many Kuiper belt objects lie in stable orbital resonances with Neptune, and more than a dozen Kuiper belt objects (including 2001KX76, the largest discovered so far), have the *same* period and semimajor axis as Pluto itself (and are nicknamed "Plutinos").

All in all, Pluto stands apart from Kuiper belt objects only in its size. But Pluto may not have been so unusual in the early solar system. Recall that Neptune's moon Triton is quite similar to (and larger than) Pluto, and its orbit indicates that it must be a captured object. Thus, Triton would once have been a "planet" orbiting the Sun. Giant impacts of other large icy objects may have created Charon and might even have given Uranus its large tilt. It remains possible that we will someday discover other objects with sizes similar to that of Pluto, though probably farther away or considerably darker. As of 2001, only a small percentage of the sky had been thoroughly searched for large Kuiper belt objects.

TIME OUT TO THINK *What is your definition of a planet? Be as rigorous as possible. Exactly how many planets are there, according to your criteria? Does your list include Pluto, Charon, Ceres or other asteroids, or Triton (a captured cousin of Pluto)? How big would a new Kuiper belt object have to be to meet your definition? Try to convince a classmate of your definition. Is your definition useful?*

of their surface markings (Figure 10.11). Charon's diameter is more than half Pluto's, and its mass is about one-eighth the mass of Pluto. Furthermore, Charon orbits only 20,000 kilometers from Pluto. (For comparison, our Moon has a mass $\frac{1}{80}$ that of Earth and orbits 400,000 kilometers away.) Charon's large relative size has caused some scientists to speculate that it, like our Moon, formed in a giant impact. Some astronomers argue that Pluto and Charon qualify as a "double planet," but others argue that neither is large enough to qualify as a planet at all.

Pluto currently has a thin atmosphere of nitrogen and other gases formed by sublimation of surface ices. However, the atmosphere is steadily thinning because Pluto's elliptical orbit is now carrying it farther from the Sun. (Pluto will not reach aphelion until 2113.) As Pluto recedes, its atmospheric gases are refreezing onto its surface.

Despite the cold, the view from Pluto would be stunning. Charon would dominate the sky, appearing almost 10 times larger in angular size than our Moon. Pluto and Charon's mutual tidal pulls long ago made them rotate synchronously with each other (see Figure 5.12). Thus, Charon is visible from only one side of Pluto and would hang motionless in the sky, always showing the same face. Moreover, the synchronous rotation means that Pluto's "day" is the same length as Charon's "month" (orbital period) of 6.4 Earth days. Charon would neither rise nor set in Pluto's skies but instead would cycle through its phases in place. The Sun would appear more than a

Planet or not, Pluto remains the largest known object in the solar system never visited by spacecraft. Close-up observations of its surface, atmosphere, and unusual moon are sure to teach us as much about icy bodies as past missions did about rocky bodies. NASA's current goal is to send a mission to Pluto before its atmosphere completely refreezes onto the surface, but budget constraints may keep the mission from becoming a reality. (Check the text Web site for updates on the status of a Pluto mission.)

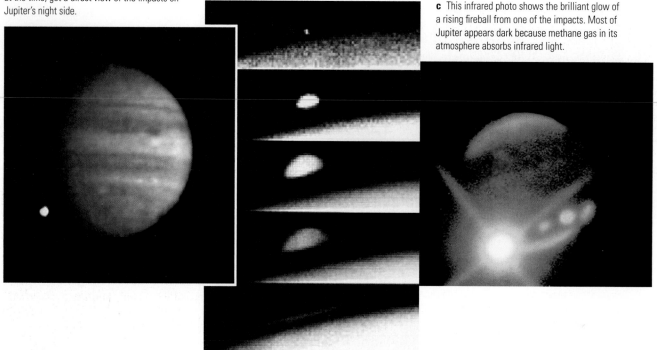

a The Galileo spacecraft, on its way to Jupiter at the time, got a direct view of the impacts on Jupiter's night side.

b In under 20 minutes, the Hubble Space Telescope observed an impact plume rise thousands of kilometers above Jupiter's clouds and then collapse back down.

c This infrared photo shows the brilliant glow of a rising fireball from one of the impacts. Most of Jupiter appears dark because methane gas in its atmosphere absorbs infrared light.

FIGURE 10.12 The SL9 impacts allowed astronomers their most direct view ever of cosmic collisions.

10.6 Cosmic Collisions: Small Bodies Versus the Planets

The hordes of small bodies orbiting the solar system are slowly shrinking in number through collisions with the planets and ejection from the solar system. Many more must have roamed the solar system in the days of the early bombardment, when most impact craters were formed [Section 7.5]. But there are still plenty left, and cosmic collisions still occur on occasion, with important ramifications for Earth as well as for other planets.

Shoemaker–Levy 9 Impacts on Jupiter

We usually think about impacts in the context of solid bodies because such impacts leave long-lasting scars in the form of impact craters. But impacts must be even more common on the jovian planets than on the terrestrial planets because of their larger size and stronger gravity, although no impact craters can form on gaseous worlds. It's estimated that a major impact on Jupiter occurs about once every 1,000 years. We were privileged to witness one in 1994.

Comet-hunters Gene and Carolyn Shoemaker, along with their colleague David Levy, came across an unusual object that turned out to be a string of comet nuclei all in a row. The reason for the odd appearance of *Shoemaker–Levy 9*, or *SL9* for short, became clear when astronomers calculated its orbit backward in time. The calculations showed that it had passed very close to Jupiter only a few months earlier (in July 1992). Because it had passed well within Jupiter's Roche tidal zone [Section 9.6], it apparently was ripped apart by tidal forces. Prior to that, SL9 was probably a single comet nucleus orbiting Jupiter, unnoticed by human observers.

The orbital calculations also showed that, thanks to gravitational nudges from the Sun, SL9 was on a collision course with Jupiter. Astronomers had more than a year to plan for observations of the impacts due in July 1994. The impacts of the 22 identified nuclei would take almost a week, and virtually every telescope on Earth and in space would be ready and watching. The only uncertainty was whether or not anything would happen. Even the best Hubble Space Telescope images were unable to determine whether the comet nuclei were massive kilometer-size chunks or merely insubstantial puffs of dust.

The answer came within minutes of the first impact: Infrared cameras recorded an intense fireball of hot gas rising thousands of kilometers above the impact site, which lay just barely on Jupiter's far side (Figure 10.12). As Jupiter's rapid rotation carried the scene into view, the collapsing plume smashed down

d This infrared photo shows high clouds created by several impacts. These clouds reflect infrared light from the Sun because they lie high above the methane gas that absorbs infrared light.

e In this visible-light photo from the Hubble Space Telescope, the high, dusty clouds left by the impacts appear as dark "scars" on Jupiter.

f This Hubble Space Telescope photo shows how the impact scars became smeared due to atmospheric winds. The scars disappeared completely within a few months.

g Artist's conception of the impacts viewed from the surface of Io.

on Jupiter's atmosphere, leading to another episode of heating and more infrared glow. News of each subsequent impact spread around the world rapidly via the Internet, and everyone waited anxiously to hear each observatory report. The Hubble Space Telescope caught impact plumes in action as they rose into sunlight and collapsed back down, leaving dust clouds high in Jupiter's stratosphere. The impact scars lingered for months, but Jupiter has recovered—for now.

Table 10.1 Major Annual Meteor Showers

Shower Name	Approximate Date
Quadrantids	January 3
Lyrids	April 22
Eta Aquarids	May 5
Delta Aquarids	July 28
Perseids	August 12
Orionids	October 22
Taurids	November 3
Leonids	November 17
Geminids	December 14
Ursids	December 23

Comet Tails and Meteor Showers

Far smaller impacts happen on Earth all the time, lighting up the sky as *meteors.* If you watch the sky on a clear night, you'll typically see a few meteors each hour. Most meteors are created by single pieces of comet dust, each no larger than a pea, that enter our atmosphere at speeds of up to 250,000 km/hr (70 km/s). The particles and the surrounding gas are heated so much that we see a brief but brilliant flash as they are vaporized high in the atmosphere. An estimated 25 million meteors occur worldwide every day, adding hundreds of tons of comet dust to the Earth daily.

Comet orbits tend to be filled with small particles ejected as the comets pass near the Sun. When Earth passes through a comet orbit, the particles striking our atmosphere produce a *meteor shower,* during which we may see dozens of meteors per hour. Earth's orbit intersects many comet orbits over the course of the year, and each produces a meteor shower (Table 10.1). The most famous meteor shower, the *Perseids,* occurs every August when the Earth passes through the orbit of comet Swift–Tuttle. Although meteors all strike the Earth on parallel tracks, meteor showers appear to radiate from a particular direction in the sky. You may have experienced a similar illusion while driving into a blizzard at night: The snowflakes seem to diverge from a single direction that

FIGURE 10.13 Snowflakes and meteor showers.

a Driving into a blizzard.

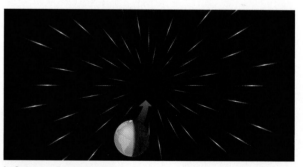

b Geometry of a meteor shower: The Earth is moving through a swarm of comet debris.

c Meteors photographed during the intense Leonid meteor shower in 1966. Star images trailed across the photo from lower left to upper right during the time exposure. Meteors appear to streak away from the top center of the frame. Note one meteor at top center made a dot instead of a streak; it's headed straight for the camera.

a Meteor Crater in Arizona was created about 50,000 years ago by an asteroid impact. The crater is more than a kilometer across and almost 200 meters deep, but the asteroid that made it was only about 50 meters across.

FIGURE 10.14 Impacts on Earth.

b The arrow points to a layer of sediment laid down by the impact 65 million years ago. At the time, the rock layers above the arrow did not exist. Dust from the impact and soot from global wildfires settled down through the atmosphere onto the seafloor that once occupied this location in Colorado.

depends on the combination of your speed and the wind speed (Figure 10.13a,b). (This illusion occurs even before the air flow around the car deflects the motion of the snowflakes.) Meteor showers get their name from the constellation from which they appear to radiate. The Perseids get their name because they appear to radiate from the constellation Perseus. (Figure 10.13c shows the Leonids from Leo.) The Earth runs into more meteors on the side facing in the direction of Earth's orbital motion (just as more snow hits the front windshield while the car is moving), so the predawn sky is the best place to watch for meteors.

TIME OUT TO THINK *The associated "comet" for the Geminid meteor shower is an object called Phaethon, which is classified as an asteroid because we've never seen a coma or tail associated with it. How is it possible that Phaethon looks like an asteroid today but once shed the particles that create the Geminid meteor shower?*

Impacts and Mass Extinctions

Meteorites and impact craters bear witness to the fact that much larger impacts occasionally occur on Earth. Meteor Crater in Arizona (see Figure 10.14a) formed about 50,000 years ago when a metallic impactor

roughly 50 meters across crashed to Earth with the explosive power of a 20-megaton hydrogen bomb. Although the crater is only a bit more than 1 kilometer across, an area covering hundreds of square kilometers was probably battered by the blast and ejecta. Far bigger impacts have occurred, sometimes with catastrophic consequences for life on Earth.

Collecting geological samples in Italy in 1978, the father–son team of Luis and Walter Alvarez discovered a thin layer of dark sediments that had apparently been deposited 65 million years ago—about the same time that the dinosaurs and many other organisms suddenly became extinct. Subsequent studies found similar sediments deposited at the same time at many sites around the world (Figure 10.14b). Careful analysis showed this worldwide sediment layer to be rich in the element iridium, which is rare on Earth's surface. But iridium is common in primitive meteorites, which led the Alvarezes to a stunning conclusion: The extinction of the dinosaurs was caused by the impact of an asteroid or comet. This conclusion was not immediately accepted and still generates some controversy, but it now seems clear that a major impact coincided with the death of the dinosaurs. Moreover, while the dinosaurs were the most famous victims of this **mass extinction**, it seems that up to 99% of all living things were killed and that 75% of all *species* living on Earth were wiped out at that time.

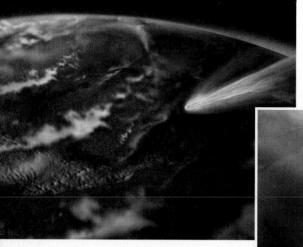

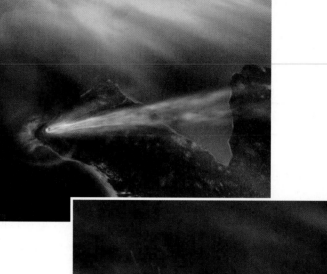

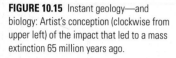

FIGURE 10.15 Instant geology—and biology: Artist's conception (clockwise from upper left) of the impact that led to a mass extinction 65 million years ago.

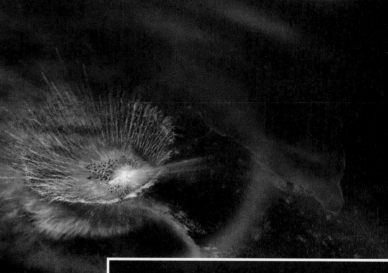

How could an impact lead to mass extinction? The amount of iridium deposited worldwide suggests that the impactor must have been about 10 kilometers across. After a decade-long search, scientists identified what appears to be the impact crater from the event. Located off the coast of Mexico's Yucatán peninsula, it is 200 kilometers across, which is close to what we expect for a 10-kilometer impactor, and dates to 65 million years ago. The impact almost immediately sent a shower of debris raining across much of nearby North America and generated huge waves that may have sloshed more than 1,000 kilometers inland (Figure 10.15). Many North American species thus may have been wiped out shortly after impact. For the rest of the world, death may have come more slowly. Heat from the impact and returning ejecta probably ignited wildfires in forests around the world. Evidence of such wildfires is found in the large amount of soot that is also present in the iridium-rich sediments from 65 million years ago. The impact also sent huge quantities of dust high into the stratosphere, where it remained for several years, blocking out sunlight, cooling the surface, and affecting atmospheric chemistry. Plants died for lack of sunlight, and effects propagated through the food chain.

Perhaps the most astonishing fact is not that 75% of all species died, but that 25% survived. Among the survivors were a few small, rodentlike mammals. These mammals may have survived because they lived in underground burrows and managed to store enough food to outlast the long spell of cold, dark

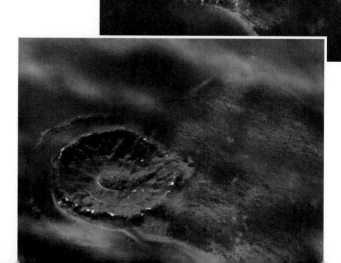

days. Small mammals had first arisen at about the same time as the dinosaurs, more than 100 million years earlier. But the sudden disappearance of the dominant dinosaurs made these mammals dominant. With an evolutionary path swept wide open, they rapidly evolved into a large assortment of much larger mammals—including us.

The event 65 million years ago was but one of at least a dozen mass extinctions in Earth's distant past. An even larger extinction event 250 million years ago has also been linked to an impact and the ensuing catastrophic effects. Could a similar event occur in the future, wiping out all that we have accomplished?

The Asteroid Threat: Real Danger or Media Hype?

The first step in analyzing the threat of future impacts is to examine the past evidence more thoroughly. Only one significant impact has clearly occurred in modern times. In 1908, an unusual explosion occurred in a sparsely inhabited region of Siberia. (The impact is known as the *Tunguska event.*) Entire forests were flattened and set on fire, and air blasts knocked over people, tents, and furniture up to 200 kilometers away (Figure 10.16). Seismic disturbances were recorded at distances up to 1,000 kilometers, and atmospheric pressure fluctuations were detected almost 4,000 kilometers away. Scientific studies were delayed for almost 20 years (due to the remoteness of the impact area and the politics of the time), but investigators eventually found meteoritic dust at the site—but no impact crater. These wide-reaching effects were apparently caused by a weak, stony meteorite only 30 meters across that exploded in the air before reaching the surface. Objects of this size probably strike our planet every century or so, usually over the ocean. But the death toll could be quite high if such an object struck a densely populated area.

Another way to gauge the threat of impacts is to look at asteroids that might strike Earth. The largest known *Earth-approaching asteroid* (Eros, the target of the NEAR mission, Figure 10.4d) is about 40 kilometers long. Hundreds of smaller Earth-approaching asteroids are known, down to sizes of tens of meters across. Orbital calculations show that none of these known objects will impact Earth in the foreseeable future. However, the vast majority of Earth-approaching asteroids probably have not yet been detected. Astronomers estimate that some 1,000 undiscovered members of this class may be larger than 1 kilometer across, and another 100,000 may be larger than 100 meters across. Many search programs are under way, detecting hundreds of objects every year. NASA has set a goal of detecting 90% of all large Earth-approaching asteroids by 2010. Comets also pose a

FIGURE 10.16 Damage from the 1908 impact over Tunguska, Siberia.

threat, particularly because their rapid plunge from the outer solar system gives little warning.

By combining observations of asteroids with information from past impact craters, we can estimate the frequency of impacts of various sizes (Figure 10.17). Larger impacts are obviously more devastating but thankfully are rare. Impactors a few meters across probably break up high in the Earth's atmosphere every few days, with each liberating the energy of an atomic bomb. However, the effects are dissipated before reaching the surface, so we do not notice them. Hundred-meter impactors that forge craters similar to Meteor Crater probably strike only about every 10,000 years or so. Impacts large enough to cause mass extinctions occur tens of millions of years apart.

Thus, while it seems a virtual certainty that the Earth will be battered by many more large impacts, the chance of a major impact happening in our lifetime is quite small. Nevertheless, the chance is not zero, and until we know the orbits of *all* large Earth-approaching asteroids we cannot predict when the next impact might occur. Some people advocate a thorough search to find all potentially dangerous asteroids, a relatively cheap proposition. But others wonder whether the information would be useful: If you knew an asteroid would hit next year, could you do anything about it? Schemes to save Earth by using nuclear weapons to demolish or divert an asteroid abound, but no one knows whether current technol-

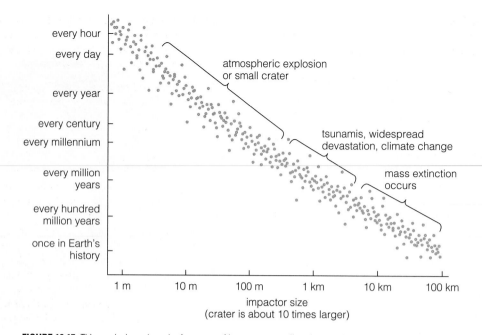

FIGURE 10.17 This graph shows how the frequency of impacts—as well as the magnitude of their effects—depends on the size of the impactor; note that smaller impacts are much more frequent than larger ones.

ogy is up to the task. For the time being, we can only hope that our number does not come up soon.

While some people see danger in Earth-approaching asteroids, others see opportunity because Earth-approaching asteroids bring valuable resources tantalizingly close to Earth. Iron-rich asteroids are particularly enticing, since they probably contain many precious metals that mostly sank to the core on Earth. In the not-too-distant future, it may prove technically feasible and financially profitable to mine metals from asteroids and bring these resources to Earth. It may also be possible to gather fuel and water from asteroids for use in missions to the outer solar system.

THE BIG PICTURE

In this chapter, we focused on the solar system's smallest objects and found that they can have big consequences. Asteroids, comets, and meteorites teach us much about the evolution of the solar system. And, in the case of the impact 65 million years ago, one such object may have made our existence possible. Keep in mind the following "big picture" ideas as you continue your studies:

- The smallest bodies in the solar system—asteroids and comets—are our best evidence of how the solar system formed.

- The small bodies are subjected to the gravitational whims of the planets, particularly the jovian planets. The subtleties of resonances play a major role in sculpting the outer solar system.

- The interplay of large and small bodies brings us meteorites to teach us our origins, comets and meteor showers to light up the sky, and impacts that can alter or obliterate life as we know it.

- Pluto is called the ninth planet, but it bears much more similarity to the thousands of Kuiper belt comets than to the other eight planets.

Review Questions

1. Briefly explain why comets, asteroids, and meteorites are so useful in terms of helping us understand the formation and development of our solar system.

2. Define and distinguish among each of the following: *asteroid, comet, meteor, meteorite.*

3. How does the largest asteroid compare in size to the terrestrial worlds? How does the total mass of all the asteroids compare to the mass of a terrestrial world?

4. What is the *asteroid belt,* and where is it located?

5. Briefly describe how orbital resonances with Jupiter have affected the asteroid belt.

6. What happens to asteroids that are kicked out of the asteroid belt by a gravitational encounter or a collision? In this context, describe the origin of the Earth-approaching asteroids.

7. Distinguish between *primitive meteorites* and *processed meteorites* in terms of composition. How do the origins of these two groups of meteorites differ?

8. How is the study of a processed meteorite an opportunity to look at a piece of a "dissected planet"?

9. How did Halley's comet get its name? Can Kepler's laws be applied to comets?

10. Describe the anatomy of a comet.

11. Briefly describe the composition of the nucleus of Halley's comet. How does gas escape from the interior of the nucleus into space? Why do the tails point away from the Sun?

12. Why can't an active comet last forever? What happens to comets after many passes near the Sun?

13. Describe the Kuiper belt and the Oort cloud in terms of location and comet orbits.

14. Where did comets that are now in the Oort cloud originally form? How did they end up in the Oort cloud?

15. Describe the origin of the Kuiper belt comets. What do we know about the sizes of Kuiper belt comets?

16. Describe Pluto and Charon. How were they discovered? Why won't Pluto collide with Neptune even though their orbits cross? How did Charon probably form?

17. Briefly summarize the evidence suggesting that Pluto is a Kuiper belt comet.

18. Describe the impact of comet Shoemaker–Levy 9 on Jupiter. How was the impact studied?

19. Explain how meteor showers are linked to comets. Why do meteor showers seem to originate from a particular location in the sky?

20. Describe the evidence suggesting that the *mass extinction* that killed off the dinosaurs was caused by the impact of an asteroid or comet. How did the impact lead to the mass extinction?

21. How often should we expect impacts of various sizes on Earth? How could we alleviate the risk of major impacts?

Discussion Questions

1. *Impact Risk.* How can we compare the risk of death from an impact to the risk of death from other hazards? One proposed way to evaluate the risk is to multiply the probability of an impact by the number of people it would kill. For example, the probability of a major impact that could kill half the world's population, or some 3 billion people, is estimated to be about 0.0000001 (10^{-7}) in any given year. If we multiply this low probability by the 3 billion people that would be killed, we find the "average" risk from such an impact to be $10^{-7} \times (3 \times 10^9) = 300$ people killed per year—about the same as the average number of people killed in airplane crashes. Do you think it is valid to say that the risk of being killed by a major impact is about the same as the risk of being killed in an airplane crash? Should we pay to limit this risk, as we pay to limit airline crashes? Defend your opinions.

2. *Rise of the Mammals.* Suppose the impact 65 million years ago had not occurred. How do you think our planet would be different? For example, do you think that mammals would still have eventually come to dominate the Earth? Would we be here? Defend your opinions.

Problems

Surprising Discoveries? For **problems 1–8**, suppose we made the discoveries described. (These are *not* real discoveries.) Decide whether the discovery should be considered reasonable or surprising. Explain.

1. A small asteroid that orbits within the asteroid belt has an active volcano.

2. Scientists discover a meteorite that, based on radioactive dating, is 7.9 billion years old.

3. An object that resembles a comet in size and composition is discovered to be orbiting in the inner solar system.

4. Studies of a large object in the Kuiper belt reveal that it is made almost entirely of rocky (as opposed to icy) material.

5. Astronomers discover a previously unknown comet that will produce a spectacular display in Earth's skies about 2 years from now.

6. A mission to Pluto finds that it has lakes of liquid water on its surface.

7. Geologists discover a crater from a 5-km object that impacted the Earth more than 100 million years ago.

8. Archaeologists learn that the fall of ancient Rome was caused in large part by an asteroid impact in southern Africa.

9. *Orbital Resonances*. How would our solar system be different if orbital resonances had never been important (for example, if Jupiter and asteroids continued to orbit the Sun but did not affect each other through resonances)? Describe at least five ways in which our solar system would be different. Which differences do you consider superficial, and which differences more profound? Consider the effects of orbital resonances discussed both in this chapter and in Chapter 9.

10. *Life Story of an Iron Atom.* Imagine that you are an iron atom in an iron meteorite recently fallen to Earth. Tell the story of how you got here, beginning from the time when you were in a gaseous state in the solar nebula 4.6 billion years ago. Include as much detail as possible. Your story should be scientifically accurate but also creative and interesting.

11. *The "Near Miss" of Toutatis.* The 5-km asteroid Toutatis passed a mere 3 million km from the Earth in 1992.

    a. In one or two paragraphs, describe what would have happened to the Earth if Toutatis had hit it.

    b. Suppose Toutatis was destined to pass *somewhere* within 3 million km of Earth. Calculate the probability that this "somewhere" would have meant that it slammed into Earth. Based on your result, do you think it is fair to say that the 1992 passage was a "near miss"? Explain. (*Hint:* You can calculate the probability by considering an imaginary dartboard of radius 3 million km on which the bull's-eye has the Earth's radius of 6,378 km.)

12. *Project: Tracking a Meteor Shower.* Armed with an expendable star chart and a flashlight covered in red plastic, set yourself up comfortably to watch a meteor shower listed in Table 10.1. Each time you see a meteor, record its path on your star chart. Record at least a dozen, and try to determine the *radiant* of the shower—the point in the sky from which the meteors appear to radiate. Also record the direction of the Earth's motion, which you can find by determining which sign of the zodiac rises at midnight.

13. *Project: Dirty Snowballs.* If there is snow where you live or study, make a filthy snowball. (The ice chunks that form behind tires work well.) How much dirt does it take to darken snow? Find out by allowing your dirty snowball to melt and measuring the approximate proportions of water and dirt afterward.

Web Projects

Find useful links for Web projects on the text Web site.

1. *The NEAR Mission.* Learn about the NEAR mission to the asteroid Eros. How did the mission reach Eros? How did it end? What did it accomplish? Write a one- to two-page summary of your findings.

2. *Stardust.* Learn about NASA's Stardust mission (launched in 1999) to bring back cometary material to Earth. What is the current status of the spacecraft? What do we hope to learn after it returns? Write a report about the mission status and its science.

3. *Asteroid and Comet Missions.* Learn about another proposed space mission to study asteroids or comets. Write a short report about the mission plans, goals, and prospects for success.

4. *Pluto Express.* NASA has gone back and forth for years about a mission to Pluto, the only planet not yet visited by spacecraft. Find the latest on plans for a mission to Pluto. Does the mission seem likely to take place any time soon? Write a short report about the current status of a Pluto mission. Discuss the scientific goals of such a mission and the political realities that may dictate whether it occurs.

Looking outward to the blackness of space, sprinkled with the glory of a universe of lights, I saw majesty—but no welcome. Below was a welcoming planet. There, contained in the thin, moving, incredibly fragile shell of the biosphere is everything that is dear to you, all the human drama and comedy. That's where life is; that's where all the good stuff is.

LOREN ACTON, U.S. ASTRONAUT

CHAPTER 11

Planet Earth
And Its Lessons on Life in the Universe

Perhaps you've heard the Earth described as the "third rock from the Sun." At first glance, the perspective of comparative planetology may seem to support this dispassionate view of the Earth. However, a deeper comparison between Earth and its neighbors reveals a far more remarkable planet.

Earth is unique in the solar system in many ways. It is the only planet with surface oceans of liquid water and substantial amounts of oxygen in its atmosphere, and its geological activity is more diverse than that of any other terrestrial world. Most important, Earth is the only world known to harbor life, which has played a considerable role in shaping the Earth's surface and atmosphere.

So far, we have found that our theories of solar system formation and planetary evolution explain most of the general features of our solar system. In this chapter, we will apply and extend what we've learned in order to form a coherent picture of the combined geological, atmospheric, and biological evolution of our planet. We'll also discuss a few important lessons that the solar system teaches us about our future on Earth and that the Earth teaches us about the possibility of life elsewhere in the universe.

11.1 How Is Earth Different?

Take a look at Figure 11.1. The abundant white clouds overlying the expanse of oceans show a planet unlike any other in our solar system. It is, of course, our own planet Earth.

In Chapter 8, we discussed many features of the Earth's geology and atmosphere. We found many similarities between the Earth and other terrestrial worlds, and also some important differences. For example, the Earth's surface is shaped by the same four geological processes (impact cratering, volcanism, tectonics, and erosion) that shape other worlds, but Earth is the only planet on which the lithosphere is clearly broken into plates that move around in what we call plate tectonics. The Earth's atmosphere shows even more substantial differences from the atmospheres of its neighbors: It is the only planet with significant atmospheric oxygen and the only terrestrial world with an ultraviolet-absorbing stratosphere.

But the greatest differences between Earth and other worlds lie in two features totally unique to Earth. First, the surface of the Earth is covered by huge amounts of water. Oceans cover nearly three-fourths of the Earth's surface, with an average depth of about 3 kilometers (1.8 miles). Water is also significant on land, where it flows through streams and rivers, fills lakes and underground water tables, and sometimes lies frozen in glaciers. Frozen water covers nearly the entire continent of Antarctica in the form of the southern ice cap. The northern ice cap sits atop the Arctic Ocean and covers the large island of Greenland. Water plays such an important role on Earth that some scientists treat it as a distinct planetary layer, called the **hydrosphere**, between the lithosphere and the atmosphere.

The second totally unique feature of Earth is its diversity of life. We find life nearly everywhere on Earth's surface, throughout the oceans, and even underground. The layer of life on Earth is sometimes called the **biosphere**. As we will see shortly, the biosphere helps shape many of the Earth's physical characteristics. For example, the biosphere explains the presence of oxygen in Earth's atmosphere. Without life, Earth's atmosphere would be very different.

Comparative planetology compels us to explore these unique properties of Earth closely. If we find that the formation and evolution of our Earth violated the rules we've derived for other planets—if Earth is "irregular"—life might be unique to Earth. If Earth does follow the rules—if Earth is "regular" or even "standard"—then life may be common in the universe.

11.2 Our Unique Geology

It's not surprising that Earth shows much more geological activity than Mercury or the Moon, since those small worlds have cooled since their formation and are now geologically "dead." But why have Earth and Venus ended up so different, despite very similar sizes? And, although Earth is significantly larger than Mars, both planets probably once had similar surface conditions that allowed liquid water to flow. Why did Earth remain hospitable, while Mars is now dry and barren? These mysteries warrant a closer look.

Figure 11.2 shows shaded relief maps comparing the surfaces of Venus, Earth, and Mars. While the three planets have many superficial similarities, they also have important differences. We find fewer im-

FIGURE 11.1 Dawn over the Atlantic Ocean, photographed from the Space Shuttle.

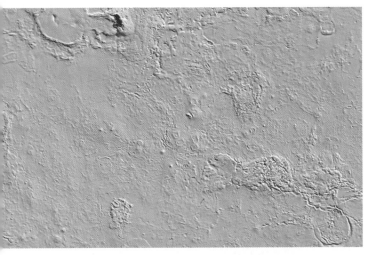

Venus

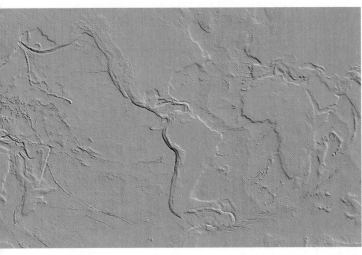

Earth

Mars

FIGURE 11.2 These shaded relief maps compare the surfaces of Venus, Earth, and Mars; the map for Earth shows the seafloor as it would appear if there were no oceans. Despite superficial similarities, there are important differences. For example, there are fewer impact craters on Earth or Venus than on Mars. Also note the unique distinction between the two kinds of surface (seafloor and continent) on Earth.

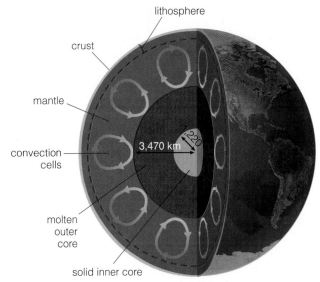

FIGURE 11.3 Internal structure of the Earth, based on the study of seismic waves.

pact craters on Venus than on Mars and fewer still on Earth. Most of Earth's volcanic mountains are steep-sided and therefore must have been made from a much more viscous lava than that found on the other planets. And Earth shows far more evidence of tectonic activity—surface reshaping driven by internal stresses. To our current knowledge, Earth is the only terrestrial planet on which the lithosphere is split into distinct *plates*. To understand these and other unique features of Earth's geology, we must look in more detail at how the four geological processes affect our planet.

Before we discuss the geological processes themselves, remember that they depend in many ways on a planet's interior structure. We know much more about Earth's interior structure than we do about that of any other planet, because geologists have created a three-dimensional "picture" of the interior by analyzing the propagation of seismic waves from earthquakes (Figure 11.3). The Earth's thin crust and the uppermost portion of the mantle make up a relatively cool and rigid *lithosphere* about 100 km thick [Section 8.2]. Below the lithosphere, the mantle is warm and partially molten in places; it supplies the magma for volcanic eruptions. Deeper in the mantle, higher pressures force the rock into the solid phase despite even higher temperatures. But even in solid form the mantle slowly flows, and convection continually carries heat up from below. The Earth's metallic core underlies the lower mantle. The outer region of the core is molten, but high pressures in the inner core force the metal into solid form. Thus, the Earth has all the ingredients needed for exciting geology: a hot interior to supply energy for volcanism, convection to help generate tectonic stresses, and a relatively thin lithosphere that does not inhibit geological activity.

The Four Geological Processes on Earth

The first of the four geological processes, *impact cratering*, should have occurred at roughly similar rates on all the terrestrial worlds. All were subject to the intense early bombardment during the solar system's first few hundred million years. Four billion years ago, the Earth's surface may have been as saturated with craters as the densely cratered lunar highlands are today [Section 8.3]. We find no 4-billion-year-old craters on Earth now, so these ancient impacts must have been erased by other geological processes. The impact rate fell substantially after the early bombardment, but occasional impacts still occur—sometimes with devastating consequences [Section 10.6]. The effects of erosion make impact craters surprisingly difficult to identify on Earth. Nevertheless, remnants of more than 100 impact craters have been found. Large impacts should have occurred more or less uniformly over the Earth's surface, including both continents and seafloors, because large impactors are not stopped by the oceans. Thus, the relative absence of large craters on the seafloor suggests that seafloor crust is much younger than is continental crust: it formed later than most of the impacts.

The rampant *erosion* on Earth arises primarily from processes involving water, but atmospheric winds also contribute. Wind tears away at geological features created by other processes, blows dust and sand across the globe, and builds features such as sand dunes. Note that wind erosion is more effective on Earth than on either Venus or Mars: Venus has very little surface wind because of its slow rotation, and wind on Mars does little damage because of the low atmospheric pressure. On Earth, water contributes to erosion through a remarkable variety of processes. Perhaps the most obvious are rain and rivers breaking down mountains, carving canyons, and transporting sand and silt across continents to deposit them in the sea. But water erosion occurs on microscopic levels too: Water seeps into cracks and crevices in rocks and breaks them down from the inside, especially when this water freezes and expands. Further examples of water-erosion processes include the slow movement of glaciers, the flow of underground rivers, and the pounding of the ocean surf.

Volcanism and *tectonics* are also very important processes on Earth. The many active volcanoes prove that volcanism still reshapes the surface, and many small-scale cliffs and valleys offer evidence of tectonics similar to that found on other terrestrial worlds. But plate tectonics may be unique to Earth. So, to gain a deeper understanding of our planet's closely linked volcanism and tectonics, we must study plate tectonics in more detail. We will also see how plate tectonics may play a role in Earth's biology.

Plate Tectonics

If you cut up a map and rearrange the continents, you'll find that they fit together in surprising ways. For example, the east coast of South America fits quite nicely into the west coast of Africa. You might chalk this up to coincidence, but we also find similar types of distinct rocks and rare fossils in eastern South America and western Africa—suggesting that these two regions were once near each other. Based on such evidence, in 1912 German meteorologist and geologist Alfred Wegener proposed the idea of *continental drift*: that the continents gradually drift across the surface of the Earth, over time scales of tens of millions of years.

For decades after Wegener made his proposal, most geologists rejected the idea that continents could move, and many ridiculed the idea openly. However, the idea of continental drift gained favor in the 1950s as supporting evidence began to accumulate. Mapping of the seafloors revealed surprising structures (Figure 11.4): high *mid-ocean ridges* extending a total length of about 60,000 km along the ocean floors and *trenches* in which the ocean depth can reach more than 8 kilometers. Geologists soon recognized that these features represented boundaries between **plates**—pieces of the lithosphere that apparently float upon the denser mantle below. Convection cells in the upper mantle move the plates around the surface, forcing them together, apart, or sideways at their boundaries—with profound geological consequences. This discovery offered a model to explain how continents could drift about, and Wegener's idea finally gained acceptance under the new name **plate tectonics**. (Recall that tectonics is geological activity driven by internal stresses, so *plate tectonics* refers to the motion of plates driven by internal stresses.) Today we know that the Earth's lithosphere is broken into more than a dozen plates. (See figure 11.4.) Most major earthquakes and volcanic eruptions occur along plate boundaries (Figure 11.5).

TIME OUT TO THINK *Study the plate boundaries in Figure 11.4. Based on this diagram, explain why the West Coast states of California, Oregon, and Washington are more prone to earthquakes and volcanoes than other parts of the United States.*

Over millions of years, plate tectonics acts like a giant conveyor belt for the Earth's crust, carrying rock up from the mantle, transporting it across the seafloor, and then returning it down into the mantle (Figure 11.6). The mid-ocean ridges mark **spreading centers** between plates—places where hot mantle material rises upward and then spreads sideways, pushing the plates apart. Worldwide along the mid-ocean ridges, new crust covers an area of about

FIGURE 11.4 The discovery of mid-ocean ridges and deep trenches provided important evidence for continental drift. The relief map uses color to show elevation, progressing from blue (lowest) to red (highest). Solid lines show plate boundaries, and arrows represent directions in which the plates are moving. Important geological features discussed later in the text are identified.

FIGURE 11.5 Earthquakes occur when plates slip violently along a plate boundary. This photo shows damage from the 1995 earthquake in Kobe, Japan.

FIGURE 11.6 Plate tectonics acts like a giant conveyor belt, driven by mantle convection, that carries rock up from the mantle at mid-ocean ridges, transports it across the seafloor, and returns it down into the mantle at ocean trenches.

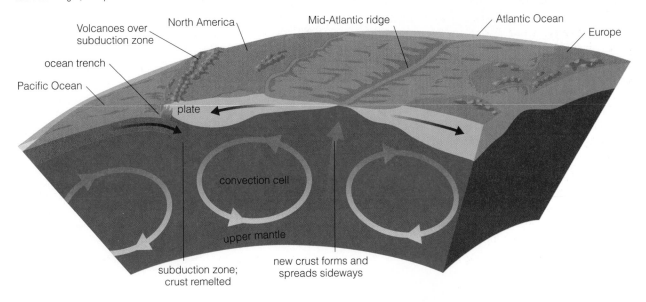

2 square kilometers every year—enough to replace the entire seafloor within a geologically short time of about 200 million years. Meanwhile, as new crust spreads over the surface, older crust must be returned to the mantle. This return occurs at the locations of the deep ocean trenches, which mark places where one plate slides under another in a process called **subduction**. The conveyor-like process of plate tectonics is undoubtedly driven by convection, although the precise nature of the convection remains uncertain.

Overall, the plates move across the Earth's surface at speeds of a few centimeters per year—about the same speed at which your fingernails grow—and we can use this motion to project the locations of the continents millions of years into the past or future (Figure 11.7). Over the past billion years, the continents have slammed together, pulled apart, spun around, and changed places on the globe. Central Africa once lay at Earth's South Pole, and Antarctica was once nearer the equator. At various times, such as was the case about 200 million years ago, the continents were all merged into a single giant continent, often called *Pangaea* (which means "all lands"). The current arrangement of the continents is no more permanent than any other.

Most of Earth's interesting geology takes place near plate boundaries (see Figure 11.4). Where one plate subducts below another, the rock of the descending plate melts, rises, and fuels tremendous volcanoes. Alaska's Aleutian island chain and the islands of Japan and the Philippines are all located where one seafloor plate subducts below another.

Earth's continents all originally formed this way. Volcanoes also occur where seafloor subducts below continental crust, such as Mount St. Helens in the Pacific Northwest, California's Sierra Nevada, and the Andes range of South America. The lava made from remelted seafloor is viscous ("sticky") and makes steep-sided volcanoes. (Some volcanoes, such as those that make up the Hawaiian Islands, occur in the middle of a plate, where molten material rises directly from the mantle and makes less steep volcanoes.)

Tectonic forces continue to reshape the continents. When two continents collide, neither can subduct, and the resulting crunch makes huge mountains, as in the case of the Himalayas (Figure 11.8b). The Appalachians were formed by a similar process and were once as tall as the Himalayas. The forces of plate tectonics can also tear a continent apart, as is happening today in the East African rift zone and with the Arabian peninsula (Figure 11.8a).

Summary of Earth's Geology

Two fundamental differences between the geology of Earth and that of other terrestrial worlds stand out:

1. Rampant erosion driven by the action of plentiful water (a direct result of Earth's atmospheric evolution, as we will see).

2. A different style of tectonics—plate tectonics—that recycles crust, drives high-viscosity volcanism to create continental crust, and causes frequent earthquakes.

FIGURE 11.7 Past, present, and predicted future arrangements of the Earth's continents.

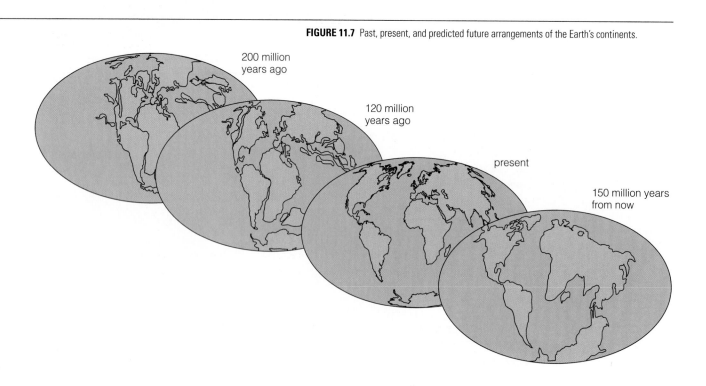

200 million years ago

120 million years ago

present

150 million years from now

a Tectonic forces have torn the Arabian peninsula from Africa. (Photo from Space Shuttle.)

FIGURE 11.8 Tectonic forces on continent.

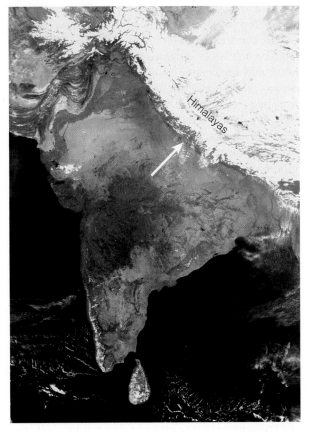

b The Himalayas are still slowly growing as the Indian plate (carrying India) pushes into the Eurasian plate (carrying most of the rest of Asia). (Satellite photo.)

Together, these two processes have reshaped the Earth's surface beyond recognition from its original state billions of years ago. They make the Earth's surface very young, perhaps the youngest in the solar system besides that of Io [Section 9.5].

11.3 Our Unique Atmosphere

In Chapter 8, we learned that the superficial differences in the atmospheres of the terrestrial worlds are fairly easy to understand. For example, the Moon and Mercury lack substantial atmospheres because of their small sizes, and a very strong greenhouse effect causes the high surface temperature on Venus. We also learned how solar heating, seasonal changes, and the Coriolis effect that comes from rotation lead to differing global circulation patterns on different worlds. But explaining how the compositions of the atmospheres came to be so different is more challenging, especially when we consider the cases of Venus, Earth, and Mars. Outgassing produced all three of these atmospheres, and the same volatiles—primarily water and carbon dioxide—were outgassed in all three cases. How, then, did Earth's atmosphere end up so different? In particular, any study of Earth's atmosphere must address four major questions:

1. Why did Earth retain most of its water—in the form of the oceans and other components of the hydrosphere—while Venus and Mars lost theirs?

2. Why does Earth have so little carbon dioxide (CO_2) in its atmosphere, when Earth should have outgassed about as much of it as Venus?

3. Why does Earth have so much more oxygen (O_2) than Venus or Mars? Where did it come from, and how does this highly reactive gas remain present in the atmosphere?

4. Why does Earth have an ultraviolet-absorbing stratosphere, while Venus and Mars do not?

To answer these questions, we must look into the history of Earth's atmosphere. The answers to all four questions turn out to be closely connected.

Recall that water ice could not condense in the region of the solar nebula where the terrestrial planets formed. How, then, did the water outgassed from volcanoes get here in the first place? (Hint: See Section 7.5.)

Water and the Origin of the Hydrosphere

On a basic level, it's fairly easy to explain why the Earth retains so much water while Venus and Mars do not. As we learned in Chapter 8, Venus probably lost an "ocean-full" of water as solar ultraviolet photons split apart water molecules and the hydrogen atoms subsequently escaped to space. Mars probably lost some of its water in a similar way, and the rest is frozen at the polar caps and under the surface.

Let's look a little more closely at the sequence of events that led to these different histories. Picture a generic terrestrial planet, billions of years ago, with erupting volcanoes outgassing carbon dioxide, water vapor, and small amounts of nitrogen. As the outgassing continued, the water vapor might have either condensed into oceans, frozen out as snow, or remained gaseous in the atmosphere. In the case of Venus, proximity to the Sun made the planet warm enough to keep all its water gaseous in the atmosphere, even before it had a strong greenhouse effect. Because water vapor itself is a greenhouse gas, this caused significant greenhouse warming and ensured that additional water vapor would also remain gaseous and warm the planet further. This tendency of the greenhouse effect to reinforce itself is called the **runaway greenhouse effect**. At the other extreme, Mars's temperature was eventually low enough for the water vapor to freeze out of the atmosphere, resulting in thick polar caps.

On Earth, moderate temperatures allowed most of the water vapor to condense as rain, leading to the accumulation of liquid water in the oceans. A small amount of water vapor remained in the atmosphere and caused greenhouse warming, but not enough to create a runaway greenhouse effect. Additional water vapor released from volcanoes therefore also condensed, gradually forming the oceans and the other components of the hydrosphere. Thus, Venus lost its water because it was too hot, Mars lost its water because it was too cold, and Earth retained its water because conditions were "just right."

Where Is All the CO_2?

Measurements of outgassing by active volcanoes and estimates of past volcanic activity suggest that Earth must have outgassed nearly as much carbon dioxide throughout its history as Venus. But Venus today has an atmosphere with a CO_2 concentration of 96%, a surface pressure 90 times greater than that on Earth, and about 500°C of greenhouse warming. In contrast, the proportion of carbon dioxide in Earth's atmosphere is much less than 1%; in fact, we usually measure CO_2 concentration on Earth in units of *parts per million*. (For example, 300 parts per million means $300 \div 1,000,000 = 0.0003 = 0.03\%$.) Clearly, some atmospheric loss process must have removed nearly all of the outgassed CO_2 on Earth.

The secret of Earth's CO_2 is wrapped up in the history of our oceans. The primary process that removes atmospheric CO_2 on Earth involves liquid water. Carbon dioxide can dissolve in water, and the oceans actually contain about 60 times more carbon dioxide than the atmosphere (still a very small amount compared to the total amount that must have been outgassed). Most of the carbon dioxide is locked up in rocks on the seafloor through chemical reactions. Rainfall erodes silicate rocks on the Earth's surface and carries the eroded minerals to the oceans. There the minerals react with dissolved carbon dioxide to form **carbonate** minerals, which fall to the ocean floor, building up thick layers of carbonate rock such as *limestone*. This is part of the **carbonate–silicate cycle** (Figure 11.9). Because the process requires liquid water, it operates only on Earth, where it has removed about as much CO_2 from Earth's atmosphere as now remains in Venus's atmosphere. (If Venus's CO_2 could somehow be removed, the remaining atmosphere would have about as much nitrogen as Earth's.) Had Earth been slightly closer to the Sun, its warmer temperature might have evaporated the oceans, leaving the CO_2 in the atmosphere and causing a runaway greenhouse effect as on Venus. Conditions might not have been so conducive to "life as we know it."

The Earth today is just warm enough for liquid water and life thanks to a moderate greenhouse effect caused by the presence of water vapor and carbon dioxide in our atmosphere. Our good fortune might seem to be based on atmospheric properties alone, but in fact our luck lies deeper—with Earth's geological processes. Much of the CO_2 is injected into the atmosphere by volcanoes near subduction zones: Carbonate rocks carried beneath the seafloor by subduction melt and release their CO_2 (see Figure 11.9). The balance between the rate at which carbonate rocks form in the oceans and the rate at which carbonate rocks melt in subduction zones largely determines the amount of CO_2 in Earth's atmosphere. If plate tectonics—especially subduction—stopped on Earth, or if it occurred at a substantially different rate, the amount of atmospheric CO_2 would undoubtedly be different. The eventual effect on Earth's temperature might prove fatal to us.

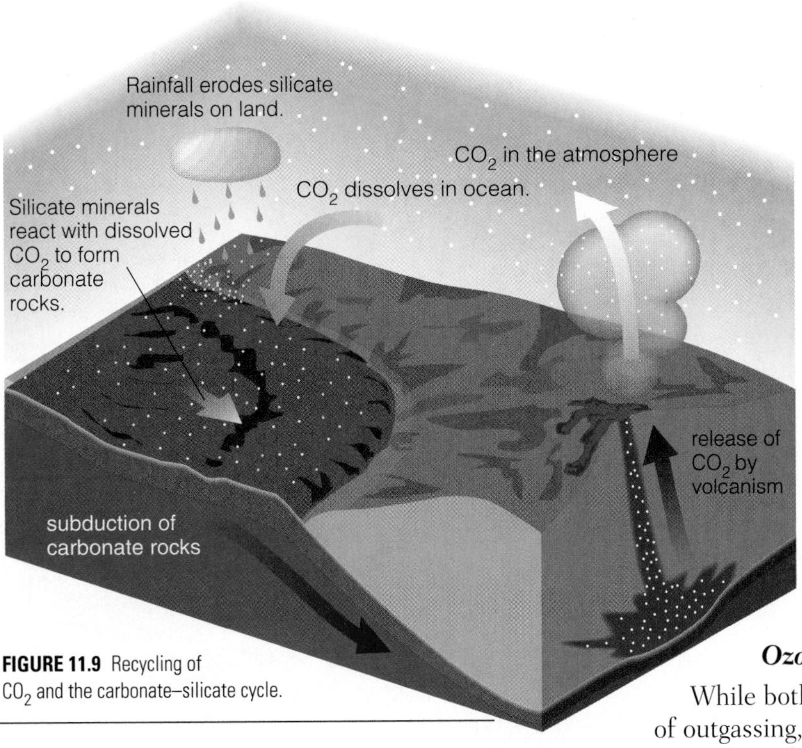

Rainfall erodes silicate minerals on land.

CO_2 in the atmosphere

CO_2 dissolves in ocean.

Silicate minerals react with dissolved CO_2 to form carbonate rocks.

release of CO_2 by volcanism

subduction of carbonate rocks

FIGURE 11.9 Recycling of CO_2 and the carbonate–silicate cycle.

The balance must have been different in Earth's past. Careful studies of ancient rocks show that liquid water was plentiful throughout most of Earth's history, so Earth's temperature has apparently been quite stable for billions of years. However, stars like the Sun increase in brightness as they age, and billions of years ago our Sun was probably about 30% fainter than it is today [Section 12.3]. With Earth's current abundance of greenhouse gases, our planet would have been frozen over. Thus, greenhouse gases must have been more abundant early in Earth's history. Venus, on the other hand, might have experienced hospitable conditions if it possessed "only" its current inventory of greenhouse gases—but it probably also had more greenhouse gases in its early history, leading to its runaway greenhouse effect. On Mars, extra greenhouse gases once provided the warm temperatures needed for liquid water to flow, but Mars lost so much of this gas that it became colder even as the Sun grew brighter.

Geological evidence is mounting for an unusual stage in Earth's history that highlights the importance of the carbonate–silicate cycle. The evidence suggests that some 600–700 million years ago glaciers resided in regions known to lie near the equator at that time. If this really occurred, the global temperature must have dipped enough for the oceans to start to freeze. Because ice reflects light much better than water, less sunlight would be absorbed by the Earth, leading quickly to what some geologists call "snowball Earth." How did the Earth recover from this "snowball" phase? After all, once covered with highly reflective ice, the planet must have become very cold. The

answer lies in the carbonate–silicate cycle (Figure 11.9). If the oceans froze and snow fell instead of rain, the cycle was broken: CO_2 was no longer pulled from the atmosphere. It therefore built up in the atmosphere, warming the planet until the oceans melted and the carbonate–silicate cycle was restored.

TIME OUT TO THINK *How would the carbonate–silicate cycle be affected if plate tectonics did not occur? Without plate tectonics, could Earth recover from a "snowball" phase?*

The Origin of Oxygen, Ozone, and the Stratosphere

While both water and carbon dioxide are products of outgassing, molecular oxygen (O_2) is not. In fact, no geological process can explain the great abundance of oxygen (about 20%) in Earth's atmosphere. Moreover, oxygen is a highly reactive chemical that would rapidly disappear from the atmosphere if it were not continuously resupplied. Fire, rust, and the discoloration of freshly cut fruits and vegetables are everyday examples of **oxidation**—chemical reactions that remove oxygen from the atmosphere. Similar reactions between oxygen and surface materials (especially iron-bearing minerals) give rise to the reddish appearance of many of Earth's rock layers, such as those in Arizona's Grand Canyon. Thus, we must explain not only how oxygen got into the Earth's atmosphere in the first place, but also how the amount of oxygen remains relatively steady even while oxidation reactions tend to bind oxygen into rocks at a rapid rate.

The answer to the oxygen mystery is *life*. The process that supplies oxygen to the atmosphere is *photosynthesis*, which converts CO_2 to O_2. The carbon becomes incorporated into amino acids, proteins, and other components of living organisms. Today, plants and single-celled photosynthetic organisms return oxygen to the atmosphere in approximate balance with the rate at which animals and oxidation reactions consume oxygen, so the oxygen content of the atmosphere stays relatively steady. Earth originally developed its oxygen atmosphere when photosynthesis added oxygen at a rate greater than these processes could remove it from the atmosphere.

Life and oxygen also explain the presence of Earth's ultraviolet-absorbing stratosphere. In the upper atmosphere, chemical reactions involving solar ultraviolet light transform some of the O_2 into molecules

b Stromatolite fossils (up to 3.5 billion years old).

a Microscopic fossil bacteria.

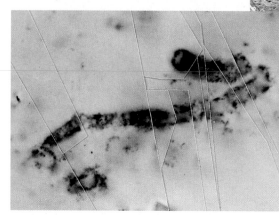

c (below) Living stromatolites today. The biological origin of fossil stromatolites was questioned until living examples were found.

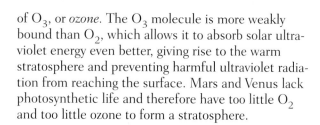

FIGURE 11.10 Ancient life.

of O_3, or *ozone*. The O_3 molecule is more weakly bound than O_2, which allows it to absorb solar ultraviolet energy even better, giving rise to the warm stratosphere and preventing harmful ultraviolet radiation from reaching the surface. Mars and Venus lack photosynthetic life and therefore have too little O_2 and too little ozone to form a stratosphere.

Summary of Earth's Atmosphere

Venus, Earth, and Mars all began on similar paths, releasing similar gases into their atmospheres by outgassing. But only Earth had conditions "just right" to support liquid oceans throughout its history. The oceans helped remove carbon dioxide from the atmosphere, ensuring that the greenhouse effect on Earth remained just strong enough to keep conditions hospitable for life, but not so strong as to create a runaway greenhouse effect like that on Venus.

But the physical connections between the Earth's geology and its atmosphere cannot fully explain the conditions on Earth today. In particular, the presence of oxygen and ozone can be explained only as products of life. Thus, to complete our understanding of the unique place of Earth in the solar system, we must now turn our attention to the role of life on Earth.

11.4 Life

Geologists and biologists have developed a fairly clear picture of the history of life on Earth, at least in broad outline. As we trace this history through time, you should ask yourself three key questions:

1. How did Earth's physical conditions affect the development of life?

2. How has life influenced the physical characteristics of Earth?

3. What can we learn about the prospects for finding life elsewhere in the solar system or universe?

The Origin of Life (4.4–3.5 billion years ago)

Life arose on Earth at least 3.5 billion years ago, and possibly much earlier than that. We've found recognizable fossils in rocks as old as 3.5 billion years—microscopic fossils of single-celled bacteria and larger fossils of bacterial "colonies" called *stromatolites* (Figure 11.10). But indirect evidence from carbon isotopes suggests that life originated at least 350 million years earlier (3.85 billion years ago), even though such ancient rock samples are too contorted for fossils to remain recognizable. All life and all fossils tested to date show the same characteristic ratio of two carbon isotopes, carbon-12 and carbon-13. Rocks as old as 3.85 billion years show the same ratio, strongly suggesting that these rocks contain remnants of life. Beyond 3.85 billion years, we cannot yet determine whether life existed, since rocks more ancient than this no longer exist on Earth's surface. It is conceivable that life began as early as 4.4 billion years ago.

TIME OUT TO THINK *You may have noticed that, while we've been talking about life, we haven't actually defined the term. In fact, it's surprisingly difficult to draw a clear boundary between life and nonlife. How would you define life? Explain your reasoning.*

If life did originate much before 3.85 billion years ago, it may have been in for a rough time. The transition from intolerable conditions during planetary formation to more hospitable conditions was a gradual one. The early bombardment that continued after the end of accretion gradually died away over a period of a few hundred million years. But large impacts capable of sterilizing the planet by temporarily boiling all the oceans may have occurred as late as 4.0–3.8 billion years ago. Some biologists speculate that life may have arisen several times in the Earth's early history, only to be extinguished by the hostile conditions. If so, life might have turned out very different if one less—or one more—sterilizing impact had occurred.

An even deeper question concerns not just *when* life arose, but *how.* No one knows the precise answer to this question, but genetic testing makes us confident of one thing: Whether life arose once or multiple times, one particular type of organism came to dominate the entire Earth. That is, every organism living today apparently developed from a single ancestor (Figure 11.11).

Genetic testing also allows at least reasonable guesses as to which living organisms most resemble the common ancestor of all life. Surprisingly, the answer appears to be organisms living in the deep oceans around seafloor volcanic vents called *black smokers* (after the dark, mineral-rich water that flows out of them) and in hot springs in places like Yellowstone (Figure 11.12). These organisms thrive in temperatures as high as 125°C. (The high pressures at the seafloor prevent the water from boiling until it reaches 450°C.) Unlike most life at the Earth's surface, which depends on sunlight, the ultimate energy source for these organisms is chemical reactions in water volcanically heated by the internal heat of the Earth itself.

Of course, knowing that all life shares a common ancestor still does not tell us how that ancestor first arose. The step from simple chemical building blocks to our common DNA-bearing ancestor (from nonlife to life) is huge, to say the least. No fossils record the transition, nor has it ever been duplicated in the laboratory.

The bare necessities for "life as we know it"— chemicals, energy, and water—were undoubtedly available on the early Earth. The building blocks of life, including amino acids and nucleic acids, are composed primarily of carbon, oxygen, nitrogen, and hydrogen—chemicals that were readily available. Energy to fuel chemical reactions was present on the

FIGURE 11.11 The tree of life, showing evolutionary relationships according to modern biology as of 2001. Note that just two small branches represent *all* plant and animal species.

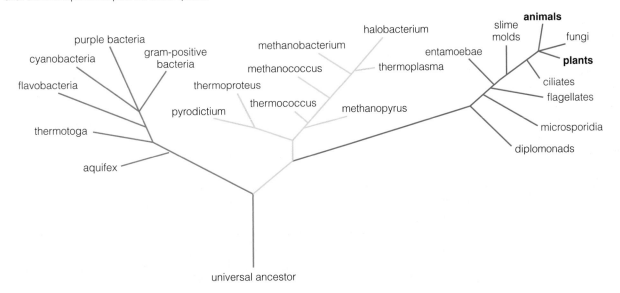

a Life around a black smoker deep beneath the ocean surface.

FIGURE 11.12 Life in hot water.

b This aerial photo shows a hot spring in Yellowstone National Park that is filled with colorful bacterial life. The path in the lower left gives an indication of scale.

surface in the form of lightning and ultraviolet light from the Sun and was also present in the oceans in the heated water near undersea volcanoes. We know that oceans were present, because we find ancient carbonate rocks that must have formed in water.

Laboratory experiments demonstrate that the chemical building blocks of life could have formed spontaneously and rapidly in the conditions that prevailed early in Earth's history. In these experiments, researchers replicate the chemical conditions of the early Earth by, for example, including the types of ingredients outgassed by volcanoes and "sparking" the mixture with electricity to simulate lightning or other energy sources. Within just a few days, the mixture spontaneously becomes rich with amino acids and nucleic acids. Such chemical reactions were undoubtedly an important source of organic molecules on Earth, but they may not have been the only source—some meteorites contain complex organic molecules.

With the building blocks for life plentiful, it may only have been a matter of time before chemical reactions created a molecule that could make a copy of itself. Such a *self-replicating molecule* might have spread quickly through regions with hospitable conditions. If life really began in this way, then some ancient self-replicating molecule was the ancestor of modern DNA—and of all life on Earth.

Early Evolution in the Oceans (3.5–2.0 billion years ago)

Regardless of its origin, life soon thrived in the oceans in the form of single-celled organisms, tapping a variety of energy sources including sunlight (i.e., through photosynthesis). Individual organisms that survived and reproduced passed on copies of their DNA to the next generation. However, the transmission of DNA from one generation to the next is not always perfect. **Mutations**—errors in the copying process—can change the arrangement of chemical bases in a strand of DNA, making the genetic information in a new cell slightly different from that of its parent. Mutations can be caused by many factors, including high-energy *cosmic rays* from space, ultraviolet light from the Sun, particles emitted by the decay of radioactive elements in the Earth, and various toxic chemicals. Most mutations are lethal, killing the cell in which the mutation occurs. However, some mutations may make a cell better able to survive in its surroundings, and the cell then passes on this improvement to its offspring, a process called **natural selection**.

Fossils show that evolution progressed remarkably slowly for most of Earth's history. For at least a billion years after life first arose, the most complex life-forms were still single-celled. Some 2 billion years ago, the land was still inhospitable because of the lack of a protective ozone layer. Continents much like those today were surrounded by oceans teeming with life, but the land itself was probably as barren as Mars is today, despite pleasant temperatures and plentiful rainfall.

Altering the Atmosphere (beginning about 2 billion years ago)

The process of photosynthesis appears to have developed quite early in the history of life. Over billions of years, the abundant single-celled organisms in the ocean pulled carbon dioxide from the atmosphere and put back oxygen. Oxygen is a highly reactive gas, and for a long time, nearly all of it was pulled back out of the atmosphere by reactions with surface rocks. However, by about 2 billion years ago, the oxidation of surface rocks was fairly complete, and oxygen began to accumulate in the atmosphere. Today it constitutes about 20% of the atmosphere.

The buildup of oxygen had two far-reaching effects for life on Earth. First, it made possible the development of oxygen-breathing *animals*. No one knows precisely how or when the first oxygen-breathing organism appeared, but for that animal the world was filled with food. You can imagine how quickly these creatures must have spread around the world, and the fact that other organisms could now be eaten changed the "rules of the game" for evolution.

The second major effect of the oxygen buildup was the formation of the ozone layer, which made it safe for life to move onto the land. Although it took more than a billion years, natural selection eventually led to plants that could survive and thrive on land, and animals followed soon after, taking advantage of this new food source.

This epoch highlights the ability of life to fundamentally alter a planet. Not only did life change the Earth's atmosphere, but in doing so it made two other kinds of life possible: oxygen-breathing animals and land organisms.

An Explosion of Diversity (beginning about 0.54 billion years ago)

As recently as 540 million years ago, most life-forms were still single-celled and tiny. Note that life had already been present on Earth for at least 3 billion years by that time, demonstrating the remarkable slowness with which evolution progressed for most of Earth's history. But the fossil record reveals a dramatic diversification of life beginning about 540 million years ago. We can trace the origin of most of today's plant and animal species—from insects to trees to vertebrates—back to this period. Apparently, once life began to diversify, the increased competition among the many more complex species accelerated the process of natural selection and the evolution of new species.

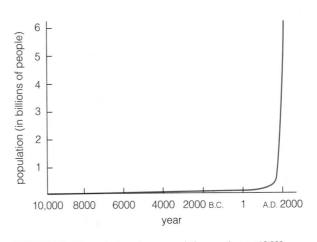

FIGURE 11.13 This graph shows human population over the past 12,000 years. Note the tremendous population growth that has occurred in just the past few centuries.

Some species have been more successful and more adaptable than others. Dinosaurs dominated the landscape of the Earth for more than 100 million years. But their sudden demise 65 million years ago paved the way for the evolution of large mammals—including us. The earliest humans appeared on the scene only a few million years ago (after 99.9% of Earth's history), and our few centuries of industry and technology have come after 99.99999% of Earth's history. Our population has been growing exponentially for the past few centuries, roughly doubling in just the past 40 years (Figure 11.13).

The tremendous growth of the human population suggests a chilling analogy: Exponential growth is also the mark of a cancerous tumor—which may seem very successful at surviving while it is growing but ultimately dies when it kills its host. Is it possible that our tremendous success is killing our host, the Earth? We can gain some perspective on this question by looking at a few lessons that the solar system teaches us about our relationship with the Earth.

TIME OUT TO THINK *The pace of change in human existence has accelerated since the advent of modern humans a few million years ago: Civilization arose about 10 thousand years ago, the industrial revolution began just a couple of centuries ago, and the computer era started just a few decades ago. Do you think the pace of change can continue to accelerate? What do you think the next stage will be?*

11.5 Lessons from the Solar System

The United States and other nations have carried out the exploration of the solar system for many reasons, including scientific curiosity, national pride, and technological advancement. Perhaps the most important return on this investment was one not expected: an improved understanding of our own planet and an enhanced ability to understand some very important global issues. In this section, we explore three particularly important issues about which comparative planetology offers important insights: global warming, ozone depletion, and mass extinctions.

Global Warming

Venus stands as a searing example of the effects of large amounts of atmospheric carbon dioxide. Earth remains habitable only because natural processes have locked up most of its carbon dioxide on the seafloor. However, humans are now tinkering with this difference between the two planets by adding carbon dioxide and other greenhouse gases to the Earth's atmosphere. The primary way we release carbon dioxide is through the burning of fossil fuels (coal, natural gas, and oil)—the carbon-rich remains of plants buried millions of years ago. But other human actions, such as deforestation and the subsequent burning of trees, also affect the carbon dioxide balance of our atmosphere. The inescapable result is that the amount of atmospheric carbon dioxide has risen steadily since the dawn of the industrial age—and continues to rise (Figure 11.14).

There is a growing scientific consensus that the average temperature of the Earth has warmed slightly (about 0.5°C) over the past 50 years—although there is much less agreement about why. Furthermore, glaciers, ice sheets, and even the snows of Mount Kilimanjaro are in retreat, while ocean levels are slowly rising. It's likely that the cause is the human addition of greenhouse gases to the atmosphere, but it's also possible that the change is natural and would have occurred even without human activity. Some scientists even suggest that the Sun itself may be warming and thus contributing to the global temperature rise.

The question of whether human activity is inducing global warming is not merely academic: The consequences of a significant continued warming could be widespread and disastrous. Some simulations predict enough warming to melt polar ice and flood the Earth's densely populated coastal regions. Such warming would also increase evaporation from the oceans, tending to make storms of all types both more frequent and more destructive: Coastal regions would be hit by more hurricanes, intense thunderstorms and associated tornadoes would strike more frequently, and even winter blizzards would be more severe and more damaging. Rainfall patterns would also shift, and much of the world's current cropland might become unusable. The most ominous possibility is the collapse of entire ecosystems. If a regional climate changes more rapidly than local species can adapt or migrate, these species might become extinct. Given the complex interactions of the Earth's biosphere, hydrosphere, and atmosphere, the consequences of such ecosystem changes are impossible to foresee.

Given the current uncertainties, it is impossible for anyone to know for sure whether global warming will even be a serious problem in the coming century, let alone make specific predictions of its severity. Our studies of the planets show that surface temper-

FIGURE 11.14 These data, collected for many years on Mauna Loa (Hawaii), show the increase in atmospheric carbon dioxide concentration. Yearly wiggles represent seasonal variations in the concentration, but the long-term trend is clearly upward. The concentration is measured in parts per million (ppm).

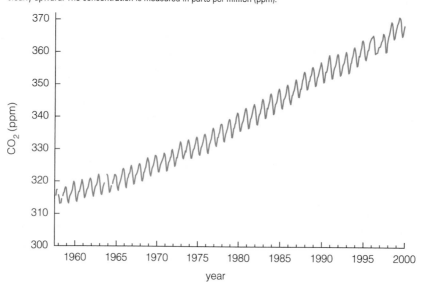

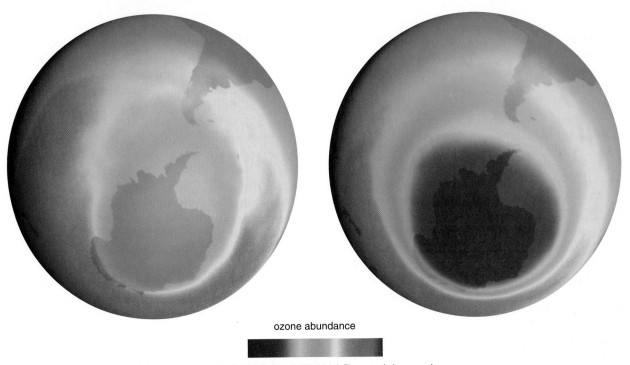

FIGURE 11.15 The extent of the ozone hole over Antarctica in 1979 (left) and 1998 (right). The ozone hole grew substantially between 1979 and 1998. Red indicates normal ozone concentration, and blue indicates severe depletion.

 atures depend on many factors and that seemingly minor changes can alter climates in surprising ways. Nevertheless, the fact that we cannot fully explain the current climate conditions on Earth suggests that we should be very careful about tampering with them.

TIME OUT TO THINK *The most obvious way to prevent global warming is to stop burning the fossil fuels that add greenhouse gases to the atmosphere. But much of the world economy depends on energy produced from fossil fuels. Given the uncertainties about the effects of global warming in the next hundred years, what if any changes do you think are justified at present? What specific proposals would you make if you were a world leader?*

Ozone Depletion

For 2 billion years, the Earth's ozone layer has shielded the surface from hazardous ultraviolet radiation, but in the past two decades humans have begun to destroy the shield. The magnitude of the problem was not recognized until the discovery of an **ozone hole** over Antarctica in the mid-1980s (Figure 11.15). The "hole" is a place where the concentration of ozone in the stratosphere is dramatically lower than it would naturally be. The ozone hole appears for a few months during each Antarctic spring and has gradually wors-

ened over the years. Since its discovery, satellite measurements have suggested that **ozone depletion** has caused ozone levels to fall by as much as a few percent worldwide. If ozone levels continue to decline, more solar ultraviolet radiation will reach the surface. Humans will face substantially greater risk from skin cancer, and plants and animals will suffer more genetic damage from mutations induced by ultraviolet radiation—with unknown consequences for the biosphere as a whole.

The apparent cause of the ozone depletion is human-made chemicals known as CFCs (chlorofluorocarbons), widely used in air conditioners and refrigerators and for many other purposes. Indeed, CFCs once seemed like an almost ideal chemical: They are useful in many industries, cheap and easy to produce, and chemically inert (i.e., they do not burn, break down, or react with anything on the Earth's surface). Ironically, their inertness is also their downfall; because CFCs are gases and are not destroyed by chemical reactions in the lower atmosphere, they eventually rise intact into the stratosphere. There they are broken down by the Sun's ultraviolet light, and one of the by-products is chlorine—a very reactive gas that catalytically destroys ozone without being consumed in the process.

Mars and Venus each played an important role in our understanding of ozone and chlorine. Mars lacks an ozone layer, so ultraviolet radiation reaches the surface. As a result, the surface is sterile: Terrestrial life could not survive on the surface, and the building blocks of life would be torn apart when exposed to the Sun. But it was the study of Venus that first alerted scientists to the dangers on Earth. Chlorine is a minor but important ingredient of Venus's atmosphere, and models of its chemical cycles showed that chlorine could rapidly destroy weak molecules such as ozone. When the models were altered to apply to Earth's atmosphere, CFCs and ozone depletion were connected.

This tale of three planets may have a happy ending, thanks to the awareness it brought of the dangers posed by ozone destruction. Today, international treaties ban the production of CFCs, although previously existing CFCs may still be used and recycled. If these treaties are obeyed, continued ozone destruction probably will not be a serious problem in the future, but the ozone hole will probably take 50 years to recover.

TIME OUT TO THINK *You've just learned that your car air conditioner is broken and will cost $200 to fix. Then you hear about a shop that can fix it for only $100—undoubtedly by violating the laws concerning CFCs. Is the small benefit to our ozone layer worth the extra $100?*

Mass Extinctions

When geology was a young science, its practitioners thought that entire mountain ranges and huge valleys were formed suddenly by cataclysmic events. Later geologists denounced this *catastrophism* and replaced it with *uniformitarianism,* which holds that geological change occurs gradually by processes acting over very long periods. Uniformitarianism certainly holds for three of our four geological processes, but impact cratering represents the "instant geology" of catastrophism. Many geologists initially rejected the importance of impact cratering as a geological process, but they were eventually convinced by the unambiguous evidence of impacts on the Moon, by astronomical studies of comets and asteroids, and by careful analysis of the few impact craters still present on Earth. Geology today is a combination of uniformitarianism and catastrophism.

Scientific ideas about evolution are undergoing a similar transformation. Biologists once assumed that evolution always proceeded gradually, with new species arising slowly and others occasionally becoming extinct. It now appears that the vast majority of all species in Earth's history died out in sudden *mass extinctions*—for which impacts are the prime suspect. The mass extinction that killed off the dinosaurs is only the most famous of these events [Section 10.6]; the fossil record shows evidence of at least a dozen other mass extinctions, some with even greater species loss than in the dinosaur event. While some species died out from direct effects of the impact, most probably died due to the disappearance from their ecosystem of species on which they depended. Furthermore, the survival or extinction of species appears to be random during mass extinctions: Surviving species show no discernible advantage over those that perish. In a sense, mass extinctions clear the blackboard, allowing evolution to get a fresh start with a new set of species. If we re-formed the Earth from scratch, life (and even intelligent life) might well arise again, but there is no scientific basis for thinking that evolution would follow the same course.

The story of mass extinctions teaches us about the danger of large impacts, but it also teaches a potentially more important lesson: Even without an impact, the Earth may be undergoing a mass extinction right now. In the normal course of evolution, the rate of species extinction is fairly low—perhaps one species lost per century. According to some estimates, human activity is now driving species to extinction so rapidly that half of today's species may be lost by the end of the twenty-first century. It is not just that humans are hunting and fishing many species to extinction. Far more are lost due to habitat destruction and pollution. And when a few species are lost by direct human influence, entire ecosystems that depend on those species may collapse. If global warming alters the climate significantly or if ozone depletion leads to more genetic damage in plants and animals, the rate of species extinction might increase even more. In any case, the "event" of losing half the world's species in just a few hundred years certainly qualifies as a mass extinction on a geological time scale, and the biological effects could be as dramatic as those of any impact. Are we unwittingly clearing the way for a new set of dominant species?

11.6 Lessons from Earth: Life in the Solar System and Beyond

Just as comparative planetology teaches us much about the Earth, the study of the Earth can teach us much about other worlds. In particular, it teaches us about the conditions for life and may help us answer what is surely one of the deepest philosophical questions of all time: Are we alone in the universe? It is fitting to close our study of the solar system by extending our understanding of biology to the planets and beyond. The study of the possibility of life elsewhere in the universe, called *astrobiology,* is a logical extension of comparative planetology.

In the time between the Copernican revolution [Section 3.4] and the space age, many people expected the other planets in our solar system to be Earth-like and to harbor intelligent life. In fact, a reward was supposedly once offered for the first evidence of intelligent life on another planet *other than Mars*. Venus, a bit closer to the Sun than Earth, was often pictured as a tropical paradise. Such expectations were dashed by the bleak images of Mars returned by spacecraft and the discovery of the runaway greenhouse effect on Venus. Many scientists began to believe that only Earth has the right conditions for life, intelligent or otherwise. Recently, the pendulum has begun to swing the other way, spurred primarily by two developments. First, biologists are learning that life thrives under a much wider range of conditions than once imagined. Second, planetary scientists are developing a much better understanding of conditions on other worlds. It now seems quite likely that conditions in at least some places on other worlds might be conducive to life.

The Hardiness, Diversity, and Probability of Life

Even on Earth, biologists long assumed that many environments were uninhabitable. But recent discoveries have found life surviving in a remarkable range of conditions. The teeming life surviving at temperatures as high as 125°C near underwater volcanic vents and in hot springs is only one of many surprises. Biologists have found microorganisms living deep inside rocks in the frozen deserts of Antarctica and inside basaltic rocks buried more than a kilometer underground. Some bacteria can even survive radiation levels once thought lethal—apparently, they have evolved cellular machinery that repairs mutations as fast as they occur. The newly discovered diversity of microscopic life has forced scientists to redraw the "tree of life" (see Figure 11.11), crowding familiar plants and animals into one corner. The majority of these microorganisms need neither sunlight, oxygen, nor "food" in the form of other organisms. Instead, they tap a variety of chemical reactions for their survival.

The new view of terrestrial biology forces us to rethink the possibility of life elsewhere. First, life on Earth thrives at extremes of temperature, pressure, and atmospheric conditions that overlap conditions found on other worlds. The Antarctic valleys, for example, are as dry and cold as certain parts of Mars, and the conditions found in terrestrial hot springs may have been duplicated on a number of planets and moons at certain times in the past. Second, life harnesses energy sources readily available on other planets. The basalt-dwelling bacteria, for example, would probably survive if they were transplanted to Mars. Third, life on Earth has evolved from a common ancestor into every imaginable ecological niche (and some unimaginable ones as well). The diversification of life on Earth has basically tested the limits of our planet, and there is no reason to doubt that it would do so on other planets. Thus, if life ever had

a foothold on any other planet in the past, some organisms might still survive in surprising ecological niches today even if the planet has undergone substantial changes. The only real question is whether life ever got started elsewhere in the first place.

What is the probability of life arising from non-living ingredients? This is probably the greatest unknown in astrobiology. The fact that we exist and are asking the question does not tell us the probability; it merely tells us that it happened once. But the rapidity with which life arose on Earth may provide a clue. As we've discussed, we find fossil evidence for life dating almost all the way back to the end of the period of early bombardment in the solar system, suggesting that life arises easily and perhaps inevitably under the right conditions. In that case, we must search the solar system for those conditions, past or present.

TIME OUT TO THINK *The preceding discussion implies that the rapid appearance of life on Earth means that life is highly probable. Do you agree with this logic? What alternative conclusions could you reach?*

Life in the Solar System

Speculation about life in the solar system usually begins with Mars, for good reason. Before it dried out billions of years ago, its early atmosphere gave the surface hospitable conditions that rivaled those on Earth, with ample running water, the necessary raw chemical ingredients for life, and a variety of familiar energy sources. Many of Earth's organisms would have thrived under early Martian conditions, and some could even survive in places in today's Martian environment. Our first attempt to search for life on Mars came with the Viking missions to Mars in the 1970s, which included two landers equipped to search for the chemical signs of life. No life was found. But the landers sampled only two locations on the planet and tested soils only very near the surface. If life once existed on Mars, it either has become extinct or is hiding in other locations.

Today, a renewed debate about Martian life is under way, thanks in part to the study of a Martian meteorite found in Antarctica in 1984. The meteorite apparently landed in Antarctica 13,000 years ago, following a 16-million-year journey through space after being blasted from Mars by an impact. The rock itself dates to 4.5 billion years ago, indicating that it solidified shortly after Mars formed and therefore was present during the time when Mars was warmer and wetter. Painstaking analysis of the meteorite reveals indirect evidence of past life on Mars, including layered carbonate minerals and complex molecules (called polycyclic aromatic hydrocarbons), both

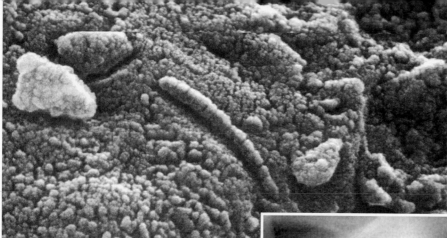

a Microscopic view of seemingly lifelike structures in a Martian meteorite.

b A comparison of microscopic chains of magnetic crystals from Earth (top) and Mars (bottom).

FIGURE 11.16 Evidence for life on Mars?

of which are associated with life when they are found in Earth rocks. Even more intriguing, highly magnified images of the meteorite reveal eerily lifelike forms (Figure 11.16a). These forms bear a superficial resemblance to terrestrial bacteria, although they are about a hundred times smaller—about the same size as recently discovered terrestrial "nanobacteria" and viruses. Magnetic crystals have also been found within the rock—crystals that on Earth are only made by bacteria (Figure 11.16b). Nevertheless, many scientists dispute the conclusion that these features suggest the past existence of life on Mars, claiming that nonbiological causes can also explain many of the meteorite's unusual features.

Future missions to Mars will search for life using more sophisticated techniques than those used by the Viking missions. The best place to look for fossil remains of extinct life is probably in ancient valley bottoms or sedimentary rocks of dried-up lake beds (Figure 8.23d). If any hot springs surround Mars's not-quite-dormant volcanoes, they may be a good place to look for surviving life. A thorough search for Martian life will probably require the return of rock samples to Earth or human exploration of the planet.

Martian meteorites also remind us that the planets may occasionally exchange rocks dislodged by major impacts. The harsh conditions under which some life on Earth exists suggest that living organisms might survive such impacts and even survive the journey from one planet to another. Earth's basalt-dwelling bacteria, for example, could probably survive an impact-cratering event, the ensuing millions of years in space, and a violent impact on Mars. If a meteorite from Earth once landed in hospitable conditions on Mars, it might have introduced life to Mars or wiped out life already present. In a sense, Earth, Venus, and Mars have been "sneezing" on one another for billions of years. Life could conceivably have originated on any of these three planets and been transported to the others.

TIME OUT TO THINK *Suppose we someday discover living organisms on Mars. How will we be able to tell whether these organisms share a common ancestor with living organisms on Earth?*

Besides Mars, the best places to search for life probably are the satellites of the jovian planets. Several of them may meet the requirements of having liquid water, appropriate chemicals, and energy sources for life. In particular, Europa appears to have a planet-wide ocean beneath its icy crust [Section 9.5], the ice and rock from which Europa formed undoubtedly included the necessary chemicals, and its internal heating might lead to undersea volcanic vents. Thus, the real question about Europa may be whether it is possible that its ocean has existed for billions of years *without* developing life. NASA is considering missions to search for signs of life on Europa. The most ambitious involves landing a robotic spacecraft on the surface that will melt its way through the icy crust to reach the ocean below.

Another enticing place to look for life past or present is Saturn's moon Titan. We've already found evidence of complex chemical reactions on Titan, many involving the same elements used by life on Earth. Liquid water and energy are in shorter supply, but both might have been supplied by impacts that heated and melted the icy surface early in Titan's history. Life might have arisen in a slushy pond during the brief period that it remained liquid and might have survived after the pond froze. Unlike the case on Earth, where early impacts probably sterilized the planet, impacts on Titan may have made life possible.

Despite our new awareness of the diversity of life on Earth, we may still be underestimating the range of conditions in which life can exist. Could life arise in liquids other than water, or possibly in an atmosphere? Some people have speculated that life could develop in the clouds of Jupiter or other jovian planets. Might life be based on different elements than those used on Earth? Are there other energy sources that we have not considered? Is our definition of life too narrow? Clearly, it will be a long time before we know all the places where life might exist even within our solar system.

Life Around Other Stars

Only a few places in our solar system seem hospitable to life, but many more hospitable worlds may be orbiting some of the hundred billion other stars in the Milky Way Galaxy or stars in some of the billions of other galaxies in the universe. Might some of the stars be orbited by planets that are as hospitable as our own Earth?

Interstellar clouds throughout the galaxy contain the basic chemical ingredients of life—carbon, oxygen, nitrogen, and hydrogen, as well as other elements. All stars are born from such interstellar clouds, so it seems likely that other solar nebulae should have given rise to planetary systems, some similar to our own [Section 7.6]. Terrestrial planets anywhere will receive light from their parent star, and if they are large enough, like Venus or Earth, they will have plenty of internal heat. Thus, the chemicals and energy sources for life should be present on many planets throughout the universe.

Planets with liquid water on their surfaces may be rarer, particularly if we consider only planets where oceans can endure for billions of years. Even if a planet is large enough to outgas substantial quantities of water and retain its atmosphere, the stories of Venus and Mars tell us that oceans are not guaranteed. To keep oceans for billions of years, the planet must lie within a range of distances from its star, sometimes called the **habitable zone**, that has temperatures just right for liquid water. How big is the habitable zone in our own solar system? We know that Venus is too close to the Sun, so this zone must begin outside the orbit of Venus. Mars is a borderline case: If it had been large enough to retain its atmosphere and sustain a stronger greenhouse effect, Mars might still be habitable today. Overall, the habitable zone around our star probably ranges from 0.8 to 1.5 AU. This zone may be broader around brighter stars and narrower around fainter stars. Computer models of solar system formation suggest that one or more terrestrial planets will usually form within a star's habitable zone, as long as the star is not part of a binary star system. Many of the extrasolar planets discovered recently lie within their star's habitable zone [Section 7.6]. Although the planets themselves are thought to be jovian—and therefore not habitable—moons orbiting the planets could be habitable. Even moons outside the habitable zone could harbor life if other energy sources (such as tidal heating) are available.

The bottom line is that, according to our theories of solar system formation, planets with all the necessities for life should be quite common in the universe. The only major question is whether these ingredients combine to form life. The fact that life arose very early in Earth's history suggests that it may be very easy to produce life under Earth-like conditions, but we will not know for sure unless and until we find other life-bearing planets. NASA is currently developing plans for orbiting telescopes that may be able to detect ozone in the spectra of planets around other stars—and, at least in our solar system, substantial ozone implies life. In addition, radio astronomers are searching the skies in hopes of receiving a signal from some extraterrestrial civilization. Perhaps, in a decade or two, we will discover unmistakable evidence of life. On that day, if it comes, we will know that we are not alone.

TIME OUT TO THINK *Consider the following statements: (1) We are the only intelligent life in the entire universe. (2) Earth is one of many planets inhabited by intelligent life. Which do you think is true? Do you find either philosophically troubling?*

THE BIG PICTURE

Humans have observed the Sun, the Moon, and the planets for thousands of years, but only recently did we learn that these other worlds have much to teach us about our own Earth. Through our study of solar system formation and comparative planetology, we have learned to look at Earth from a very new perspective. Keep in mind the following "big picture" ideas:

- Earth has been shaped by the same geological and atmospheric processes that shaped the other terrestrial worlds. Earth is not a special case from a planetary point of view but rather a place where natural processes led to conditions conducive to life.

- Most of Earth's unique features can be traced to the fact that abundant water has remained liquid throughout our planet's history—thanks to our distance from the Sun and the size of our planet.

- Life arose early on Earth and played a crucial role in shaping our planet's history. The abundant oxygen in our atmosphere is just one of many phenomena that demonstrate how life can transform a planet.

- Humans are ideally adapted to the Earth today, but there is no guarantee that the Earth will remain as hospitable in the future. The study of our solar system teaches us how planets can change.

- The conditions necessary for "life as we know it" are probably common in the universe and may even be found in our own solar system. But so far, we have no proof that life exists elsewhere.

Review Questions

1. Briefly summarize how Earth is different from the other terrestrial planets. In particular, define *hydrosphere* and *biosphere*.

2. Briefly describe the interior structure of the Earth, including its molten outer core and its lithosphere.

3. Do we find evidence of impact cratering on Earth? Why do we find fewer craters on the seafloor than on the continents?

4. Why is erosion more important on Earth than on Venus or Mars? Describe how erosion affects the Earth's geology.

5. Briefly describe the conveyor-like process of plate tectonics. What are *spreading centers* and *subduction zones?*

6. Describe how subduction has created the continents over billions of years. Give examples of how this process has affected Japan, the Philippines, and the western United States.

7. Briefly describe how tectonic stresses can create rift valleys and tall mountain ranges. Give examples of places that have been affected in each of these ways.

8. What is a *runaway greenhouse effect?* Why did it occur on Venus but not on Earth?

9. Describe how the *carbonate–silicate cycle* helps maintain the relatively small CO_2 content of our atmosphere.

10. What are *oxidation* reactions? If there were no life on Earth, would Earth's atmosphere still contain significant amounts of oxygen or ozone?

11. Approximately when did life arise on Earth? How do we know?

12. Briefly describe how life gradually altered the Earth's atmosphere until it reached its current state.

13. What is global warming? Briefly describe some of the potential dangers of global warming.

14. Briefly describe the phenomena and causes of the Antarctic *ozone hole* and worldwide *ozone depletion.*

15. Is the Earth currently undergoing a mass extinction? Explain.

16. Summarize why we now think that life can survive in a much broader range of conditions than we did just a few decades ago. Why is this important to the prospect of finding life elsewhere?

17. Describe the evidence from Martian meteorites suggesting that life may once have existed on Mars. Explain how life might have originated on one terrestrial planet and been transferred to others.

18. Why are Europa and Titan considered prospects for harboring life?

19. What is a *habitable zone?* Using this idea, briefly describe the prospect of finding life on planets around other stars.

Discussion Questions

1. *Cancer of the Earth?* In the text, we discussed how the spread of humans over the Earth resembles, at least in some ways, the spread of cancer in a human body. Cancers end up killing themselves because of the damage they do to their hosts. Do you think we are in danger of killing ourselves through our actions on the Earth? If so, what should we do to alleviate this danger? Overall, do you think the cancer analogy is valid or invalid? Defend your opinions.

2. *Contact.* Suppose we discover definitive evidence for microbial life on Mars or Europa. Would this discovery alter your view of our place in the universe? If so, how? What if we made contact with an intelligent species from another world? Do you think it is likely that either kind of life exists elsewhere in the universe? Do you think either kind will be discovered in your lifetime? Explain.

Problems

Surprising Discoveries? For **problems 1–8**, suppose we made the discoveries described. (These are *not* real discoveries.) Decide whether the discovery should be considered reasonable or surprising. Explain. (In some cases, either view can be defended.)

1. A fossil of an organism that died more than 300 million years ago, found in the crust near a mid-ocean ridge.

2. Evidence that fish once swam in a region that is now high on a mountaintop.

3. A "lost continent" on which humans had a great city just a few thousand years ago but that now resides deep underground near a subduction zone.

4. A planet in another solar system that has an Earth-like atmosphere but no life.

5. A planet in another solar system that has an ozone layer but no ordinary oxygen (O_2) in its atmosphere.

6. Evidence that the early Earth had more carbon dioxide in its atmosphere than the Earth does today.

7. Discovery of life on Mars that also uses DNA as its generic molecule and that uses a genetic code very similar to that used by life on Earth.

8. Evidence of photosynthesis occurring on a planet in another solar system that lies outside that solar system's habitable zone.

9. *Change in Formation Properties.* Consider Earth's four formation properties of size, distance, composition, and rotation rate. Choose one property, and suppose it had been different (e.g., smaller size, greater distance). Describe how this change might have affected Earth's subsequent history and the possibility of life on Earth.

10. *Defining Life.* Write a definition of life and explain the basis of your definition in a few sentences. Then evaluate whether each of the following three cases meets your definition. Explain why or why not in a paragraph for each case. (i) The first self-replicating molecule on Earth lies near the boundary of life and nonlife. Does it meet your definition of life? (ii) Modern-day viruses are essentially packets of DNA encased in a microscopic shell of protein. Viruses cannot replicate themselves; instead, they reproduce by infecting living cells and "hijacking" the cell's reproduction machinery to make copies of the viral

DNA and proteins. Are viruses alive? (iii) Imagine that humans someday travel to other stars and discover a planet populated by what appear to be robots programmed to mine metal, refine it, and assemble copies of themselves. Examination of fossil "robots" shows that they have improved, perhaps because cosmic rays have caused errors in their programs. Is this race of robots alive? Would your answer depend on whether another race had built the first robots?

11. *Ozone Signature.* Suppose a powerful future telescope is able to take a spectrum of a terrestrial planet around another star. The spectrum reveals the presence of significant amounts of ozone. Why would this discovery strongly suggest the presence of life on this planet? Would it tell us whether the life is microscopic or more advanced? Summarize your answers in one or two paragraphs.

12. *Explaining Ourselves to the Aliens.* Imagine that someday we make contact with intelligent aliens and even learn to communicate. Even so, many facets of life we take for granted may be utterly incomprehensible to them: music (perhaps they have no sense of hearing), money (perhaps there has never been a need), love (perhaps they don't feel this emotion), meals (perhaps they photosynthesize). Write a page attempting to explain one of these concepts, or another of your choosing. Remember that even the words in your explanation require definition; start at the lowest possible level.

Web Projects

Find useful links for Web projects on the text Web site.

1. *Life on Mars.* Find the latest information regarding the controversy over evidence for life in Martian meteorites. Write a one- to two-page report summarizing the current state of the controversy and your own opinion as to whether life once existed on Mars.

2. *Search for Life.* Learn about a proposed mission to search for life on Mars, Europa, or elsewhere in our solar system or beyond. Write a one- to two-page summary of the mission and its prospects for success.

3. *Human Threats to the Earth.* Write an in-depth research report, three to five pages in length, about current understanding and controversy regarding one of the following issues: global warming, ozone depletion, or the loss of species on Earth due to human activity. Be sure to address both the latest knowledge about the issue and proposals for alleviating any dangers associated with it. End your report by making your own recommendations about what, if anything, needs to be done to prevent damage to the Earth.

4. *Local Geology.* Write a one- to two-page report about the geology of an area you know well—perhaps the location of your campus or your hometown. Which of the four geological processes has played the most important role, in your opinion? Has plate tectonics played a direct role in shaping the area?

PART IV
Stellar Alchemy

CHAPTER 12

Our Star

Today, astronomy involves the study of the entire universe, but the root of the word *astronomy* originally came from the Greek word for "star." Although we have learned a lot about the universe up to this point in the book, only now do we turn our attention to the study of the stars, the namesakes of our subject.

When we think of stars, we usually think of the beautiful points of light visible on a clear night. But the nearest and most easily studied star is visible only in the daytime—our Sun. Of course, the Sun is important to us in many more ways than as an object for astronomical study. The Sun is the source of virtually all light, heat, and energy reaching the Earth, and life on Earth's surface could not survive without it.

In this chapter, we will study the Sun in some depth. We will learn how the Sun makes life possible on Earth. Equally important, we will study our Sun as a star, so that in subsequent chapters we can more easily understand stars throughout the universe.

12.1 Why Does the Sun Shine?

Ancient peoples recognized the vital role of the Sun in their lives. Some worshiped the Sun as a god, and others created elaborate mythologies to explain its daily rise and set. Only recently, however, have we learned how the Sun provides us with light and heat.

Most ancient thinkers viewed the Sun as some type of fire, perhaps a lump of burning coal or wood. The Greek philosopher Anaxagoras (c. 500–428 B.C.) imagined the Sun to be a very hot, glowing rock about the size of the Greek peninsula of Peloponnesus (comparable in size to Massachusetts), making him one of the first people in history to believe that the heavens and the Earth are made from the same types of materials.

By the mid-1800s, the size and distance of the Sun were reasonably well known, and scientists seriously began to address the question of how the Sun shines. Two early ideas held either that the Sun was a cooling ember that had once been much hotter or that the Sun generated energy from some type of chemical burning similar to the burning of coal or wood. Although simple calculations showed that a cooling or chemically burning Sun could shine for a few thousand years—an age that squared well with biblically based estimates of the age of the Earth that were popular at the time—these ideas suffered from fatal flaws. If the Sun were a cooling ember, it would have been much hotter just a few hundred years earlier, making it too hot for civilization to have existed. Chemical burning was ruled out because it cannot generate enough energy to account for the rate of radiation observed from the Sun's surface.

A more plausible hypothesis of the late 1800s suggested that the Sun generates energy by contracting in size, a process called **gravitational contraction**. If the Sun were shrinking, it would constantly be converting gravitational potential energy into thermal energy, thereby keeping the Sun hot. Because of its large mass, the Sun would need to contract only very slightly each year to maintain its temperature—so slightly that it would be unnoticeable. Calculations showed that the Sun could shine for up to about 25 million years generating energy by gravitational contraction. However, geologists of the late 1800s had already established the age of the Earth to be far older than 25 million years, leaving astronomers in an embarrassing position.

Only after Einstein published his special theory of relativity, which included his discovery of $E = mc^2$, did the true energy-generation mechanism of the Sun become clear. We now know that the Sun generates energy by *nuclear fusion,* a source so efficient that the Sun can shine for about 10 billion years. Because the Sun is only 4.6 billion years old today, we expect it to keep shining for some 5 billion years to come.

FIGURE 12.1 An acrobat stack is in gravitational equilibrium: The lowest person supports the most weight and feels the greatest pressure, and the overlying weight and underlying pressure decrease for those higher up.

Our current model of solar-energy generation by nuclear fusion means that the Sun's size is generally stable, maintained by a balance between the competing forces of gravity pulling inward and pressure pushing outward. This balance is called **gravitational equilibrium** (also referred to as *hydrostatic equilibrium*). It means that, at any point within the Sun, the weight of overlying material is supported by the underlying pressure. A stack of acrobats provides a simple example of this balance (Figure 12.1). The bottom person supports the weight of everybody above him, so the pressure on his body is very great. At each higher level, the overlying weight is less, so the pressure decreases. Gravitational equilibrium in the Sun means that the pressure increases with depth, making the Sun extremely hot and dense in its central core (Figure 12.2).

TIME OUT TO THINK *The Earth's atmosphere is also in gravitational equilibrium, with the weight of upper layers supported by the pressure in lower layers. Use this idea to explain why the air gets thinner at higher altitudes.*

Interestingly, while gravitational contraction is not an important energy-generation mechanism in the

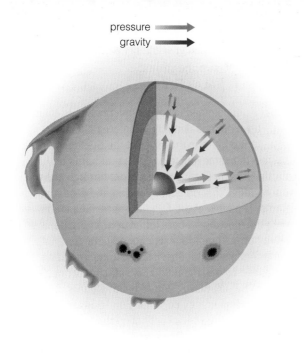

pressure →
gravity →

FIGURE 12.2 Gravitational equilibrium in the Sun: At each point inside, the pressure pushing outward balances the weight of the overlying layers.

Sun today, it was important in the distant past and will be important again in the distant future. Our Sun was born from a collapsing cloud of interstellar gas. The contraction of the cloud released gravitational potential energy, raising the interior temperature higher and higher. When the central temperature and density reached the values necessary to sustain nuclear fusion, the newly released energy provided enough pressure to halt the collapse. That is, the onset of fusion brought the Sun into gravitational equilibrium. About 5 billion years from now, when the Sun finally exhausts its nuclear fuel, the internal pressure will drop, and gravitational contraction will begin once again. As we will see later, some of the most important and spectacular processes in astronomy hinge on this ongoing "battle" between the crush of gravity and a star's internal sources of pressure.

Thus, the answer to the question "Why does the Sun shine?" is that about 4.6 billion years ago *gravitational contraction* made the Sun hot enough to sustain nuclear fusion in its core. Ever since, energy liberated by fusion has maintained the Sun's *gravitational equilibrium* and kept the Sun shining steadily, supplying the light and heat that sustain life on Earth.

12.2 Plunging to the Center of the Sun: An Imaginary Journey

In the rest of this chapter, we will discuss in detail how the Sun produces energy and how that energy travels to Earth. But first, to get a "big picture" view of the Sun, let's imagine you have a spaceship that can somehow withstand the immense heat and pressure of the solar interior and take an imaginary journey from Earth to the center of the Sun.

As you begin your voyage from Earth, the Sun appears as a whitish ball of glowing gas. With spectroscopy [Section 6.5], you verify that the Sun's mass is 70% hydrogen and 28% helium. Heavier elements make up the remaining 2%. The total power output of the Sun, called its **luminosity**, is an incredible 3.8×10^{26} watts. That is, every second, the Sun radiates a total of 3.8×10^{26} joules of energy into space (recall that 1 watt = 1 joule/s). If we could somehow capture and store just 1 second's worth of the Sun's luminosity, it would be enough to meet current human energy demands for roughly the next 500,000 years! Of course, only a tiny fraction of the Sun's total energy output reaches Earth, with the rest dispersing in all directions into space. Most of this energy is radiated in the form of visible light, but once you leave the protective blanket of the Earth's atmosphere you'll encounter significant amounts of other types of solar radiation, including dangerous ultraviolet and X rays. You'll need substantial shielding on your spaceship to prevent serious radiation burns from these high-energy forms of light.

Through a telescope, you can see that the Sun seethes with churning gases, and at most times you'll see at least a few **sunspots** blotching its surface (Figure 12.3). If you focus your telescope solely on a sunspot, you'll find that it is blindingly bright; sunspots appear dark only in contrast to the even brighter solar surface that surrounds them. A typical sunspot is large enough to swallow the entire Earth, dramatically illustrating that the Sun is immense by any earthly standard: The Sun's radius is nearly 700,000 kilometers, and its mass is 2×10^{30} kilograms—about 300,000 times more massive than the Earth.

FIGURE 12.3 This photo of the visible surface of the Sun shows several dark sunspots.

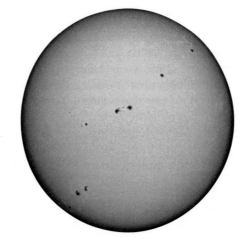

Table 12.1 Basic Properties of the Sun

Radius (R_{Sun})	696,000 km (about 109 times the radius of the Earth)
Mass (M_{Sun})	2×10^{30} kg (about 300,000 times the mass of the Earth)
Luminosity (L_{Sun})	3.8×10^{26} watts
Composition (by percentage of mass)	70% hydrogen, 28% helium, 2% heavier elements
Rotation rate	27 days (equator) to 31 days (poles)
Surface temperature	5,800 K (average); 4,000 K (sunspots)
Core temperature	15 million K

You can determine that the Sun is rotating by watching the sunspots day after day. If you watch very carefully, you may notice that sunspots near the solar equator circle the Sun faster than those at higher solar latitudes. This observation reveals that, unlike a spinning ball, the entire Sun does *not* rotate at the same rate. Instead, the solar equator completes one rotation in about 27 days, and the rotation period increases with latitude to about 31 days near the solar poles. Table 12.1 summarizes some of the basic properties of the Sun.

TIME OUT TO THINK *As a brief review, describe how we measure the mass of the Sun using Newton's version of Kepler's third law. (Hint: Look back at Chapter 5.)*

As you and your spaceship continue to fall toward the Sun, you notice an increasingly powerful headwind exerting a bit of drag on your descent. This headwind, called the **solar wind**, is created by the energy of photons, ions, and subatomic particles flowing outward from the solar surface. The solar wind helps shape the magnetospheres of planets and blows off the material that forms the tails of comets [Section 10.4].

A few million kilometers above the solar surface, you enter the solar **corona**, the tenuous uppermost layer of the Sun's atmosphere (Figure 12.4). Here you find the temperature to be astonishingly high—about 1 million Kelvin. This region emits most of the Sun's X rays. However, the density here is so low that your spaceship feels relatively little heat despite the million-degree temperature [Section 4.2]. Nearer the surface, the temperature suddenly drops to about 10,000 K in the **chromosphere**, the primary source of the Sun's ultraviolet radiation. At last, you plunge

Common Misconceptions: The Sun Is Not on Fire

We are accustomed to saying that the Sun is "burning," a way of speaking that conjures up images of a giant bonfire in the sky. But the Sun does not burn in the same sense as a fire that burns on Earth. Fires on Earth generate light through chemical changes that consume oxygen and produce a flame. The glow of the Sun has more in common with the glowing embers left over after the flames have burned out. Much like hot embers, the Sun's surface shines with the visible thermal radiation produced by any object that is sufficiently hot [Section 6.3]. However, whereas hot embers quickly stop glowing as they cool, the Sun keeps shining because its surface is kept hot by the energy rising from the Sun's core. Because this energy is generated by nuclear fusion, we sometimes say that it is the result of "nuclear burning"—a term that suggests nuclear changes in much the same way that "chemical burning" suggests chemical changes (such as consumption of oxygen). Nevertheless, while it is reasonable to say that the Sun undergoes nuclear burning in its core, it is not accurate to speak of any kind of burning on the Sun's surface, where light is produced primarily by thermal radiation.

through the visible surface of the Sun, called the **photosphere**, where the temperature averages just under 6,000 K. Although the photosphere looks like a well-defined surface from Earth, it consists of gas far less dense than the Earth's atmosphere.

Throughout the solar atmosphere, you notice that the Sun has its own version of weather, in which conditions at a particular altitude differ from one region to another. Some regions of the chromosphere and corona are particularly hot and bright, while other regions are cooler and less dense. In the photosphere, sunspots are cooler than the surrounding surface, though they are still quite hot and bright by earthly standards. In addition, your compass goes crazy as you descend through the solar atmosphere, indicating that solar weather is shaped by intense magnetic fields. Occasionally, huge magnetic storms occur, shooting hot gases far into space.

Up to this point in your journey, you have seen the Earth and the stars behind you, but as you slip beneath the photosphere, blazing light engulfs you. You are inside the Sun, and your spacecraft is tossed about by incredible turbulence. If you can hold steady long enough to see what is going on around you, you'll

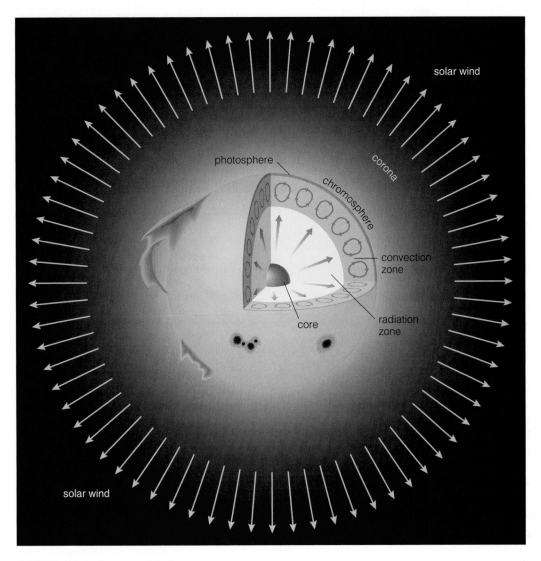

FIGURE 12.4 The basic structure of the Sun.

notice spouts of hot gas rising upward, surrounded by cooler gas cascading down from above. You are in the **convection zone**, where energy generated in the solar core travels upward, transported by the rising of hot gas and falling of cool gas called *convection* [Section 8.2]. With some quick thinking, you may realize that the photosphere above you is the top of the convection zone and that convection is the cause of the Sun's seething, churning appearance.

As you descend through the convection zone, the surrounding density and pressure increase substantially, along with the temperature. Soon you reach depths at which the Sun is far denser than water; nevertheless, it is still a *gas*—more specifically, a *plasma* of positively charged ions and free electrons—because each particle moves independently of its neighbors [Section 4.3].

About a third of the way down to the center, the turbulence of the convection zone gives way to the stabler plasma of the **radiation zone**, where energy is carried outward primarily by photons of light. The temperature is now almost 10 million K, and your spacecraft is bathed in X rays trillions of times more intense than the visible light at the solar surface. No real spacecraft could survive, but your imaginary one keeps plunging straight down to the solar **core**. There you finally find the source of the Sun's energy: nuclear fusion transforming hydrogen into helium. At the Sun's center, the temperature is about 15 million K, the density is more than 100 times that of water, and the pressure is 200 billion times that on the surface of the Earth.

With your journey complete, it's time to turn around and head back home. We'll continue this chapter by studying fusion in the solar core and then tracing the flow of the energy generated by fusion as it moves outward through the Sun.

12.3 The Cosmic Crucible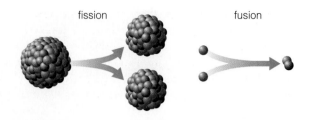

The prospect of turning common metals like lead into gold enthralled medieval alchemists, but their get-rich-quick schemes never managed to work. There is no easy way to turn other elements into gold, but it *is* possible to transmute one element or isotope into another. If a nucleus gains or loses protons, its atomic number changes and it becomes a different element. If it gains or loses neutrons, its atomic mass changes and it becomes a different isotope [Section 4.3]. The process of splitting a nucleus so that it loses protons or neutrons is called **nuclear fission**. The process of combining nuclei to make a nucleus with a greater number of protons or neutrons is called **nuclear fusion** (Figure 12.5). Human-built nuclear power plants rely on nuclear fission of uranium or plutonium. The nuclear power plant at the center of the Sun relies on nuclear fusion, turning hydrogen into helium.

FIGURE 12.5 Nuclear fission splits a nucleus into smaller pieces, while nuclear fusion combines smaller nuclei into a larger nucleus.

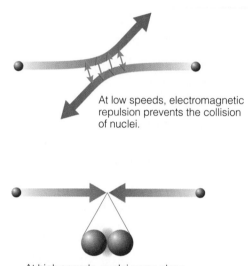

At low speeds, electromagnetic repulsion prevents the collision of nuclei.

At high speeds, nuclei come close enough for the strong nuclear force to bind them together.

FIGURE 12.6 Positively charged nuclei can fuse only if a high-speed collision brings them close enough for the strong nuclear force to come into play.

Nuclear Fusion

The 15-million-K plasma in the solar core is like a "soup" of hot gas, with bare, positively charged atomic nuclei (and individual negatively charged electrons) whizzing about at extremely high speeds. At any one time, some of these nuclei are on high-speed collision courses with each other. In most cases, electromagnetic forces deflect the nuclei, preventing actual collisions, because positive charges repel one another. However, if nuclei collide with sufficient energy, they can stick together to form a heavier nucleus (Figure 12.6).

Sticking positively charged nuclei together is not easy. The **strong force**, which binds protons and neutrons together in atomic nuclei, is the only force in nature that can overcome the electromagnetic repulsion between two positively charged nuclei. In contrast to gravitational and electromagnetic forces, which drop off gradually as the distances between particles increase (by an inverse square law), the strong force is more like glue or Velcro: It overpowers the electromagnetic force over very small distances but is insignificant when the distances between particles exceed the typical sizes of atomic nuclei. The trick to nuclear fusion therefore is to push the positively charged nuclei close enough together for the strong force to outmuscle electromagnetic repulsion.

The high pressures and temperatures in the solar core are just right for fusion of hydrogen nuclei into helium nuclei. The high temperature is important because the nuclei must collide at very high speeds if they are to come close enough together to fuse. The higher the temperature, the harder the collisions, making fusion reactions more likely at higher temperatures. The high pressure of the overlying layers is necessary because without it the hot plasma of the solar core would simply explode into space, shutting off the nuclear reactions.

Hydrogen Fusion in the Sun: The Proton–Proton Chain

Recall that hydrogen nuclei are nothing more than individual protons, while the most common form of helium consists of two protons and two neutrons. Thus, the overall hydrogen fusion reaction in the Sun is:

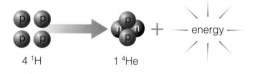

$4\ ^1H$ $1\ ^4He$

The sequence of steps that occurs in the Sun is called the **proton–proton chain** because it begins with collisions between individual protons (hydrogen nuclei). Each resulting helium-4 nucleus has a mass that is slightly less (by about 0.7%) than the combined mass of the four protons that created it. Overall, fusion in the Sun converts about 600 million tons of hydrogen into 596 million tons of helium every second; the "missing" 4 million tons of matter becomes energy in accord with Einstein's formula $E = mc^2$. About 2% of this energy is carried off by *neutrinos*, minuscule particles with very little mass. The rest of the energy emerges as kinetic energy of the nuclei and radiative energy in the form of gamma rays. As we will see, this energy slowly percolates to the solar surface, eventually emerging as the sunlight that bathes the Earth.

The Solar Thermostat

The rate of nuclear fusion in the solar core, which determines the energy output of the Sun, is very sensitive to temperature. A slight increase in temperature would mean a much higher fusion rate, and a slight decrease in temperature would mean a much lower fusion rate. If the Sun's rate of fusion varied erratically, the effects on the Earth might be devastating. Fortunately, the Sun's central temperature is steady thanks to gravitational equilibrium—the balance between the pull of gravity and the push of internal pressure.

Suppose the core temperature rose very slightly. The rate of nuclear fusion would soar, generating excess energy that would increase the pressure. The push of this pressure would temporarily exceed the pull of gravity, causing the core to expand and cool. With this cooling, the fusion rate would drop back down. The expansion and cooling would continue until gravitational equilibrium was restored, at which point the fusion rate would be back at its original value. An opposite process restores the normal fusion rate if the core temperature drops. A decrease in core temperature would lead to decreased nuclear burning, a drop in the central pressure, and contraction of the core. As the core shrank, its temperature would rise until the burning rate returned to normal.

The response of the core pressure to changes in the nuclear fusion rate is essentially a *thermostat* that keeps the Sun's central temperature steady. Any change in the core temperature is automatically corrected by the change in the fusion rate and the accompanying change in pressure.

While the processes involved in gravitational equilibrium prevent erratic changes in the fusion rate, they also ensure that the fusion rate gradually rises over billions of years. Theoretical models indicate that the Sun's core temperature should have increased enough to raise its fusion rate and the solar luminosity by about 30% since the Sun was born 4.6 billion years ago. The gradual increase in the solar luminosity poses a bit of a puzzle, because geological evidence shows that the Earth's temperature has remained fairly steady since the Earth finished forming over 4 billion years ago. Studying how the Sun changes with time and how the Earth responds to these changes may teach us much about how the Earth's physical and biological systems regulate the surface temperature.

"Observing" the Solar Interior

We cannot see inside the Sun, so you may be wondering how we can know so much about what goes on underneath its surface. The primary way we learn about the interior of the Sun and other stars is by creating *mathematical models* that use the laws of physics to predict the internal conditions. A basic model uses the Sun's observed composition and mass as inputs to equations that describe gravitational equilibrium and the solar thermostat. With the aid of a computer, we can use the model to calculate the Sun's temperature, pressure, and density at any depth. We can then predict the rate of nuclear fusion in the solar core by combining these calculations with knowledge about nuclear fusion gathered in laboratories here on Earth. Remarkably, such models correctly "predict" the radius, surface temperature, luminosity, age, and many other properties of the Sun. However, current models do not predict *everything* about the Sun correctly, so scientists are constantly working to discover what is missing from them. The fact that the models successfully predict so many observed characteristics of the Sun gives us confidence that they are on the right track and that we really do understand what is going on inside the Sun.

A second way to learn about the inside of the Sun is to observe "sun quakes"—vibrations of the Sun that are similar to the vibrations of the Earth caused by earthquakes, although they are generated very differently. Earthquakes occur when the Earth's crust suddenly shifts, generating *seismic waves* that propagate through the Earth's interior. We can learn about the Earth's interior by recording seismic waves on the Earth's surface with seismographs. Sun quakes result from waves of pressure (sound waves) that propagate deep within the Sun at all times. These waves cause the solar surface to vibrate when they reach it. Although we cannot set up seismographs on the Sun, we can detect the vibrations of the surface by measuring Doppler shifts [Section 6.4]: Light from portions of the surface that are rising toward us is slightly blueshifted, while light from portions that are falling away from us is slightly redshifted. The vibrations

FIGURE 12.7 Vibrations on the surface of the Sun can be detected by Doppler shifts. In this schematic representation, red indicates falling gas, and blue indicates rising gas.

are relatively small but measurable (Figure 12.7). In principle, we can deduce a great deal about the solar interior by carefully analyzing these vibrations. (By analogy to seismology on Earth, this type of study of the Sun is called *helioseismology* [recall that *helios* means "sun"].) Results to date confirm that mathematical models of the solar interior are on the right track and at the same time provide data that can be used to improve the models further.

The Solar Neutrino Problem

Another way to study the solar interior is to observe the neutrinos coming from fusion reactions in the core. Don't panic, but as you read this sentence about a thousand trillion solar neutrinos will zip through your body. Fortunately, they won't do any damage, because neutrinos rarely interact with anything. Neutrinos created by fusion in the solar core

fly quickly through the Sun as if passing through empty space. In fact, while an inch of lead will stop an X ray, stopping an average neutrino would require a slab of lead more than a light-year thick! Clearly, counting neutrinos is dauntingly difficult, because virtually all of them stream right through any detector built to capture them.

TIME OUT TO THINK *Is the number of solar neutrinos zipping through our bodies significantly lower at night? (Hint: How does the thickness of the Earth compare with the thickness of a slab of lead needed to stop an average neutrino?)*

Nevertheless, neutrinos *do* occasionally interact with matter, and it is possible to capture a few solar neutrinos with a large enough detector such as the "Homestake experiment," located 1,500 meters underground in a South Dakota gold mine (Figure 12.8). The designers of this experiment expected to capture an average of just one neutrino per day. This predicted capture rate was based on measured properties of chlorine nuclei and models of nuclear fusion in the Sun. However, over a period of more than two decades, neutrinos were captured only about once every 3 days on average. That is, the Homestake experiment detected only about one-third of the predicted number of neutrinos. This disagreement between model predictions and actual observations is called the **solar neutrino problem**.

FIGURE 12.8 This tank of dry-cleaning fluid (visible underneath the catwalk), located deep within South Dakota's Homestake mine, was a solar neutrino detector. The chlorine nuclei in the cleaning fluid turned into argon nuclei when they captured neutrinos from the Sun.

FIGURE 12.9 The Super-Kamiokande experiment in Japan is one of the world's premier neutrino detectors.

TIME OUT TO THINK *Although the observed number of neutrinos falls short of theoretical predictions, experiments like Homestake have shown that at least some neutrinos are coming from the Sun. Explain why this provides direct evidence that nuclear fusion really is taking place in the Sun right now.*

The shortfall of neutrinos found with the Homestake experiment has led to many more recent attempts to detect solar neutrinos using more sophisticated detectors (Figure 12.9). These experiments have generally found fewer neutrinos than current models of the Sun predict. This discrepancy between model and experiment probably means one of two things: Either something is wrong with our models of the Sun, or something is missing in our understanding of how neutrinos behave. For the moment, many physicists and astronomers are betting that we understand the Sun just fine and that the discrepancy has to do with the neutrinos themselves. Early results from the Sudbury Neutrino Observatory in Canada, a new detector designed to search for all types of neutrinos, suggests that many solar neutrinos actually change type before reaching Earth, making themselves much harder to detect. However, the observations are ongoing, and it will probably be several more years before this possible solution to the solar neutrino problem can be definitively confirmed.

a Scientists inspecting individual detectors within Super-Kamiokande.

12.4 From Core to Corona

Energy liberated by nuclear fusion in the Sun's core must eventually reach the solar surface, where it can be radiated into space. The path that the energy takes to the surface is long and complex. In this section, we follow the long path that energy travels after its production in the core.

The Path Through the Solar Interior

In Chapter 6, we discussed how atoms can absorb or emit photons. In fact, photons can also interact with any charged particle, and a photon that "collides"

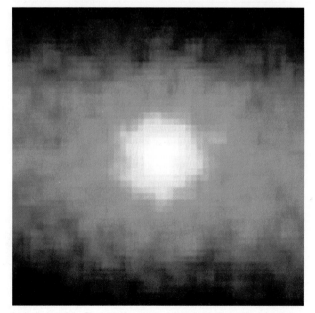

b An image of the Sun based on neutrinos detected by Super-Kamiokande.

FIGURE 12.10 A photon in the solar interior bounces randomly among electrons, slowly working its way outward over about 1 million years.

with an electron can be deflected into a completely new direction. Deep in the solar interior, the plasma is so dense that the gamma-ray photons resulting from fusion travel only a fraction of a millimeter before colliding with an electron. Because each collision sends the photon in a random new direction, the photon bounces around the core in a haphazard way, sometimes called a *random walk* (Figure 12.10). With each random bounce, the photon drifts farther and farther, on average, from its point of origin. Along the way, the photons exchange energy with their surroundings; because the surrounding temperature declines as the photons move outward through the Sun, they are gradually transformed from gamma rays to photons of lower energy. (Because energy must be conserved, each gamma-ray photon becomes many lower-energy photons.) By the time the energy of fusion reaches the surface, the photons are primarily visible light.

This random bouncing of photons is the primary way by which energy moves outward through the *radiation zone,* which stretches from the core to about 70% of the Sun's radius (see Figure 12.4). Above this point, where the temperature has dropped to about 2 million K, the solar plasma absorbs photons more readily (rather than just bouncing them around). This is the beginning of the solar *convection zone,* where the build-up of heat resulting from photon absorption causes bubbles of hot plasma to rise upward in the process known as **convection** [Section 8.2]. Convection occurs because hot gas is less dense than cool gas. Like a hot-air balloon, a hot bubble of solar plasma rises upward through the cooler plasma above it. Meanwhile, cooler plasma from above slides around the rising bubble and sinks to lower layers, where it is heated. The rising of hot plasma and sinking of cool plasma form a cycle that transports energy outward from the top of the radiation zone to the solar surface (Figure 12.11a).

The Solar Surface

The Earth has a solid crust, so its surface is well defined. In contrast, the Sun is made entirely of gaseous plasma. Defining where the surface of the Sun begins is therefore something like defining the surface of a cloud: From a distance it looks quite distinct, but up close the surface is fuzzy, not sharp. We generally define the solar surface as the layer that appears distinct from a distance. This is the layer we identified as the *photosphere* when we took our imaginary journey into the Sun. More technically, the photosphere is the layer of the Sun from which pho-

FIGURE 12.11 Convection transports energy outward in the Sun's convection zone.

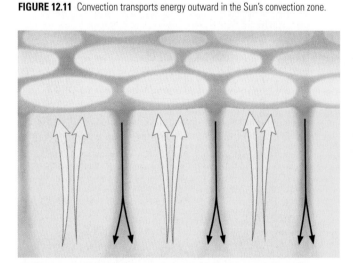

a This schematic diagram shows how hot gas (white arrows) rises while cooler gas (black arrows) descends around it. Bright spots appear on the solar surface in places where hot gas is rising from below, creating the granulated appearance of the solar photosphere.

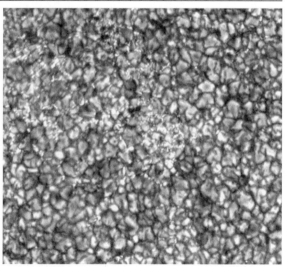

b Granulation is evident in this photo of the Sun's surface. Each bright granule is the top of a rising column of gas. At the darker lines between the granules, cooler gas is descending below the photosphere.

tons finally escape into space after the million-year journey of solar energy outward from the core. Most of the energy produced by fusion in the solar core ultimately leaves the photosphere as thermal radiation [Section 6.3]. The average temperature of the photosphere is about 5,800 K, corresponding to a thermal radiation spectrum that peaks in the green portion of the visible spectrum, with substantial energy coming out in all colors of visible light.

Although the average temperature of the photosphere is 5,800 K, actual temperatures vary significantly from place to place. The photosphere is marked throughout by the bubbling pattern of **granulation** produced by the underlying convection (Figure 12.11b). Each *granule* appears bright in the center, where hot gas bubbles upward, and dark around the edges, where cool gas descends. If we made a movie of the granulation, we'd see it bubbling rather like a pot of boiling water. Just as bubbles in a pot of boiling water burst on the surface and are replaced by new bubbles, each granule lasts only a few minutes before being replaced by other granules bubbling upward.

Sunspots and Magnetic Fields

Sunspots are the most striking features on the solar surface. The temperature of the plasma in sunspots is about 4,000 K, significantly cooler than the 5,800-K plasma of the surrounding photosphere. If you think about this for a moment, you may wonder how sunspots can be so much cooler than their surroundings. What keeps the surrounding hot plasma from heating the sunspots?

Something must be preventing hot plasma from entering the sunspots, and that "something" turns out to be magnetic fields. Strong magnetic fields can alter the energy levels in atoms and ions and therefore can alter the spectral lines they produce. More specifically, magnetic fields cause some spectral lines to split into two or more closely spaced lines (Figure 12.12). Thus, scientists can map magnetic fields on the Sun by studying the spectral lines in light from different parts of the solar surface. Such maps reveal sunspots to be regions of intense magnetic fields.

Magnetic fields are invisible, but in principle we could visualize a magnetic field by laying out many compasses; each compass needle would point to local magnetic north. We can represent the magnetic field by drawing a series of lines, called **magnetic field lines**, connecting the needles of these imaginary compasses (Figure 12.13a). The strength of the magnetic field is indicated by the spacing of the lines: Closer lines mean a stronger field (Figure 12.13b). Because these imaginary field lines are so much

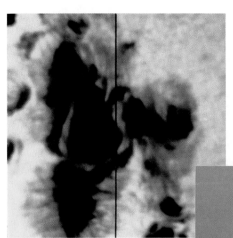

a The vertical line on this photo of a sunspot and its surrounding region represents the line along which light was dispersed into a spectrum.

b A portion of this spectrum is shown with a spectral line that is a single line outside the sunspot region but splits into three parts under the influence of the sunspot's strong magnetic field.

FIGURE 12.12 Sunspots are regions of intense magnetic activity.

easier to visualize than the magnetic field itself, we usually discuss magnetic fields by talking about how the field lines would appear. Charged particles, such as the ions and electrons in the solar plasma, follow paths that spiral along the magnetic field lines (Figure 12.13c). Thus, the solar plasma can move freely *along* magnetic field lines but cannot easily move perpendicular to them.

The magnetic field lines act somewhat like elastic bands, twisted into contortions and knots by turbulent motions in the solar atmosphere. Sunspots occur where the most taut and tightly wound magnetic fields poke nearly straight out from the solar interior (Figure 12.14). Note that sunspots tend to occur in pairs connected by a loop of magnetic field lines. These tight magnetic field lines suppress convection within each sunspot and prevent surrounding plasma from sliding sideways into the sunspot. With hot plasma unable to enter the region, the sunspot plasma becomes cooler than that of the rest of the photosphere.

The magnetic field lines connecting two sunspots often soar high above the photosphere, through the chromosphere, and into the corona. These vaulted loops of magnetic field sometimes appear as

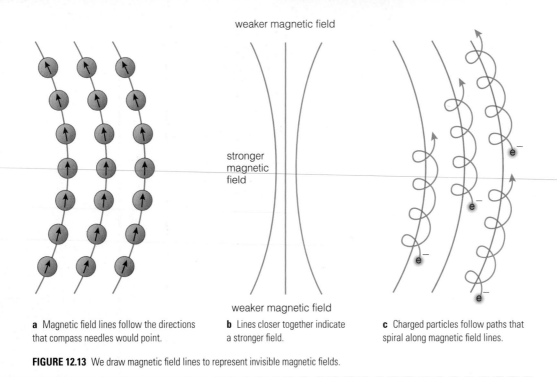

weaker magnetic field

stronger magnetic field

weaker magnetic field

a Magnetic field lines follow the directions that compass needles would point.

b Lines closer together indicate a stronger field.

c Charged particles follow paths that spiral along magnetic field lines.

FIGURE 12.13 We draw magnetic field lines to represent invisible magnetic fields.

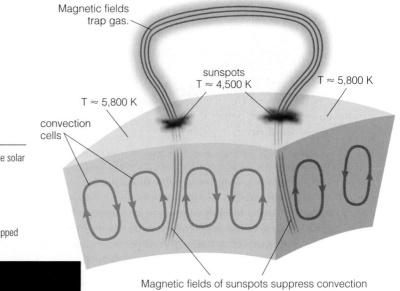

FIGURE 12.14 Loops of magnetic field lines can arch high above the solar surface.

Magnetic fields trap gas.

sunspots T ≈ 4,500 K

T ≈ 5,800 K

T ≈ 5,800 K

convection cells

Magnetic fields of sunspots suppress convection and prevent surrounding plasma from sliding sideways into sunspot.

b Pairs of sunspots are connected by tightly wound magnetic field lines.

a This X-ray photo (from NASA's TRACE mission) shows hot gas trapped within looped magnetic field lines.

solar prominences, in which the field traps gas that may glow for days or even weeks. Some prominences rise to heights of more than 100,000 kilometers above the Sun's surface (Figure 12.15).

The most dramatic events on the solar surface are **solar flares**, which emit bursts of X rays and fast-moving charged particles into space (Figure 12.16). Flares generally occur in the vicinity of sunspots, leading us to suspect that they occur when the magnetic field lines become so twisted and knotted that they can bear the tension no longer. The magnetic field lines suddenly snap like tangled elastic bands twisted beyond their limits, releasing a huge amount of energy. This energy heats the nearby plasma to 100 million K over the next few minutes to few hours, generating X rays and accelerating some of the charged particles to nearly the speed of light.

The Chromosphere and Corona

The high temperatures of the chromosphere and corona perplexed scientists for decades. After all, temperatures gradually decline as we move outward from the core to the top of the photosphere. Why should this decline suddenly reverse? Some aspects of this atmospheric heating remain a mystery today, but we have at least a general explanation: The Sun's strong magnetic fields carry energy upward from the churning solar surface to the chromosphere and corona. More specifically, the rising and falling of gas in the convection zone probably shake magnetic field lines beneath the solar surface. This shaking generates waves along the magnetic field lines that carry energy upward to the solar atmosphere. Precisely how the waves deposit their energy in the chromosphere and corona is not known, but the waves agitate the low-density plasma of these layers, somehow heating them to high temperatures. Much of this heating appears to happen near where the magnetic field lines emerge from the Sun's surface.

Note that, according to this model of solar heating, the same magnetic fields that keep sunspots cool make the overlying plasma of the chromosphere and

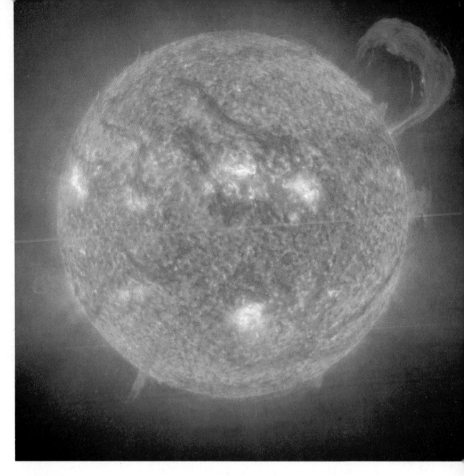

FIGURE 12.15 A gigantic solar prominence erupts from the solar surface at the upper right of this ultraviolet-light photo (from the SOHO mission). The gas within this prominence, which is over 20 times the size of Earth, is quite hot but still cooler than the million-degree gas of the surrounding corona.

FIGURE 12.16 This photo (from TRACE) of ultraviolet light emitted by hydrogen atoms shows a solar flare erupting from the Sun's surface.

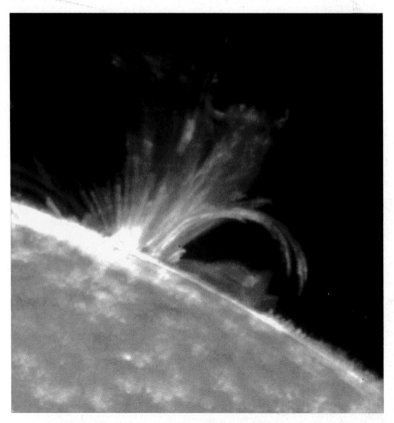

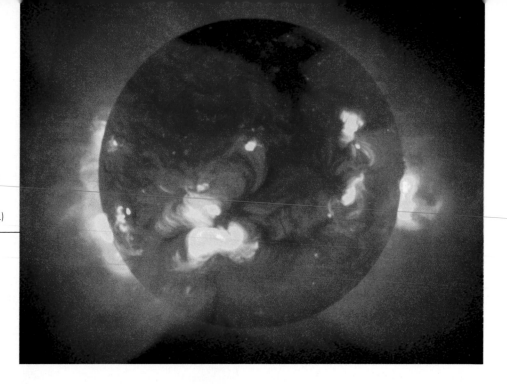

FIGURE 12.17 An X-ray image of the Sun reveals the corona: Brighter regions of this image correspond to regions of stronger X-ray emission. (From the Yohkoh space observatory.)

corona hot. We can test this idea observationally. The gas of the chromosphere and corona is so tenuous that we cannot see it with our eyes except during a total eclipse, when we can see the faint visible light scattered by electrons in the corona [Section 2.3]. However, the roughly 10,000-K plasma of the chromosphere emits strongly in the ultraviolet, and the million-K plasma of the corona is the source of virtually all X rays coming from the Sun. Figure 12.17 shows an X-ray image of the Sun: As the solar heating model predicts, the brightest regions of the corona tend to be directly above sunspot groups.

Some regions of the corona, called **coronal holes**, barely show up in X-ray images. More detailed analyses show that the magnetic field lines in coronal holes project out into space like broken rubber bands, allowing particles spiraling along them to escape the Sun altogether. These particles streaming outward from the corona constitute the *solar wind,* which blows through the solar system at an average speed of about 500 kilometers per second and has important effects on planetary surfaces, atmospheres, and magnetospheres [Section 8.4]. Somewhere beyond the orbit of Neptune, the pressure of interstellar gas must eventually halt the solar wind.

12.5 Solar Weather and Climate

Short-lived phenomena like sunspots, prominences, and flares constitute what we call *solar weather* or **solar activity.** You know from personal experience that the Earth's weather is notoriously unpredictable, and the same is true for the Sun: We cannot predict precisely when or where a particular sunspot or flare

will appear. The Earth's *climate,* on the other hand, is quite regular from season to season. So it is with the Sun, where despite day-to-day variations the general nature and intensity of solar activity follow a predictable cycle.

The Sunspot Cycle

Long before we realized that sunspots were magnetic disturbances, astronomers had recognized patterns in sunspot activity. The most notable pattern is the number of sunspots visible on the Sun at any particular time. Thanks to telescopic observations of the Sun recorded by astronomers since the 1600s, we know that the number of sunspots gradually rises and falls in a **sunspot cycle** with a period of about 11 years (Figure 12.18a). At the time of **solar maximum,** when sunspots are most numerous, we may see dozens of sunspots on the Sun at one time. In contrast, we see few if any sunspots at the time of **solar minimum.** Note that the sunspot cycle is not perfectly predictable: The interval between solar maxima is sometimes as long as 15 years or as short as 7 years. More dramatically, sunspot activity virtually ceased between the years 1645 and 1715, a period sometimes called the *Maunder minimum* (after E. W. Maunder, who identified it in historical sunspot records). The frequency of prominences and flares also follows the sunspot cycle, with these events being most common at solar maximum and least common at solar minimum.

Another feature of the sunspot cycle is a gradual change in the solar latitudes at which individual sunspots form and dissolve (Figure 12.18b). As a cycle begins at solar minimum, sunspots form primarily at mid-latitudes (30° to 40°) on the Sun. The sunspots tend to form at lower latitudes as the cycle progresses, appearing very close to the solar equator as the next solar minimum approaches.

A less obvious feature of the sunspot cycle is that at each solar minimum something peculiar happens to the Sun's magnetic field. The field lines connecting all pairs of sunspots (see Figure 12.14) tend to point in the same direction throughout an 11-year solar cycle (within each hemisphere); for example, all compass needles might point from the easternmost sunspot to the westernmost sunspot in a pair. However, as the cycle ends at solar minimum, the magnetic field reverses: In the subsequent solar cycle, the field lines connecting pairs of sunspots point in the opposite direction. Apparently, the entire magnetic field of the Sun flip-flops every 11 years. These magnetic reversals hint that the sunspot cycle is related to the generation of magnetic fields on the Sun. They also tell us that the *complete* magnetic cycle of the Sun, called the *solar cycle,* really averages 22 years, since it takes two 11-year cycles before the magnetic field is back the way it started.

What Causes the Sunspot Cycle?

The causes of the Sun's magnetic fields and the sunspot cycle are not well understood, but we believe we know the general nature of the processes involved. Convection is thought to dredge up weak magnetic fields generated in the solar interior, amplifying them as they rise. The Sun's **differential rotation**—faster at its equator than near its poles—then stretches and shapes these fields.

Imagine what happens to a magnetic field line that originally runs along the Sun's surface directly from the north pole to the south pole. At the equator, the field line circles the Sun in 27 days, but at higher latitudes the field line lags behind. Gradually, differential rotation winds the field line more and more tightly around the Sun (Figure 12.19). This process, operating at all times over the entire Sun, produces the contorted field lines that generate sunspots and other solar activity.

Making more specific statements about how the Sun's magnetic field develops and changes in time requires the aid of sophisticated computer models,

FIGURE 12.18 Sunspot cycle during the past century.

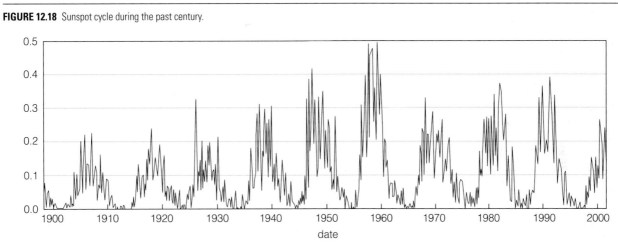

a This graph shows how the number of sunspots on the Sun changes with time. The numbers on the vertical axis indicate the percentage of the Sun's visible hemisphere covered by sunspots. Note the approximately 11-year cycle.

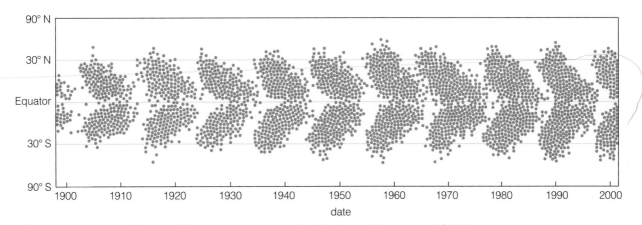

b This graph shows how the latitudes at which sunspot groups appear tend to shift during a single sunspot cycle. Each dot represents a group of sunspots and indicates the year (horizontal axis) and latitude (vertical axis) at which the group appeared.

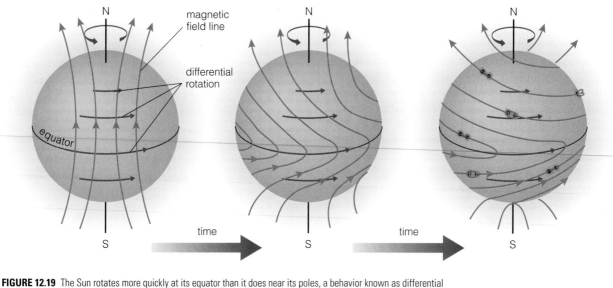

FIGURE 12.19 The Sun rotates more quickly at its equator than it does near its poles, a behavior known as differential rotation. Because gas circles the Sun faster at the equator, it drags the Sun's north-south magnetic field lines into a more twisted configuration. The magnetic field lines linking pairs of sunspots, depicted here as green and black blobs, trace out the directions of these stretched and distorted field lines.

which have successfully replicated some features of the sunspot cycle, such as changes in the number and latitude of sunspots and the magnetic field reversals that occur about every 11 years. However, much still remains mysterious, including why the period of the sunspot cycle varies and why solar activity is different from one cycle to the next.

Solar Activity and the Earth

During solar maximum, solar flares and other forms of solar activity feed large numbers of highly energetic charged particles (protons and electrons) into the solar wind. Do these forms of solar weather affect the Earth? In at least some ways, the answer is a definitive yes.

The solar wind continually carries charged particles from the solar corona to the Earth, creating the shimmering light of the *aurora*. Particles streaming from the Sun after the occurrence of solar flares or other major solar storms can also have practical impacts on society. For example, these particles can hamper radio communications, disrupt electrical power delivery, and damage the electronic components in orbiting satellites. During a particularly powerful magnetic storm on the Sun in March 1989, the U.S. Air Force temporarily lost track of over 2,000 satellites, and powerful currents induced in the ground circuits of the Quebec hydroelectric system caused it to collapse for more than 8 hours.

Satellites in low-Earth orbit are particularly vulnerable during solar maximum, because the increase in solar X rays and energetic particles heats the Earth's upper atmosphere, causing it to expand. The density of the gas surrounding low-flying satellites therefore rises, exerting drag that saps their energy and angular momentum. If this drag proceeds unchecked, the satellites ultimately plummet back to Earth. Satellites in low orbits, including the Hubble Space Telescope, therefore require occasional boosts to prevent them from falling out of the sky.

Connections between solar activity and the Earth's climate are much less clear. The period from 1645 to 1715, when solar activity seems to have virtually ceased, was a time of exceptionally low temperatures in Europe and North America known as the *Little Ice Age*. Did the low solar activity cause these low temperatures, or was their occurrence a coincidence? No one knows for sure. Some researchers have claimed that certain weather phenomena, such as drought cycles or frequencies of storms, are correlated with the 11- or 22-year cycle of solar activity. However, the data supporting these correlations are weak in many cases, and even real correlations may be coincidental.

Part of the difficulty in linking solar activity with climate is that no one understands how the linkage might work. Although emissions of ultraviolet light, X rays, and high-energy particles increase substantially from solar minimum to solar maximum, the total luminosity of the Sun barely changes at all. (The Sun becomes slightly brighter during solar maximum, but only by about 0.1%.) Thus, if solar activity really is affecting Earth's climate, it must be through some very subtle mechanism. For example, perhaps the expansion of the Earth's upper atmosphere that occurs with solar maximum somehow causes changes in weather.

The question of how solar activity is linked to the Earth's climate is very important, because we need to know whether global warming is affected by solar activity in addition to human activity. Unfortunately, for the time being at least, we can say little about this question.

THE BIG PICTURE

In this chapter, we have examined our Sun, the nearest star. When you look back at this chapter, make sure you understand these "big picture" ideas:

■ The ancient riddle of why the Sun shines is now solved. The Sun shines with energy generated by fusion of hydrogen into helium in the Sun's core. After a million-year journey through the solar interior and an 8-minute journey through space, a small fraction of this energy reaches the Earth and supplies sunlight and heat.

■ Gravitational equilibrium, the balance between pressure and gravity, determines the Sun's interior structure and maintains its steady nuclear burning rate. If the Sun were not so steady, life on Earth might not have been possible.

■ The Sun's atmosphere displays its own version of weather and climate, governed by solar magnetic fields. Solar weather has important influences on the Earth.

■ The Sun is important not only as our source of light and heat, but also because it is the only star near enough for us to study in great detail. In the coming chapters, we will use what we've learned about the Sun to help us understand other stars.

Review Questions

1. Briefly describe how *gravitational contraction* generates energy and when it was important in the Sun's history.

2. What two forces are balanced in *gravitational equilibrium*? How does gravitational equilibrium determine the pressure at any depth within a planet or star?

3. What is the Sun made of?

4. Briefly describe the Sun's luminosity, mass, radius, and average surface temperature.

5. Briefly describe the distinguishing features of each of the Sun's major layers shown in Figure 12.4.

6. What are *sunspots*? Why do they appear dark in pictures of the Sun?

7. Distinguish between *nuclear fission* and *nuclear fusion*. Which one is used in nuclear power plants? Which one is used by the Sun?

8. Why does fusion require high temperatures and pressures?

9. How does the idea of gravitational equilibrium explain why the Sun's core temperature and fusion rate are self-regulating, like a thermostat?

10. Explain how mathematical models allow us to predict conditions inside the Sun. How can we be confident that the models are on the right track?

11. How are "sun quakes" similar to earthquakes? How are they different? Describe how we can observe them and how they help us learn about the solar interior.

12. What is the *solar neutrino problem*?

13. What is *convection*? Describe how convection transports energy through the Sun's convection zone.

14. Describe the appearance and temperature of the Sun's *photosphere*. What is *granulation*? How would granulation appear in a movie?

15. Briefly describe how *magnetic field lines* are related to what we would see if we placed compasses in a magnetic field.

16. How are magnetic fields related to sunspots? Describe how magnetic fields explain each of the following: the fact that sunspot plasma is cool, the existence and appearance of *prominences*, and the dramatic explosions of *solar flares*.

17. Why is the *chromosphere* best viewed with ultraviolet telescopes? Why is the *corona* best viewed with X-ray telescopes?

18. Describe the current theory of how the chromosphere and corona are heated. What evidence supports it?

19. What are *coronal holes*? How are they related to the solar wind?

20. What do we mean by *solar activity*? Explain how it is similar to weather on short time scales and to climate on longer time scales.

21. Describe the *sunspot cycle*. What do we mean by *solar maximum* and *solar minimum*? Explain why the complete cycle is really 22 years rather than 11 years.

22. Describe a few known ways solar activity affects the Earth. Why is it so difficult to establish links between solar activity and the Earth's climate?

Discussion Question

1. *The Sun and Global Warming.* One of the most pressing environmental issues on Earth concerns the extent to which human emissions of greenhouse gases are warming our planet. Some people claim that part or all of the observed warming over the past century may be due to changes on the Sun, rather than to anything humans have done. Discuss how a better understanding of the Sun might help us understand the threat posed by greenhouse gas emissions. Why is it so difficult to develop a clear understanding of how the Sun affects the Earth's climate?

Problems

Sensible Statements? For **problems 1–7**, decide whether the statement is sensible and explain why it is or is not.

1. Before Einstein, gravitational contraction appeared to be a perfectly plausible mechanism for solar energy generation.

2. A sudden temperature rise in the Sun's core is nothing to worry about, because conditions in the core will soon return to normal.

3. If fusion in the solar core ceased today, worldwide panic would break out tomorrow as the Sun began to grow dimmer.

4. Astronomers have recently photographed magnetic fields churning deep beneath the solar photosphere.

5. Neutrinos probably can't harm me, but just to be safe I think I'll wear a lead vest.

6. If you want to see lots of sunspots, just wait for solar maximum!

7. News of a major solar flare today caused concern among professionals in the fields of communications and electrical power generation.

8. *Differential Rotation.* The Sun and Jupiter exhibit differential rotation, but Earth does not.
   a. Explain clearly what we mean by *differential rotation.*
   b. Briefly explain why it is possible for jovian planets to rotate differentially, but it is not possible for terrestrial planets to do so.
   c. Would you expect differential rotation to be common among other stars? Why or why not?

9. *The Lifetime of the Sun.* The total mass of the Sun is about 2×10^{30} kg, of which about 75% was hydrogen when the Sun formed. However, only about 13% of this hydrogen ever becomes available for fusion in the core; the rest remains in layers of the Sun where the temperature is too low for fusion.
   a. Based on the given information, calculate the total mass of hydrogen available for fusion over the lifetime of the Sun.
   b. Combine your results from part (a) and the fact that the Sun fuses about 600 billion kg of hydrogen each second to calculate how long the Sun's initial supply of hydrogen can last. Give your answer in both seconds and years.
   c. Given that our solar system is now about 4.6 billion years old, when will we need to worry about the Sun running out of hydrogen for fusion?

Web Projects

Find useful links for Web projects on the text Web site.

1. *Current Solar Activity.* Daily information about solar activity is available at NASA's Web site sunspotcycle.com. Where are we in the sunspot cycle right now? When is the next solar maximum or minimum expected? Have there been any major solar storms in the past few months? If so, did they have any significant effects on Earth? Summarize your findings in a one- to two-page report.

2. *Solar Observatories in Space.* Visit NASA's Web site for the Sun–Earth connection and explore some of the current and planned space missions designed to observe the Sun. Choose one mission to study in greater depth, and write a one- to two-page report on the mission status and goals and what it has taught or will teach us about the Sun.

3. *Sudbury Neutrino Observatory.* Visit the Web site for the Sudbury Neutrino Observatory (SNO) and learn how it may help solve the solar neutrino problem. Write a one- to two-page report describing the observatory, any recent results, and what we can expect from it in the future.

"All men have the stars," he answered, "but they are not the same things for different people. For some, who are travelers, the stars are guides. For others they are no more than little lights in the sky. For others, who are scholars, they are problems. For my businessman they were wealth. But all these stars are silent. You—you alone—will have the stars as no one else has them."

ANTOINE DE SAINT-EXUPERY,
FROM *THE LITTLE PRINCE*

CHAPTER 13
Stars

On a clear, dark night, a few thousand stars are visible to the naked eye. Many more become visible through binoculars, and with a powerful telescope we can see so many stars that we could never hope to count them. Like individual people, each individual star is unique. Like the human family, all stars share much in common.

Today, we know that stars are born from clouds of interstellar gas, shine brilliantly by nuclear fusion for millions or billions of years, and then die, sometimes in dramatic ways. This chapter outlines how we study and categorize stars and how we have come to realize that stars, like people, change over their lifetime.

13.1 Snapshot of the Heavens

Imagine that an alien spaceship flies by Earth on a simple but short mission: The visitors have just 1 minute to learn everything they can about the human race. In 60 seconds, they will see next to nothing of each individual person's life. Instead, they will obtain a collective "snapshot" of humanity that shows people from all stages of life engaged in their daily activities. From this snapshot alone, they must piece together their entire understanding of human beings and their lives, from birth to death.

We face a similar problem when we look at the stars. Compared with stellar lifetimes of millions or billions of years, the few hundred years humans have spent studying stars with telescopes is rather like the aliens' 1-minute glimpse of humanity. We see only a brief moment in any star's life, and our collective snapshot of the heavens consists of such frozen moments for billions of stars. From this snapshot, we try to reconstruct the life cycles of stars while also analyzing what makes one star different from another.

Thanks to the efforts of hundreds of astronomers studying this snapshot of the heavens, stars are no longer mysterious points of light in the sky. We now know that all stars form in great clouds of gas and dust. All stars begin their life with roughly the same chemical composition: About three-quarters of the star's mass at birth is hydrogen, and about one-quarter is helium, with no more than about 2% consisting of elements heavier than helium. During most of any star's life, the rate at which it generates energy depends on the same type of balance between the inward pull of gravity and the outward push of internal pressure that governs the rate of fusion in our Sun. Despite these similarities, stars appear different from one another for two primary reasons: They differ in mass, and we see different stars at different stages of their lives.

The key that finally unlocked these secrets of stars was an appropriate classification system. Before the twentieth century, humans classified stars primarily by their brightness and location in our sky. Today, astronomers classify a star primarily according to its *luminosity* and *surface temperature*. Our task in this chapter is to learn how this extraordinarily effective classification system reveals the true natures of stars and their life cycles. We begin by investigating how to determine a star's luminosity, surface temperature, and mass.

13.2 Stellar Luminosity

A star's **luminosity** is the total amount of power it radiates into space, which can be stated in *watts* [Section 12.2]. For example, the Sun's luminosity is

3.8×10^{26} watts. We cannot measure a star's luminosity directly, because its brightness in our sky depends on its distance as well as its true luminosity. For example, our Sun and Alpha Centauri A (the brightest of the three stars in the Alpha Centauri system [Section 1.2]) are similar in luminosity. But Alpha Centauri A is a feeble point of light in the night sky, while our Sun provides enough light and heat to sustain life on Earth. The difference in brightness arises because Alpha Centauri A is about 270,000 times farther from Earth than is the Sun.

More precisely, we define the **apparent brightness** of any star in our sky as the amount of light reaching us *per unit area*. (A more technical term for apparent brightness is *flux*.) The apparent brightness of any light source obeys an *inverse square law* with distance, similar to the inverse square law that describes the force of gravity [Section 5.3]. If we viewed the Sun from twice the Earth's distance, it would appear dimmer by a factor of $2^2 = 4$. If we viewed it

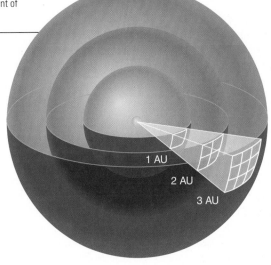

from 10 times the Earth's distance, it would appear $10^2 = 100$ times dimmer. From 270,000 times the Earth's distance, it would look like Alpha Centauri A—dimmer by a factor of $270,000^2$, or about 70 billion.

Figure 13.1 shows why apparent brightness follows an inverse square law. The same total amount of light must pass through each imaginary sphere surrounding the star. If we focus our attention on the light passing through a small square on the sphere located at 1 AU, we see that the same amount of light must pass through *four* squares of the same size on the sphere located at 2 AU. Thus, each square on the sphere at 2 AU receives only $\frac{1}{2^2} = \frac{1}{4}$ as much light as the square on the sphere at 1 AU. Similarly, the same amount of light passes through *nine* squares of the same size on the sphere located at 3 AU. Thus, each of these squares receives only $\frac{1}{3^2} = \frac{1}{9}$ as much light as the square on the sphere at 1 AU. Generalizing, we see that the amount of light received per unit area decreases with increasing distance by the square of the distance—an inverse square law.

This inverse square law leads to a very simple and important formula relating the apparent brightness, luminosity, and distance of any light source. We will call it the **luminosity–distance formula**:

$$\text{apparent brightness} = \frac{\text{luminosity}}{4\pi \times (\text{distance})^2}$$

Note that, because the standard units of luminosity are watts, the units of apparent brightness are *watts per square meter*. Because we can always measure the apparent brightness of a star, this formula provides a way to calculate a star's luminosity if we can first measure its distance or to calculate a star's distance if we somehow know its luminosity. (The luminosity–distance formula is strictly correct only if interstellar dust does not absorb or scatter the starlight along its path to Earth.)

Although watts are the standard units for luminosity, it's often more meaningful to describe stellar luminosities in comparison to the Sun by using units of **solar luminosity**: $L_{\text{Sun}} = 3.8 \times 10^{26}$ watts. For example, Proxima Centauri, the nearest of the three stars in the Alpha Centauri system and hence the nearest star besides our Sun, is only about 0.0006 times as luminous as the Sun, or $0.0006 L_{\text{Sun}}$. Betelgeuse, the bright left-shoulder star of Orion, has a luminosity of $38,000 L_{\text{Sun}}$, meaning that it is 38,000 times more luminous than the Sun.

Measuring the Apparent Brightness

We can measure a star's apparent brightness by using a detector that records how much energy strikes its light-sensitive surface each second. No detector can record light of all wavelengths, so we necessarily measure apparent brightness in only some small range of the complete spectrum. For example, the human eye is sensitive to visible light but does not respond to ultraviolet or infrared photons. Thus, when we perceive a star's brightness, our eyes are measuring the apparent brightness only in the visible region of the spectrum. Clearly, when we measure the apparent brightness in visible light, we can calculate only the star's *visible-light luminosity*. Similarly, when we observe a star with a spaceborne X-ray telescope, we measure only the apparent brightness in X rays and can calculate only the star's *X-ray luminosity*. We will use the terms **total luminosity** and **total apparent brightness** to describe the luminosity and apparent brightness we would measure *if* we could detect photons across the entire electromagnetic spectrum.

Measuring Distance Through Stellar Parallax

Once we have measured a star's apparent brightness, the next step in determining its luminosity is to measure its distance. The most direct way to measure the distances to stars is with *stellar parallax*, the small annual shifts in a star's apparent position caused by the Earth's motion around the Sun [Section 2.4]. Recall that you can observe parallax of your finger by holding it at arm's length and looking at it alternately with first one eye closed and then the other. Astronomers measure stellar parallax by comparing observations of a nearby star made 6 months apart (Figure 13.2).

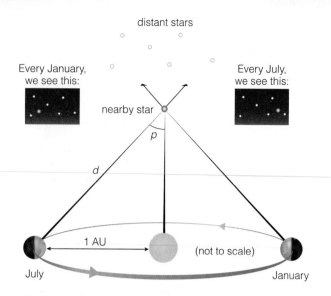

distant stars

Every January, we see this:

Every July, we see this:

nearby star

p

d

1 AU

(not to scale)

July

January

FIGURE 13.2 Parallax makes the apparent position of a nearby star shift back and forth with respect to distant stars over the course of each year. If we measure the parallax angle p in arcseconds, the distance d to the star in parsecs is $\frac{1}{p}$. The angle in this figure is greatly exaggerated: All stars have parallax angles of less than 1 arcsecond.

The nearby star appears to shift against the background of more distant stars because we are observing it from two opposite points of the Earth's orbit. The star's **parallax angle** is defined as *half* the star's annual back-and-forth shift.

Measuring stellar parallax is difficult because stars are so far away, making their parallax angles very small. Even the nearest star, Proxima Centauri, has a parallax angle of only 0.77 arcsecond. For increasingly distant stars, the parallax angles quickly become too small to measure even with our highest-resolution telescopes. Current technology allows us to measure parallax only for stars within a few hundred light-years—not much farther than what we call *local solar neighborhood* in the vast, 100,000-light-year-diameter Milky Way Galaxy.

By definition, the distance to an object with a parallax angle of 1 arcsecond is one **parsec**, abbreviated **pc**. (The word *parsec* comes from the words *parallax* and *arcsecond*.) With a little geometry and Figure 13.2, it is possible to show that:

$$1 \text{ pc} = 3.26 \text{ light-years} = 3.09 \times 10^{13} \text{ km}$$

If we use units of arcseconds for the parallax angle, a simple formula allows us to calculate distances in parsecs:

$$d \text{ (in parsecs)} = \frac{1}{p \text{ (in arcseconds)}}$$

For example, the distance to a star with a parallax angle of 1/2 arcsecond is 2 parsecs, the distance to a star with a parallax angle of 1/10 arcsecond is 10 parsecs, and the distance to a star with a parallax angle

of 1/100 arcsecond is 100 parsecs. Astronomers often express distances in parsecs or light-years interchangeably, so you should learn to convert quickly between them by remembering that 1 pc = 3.26 light-years. Thus, 10 parsecs is about 32.6 light-years; 1,000 parsecs, or 1 kiloparsec (1 kpc), is about 3,260 light-years; and 1 million parsecs, or 1 megaparsec (1 Mpc), is about 3.26 million light-years.

Enough stars have measurable parallax to give us a fairly good sample of the many different types of stars. For example, we know of over 300 stars within about 33 light-years (10 parsecs) of the Sun. About half are binary star systems consisting of two orbiting stars or multiple star systems containing three or more stars. Most are tiny, dim red stars such as Proxima Centauri—so dim that we cannot see them with the naked eye, despite the fact that they are relatively close. A few nearby stars, such as Sirius (2.6 parsecs), Vega (8 parsecs), Altair (5 parsecs), and Fomalhaut (7 parsecs), are white in color and bright in our sky, but many more of the brightest stars in the sky lie farther away. The intrinsic luminosities of stars must therefore span a wide range, since so many nearby stars appear dim while many more distant stars appear bright.

The Magnitude System

Many amateur and professional astronomers describe stellar brightness using the ancient **magnitude system** devised by the Greek astronomer Hipparchus (c. 190–120 B.C.). The magnitude system originally classified stars according to how bright they look to our eyes—the only instruments available in ancient times. The brightest stars received the designation

Hipparchus

"first magnitude," the next brightest "second magnitude," and so on. The faintest visible stars were magnitude 6. We call these descriptions **apparent magnitudes** because they compare how bright different stars *appear* in the sky. Star charts often use dots of different sizes to represent the apparent magnitudes of stars.

In modern times, the magnitude system has been extended and more precisely defined. As a result, stars can have fractional apparent magnitudes, and a few bright stars have apparent magnitudes *less than* 1 —which means *brighter* than magnitude 1. For example, the brightest star in the night sky, Sirius, has an apparent magnitude of −1.46.

The modern magnitude system also defines **absolute magnitudes** as a way of describing stellar luminosities. A star's absolute magnitude is the apparent magnitude it would have *if* it were at a distance of 10 parsecs from Earth. For example, the Sun's absolute magnitude is about 4.8, meaning that the Sun would have an apparent magnitude of 4.8 *if* it were 10 parsecs away from us—bright enough to be visible, but not conspicuous, on a dark night.

It's worth acquainting yourself with the magnitude system, because it is still commonly used. However, for the discussions in this book, it's easier to use the luminosity–distance formula, so we will avoid using magnitudes in this book.

13.3 Stellar Surface Temperature

The second basic property of stars (besides luminosity) needed for modern stellar classification is surface temperature. Measuring a star's surface temperature is somewhat easier than measuring its luminosity because the measurement is not affected by the star's distance. Instead, we determine surface temperature directly from the star's color or spectrum. One note of caution: We can measure only a star's *surface temperature,* not its interior temperature. (Interior temperatures are calculated with theoretical models [Section 12.3].) When astronomers speak of the "temperature" of a star, they usually mean the surface temperature unless they say otherwise.

A star's surface temperature determines the color of light it emits. A red star is cooler than a yellow star, which in turn is cooler than a blue star. The naked eye can distinguish colors only for the brightest stars, but colors become more evident when we view stars through binoculars or a telescope (Figure 13.3). Astronomers can determine the "color" of a star more precisely by comparing its apparent brightness as viewed through two different filters. For example, a cool star such as Betelgeuse, with a surface temperature of about 3,400 K, emits more red light than blue light and therefore looks much brighter when viewed through a red filter than when viewed through a blue filter. In contrast, a hotter star such as Sirius, with a surface temperature of about 9,400 K, emits more blue light than red light and looks much brighter through a blue filter than through a red filter.

FIGURE 13.3 This Hubble Space Telescope view through the heart of our Milky Way Galaxy reveals that stars come in many different colors.

Spectral Type

The emission and absorption lines in a star's spectrum provide an independent and more accurate way to measure its surface temperature. Stars displaying spectral lines of highly ionized elements must be fairly hot, while stars displaying spectral lines of molecules must be relatively cool [Section 6.3]. Astronomers classify stars according to surface temperature by assigning a **spectral type** determined from the spectral lines present in a star's spectrum. The hottest stars, with the bluest colors, are called spectral type O, followed in order of declining surface temperature by spectral types B, A, F, G, K, and M. The time-honored mnemonic for remembering this sequence, OBAFGKM, is "Oh Be A Fine Girl/Guy, Kiss Me!" Table 13.1 summarizes the characteristics of each spectral type.

Each spectral type is subdivided into numbered subcategories (e.g., B0, B1, . . . , B9). The larger the number, the cooler the star. For example, the Sun is designated spectral type G2, which means it is slightly hotter than a G3 star but cooler than a G1 star.

TIME OUT TO THINK *Invent your own mnemonic for the OBAFGKM sequence. To help get you thinking, here are two examples: (1) Only Bungling Astronomers Forget Generally Known Mnemonics; and (2) Only Business Acts For Good, Karl Marx.*

13.4 Stellar Masses

The most important property of a star is its mass, but stellar masses are much harder to measure than luminosities or surface temperatures. The most dependable method for "weighing" a star relies on Newton's version of Kepler's third law [Section 5.3].

This law allows us to calculate the masses of orbiting objects by measuring the period and average distance (semimajor axis) of their orbit. Note that this law can be applied only when we can measure both the orbital period and semimajor axis. Thus, unless we have identified a planet orbiting a star, we can measure stellar masses only in binary star systems in which we can determine the orbital properties of the two stars.

About half of all stars orbit a companion star of some kind, and these binary star systems fall into three classes:

- A **visual binary** is a pair of stars that we can see orbiting each other. Mizar, the second star in the handle of the Big Dipper, is one example

Table 13.1 The Spectral Sequence

Spectral Type	Example(s)	Temperature Range	Key Absorption Line Features
O	Stars of Orion's Belt	>30,000 K	Lines of ionized helium, weak hydrogen lines
B	Rigel	30,000 K–10,000 K	Lines of neutral helium, moderate hydrogen lines
A	Sirius	10,000 K–7,500 K	Very strong hydrogen lines
F	Polaris	7,500 K–6,000 K	Moderate hydrogen lines, moderate lines of ionized calcium
G	Sun, Alpha Centauri A	6,000 K–5,000 K	Weak hydrogen lines, strong lines of ionized calcium
K	Arcturus	5,000 K–3,500 K	Lines of neutral and singly ionized metals, some molecules
M	Betelgeuse, Proxima Centauri	<3,500 K	Molecular lines strong

of a visual binary (Figure 13.4). Sometimes we observe a star slowly shifting position in the sky as if it were a member of a visual binary, but its companion is too dim to be seen. Slow shifts in the position of Sirius, the brightest star in the sky, revealed it to be a binary star long before its companion was discovered (Figure 13.5).

- An **eclipsing binary** is a pair of stars that orbit in the plane of our line of sight (Figure 13.6). Each star occasionally passes in front of the other, causing an eclipse. When neither star is eclipsed, we see the combined light of both stars. But when one star eclipses the other, the apparent brightness of the system drops because some of the light is blocked from our view.

- If a binary system is neither visual nor eclipsing, we may be able to detect its binary nature by observing Doppler shifts in its spectral lines [Section 6.4]. If one star is orbiting another, it periodically moves toward us and away from us in its orbit, and its spectral lines show blueshifts and redshifts as a result of this motion (Figure 13.7). Because we detect the binary nature of such star systems by studying their spectra, they are called **spectroscopic binary** systems. Interestingly, each of the two stars we see when we view Mizar through a telescope is a spectroscopic binary.

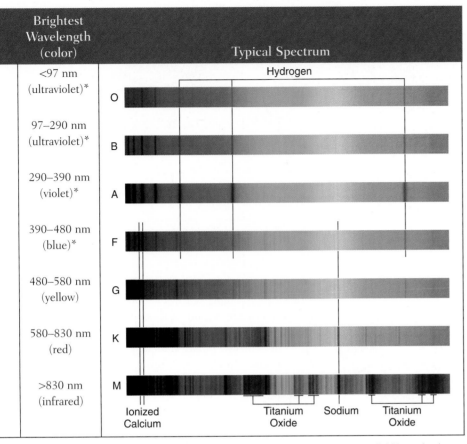

Brightest Wavelength (color)	Typical Spectrum
<97 nm (ultraviolet)*	O
97–290 nm (ultraviolet)*	B
290–390 nm (violet)*	A
390–480 nm (blue)*	F
480–580 nm (yellow)	G
580–830 nm (red)	K
>830 nm (infrared)	M

Hydrogen

Ionized Calcium   Titanium Oxide   Sodium   Titanium Oxide

*All stars above 6,000 K look more or less white to the human eye because they emit plenty of radiation at all visible wavelengths.

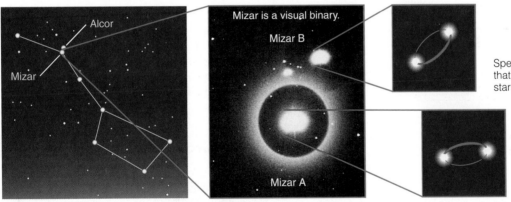

Alcor

Mizar

Mizar is a visual binary.

Mizar B

Mizar A

Spectroscopy shows that each of the visual stars is itself a binary.

FIGURE 13.4 Mizar looks like one star to the naked eye but is actually a system of four stars. Through a telescope Mizar appears to be a visual binary made up of two stars, Mizar A and Mizar B, that gradually change positions, indicating that they orbit every few thousand years. However, each of these two stars is actually a spectroscopic binary, making a total of four stars. (The star Alcor appears very close to Mizar to the naked eye but does *not* orbit it.)

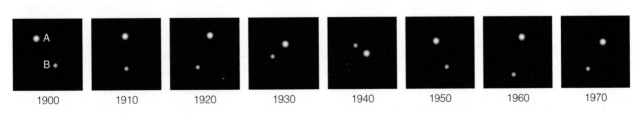

A

B

1900   1910   1920   1930   1940   1950   1960   1970

FIGURE 13.5 Each frame represents the relative positions of Sirius A and Sirius B at 10-year intervals from 1900 to 1970. The back-and-forth "wobble" of Sirius A allowed astronomers to infer the existence of Sirius B even before the two stars could be resolved in telescopic photos.

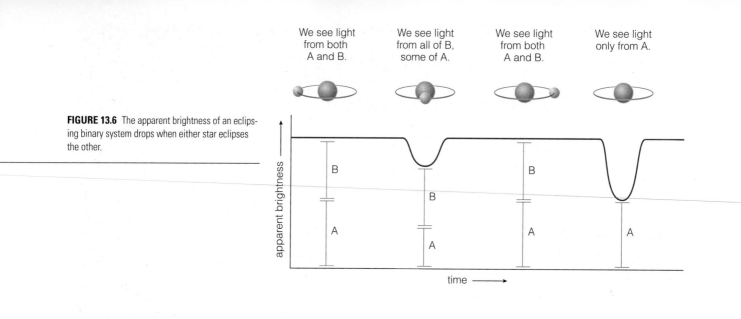

We see light from both A and B.

We see light from all of B, some of A.

We see light from both A and B.

We see light only from A.

FIGURE 13.6 The apparent brightness of an eclipsing binary system drops when either star eclipses the other.

apparent brightness

time ⟶

Star B spectrum at time 1: approaching, therefore blueshifted

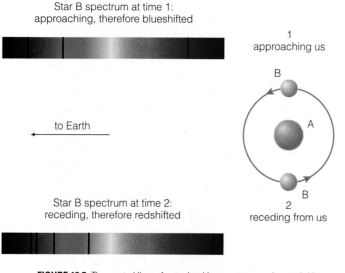

to Earth

1 approaching us

B

A

B

2 receding from us

Star B spectrum at time 2: receding, therefore redshifted

FIGURE 13.7 The spectral lines of a star in a binary system are alternately blueshifted as it comes toward us in its orbit and redshifted as it moves away from us.

us or away from us. Orbiting stars generally do not move directly along our line of sight, so their actual velocities can be significantly greater than those we measure through the Doppler effect. However, the orbits of eclipsing binary stars take them directly toward us and directly away from us once each orbit, enabling us to measure their true orbital velocities. Eclipsing binaries are therefore particularly important to the study of stellar masses. As an added bonus, eclipsing binaries allow us to measure stellar radii directly: Because we know how fast the stars are moving across our line of sight as one eclipses the other, we can determine their radii by timing how long each eclipse lasts.

TIME OUT TO THINK *Suppose two orbiting stars are moving in a plane perpendicular to our line of sight. Would the spectral features of these stars appear shifted in any way? Explain.*

13.5 The Hertzsprung–Russell Diagram

During the first decade of the twentieth century, a similar thought occurred independently to astronomers Ejnar Hertzsprung, working in Denmark, and Henry Norris Russell, working in the United States at Princeton University: Each decided to make a graph plotting stellar luminosities on one axis and spectral types on the other. Such graphs are now called **Hertzsprung–Russell (H–R) diagrams**. Soon after they began making their graphs, Hertzsprung and Russell uncovered some previously unsuspected patterns in the properties of stars. As we will see shortly, understanding these patterns and the H–R diagram is central to the study of stars.

Basics of the H–R Diagram

Figure 13.8 displays an example of an H–R diagram. Note that

Once we have identified a binary star system, we can apply Newton's version of Kepler's third law only if we can measure both the orbital period and the separation of the two stars. Measuring orbital period is fairly easy: In a visual binary, we simply observe how long each orbit takes (or extrapolate from part of an orbit); in an eclipsing binary, we measure the time between eclipses; and in a spectroscopic binary, we measure the time it takes the spectral lines to shift back and forth. In contrast, determining the average separation of the stars in a binary system is usually much more difficult: Except in rare cases, as with visual binaries, in which we can measure the separation directly, we can calculate the separation only if we know the actual orbital speeds of the stars. The primary technique for measuring stellar speeds relies on the Doppler effect, but Doppler shifts tell us only the portion of a star's velocity that is directly toward

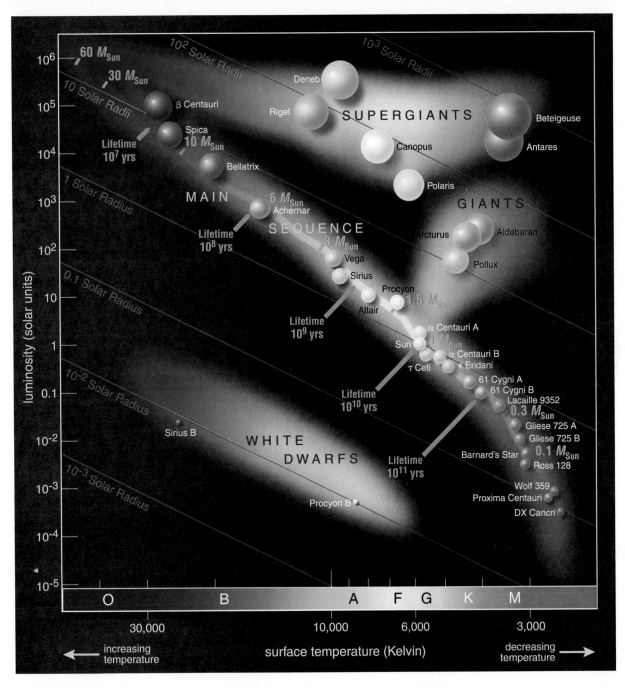

FIGURE 13.8 An H–R diagram, one of astronomy's most important tools, shows how the surface temperatures of stars (plotted along the horizontal axis) relate to their luminosities (plotted along the vertical axis). Several of the brightest stars in the sky are plotted here, along with a few of the closest to Earth. They are not drawn to scale—the diagonal lines, labeled in solar radii, indicate how large they are compared to the Sun. The lifetime and mass labels apply only to main-sequence stars (see Figure 13.9). (Star positions on this diagram are based on data from the Hipparcos satellite.)

- The horizontal axis represents stellar surface temperature, which, as we've discussed, corresponds to spectral type.

- Temperature increases *from right to left* because Hertzsprung and Russell based their diagrams on the spectral sequence OBAFGKM.

- The vertical axis represents stellar luminosity, in units of the Sun's luminosity (L_{Sun}).

- Stellar luminosities span a wide range, so we keep the graph compact by making each tick mark represent a luminosity 10 times larger than the prior tick mark.

- Each dot on the diagram represents the spectral type and luminosity of a single star. For example, the dot representing the Sun in Figure 13.8 corresponds to the Sun's spectral type, G2, and its luminosity, $1L_{Sun}$.

Because luminosity increases upward on the diagram and surface temperature increases leftward, stars near the upper left are hot and luminous. Similarly, stars near the upper right are cool and luminous, stars near the lower right are cool and dim, and stars near the lower left are hot and dim.

TIME OUT TO THINK *Explain how the colors of the stars in Figure 13.8 help indicate stellar surface temperature. Do these colors tell us anything about interior temperatures? Why or why not?*

The H–R diagram also provides direct information about stellar radii, because a star's luminosity depends on both its surface temperature and its surface area or radius. Recall that surface temperature determines the amount of power emitted by the star *per unit area*: Higher temperature means greater power output per unit area [Section 6.3]. Thus, if two stars have the same surface temperature, one can be more luminous than the other only if it is larger in size. Stellar radii therefore must increase as we go from the high-temperature, low-luminosity corner on the lower left of the H–R diagram to the low-temperature, high-luminosity corner on the upper right.

Patterns in the H–R Diagram

Figure 13.8 also shows that stars do not fall randomly throughout the H–R diagram. Instead, they fall into several distinct groups:

- Most stars fall somewhere along the **main sequence**, the prominent streak running from the upper left to the lower right on the H–R diagram. Note that our Sun is a main-sequence star.

- The stars along the top are called **supergiants** because they are very large in addition to being very bright.

- Just below the supergiants are the **giants**, which are somewhat smaller in radius and lower in luminosity (but still much larger and brighter than main-sequence stars of the same spectral type).

- The stars near the lower left are small in radius and appear white in color because of their high temperature; we call these stars the **white dwarfs**.

When classifying a star, astronomers generally report both the star's spectral type and a **luminosity class** that describes the region of the H–R diagram in which the star falls. Table 13.2 summarizes the luminosity classes. Note that luminosity class I represents supergiants, luminosity class III represents giants, luminosity class V represents main-sequence stars, and luminosity classes II and IV are intermedi-

Table 13.2 Stellar Luminosity Classes

Class	Description
I	Supergiants
II	Bright giants
III	Giants
IV	Subgiants
V	Main sequence

ate to the others. For example, the complete spectral classification of our Sun is G2 V: The G2 spectral type means it is yellow in color, and the luminosity class V means it is a main-sequence star. Betelgeuse is M2 I, making it a *red supergiant*. Proxima Centauri is M5 V—similar in color and surface temperature to Betelgeuse, but far dimmer because of its much smaller size. White dwarfs are usually designated with the letters *wd* rather than with a Roman numeral.

The Main Sequence

The common trait of main-sequence stars is that, like our Sun, they are fusing hydrogen into helium in their cores. Because stars spend the majority of their lives fusing hydrogen, most stars fall somewhere along the main sequence of the H–R diagram.

But why do main-sequence stars span such a wide range of luminosities and surface temperatures? By measuring the masses of stars in binary systems, astronomers have discovered that stellar masses increase upward along the main sequence (Figure 13.9). At the upper end of the main sequence, the hot, luminous O stars can have masses as high as 100 times that of the Sun ($100M_{Sun}$). On the lower end, cool, dim M stars may have as little as 0.08 times the mass of the Sun ($0.08M_{Sun}$). Many more stars fall on the lower end of the main sequence than on the upper end, which tells us that low-mass stars are much more common than high-mass stars.

The orderly arrangement of stellar masses along the main sequence tells us that *mass* is the most important attribute of a hydrogen-burning star. Luminosity depends directly on mass because the weight of a star's outer layers determines the nuclear burning rate in its core; more weight means the star must sustain a higher nuclear burning rate in order to maintain gravitational equilibrium [Section 12.3]. The nuclear burning rate, and hence the luminosity, is very sensitive to mass. For example, a $10M_{Sun}$ star on the main sequence is about 10,000 times more luminous than the Sun.

The relationship between mass and surface temperature is a little subtler. In general, a very luminous star must either be very large or have a very high surface temperature, or some combination of both. Stars on the upper end of the main sequence are thousands of times more luminous than the Sun but only about 10 times larger than the Sun in radius. Thus, their surfaces must be significantly hotter than the Sun's surface to account for their high luminosities. Main-sequence stars more massive than the Sun therefore have higher surface temperatures than the Sun, and those less massive than the Sun have lower surface temperatures. That is why the main sequence slices diagonally from the upper left to the lower right on the H–R diagram.

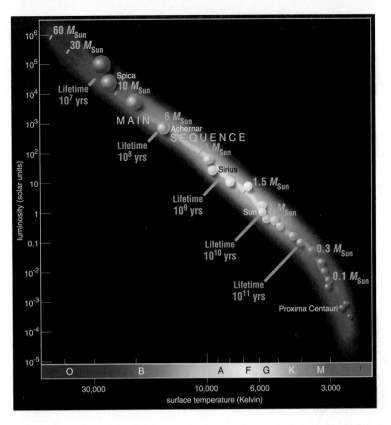

FIGURE 13.9 Along the main sequence, more massive stars are brighter and hotter but have shorter lifetimes. (Stellar masses are given in units of solar masses: 1 M_{Sun} = 2 × 10^{30} kg.)

Main-Sequence Lifetimes

A star has a limited supply of core hydrogen and therefore can remain as a hydrogen-fusing main-sequence star for only a limited time—the star's **main-sequence lifetime** (or *hydrogen-burning lifetime*). Because stars spend the vast majority of their lives fusing hydrogen into helium, we sometimes refer to the main-sequence lifetime as simply the "lifetime." Like masses, stellar lifetimes vary in an orderly way as we move up the main sequence: Massive stars near the upper end of the main sequence have *shorter* lives than less massive stars near the lower end (see Figure 13.9).

Why do more massive stars live shorter lives? A star's lifetime depends on both its mass and its luminosity. Its mass determines how much hydrogen fuel the star initially contains in its core. Its luminosity determines how rapidly the star uses up its fuel. Massive stars live shorter lives because, even though they start their lives with a larger supply of hydrogen, they consume their hydrogen at a prodigious rate.

The main-sequence lifetime of our Sun is about 10 billion years [Section 12.1]. A 30-solar-mass star has roughly 30 times more hydrogen than the Sun but burns it with a luminosity some 300,000 times greater. Consequently, its lifetime is roughly 30/300,000 = 1/10,000 as long as the Sun's—corresponding to a lifetime of only a few million years. Cosmically speaking, a few million years is a remarkably short time, which is one reason why massive stars are so rare: Most of the massive stars that have ever been born are long since dead. (A second reason is that lower-mass stars form in larger numbers than higher-mass stars [Section 14.2].) The fact that any massive stars exist tells us that stars form continuously in our gal-

axy. The massive, bright O stars in our galaxy today formed only recently and will die long before they have a chance to complete even one orbit around the center of the galaxy.

TIME OUT TO THINK *Would you expect to find life on planets orbiting massive O stars? Why or why not? (Hint: Compare the lifetime of an O star to the amount of time that passed from the formation of our solar system to the origin of life on Earth.)*

On the other end of the scale, a 0.3-solar-mass star emits a luminosity just 0.01 times that of the Sun and consequently lives roughly 0.3/0.01 = 30 times longer than the Sun. In a universe that is now between 12 and 16 billion years old, even the most ancient of these small, dim M stars still survive and will continue to shine faintly for hundreds of billions of years to come.

Giants, Supergiants, and White Dwarfs

Giants and supergiants are stars nearing the ends of their lives because they have already exhausted their core hydrogen. Surprisingly, stars grow more luminous when they begin to run out of fuel. As we will discuss in the next chapter, a star generates energy furiously during the last stages of its life as it tries to stave off the inevitable crushing force of gravity. As ever-greater amounts of power well up from the core, the outer layers of the star expand, making it into a giant or supergiant. The largest of these celestial behemoths have radii more than 1,000 times that of the Sun. If our Sun were this big, it would engulf the planets out to Jupiter.

Because they are so bright, giants and supergiants can be seen even when they are not especially close to us. Many of the brightest stars visible to the naked eye are giants or supergiants; they are often identifiable by their reddish color. Nevertheless, giants and supergiants are rarer than main-sequence stars because, in our snapshot of the heavens, we catch most stars in the act of hydrogen burning and relatively few in a later stage of life.

Giants and supergiants eventually run out of fuel entirely. A giant with a mass similar to that of our Sun ultimately ejects its outer layers, leaving behind a "dead" core in which all nuclear fusion has ceased. White dwarfs are these remaining embers of former giants. They are hot because they are essentially exposed stellar cores, but they are dim because they lack an energy source and radiate only their leftover heat into space. A typical white dwarf is no larger in size than the Earth, although it may have a mass as great as that of our Sun. (Giants and supergiants with masses much larger than that of the Sun ultimately explode, leaving behind neutron stars or black holes as corpses [Section 14.4].)

Pulsating Variable Stars

Not all stars shine steadily like our Sun. Any star that significantly varies in brightness with time is called a *variable star*. A particularly important type of variable star has a peculiar problem with achieving the proper balance between the power welling up from its core and the power being radiated from its surface. Sometimes the upper layers of such a star are too opaque; energy and pressure build up beneath the photosphere, and the star expands in size. However, this expansion puffs the upper layers outward, making them too transparent. So much energy then escapes that the underlying pressure drops, and the star contracts again. In a futile quest for a steady equilibrium, the atmosphere of such a **pulsating variable star** alternately expands and contracts, causing the star to rise and fall in luminosity. Figure 13.10 shows a typical light curve for a pulsating variable star, with the star's brightness graphed against time. Any pulsating variable star has its own particular period between peaks in luminosity, which we can discover easily from its light curve. These periods can range from as short as several hours to as long as several years.

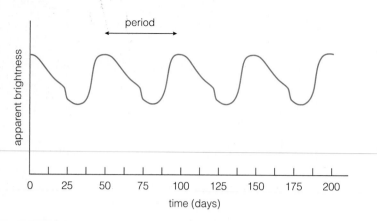

FIGURE 13.10 A typical light curve for a Cepheid variable star. Cepheids are giant, whitish stars whose luminosities regularly pulsate over periods of a few days to about a hundred days. The pulsation period of this Cepheid is about fifty days.

Most pulsating variable stars inhabit a strip (called the *instability strip*) on the H–R diagram that lies between the main sequence and the red giants (Figure 13.11). A special category of very luminous pulsating variables lies at the top of this strip: the **Cepheid variables**, or **Cepheids** (so named because the first identified star of this type was the star Delta Cephei).

Cepheids fluctuate in luminosity with periods of a few days to a few months. In 1912, Henrietta Leavitt discovered that the periods of these stars are very closely related to their luminosities: The longer the period, the more luminous the star. This **period–luminosity relation** holds because larger (and hence more luminous) Cepheids take longer to pulsate in and out in size.

Once we measure the period of a Cepheid variable, we can use the period–luminosity relation to determine its luminosity. We can then calculate its distance with the luminosity–distance formula. In fact, as we'll discuss in Chapter 17, Cepheids provide our primary means of measuring distances to other galaxies and thus teach us the true scale of the cosmos. The next time you look at the North Star, Polaris, gaze upon it with renewed appreciation. Not only has it guided generations of navigators in the Northern Hemisphere, but it is also one of these special Cepheid variable stars.

13.6 Star Clusters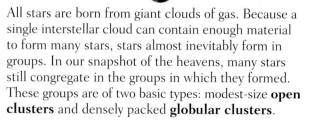

All stars are born from giant clouds of gas. Because a single interstellar cloud can contain enough material to form many stars, stars almost inevitably form in groups. In our snapshot of the heavens, many stars still congregate in the groups in which they formed. These groups are of two basic types: modest-size **open clusters** and densely packed **globular clusters**.

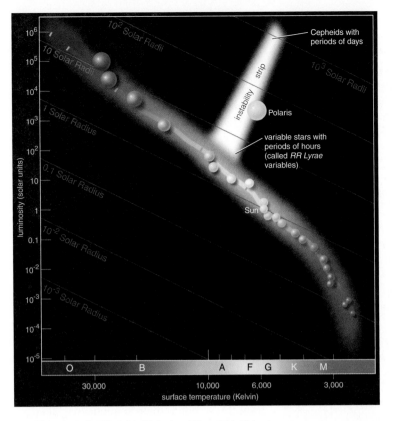

FIGURE 13.11 An H–R diagram with the instability strip highlighted.

Open clusters of stars are always found in the disk of the galaxy. They can contain up to several thousand stars and typically span about 30 light-years (10 parsecs). The most famous open cluster is the *Pleiades*, a prominent clump of stars in the constellation Taurus (Figure 13.12). The Pleiades are often called the *Seven Sisters*, although only six of the cluster's several thousand stars are easily visible to the naked eye. Other cultures have other names for this beautiful group of stars; in Japanese it is *Subaru*, which is why the logo for Subaru automobiles is a diagram of the Pleiades.

Globular clusters are found in both the halo and the disk of our galaxy. A globular cluster can contain more than a million stars concentrated in a ball typically 60 to 150 light-years across (20 to 50 parsecs). Its innermost region can have 10,000 stars packed within just a few light-years (Figure 13.13).

Star clusters are extremely useful to astronomers for two key reasons:

1. All the stars in a cluster lie at about the same distance from Earth.

2. Cosmically speaking, all the stars in a cluster formed at about the same time (i.e., within a few million years of one another).

FIGURE 13.12 A photo of the Pleiades, a nearby open cluster of stars. The most prominent stars in this open cluster are of spectral type B, indicating that the Pleiades are no more than 100 million years old, relatively young for a star cluster.

FIGURE 13.13 This globular cluster, known as M80, is over 12 billion years old. The prominent reddish stars in this Hubble Space Telescope photo are red giant stars nearing the ends of their lives.

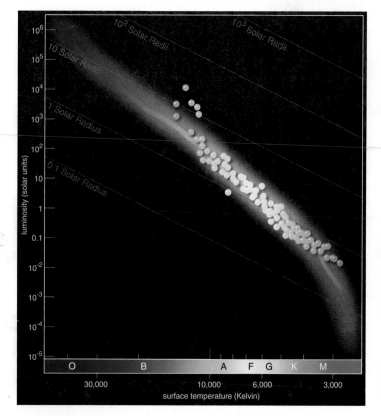

FIGURE 13.14 An H–R diagram for the stars of the Pleiades. Note that the Pleiades cluster is missing its upper main-sequence stars, indicating that these stars have already ended their hydrogen-burning lives. The main-sequence turnoff point at about spectral type B6 tells us that the Pleiades are about 100 million years old.

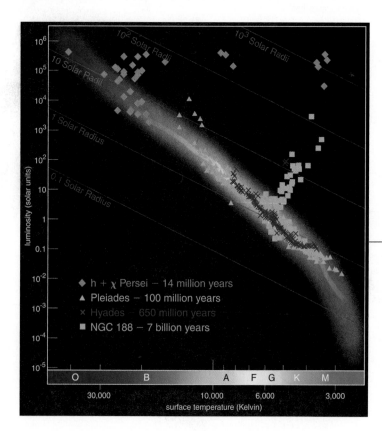

Astronomers can therefore use star clusters as laboratories for comparing the properties of stars, as yardsticks for measuring distances in the universe [Section 17.3], and as timepieces for measuring the age of our galaxy.

We can use clusters as timepieces because we can determine their ages from H–R diagrams of cluster stars. To understand how the process works, consider Figure 13.14, which shows an H–R diagram for the Pleiades. Note that most of the stars in the Pleiades fall along the standard main sequence, with one important exception: The Pleiades' stars trail away to the right of the main sequence at the upper end. That is, the Pleiades are missing the hot, short-lived O main-sequence stars. Apparently, the Pleiades are old enough for its O stars to have already ended their hydrogen-burning lives. At the same time, they are young enough for some of its B stars to still survive on the main sequence.

The precise point on the H–R diagram at which the Pleiades' main sequence diverges from the standard main sequence is called the **main-sequence turnoff** point. In this cluster, it occurs around spectral type B6. The main-sequence lifetime of a B6 star is roughly 100 million years, so this must be the age of the Pleiades. Any star in the Pleiades that was born with a main-sequence spectral type hotter than B6 had a lifetime shorter than 100 million years and hence is no longer found on the main sequence. Stars with lifetimes longer than 100 million years are still fusing hydrogen and hence remain as main-sequence stars.

FIGURE 13.15 This H–R diagram shows stars from four clusters with very different ages. Each star cluster has a different main-sequence turnoff point. The youngest cluster, h+χ Persei, still contains main-sequence O and B stars, indicating that it is only about 14 million years old. Stars at the main-sequence turnoff point for the oldest cluster, NGC 188, are only slightly more luminous and massive than the Sun, indicating an age of 7 billion years.

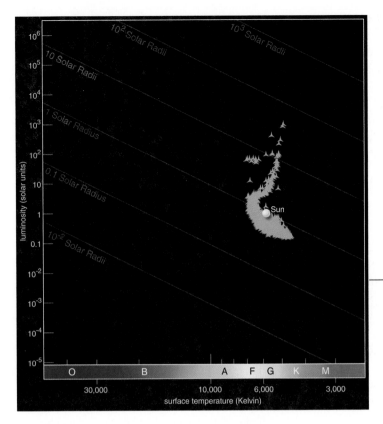

FIGURE 13.16 This H–R diagram shows stars from the globular cluster Palomar 3. The main-sequence turnoff point is in the vicinity of stars like our Sun, indicating an age for this cluster of around 10 billion years. A more technical analysis of this cluster places its age at around 12 to 14 billion years. (Stars in globular clusters tend to contain virtually no elements other than hydrogen and helium. Because of their different composition, these stars are somewhat bluer and more luminous than stars of the same mass and composition as that of our Sun.)

Over the next few billion years, the B stars in the Pleiades will die out, followed by the A stars and the F stars. Thus, if we could make H–R diagrams for the Pleiades every few million years, we would find that the main sequence gradually grows shorter. Comparing the H–R diagrams of other open clusters makes this effect more apparent (Figure 13.15). In each case, we determine the cluster's age from the lifetimes of the stars at its main-sequence turnoff point:

> age of the cluster = lifetime of stars at
>   main-sequence turnoff point

Thus, stars in a particular cluster that once resided above the turnoff point on the main sequence have already exhausted their core supply of hydrogen, while stars below the turnoff point remain on the main sequence.

TIME OUT TO THINK *Suppose a star cluster is precisely 10 billion years old. On an H–R diagram, where would you expect to find its main-sequence turnoff point? Would you expect this cluster to have any main-sequence stars of spectral type A? Would you expect it to have main-sequence stars of spectral type K? Explain. (Hint: What is the lifetime of our Sun?)*

The technique of identifying main-sequence turnoff points is our most powerful tool for evaluating the ages of star clusters. We've learned, for example, that most open clusters are relatively young and that very few are older than about 5 billion years. In contrast, the stars at the main-sequence turnoff points in globular clusters are usually less massive than our Sun (Figure 13.16). Because stars like our Sun have a lifetime of about 10 billion years and these stars have already died in globular clusters, we conclude that globular-cluster stars are older than 10 billion years. More precise studies of the turnoff points in globular clusters, coupled with theoretical calculations of stellar lifetimes, place the ages of these clusters at between 12 and 16 billion years, making them the oldest known objects in the galaxy. In fact, globular clusters place a limit on the possible age of the universe: If stars in globular clusters are 12 billion years old, then the universe must be at least this old.

THE BIG PICTURE

We have classified the diverse families of stars visible in the night sky. Much of what we know about stars, galaxies, and the universe itself is based on the fundamental properties of stars introduced in this chapter, so make sure you understand the following "big picture" ideas:

■ All stars are made primarily of hydrogen and helium, at least at the time they form. The differences between stars come about primarily because of differences in mass and age.

■ Much of what we know about stars comes from studying the patterns that appear when we plot stellar surface temperatures and luminosities in an H–R diagram. Thus, the H–R diagram is one of the most important tools of astronomers.

■ Stars spend most of their lives as main-sequence stars, fusing hydrogen into helium in their cores. The most massive stars live only a few million years, while the least massive stars will survive until the universe is many times its present age.

■ Much of what we know about the universe comes from studies of star clusters. Here again, H–R diagrams play a vital role. For example, H–R diagrams of star clusters allow us to determine their ages.

Review Questions

1. Briefly explain how we can learn about the lives of stars, despite the fact that stellar lives are far longer than human lives.

2. What basic composition are all stars born with? Given that all stars begin their lives with this same basic composition, why do stars differ from one another?

3. What do we mean by a star's *luminosity*? What are the standard units for luminosity? Explain what we mean when we measure luminosity in units of L_{Sun}.

4. What do we mean by a star's *apparent brightness*? What are the standard units for apparent brightness? Explain how a star's apparent brightness depends on its distance from us.

5. State the *luminosity–distance formula,* and explain why it is so important.

6. What do we mean by a star's visible-light luminosity or visible-light apparent brightness? By its X-ray luminosity or X-ray apparent brightness? Explain why we sometimes talk about such wavelength-specific measures, rather than *total luminosity* and *total apparent brightness.*

7. Briefly explain how we calculate a star's distance in parsecs by measuring its *parallax angle* in arcseconds.

8. What is the *magnitude system?* What are its origins? Briefly explain what we mean by the *apparent magnitude* and *absolute magnitude* of a star.

9. Briefly describe how a star's *spectral type* is related to its surface temperature. List the seven basic spectral types in order of decreasing temperature. How are the spectral types subdivided by number (e.g., A3, G5)?

10. Describe and distinguish between *visual binary* systems, *eclipsing binary* systems, and *spectroscopic binary* systems.

11. Describe how we measure stellar masses and why eclipsing binaries are so important to such measurements.

12. Draw a sketch of a basic *Hertzsprung–Russell diagram (H–R diagram).* Label the *main sequence, giants, supergiants,* and *white dwarfs.* Where on this diagram do we find stars that are cool and dim? Cool and luminous? Hot and dim? Hot and bright?

13. What do we mean by a star's *luminosity class?* On your sketch of the H–R diagram, identify the regions for luminosity classes I, III, and V.

14. Explain how mass determines the luminosity and surface temperature of a main-sequence star. How do masses differ as we look along the main sequence on an H–R diagram?

15. What do we mean by a star's *main-sequence lifetime?* Explain why more massive stars live shorter lives.

16. What are *pulsating variable stars?* Why do they vary periodically in brightness?

17. What are *Cepheid variables?* What is the *period–luminosity relation* for Cepheids, and why is it useful for distance measurements?

18. Describe in general terms how *open clusters* and *globular clusters* differ in their numbers of stars, ages, and locations in the galaxy. Why are star clusters so important to astronomers?

19. Explain why H–R diagrams look different for star clusters of different ages. How does the location of the *main-sequence turnoff point* tell us the age of the star cluster?

Discussion Question

1. *Classification.* Annie Jump Cannon and other women astronomers at Harvard University made many important contributions to the area of systematic stellar classification. Why do you think rapid advances in our understanding of stars followed quickly on the heels of their efforts? Can you think of other areas in science where huge advances in understanding followed directly from improved systems of classification?

Problems

True Statements? For **problems 1–10**, decide whether the statements are true or false and clearly explain how you know.

1. Two stars that look very different must be made out of different kinds of elements.

2. Sirius is the brightest star in the night sky, but if we moved it 10 times farther away it would look only one-tenth as bright.

3. Sirius looks brighter than Alpha Centauri, but we know that Alpha Centauri is closer because its apparent position in the sky shifts by a larger amount as the Earth orbits the Sun.

4. Stars that look red-hot have hotter surfaces than stars that look blue.

5. Some of the stars on the main sequence of the H–R diagram are not converting hydrogen into helium.

6. The smallest, hottest stars are plotted in the lower left-hand portion of the H–R diagram.

7. Stars that begin their lives with the most mass live longer than less massive stars because it takes them a lot longer to use up their hydrogen fuel.

8. Star clusters with lots of bright, blue stars are generally younger than clusters that don't have any such stars.

9. All giants, supergiants, and white dwarfs were once main-sequence stars.

10. Most of the stars in the sky are more massive than the Sun.

11. *Stellar Data.* Consider the following data table for several bright stars. Note that M_v is absolute magnitude and m_v is apparent magnitude.

Star	M_v	m_v	Spectral Type	Luminosity Class
Aldebaran	−0.2	+0.9	K5	III
Alpha Centauri A	+4.4	0.0	G2	V
Antares	−4.5	+0.9	M1	I
Canopus	−3.1	−0.7	F0	II
Fomalhaut	+2.0	+1.2	A3	V
Regulus	−0.6	+1.4	B7	V
Sirius	+1.4	−1.4	A1	V
Spica	−3.6	+0.9	B1	V

Answer each of the following questions, including a brief explanation with each answer.

  a. Which star appears brightest in our sky?

  b. Which star appears faintest in our sky?

  c. Which star has the greatest luminosity?

  d. Which star has the smallest luminosity?

  e. Which star has the highest surface temperature?

  f. Which star has the lowest surface temperature?

  g. Which star is most similar to the Sun?

  h. Which star is a red supergiant?

  i. Which star has the largest radius?

  j. Which stars have finished burning hydrogen in their cores?

  k. Among the main-sequence stars listed, which one is the most massive?

  l. Among the main-sequence stars listed, which one has the longest lifetime?

12. *Data Tables.* Study the spectral types listed in Appendix F for the 20 brightest stars and for the stars within 12 light-years. Why do you think the two lists are so different? Explain.

Web Projects

Find useful links for Web projects on the text Web site.

1. *Women in Astronomy.* Until fairly recently, men greatly outnumbered women in professional astronomy. Nevertheless, many women made crucial discoveries in astronomy throughout history. Do some research about the life and discoveries of a woman astronomer from any time period, and write a two- to three-page scientific biography.

2. *The Hipparcos Mission.* The European Space Agency's Hipparcos mission, which operated from 1989–1993, made precise parallax measurements for more than 40,000 stars. Learn about how Hipparcos allowed astronomers to measure smaller parallax angles than they could from the ground and how Hipparcos discoveries have affected our knowledge of the universe. Write a one- to two-page report on your findings.

3. *Future Parallax Missions.* Astronomers have plans for missions that will further extend our ability to measure distances through parallax. Write a one- to two-page report on one such mission, such as NASA's Space Interferometry Mission (SIM; launch planned for 2009) or the European Space Agency's GAIA mission (launch scheduled for 2012).

I can hear the sizzle of newborn stars, and know anything of meaning, of the fierce magic emerging here. I am witness to flexible eternity, the evolving past, and know I will live forever, as dust or breath in the face of stars, in the shifting pattern of winds.

JOY HARJO, *SECRETS FROM THE CENTER OF THE WORLD*

CHAPTER 14
Star Stuff

We inhale oxygen with every breath. Iron-bearing hemoglobin carries this oxygen through the bloodstream. Chains of carbon and nitrogen form the backbone of the proteins, fats, and carbohydrates in our cells. Calcium strengthens our bones, while sodium and potassium ions moderate communications of the nervous system. What does all this have to do with astronomy? The profound answer, recognized only in the latter half of the twentieth century, is that these elements were created by stars.

We've already discussed in general terms how the elements in our bodies came to exist: Hydrogen and helium were produced in the Big Bang, and heavier elements were created later by stars and scattered into space by stellar explosions. There, in the spaces between the stars, these elements mixed with interstellar gas and became incorporated into subsequent generations of stars.

In this chapter, we will discuss the origins of the elements in greater detail by delving into the lives of stars. As you read, keep in mind that no matter how far the stars may seem to be removed from our everyday lives, they actually are connected to us in the most intimate way possible: Without the births, lives, and deaths of stars, none of us would be here.

14.1 Lives in the Balance

The story of a star's life is in many ways the story of an extended battle between two opposing forces: gravity and pressure. The most common type of pressure in stars is **thermal pressure**—the familiar type of pressure that keeps a balloon inflated and that increases when the temperature or thermal energy increases.

A star can maintain its internal thermal pressure only if it continually generates new thermal energy to replace the energy it radiates into space. This energy can come from two sources: *nuclear fusion* of light elements into heavier ones and the process of *gravitational contraction,* which converts gravitational potential energy into thermal energy [Section 12.1]. These energy-production processes operate only temporarily, although in this case "temporarily" means millions or billions of years. In contrast, gravity acts eternally, and any time gravity succeeds in shrinking a star's core the strength of gravity grows. (Recall that the force of gravity inside an object grows stronger if it either gains mass or shrinks in radius [Section 5.3].) Because a star cannot rely on thermal energy to resist gravity forever, its ultimate fate depends on whether something other than thermal pressure manages to halt the unceasing crush of gravity.

The outcome of a star's struggle between gravity and pressure depends almost entirely on its birth mass. All stars are born from spinning clumps of gas, but newborn stars can have masses ranging from less than 10% of the mass of our Sun to about 100 times that of our Sun. The most massive stars live fast and die young, proceeding from birth to explosive death in just a few million years. The lowest-mass stars consume hydrogen so slowly that they will continue to shine until the universe is many times older than it is today.

Because stars come in such a great variety of masses, we can simplify our discussion of stellar lives by dividing stars into three basic groups: **low-mass stars** are born with less than about two times the mass of our Sun, or less than 2 *solar masses* ($2M_{Sun}$) of material; **intermediate-mass stars** have birth weights between about 2 and 8 solar masses; and **high-mass stars** include all stars born with masses greater than about 8 solar masses. Both low-mass and intermediate-mass stars swell into red giants near the ends of their lives and ultimately become white dwarfs. High-mass stars also become red and large in their latter days, but their lives end much more violently. We will focus most of our discussion in this chapter on the dramatic differences between the lives of low- and high-mass stars. Because the life stages of intermediate-mass stars are quite similar to the corresponding stages of high-mass stars until the very ends of their lives, we include them in our discussion of high-mass stars.

Given the brevity of human history compared to the life of any star, you might wonder how we can know so much about stellar life cycles. As with any scientific inquiry, we study stellar lives by comparing theory and observation. On the theoretical side, we use mathematical models based on the known laws of physics to predict the life cycles of stars. On the observational side, we study stars of different mass but the same age by looking in star clusters whose ages we have determined by main-sequence turnoff [Section 13.6]. Occasionally, we even catch a star in its death throes. Theoretical predictions of the life cycles of stars now agree quite well with these observations. In the remainder of this chapter, we will examine in detail our modern understanding of the life stories of stars and how they manufacture the variety of elements—the *star stuff*—that make our lives possible.

14.2 Star Birth

A star's life begins in an interstellar cloud (Figure 14.1). Star-forming clouds tend to be quite cold—typically only 10–30 K. (Recall that 0 K is absolute zero, and temperatures on Earth are around 300 K.) They also tend to be quite dense compared to the rest of the gas between the stars, although they would qualify as a superb vacuum by earthly standards. Like the galaxy as a whole, star-forming clouds are made almost entirely of hydrogen and helium.

The clouds that form stars are sometimes called **molecular clouds**, because their low temperatures allow hydrogen atoms to pair up to form hydrogen molecules (H_2). The relatively rare atoms of elements heavier than helium can also form molecules, such as carbon monoxide or water, or tiny, solid grains of dust. More important, the cold temperatures and relatively high densities allow gravity to overcome thermal pressure more readily in molecular clouds than elsewhere in interstellar space, initiating the gravitational collapse that gives birth to new stars.

From Cloud to Protostar

As a piece of a molecular cloud starts to collapse, gravitational contraction increases the cloud's thermal energy. However, the cloud quickly radiates this thermal energy away, preventing the pressure from building enough to resist gravity. During this early phase of collapse, the temperature remains below 100 K, and the cloud glows in long-wavelength infrared light (Figure 14.2).

The ongoing collapse increases the cloud's density, making it increasingly difficult for radiation to

FIGURE 14.1 A star-forming cloud of molecular hydrogen gas in the constellation Scorpius extends from the upper-right corner of this photo through the center. The cloud appears dark because dust particles within it obscure the light from more distant stars lying behind it. Blue-white blotches near the edges of the dark cloud are newly formed stars. They appear fuzzy because some of their light is reflecting off patchy gas in their vicinity.

FIGURE 14.2 This false-color picture shows infrared radiation from star-forming regions in the constellation Orion. The colors correspond to the temperature of the emitting gas: Red is cooler, and white is hotter. Star formation is most intense in the yellow-white regions, which have been heated to 60–100 K (but note that this is still quite cold). The Orion Nebula, home to many protostars, is the prominent yellow-white area near the bottom. Betelgeuse, a red supergiant, appears as a blue-white dot near the center.

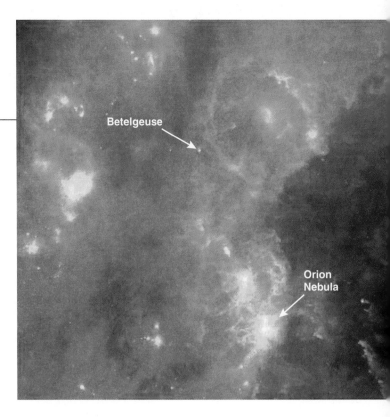

escape. Eventually, its central regions grow completely opaque, trapping much of the thermal energy produced by gravitational contraction. Because thermal energy can no longer escape easily, the cloud's internal temperature and pressure rise dramatically. This rising pressure begins to fight back against the crush of gravity, and the dense cloud fragment becomes a **protostar**—the seed from which a star will grow. A protostar looks starlike, but its core is not yet hot enough for fusion.

The law of conservation of angular momentum [Section 5.2] implies that a **protostellar disk** should encircle a forming protostar, similar to the *proto-planetary disk* from which the planets of our solar

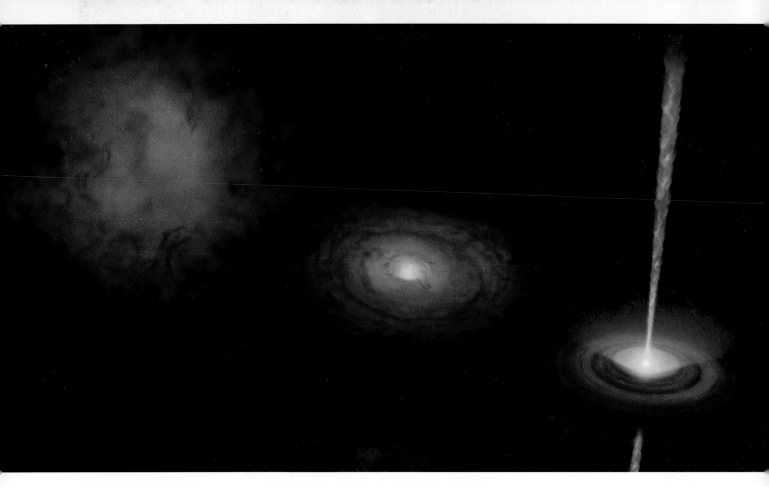

FIGURE 14.3 Artist's conception of star birth. A cloud fragment (top left) necessarily has some angular momentum, which causes it to spin faster and flatten into a protostellar disk as it collapses (center). In the late stages of collapse (bottom right), the central protostar has a strong protostellar wind and may fire jets of high-speed gas outward along its rotation axis.

system formed [Section 7.3]. A collapsing cloud fragment necessarily starts with some overall angular momentum (the sum total of the angular momenta of every constituent gas particle), although it might be unmeasurable when the fragment is large. But conservation of angular momentum ensures that the fragment spins faster and faster as it collapses (Figure 14.3). The rotating cloud flattens to form the protostellar disk as it shrinks. Protostellar disks sometimes coalesce into true planetary systems, but we do not yet know how commonly this occurs.

TIME OUT TO THINK *The term* protostellar disk *refers to any disk of material surrounding a protostar. The term* protoplanetary disk *refers to a disk of material that later produces full-fledged planets. Do you think that all protostellar disks are also protoplanetary disks? If so, why? If not, what do you think might prevent planets from forming in some protostellar disks?*

The protostellar disk probably plays a large role in eventually slowing the rotation of the protostar. The protostar's rapid rotation generates a strong magnetic field. As the magnetic field lines sweep through the protostellar disk, they transfer some of the protostar's angular momentum outward, slowing its rotation. The strong magnetic field also helps generate a strong **protostellar wind**—an outward flow of particles similar to the *solar wind* [Section 12.2]—that may carry additional angular momentum from the protostar to interstellar space.

Rotation is likely to be responsible for the formation of some binary star systems. Protostars that are unable to rid themselves of enough angular momentum spin too fast to become a stable single star and tend to split in two. Each of these two fragments can go on to form a separate star. If the stars are particularly close together, the resulting pair is called a **close binary** system, in which two stars coexist in close proximity, rapidly orbiting each other.

Observations show that the latter stages of a star's formation can be surprisingly violent. Besides the

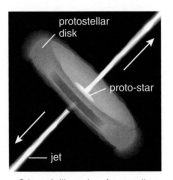

a Schematic illustration of protostellar disk–jet structure.

b Photograph of a protostellar disk and jet. We are seeing the disk edge-on, as in (a). The disk's top and bottom surfaces, illuminated by the protostar, are shown in green, and the jets emerging along the disk's axis are shown in red. The dark central layers of the disk block our view of the protostar itself.

c A wider-angle photograph of a jet emanating from a protostar (left) and ramming into surrounding interstellar gas (right).

FIGURE 14.4 Some protostars can be seen shooting jets of matter into interstellar space.

strong wind, many young stars also fire high-speed streams, or **jets**, of gas into interstellar space (Figure 14.4). No one knows exactly how protostars generate these jets, but two high-speed streams generally flow out along the rotation axis of the protostar, shooting in opposite directions. We also sometimes see blobs of material along the jets (called *Herbig–Haro objects* after their discoverers). These blobs appear to be collections of gas swept up as the jet plows into the surrounding interstellar material. Together, winds and jets play an important role in clearing away the cocoon of gas that surrounds a forming star, eventually allowing the star to emerge into view.

A Star Is Born

A full-fledged star is born when the core temperature of the protostar exceeds 10 million K—hot enough for hydrogen fusion to operate efficiently by the *proton–proton chain* [Section 12.3]. The ignition of fusion halts the protostar's gravitational contraction. The new star's interior structure stabilizes, with the thermal energy generated by fusion maintaining the

balance between gravity and pressure that we call *gravitational equilibrium* [Section 12.1]. The star is now a hydrogen-burning, *main-sequence* star [Section 13.5].

The length of time from the formation of a protostar to the birth of a main-sequence star depends on the star's mass. Remember that massive stars do everything faster. The collapse of a high-mass protostar may take only a million years or less. Collapse into a star like our Sun takes about 50 million years, and collapse into a small star of spectral type M may take more than a hundred million years. Thus, the most massive stars in a young star cluster may live and die before the smallest stars finish their prenatal period.

We can summarize the transitions that occur during star birth with a special type of H–R diagram. Recall that a standard H–R diagram shows luminosities and surface temperatures for many different stars [Section 13.5]. In contrast, this special H–R diagram shows part of a **life track** (also called an *evolutionary track*) for a single star; for reference, the diagram also shows the standard main sequence. Each point along a star's life track represents its surface temperature and luminosity at some moment during its life.

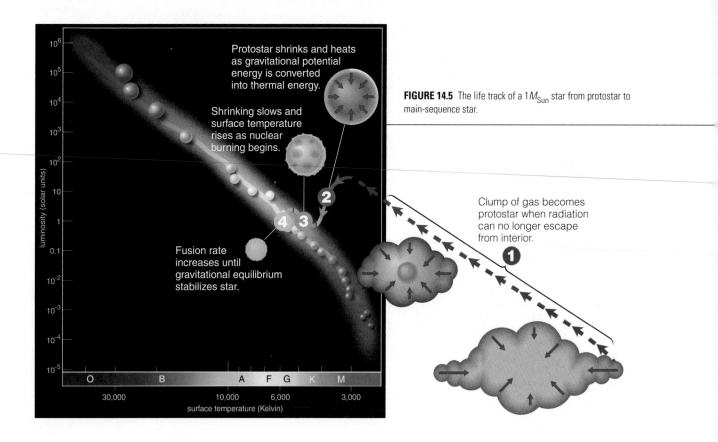

Protostar shrinks and heats as gravitational potential energy is converted into thermal energy.

Shrinking slows and surface temperature rises as nuclear burning begins.

Fusion rate increases until gravitational equilibrium stabilizes star.

Clump of gas becomes protostar when radiation can no longer escape from interior.

FIGURE 14.5 The life track of a $1M_{Sun}$ star from protostar to main-sequence star.

Figure 14.5 shows a life track for the prenatal period in the life of a $1M_{Sun}$ star like our Sun.

Stars of different masses follow a similar set of stages during their prenatal periods. Figure 14.6 illustrates life tracks on an H–R diagram for the prenatal stages of several stars of different masses.

TIME OUT TO THINK *Explain in your own words what we mean by a* life track *for a star. Why do we say that Figures 14.5 and 14.6 show only* pre-main-sequence *life tracks? In general terms, predict the appearance on Figure 14.6 of pre-mainsequence life tracks for a* $25M_{Sun}$ *star and for a* $0.1M_{Sun}$ *star.*

Stellar Birth Weights

A single group of molecular clouds can contain thousands of solar masses of gas, which is why stars generally are born in clusters. We do not yet fully understand the processes that govern the clumping and fragmentation of these clouds into protostars with a wide array of masses. However, we can observe the results. In a newly formed star cluster, stars with low masses greatly outnumber stars with high masses. For every star with a mass between 10 and 100 solar masses, there are typically 10 stars with masses between 2 and 10 solar masses, 50 stars with masses between 0.5 and 2 solar masses, and a few hundred stars with masses below $0.5M_{Sun}$. Thus, although the Sun lies nearly in the middle of the overall *range* of stellar masses, most stars in a new star cluster are less massive than the Sun. With the passing of time, the balance tilts even more in favor of the low-mass stars as the high-mass stars die away.

The masses of stars have limits. Theoretical models indicate that stars above about $100M_{Sun}$ generate power so furiously that gravity cannot contain their internal pressure. Such stars effectively tear themselves apart and drive their outer layers into space. Observations confirm the absence of stars much larger than $100M_{Sun}$—if any stars this massive existed nearby, they would be so luminous that we would easily detect them. On the other end of the scale, calculations show that the central temperature of a protostar with less than $0.08M_{Sun}$ never climbs above the 10-million-K threshold needed for efficient hydrogen fusion. Instead, a strange effect called **degeneracy pressure** halts the gravitational contraction of the core before hydrogen burning can begin.

Degeneracy pressure arises from a quantum mechanical law called the *exclusion principle,* which prevents two identical electrons (or protons or neutrons) from occupying the same place at the same time. The following analogy might help you visualize how degeneracy pressure works. Imagine that a group of people are in an auditorium filled with hun-

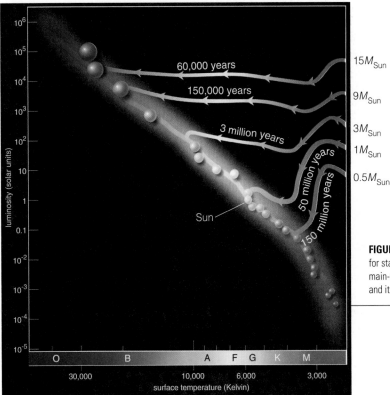

FIGURE 14.6 Life tracks from protostar to the main sequence for stars of different masses. A high-mass protostar becomes a main-sequence star much more quickly than a lower-mass star, and its luminosity does not decline as much as it collapses.

dreds of folding chairs. As long as the chairs greatly outnumber the people, it's rare that two people ever fight over the same chair. However, if the number of people is nearly equal to the number of chairs, it's hard to pack them any tighter because people aren't allowed to share seats.

In the context of this analogy, protostars with masses above $0.08M_{Sun}$ are like auditoriums with many more available chairs than people, so the people (particles) can easily squeeze into a smaller section of the auditorium. But the cores of protostars with masses below $0.08M_{Sun}$ are like auditoriums with so few chairs that the people (particles) fill nearly all of them. Because there are no extra chairs available, the people (particles) cannot squeeze into a smaller section of the auditorium. This resistance to squeezing is why degeneracy pressure halts gravitational contraction. Note that the degeneracy pressure arises *only* because particles have no place else to go. Thus, unlike thermal pressure, *degeneracy pressure has nothing to do with temperature.*

Because degeneracy pressure halts the collapse of a protostellar core with less than $0.08M_{Sun}$ before fusion becomes self-sustaining, the result is a "failed star" that slowly radiates away its internal thermal energy, gradually cooling with time. Such objects, called **brown dwarfs,** occupy a fuzzy gap between what we call a planet and what we call a star. (Note that $0.08M_{Sun}$ is about 80 times the mass of Jupiter.) Because degeneracy pressure does *not* rise and fall

with temperature, the gradual cooling of a brown dwarf's interior does not weaken its degeneracy pressure. In the constant battle of any "star" to resist the crush of gravity, brown dwarfs are winners, albeit dim ones: Their degeneracy pressure will not diminish with time, so gravity will never gain the upper hand.

14.3 Life as a Low-Mass Star

In the grand hierarchy of stars, our Sun ranks as rather mediocre. We should be thankful for this mediocrity. If the Sun had been a high-mass star, it would have lasted only a few million years, dying before life could have arisen on Earth. Instead, the Sun has shone steadily for nearly 5 billion years and will continue to do so for about 5 billion more. In this section, we investigate the lives of low-mass stars like our Sun.

Slow and Steady

Low-mass stars spend their main-sequence lives fusing hydrogen into helium in their cores slowly and steadily via the *proton–proton chain* [Section 12.3]. As in the Sun, the energy released by nuclear fusion in the core may take a million years to reach the surface, where it finally escapes into space as the star's luminosity. The energy moves outward from the core through the random bounces of photons from one electron to another and convection [Section 12.4].

Convection is the dominant process in the outer one-third of the Sun's interior. More massive stars have hotter interiors and shallower convection zones. Lower-mass stars have cooler interiors and deeper convection zones. In very low mass stars, the convection zone extends all the way down to the core. The highest-mass stars have no convection zone at all near their surface, but they have a convective core because they produce energy so furiously (Figure 14.7).

A low-mass star gradually consumes its core hydrogen, converting it into helium over a period of billions of years. In the process, the declining number of independent particles in the core (four independent protons fuse into just one independent helium nucleus) causes the core to shrink and heat very gradually, pushing the luminosity of the main-sequence star slowly upward as it ages. But the most dramatic changes occur when nuclear fusion exhausts the hydrogen in the star's core.

Red Giant Stage

The energy released by hydrogen fusion during a star's main-sequence life maintains the thermal pressure that holds gravity at bay. But when the core hydrogen is finally depleted, nuclear fusion ceases in the star's core. With no fusion to supply thermal energy, the core pressure can no longer resist the crush of gravity, and the core begins to shrink more rapidly.

Surprisingly, the star's outer layers expand outward while the core is shrinking. On an H–R dia-gram, the star's life track moves almost horizontally to the right as the star grows in size to become a **subgiant** (Figure 14.8). Then, as the expansion of the outer layers continues, the star's luminosity begins to increase substantially, and the star slowly becomes a **red giant**. For a $1M_{Sun}$ star, this process takes about a billion years, during which the star's radius increases about 100-fold and its luminosity grows by an even greater factor. (Like all phases of stellar lives, the process occurs faster for more massive stars and slower for less massive stars.) This process may at first seem paradoxical: Why does the star grow bigger and more luminous at the same time that its core is shrinking?

We can find the answer by considering the interior structure of the star. The core is now made of helium—the "ash" left behind by hydrogen fusion—but the surrounding layers still contain plenty of fresh hydrogen. Gravity shrinks both the *inert* (non-burning) helium core and the surrounding *shell* of hydrogen, and the shell soon becomes hot enough to sustain hydrogen fusion (Figure 14.9). In fact, the shell becomes so hot that this **hydrogen shell burning** proceeds at a higher rate than core hydrogen fusion did during the star's main-sequence life. The result is that the star becomes more luminous than ever before. The thermal pressure generated by the hydrogen shell burning overcomes gravity in the star's upper layers, pushing them to expand outward. What was once a fairly dim main-sequence star balloons into a luminous red giant. Note that the red giant is large on the outside but most of its mass is buried deep in a shrunken stellar core.

FIGURE 14.7 Among main-sequence stars, convection zones extend deeper in lower-mass stars. High-mass stars have convective cores but have no convection zones near their surfaces.

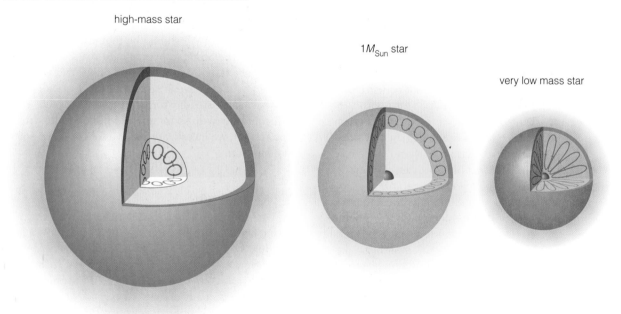

high-mass star

$1M_{Sun}$ star

very low mass star

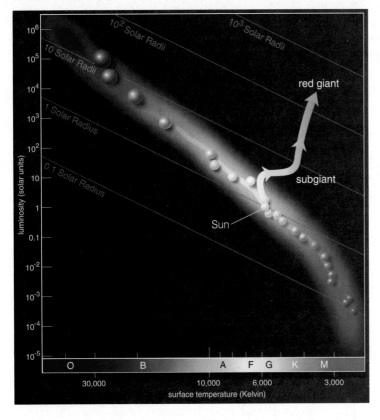

FIGURE 14.8 The life track of a $1 M_{Sun}$ star on an H–R diagram from the end of its main-sequence life until it becomes a red giant.

thing to inflate the inert core that lies underneath. Instead, newly produced helium keeps adding to the mass of the helium core, amplifying its gravitational pull and shrinking it further. The hydrogen-burning shell shrinks along with the core, growing hotter and denser. The fusion rate in the shell consequently rises, feeding even more helium ash to the core. The star is caught in a vicious circle. The core and shell continue to shrink in size and grow in luminosity, while thermal pressure continues to push the star's upper layers outward. This cycle breaks down only when the inert helium core reaches a temperature of about 100 million K, at which point helium nuclei can fuse together. Throughout the expansion phase, *stellar winds* carry away much more matter than the solar wind does for our Sun, but at much slower speeds.

TIME OUT TO THINK *Before you read on, briefly summarize why a star grows larger and brighter after it exhausts its core hydrogen. When does the growth of a red giant finally halt, and why? Would a star's red giant stage be different if the temperature required for helium fusion were around 200 million K, rather than 100 million K? Why?*

The situation grows more extreme as long as the helium core remains inert. Recall that, in a main-sequence star like the Sun, a rise in the fusion rate causes the core to inflate and cool until the fusion rate drops back down; in Chapter 12, we referred to this self-correcting process as the solar *thermostat* [Section 12.3]. But thermal energy generated in the hydrogen-burning shell of a red giant cannot do any-

Helium Burning

Recall that fusion occurs only when two nuclei come close enough together for the attractive *strong force* to overcome electromagnetic repulsion [Section 12.3]. Helium nuclei have two protons (and two neutrons) and hence a greater positive charge than the single proton of a hydrogen nucleus. The greater charge

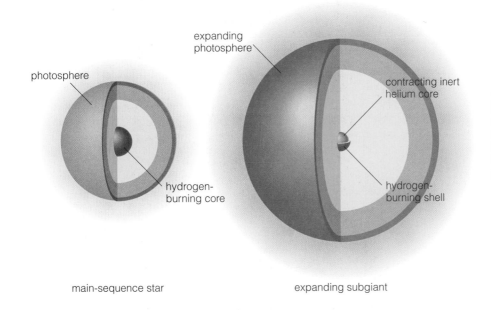

photosphere

expanding photosphere

contracting inert helium core

hydrogen-burning core

hydrogen-burning shell

main-sequence star

expanding subgiant

FIGURE 14.9 After a star ends its main-sequence life, its inert helium core contracts while hydrogen shell burning begins. The high rate of fusion in the hydrogen shell forces the star's upper layers to expand outward.

means that helium nuclei repel one another more strongly than hydrogen nuclei. **Helium fusion** therefore occurs only when nuclei slam into one another at much higher speeds than those needed for hydrogen fusion. Therefore, helium fusion requires much higher temperatures. The helium fusion process converts three helium nuclei into one carbon nucleus:

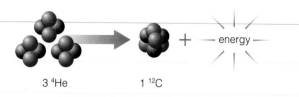

3 ^{4}He   1 ^{12}C

Energy is released because the carbon-12 nucleus has a slightly lower mass than the three helium-4 nuclei, and the lost mass becomes energy in accord with $E = mc^2$.

When helium fusion begins in a low-mass star, the energy it releases rapidly inflates the core. This core expansion pushes the hydrogen-burning shell outward, lowering its temperature and its burning rate. The result is that, even though the star now has core helium fusion *and* hydrogen shell burning taking place simultaneously (Figure 14.10), the total energy production falls from its peak during the red giant phase. The reduced total energy output of the star reduces its luminosity, thereby allowing its outer layers to contract from their peak size during the red giant phase. As the outer layers contract, the star's surface temperature also increases somewhat.

Because the helium-burning star is now smaller and hotter than it was as a red giant, its life track on the H–R diagram drops downward and to the left (Figure 14.11). The helium cores of all low-mass stars fuse helium into carbon at about the same rate, so these stars all have about the same luminosity. However, the outer layers of these stars can have different masses, depending on how much mass they lost through their stellar winds. Stars that lost more mass end up with smaller radii and higher surface temperatures and hence are farther to the left on the H–R diagram. In a cluster of stars, those stars that are currently in their helium-burning phase therefore all have about the same luminosity but differ in surface temperature.

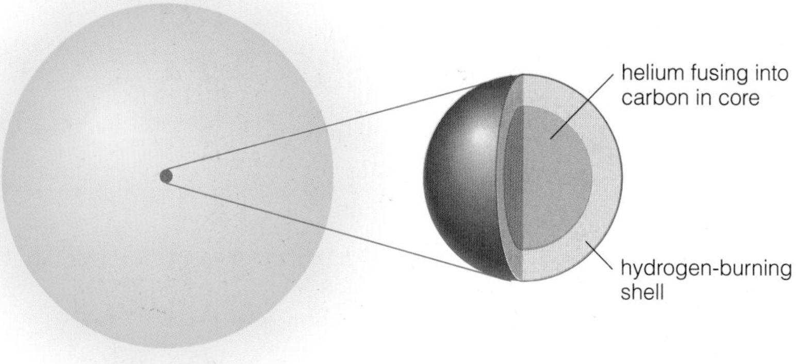

FIGURE 14.10 Core structure of a helium-burning star. Helium fusion causes the core and hydrogen-burning shell to expand and slightly cool, thereby reducing the overall energy generation rate in comparison to the rate during the red giant stage. The outer layers shrink back, so a helium-burning star is smaller than a red giant of the same mass.

Last Gasps

It is only a matter of time until a helium-burning star fuses all its core helium into carbon. The core helium in a low-mass star will run out in about a hundred million years. When the core helium is exhausted, fusion turns off, and the core begins to shrink once again under the crush of gravity.

The basic processes that changed the star from a main-sequence star into a red giant now resume, but this time it is helium fusion that ignites in a shell around an inert carbon core. Meanwhile, the hydrogen shell still burns atop the helium layer. Both shells contract along with the inert core, driving their temperatures and fusion rates much higher. The luminosity of this double-shelled star grows greater than ever, and its outer layers swell to an even huger size. On the H–R diagram, the star's life track once again turns upward (Figure 14.12).

The furious burning in the helium and hydrogen shells cannot last long—maybe a few million years or less for a $1M_{Sun}$ star. The star's only hope of extending its life lies with the carbon core, but this is a false hope in the case of low-mass stars. Carbon fusion is possible only at temperatures above about 600 million K. Before the core of a low-mass star ever reaches such a lofty temperature, degeneracy pressure halts its gravitational collapse.

For a low-mass star with a carbon core, the end is near. The huge size of the dying star means that it has a very weak grip on its outer layers. As the star's luminosity and radius keep rising, matter flows from its surface at increasingly high rates. Meanwhile, strong convection dredges up carbon from the core, enriching the surface of the star with carbon. Red giants whose atmospheres become especially carbon-rich in this way are called **carbon stars**. Carbon stars have cool, low-speed stellar winds, and the temperature of the gas in these winds drops with distance from the stellar surface. At the point at which the temperature has dropped to 1,000–2,000 K, some of the gas atoms in these slow-moving winds

helium fusing into carbon in core

hydrogen-burning shell

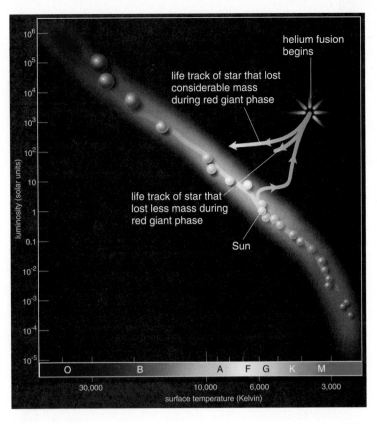

FIGURE 14.11 The onset of helium fusion. When helium fusion begins, a star's surface shrinks and heats, so the star's life track moves downward and to the left on the H–R diagram.

begin to stick together in microscopic clusters, forming small, solid particles of dust. These dust particles continue their slow drift with the stellar wind into interstellar space, where they become **interstellar dust grains**. The process of particulate formation is very similar to the formation of smoke particles in a fire. Thus, in a sense, carbon stars are the most voluminous polluters in the universe. However, this "carbon smog" is essential to life: Most of the carbon in your body (and in all life on Earth) was manufactured in carbon stars and blown into space by their stellar winds.

TIME OUT TO THINK *Suppose the universe contained only low-mass stars. Would elements heavier than carbon exist? Why or why not?*

Before a low-mass star dies, it treats us to one last spectacle. Through winds and other processes, the star ejects its outer layers into space. The result is a huge shell of gas expanding away from the inert, degenerate carbon core. The exposed core is still very hot and therefore emits intense

FIGURE 14.12 The life track of a $1 M_{Sun}$ star from main-sequence star to white dwarf. Core structure is shown at key stages.

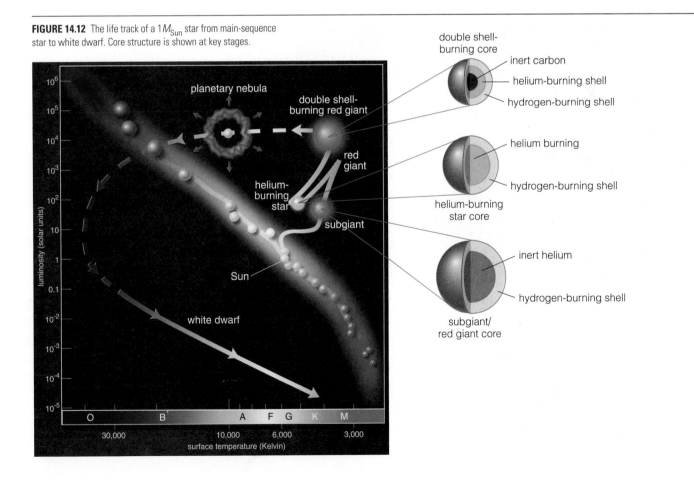

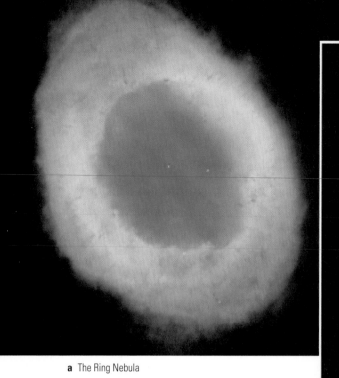

a The Ring Nebula

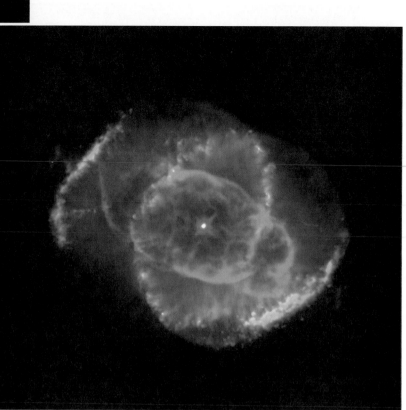

b The Cat's Eye Nebula

c The Twin Jet Nebula

d The Hourglass Nebula

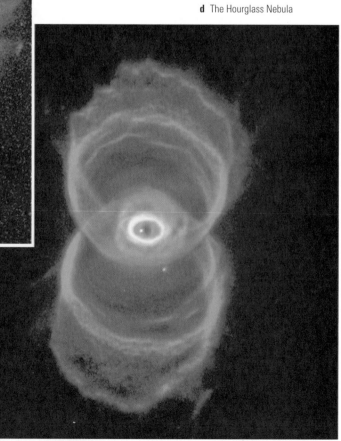

FIGURE 14.13 Planetary nebulae occur when low-mass stars in their final death throes cast off their outer layers of gas, as seen in these photos from the Hubble Space Telescope. The hot core that remains ionizes and energizes the richly complex envelope of gas surrounding it.

ultraviolet radiation that ionizes the gas in the expanding shell, causing it to glow brightly as what we call a **planetary nebula**. Despite their name, planetary nebulae have nothing to do with planets. The name comes from the fact that nearby nebulae look disk-shaped (like planets) through a small telescope. Through a larger telescope, more detail is visible. The famous Ring Nebula in the constellation Lyra is a planetary nebula; many others are also quite beautiful (Figure 14.13).

The glow of the planetary nebula fades as the exposed core cools and the ejected gas disperses into space. The nebula will disappear within a million years, leaving behind the cooling carbon core. On the H–R diagram, the life track now represents this "dead" core (see Figure 14.12). At first the life track heads to the left, because the core is initially quite hot and luminous. But the life track soon veers downward and to the right as the remaining ember cools and fades. You already know these naked, inert cores by another name: *white dwarfs* [Section 13.5].

In the ongoing battle between gravity and a star's internal pressure, white dwarfs are a sort of stalemate. As long as no mass is added to the white dwarf from some other source (such as a companion star in a binary system), neither the strength of gravity nor the strength of the degeneracy pressure that holds gravity at bay will ever change. Thus, a white dwarf is little more than a decaying corpse that will cool for the indefinite future, eventually disappearing from view as a *black dwarf*.

The Fate of Life on Earth

The evolutionary stages of low-mass stars are immensely important to the Earth, because we orbit a low-mass star—the Sun. The Sun will gradually brighten during its remaining time as a main-sequence star; it has already been doing so for the past 4 billion years with no serious consequences yet for life on Earth. However, somewhere around the year A.D. 5,000,000,000, the hydrogen in the solar core will run out. The Sun will then expand to a subgiant about three times as large and twice as bright as it is today. This increased luminosity will evaporate the oceans, perhaps leading to a runaway greenhouse effect that could raise Earth's average temperature to something similar to that of Venus and probably hotter [Section 11.3]. If any life survives, it will have to be very tolerant of heat.

Things will only get worse as the Sun grows into a red giant over the next several hundred million years. Just before helium flash, the Sun will be about 100 times larger in radius and over 1,000 times more luminous than it is today. Earth's surface temperature will exceed 1,000 K, and the oceans will long since have boiled away. By this time, any surviving humans will need to have found a new home. The Sun will shrink somewhat after helium burning begins, providing a temporary lull in incineration while the Sun spends 100 million years as a helium-burning star.

Anyone who survives the Sun's helium-burning phase will need to prepare for one final disaster. After exhausting its core helium, the Sun will expand again during its last million years. Its luminosity will soar to thousands of times what it is today, and it will grow so large that solar prominences might lap at the Earth's surface. Then it will eject its outer layers as a planetary nebula that will engulf Jupiter and Saturn and drift on past Pluto into interstellar space. If the Earth is not destroyed, its charred surface will be cold and dark in the faint, fading light of the white dwarf that the Sun has become. From then on, the Earth will be little more than a chunk of rock circling the corpse of our once-brilliant star.

14.4 Life as a High-Mass Star

Human life would be impossible without both low-mass stars and high-mass stars. The long lives of low-mass stars allow evolution to proceed for billions of years, but only high-mass stars produce the full array of elements on which life depends. Fusion of elements heavier than helium to produce elements heavier than carbon requires extremely high temperatures in order to overcome the larger electromagnetic repulsion of more highly charged nuclei. Reaching such temperatures requires an extremely strong crush of gravity on a star's core—a crush that occurs only under the immense weight of the overlying gas in high-mass stars.

In the final stages of their lives, the highest-mass stars proceed to fuse increasingly heavy elements until they have exhausted all possible fusion sources. When fusion ceases, gravity drives the core to implode suddenly, which, as we will soon see, causes the star to self-destruct in the titanic explosion we call a *supernova*. The fast-paced life and cataclysmic death of a high-mass star—surely one of the great dramas of the universe—are the topics of this section.

Brief but Brilliant

During its main-sequence life, the strong gravity of a high-mass star compresses its hydrogen core to higher temperatures than we find in lower-mass stars. You already know that the rate of fusion via the proton–proton chain increases substantially at higher temperatures, but the even hotter core temperatures of high-mass stars enable protons to slam into carbon, oxygen, or nitrogen nuclei as well as into other protons. Although carbon, nitrogen, and oxygen make up less than 2% of the material from which stars form in interstellar space, this 2% is more than enough to be useful in a stellar core. The carbon, nitrogen, and oxygen act as catalysts for hydrogen fusion, making it proceed at a far higher rate than would be possible by the proton–proton chain alone. (A *catalyst* is something that aids the progress of a reaction without being consumed in the reaction.) The lives of high-mass stars are truly brief but brilliant.

The chain of reactions that leads to hydrogen fusion in high-mass stars is called the **CNO cycle**; the letters *CNO* stand for carbon, nitrogen, and oxygen, respectively. Just as in the proton–proton chain [Section 12.3], four hydrogen nuclei go in while one helium-4 nucleus comes out. The amount of energy generated in each reaction cycle therefore is the same as in the proton–proton chain: It is equal to the difference in mass between the four hydrogen nuclei and the one helium nucleus multiplied by c^2. The CNO cycle is simply another, faster, way to accomplish hydrogen fusion.

The escalated fusion rates in high-mass stars generate remarkable amounts of power. Many more photons stream from the photospheres of high-mass stars than from the Sun, and many more photons are bouncing around inside. Although photons have no mass, they act like particles and carry *momentum* [Section 5.1], which they can transfer to anything they hit, imparting a very slight jolt. The combined jolts from the huge number of photons streaming outward through a high-mass star apply a type of pressure called **radiation pressure**.

This type of pressure can have dramatic effects on high-mass stars. In the most massive stars, radiation pressure is even more important than thermal pressure in keeping gravity at bay and can drive strong, fast-moving stellar winds.

TIME OUT TO THINK *We have now discussed three types of pressure: thermal pressure, degeneracy pressure, and radiation pressure. Before continuing, briefly review each type of pressure and describe the conditions under which each is important.*

Advanced Nuclear Burning

The exhaustion of core hydrogen in a high-mass star sets in motion the same processes that turn a low-mass star into a red giant, but the transformation proceeds much more quickly. The star develops a hydrogen-burning shell, and its outer layers begin to expand outward. At the same time, the core contracts, and this gravitational contraction releases energy that raises the core temperature until it becomes hot enough to fuse helium into carbon.

A high-mass star fuses helium into carbon so rapidly that it is left with an inert carbon core after just a few hundred thousand years or less. Once again, the absence of fusion leaves the core without a thermal energy source to fight off the crush of gravity. The inert carbon core shrinks, the crush of gravity intensifies, and the core pressure, temperature, and density all rise. Meanwhile, a helium-burning shell forms between the inert core and the hydrogen-burning shell. The star's outer layers swell again.

Up to this point, the life stories of intermediate-mass stars ($2–8M_{Sun}$) and high-mass stars ($>8M_{Sun}$) are very similar, except that all stages proceed more rapidly in higher-mass stars. However, degeneracy pressure prevents the cores of intermediate-mass stars from reaching the temperatures required to burn carbon or oxygen into anything much heavier. These stars eventually blow away their upper layers and finish their lives as white dwarfs. The rest of a high-mass star's life, on the other hand, is unlike anything that a low- or intermediate-mass star ever experiences.

The crush of gravity in a high-mass star is so overwhelming that degeneracy pressure never comes into play in the collapsing carbon core. The gravitational contraction of the core continues, and the core temperature soon reaches the 600 million K required to fuse carbon into heavier elements. Carbon fusion provides the core with a new source of energy that restores the balance versus gravity, but only temporarily. In the highest-mass stars, carbon burning may last only a few hundred years. When the core carbon is depleted, the core again begins to collapse, shrinking and heating until it can fuse a still-heavier element. The star is engaged in the final phases of a desperate battle against the ever-strengthening crush of gravity. Each successive stage of core nuclear burning proceeds more rapidly than prior stages. In a field in which we usually think in terms of millions or billions of years, we are suddenly dealing with time scales that are short even for humans.

The nuclear reactions in the star's final stages of life become quite complex, and many different reactions may take place simultaneously. The simplest sequence of fusion stages involves **helium capture**—the fusing of helium nuclei into progressively heavier elements. This process, which adds two protons with each reaction, tends to create elements with even numbers of protons.

Each time the core depletes the elements it is fusing, it shrinks and heats until it becomes hot enough for other fusion reactions. Meanwhile, a new type of shell burning ignites between the core and the overlying shells of fusion. Near the end, the star's central region resembles the inside of an onion, with layer upon layer of shells burning different elements

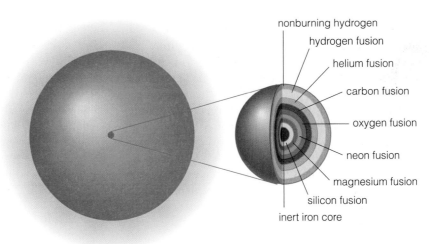

FIGURE 14.14 The multiple layers of nuclear burning in the core of a high-mass star during the final days of its life.

nonburning hydrogen
hydrogen fusion
helium fusion
carbon fusion
oxygen fusion
neon fusion
magnesium fusion
silicon fusion
inert iron core

(Figure 14.14). During the star's final few days, iron begins to pile up in the silicon-burning core.

Despite the dramatic events taking place in its interior, the high-mass star's outer appearance changes only slowly. As each stage of core fusion ceases, the surrounding shell burning intensifies and further inflates the star's outer layers. Each time the core flares up again, the outer layers may contract a bit. The result is that the star's life track zigzags across the top of the H–R diagram (Figure 14.15). In very massive stars, the core changes happen so quickly that the outer layers don't have time to respond, and the star

progresses steadily toward becoming a red supergiant. Betelgeuse, the upper-left shoulder star of Orion, is the best-known red supergiant star. Its radius is over 500 solar radii, or more than twice the distance from the Sun to Earth. We have no way of knowing what stage of nuclear burning is now taking place in Betelgeuse's core. Betelgeuse may have a few thousand years of nuclear burning still ahead, or we may be seeing it as iron piles up in its core. If the latter is the case, then sometime in the next few days we will witness one of the most dramatic events that ever occurs in the universe.

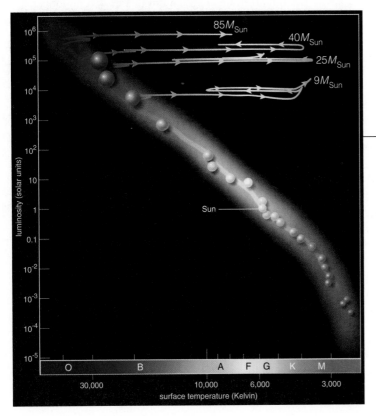

FIGURE 14.15 Life tracks on the H–R diagram from main-sequence star to red supergiant for selected high-mass stars. Labels on the tracks give the star's mass at the beginning of its main-sequence life. Because of the strong wind from such a star, its mass can be considerably smaller when it leaves the main sequence. (Based on models from A. Maeder and G. Meynet.)

Iron: Bad News for the Stellar Core

As a high-mass star develops an inert core of iron, the core continues shrinking and heating while iron continues to pile up from nuclear burning in the surrounding shells. If iron were like the other elements in prior stages of nuclear burning, this core contraction would stop when iron fusion ignited. However, iron is unique among the elements in a very important way: It is the one element from which it is *not* possible to generate any kind of nuclear energy.

To understand why iron is unique, remember that only two basic processes can release nuclear energy: *fusion* of light elements into heavier ones, and *fission* of very heavy elements into not-so-heavy ones. Recall that hydrogen fusion converts four protons (hydrogen nuclei) into a helium nucleus that consists of two protons and two neutrons. Thus, the total number of *nuclear particles* (protons and neutrons combined) does not change. However, this fusion reaction generates energy (in accord with $E = mc^2$) because the *mass* of the helium nucleus is less than the combined mass of the four hydrogen nuclei that fused to create it—despite the fact that the *number* of nuclear particles is unchanged.

In other words, fusing hydrogen into helium generates energy because helium has a lower *mass per nuclear particle* than hydrogen. Similarly, fusing three helium-4 nuclei into one carbon-12 nucleus generates energy because carbon has a lower mass per nuclear particle than helium—which means that some mass disappears and becomes energy in this fusion reaction. In fact, the decrease in mass per nuclear particle from hydrogen to helium to carbon is part of a general trend shown in Figure 14.16: The mass per

nuclear particle tends to decrease as we go from light elements to iron, which means that fusion of light nuclei into heavier nuclei generates energy. This trend reverses beyond iron: The mass per nuclear particle tends to *increase* as we look to still heavier elements. As a result, elements heavier than iron can generate nuclear energy only through fission into lighter elements. For example, uranium has a greater mass per nuclear particle than lead, so uranium fission (which ultimately leaves lead as a by-product) must convert some mass into energy.

Iron has the lowest mass per nuclear particle of all nuclei and therefore cannot release energy by either fusion or fission. Thus, once the matter in a stellar core turns to iron, it can generate no further thermal energy or pressure. The iron core's only hope of resisting the crush of gravity lies with degeneracy pressure. But iron keeps piling up in the core until degeneracy pressure can no longer support it either. What ensues is the ultimate nuclear-waste catastrophe.

TIME OUT TO THINK *How would the universe be different if hydrogen, rather than iron, had the lowest mass per nuclear particle? Why?*

Supernova

The degeneracy pressure that briefly supports the inert iron core arises because the laws of quantum mechanics prohibit electrons from getting too close together. But once gravity pushes the electrons past the quantum mechanical limit, they can no longer exist freely. The electrons disappear by combining with protons to form neutrons, releasing neutrinos in the process (Figure 14.17). The electron degeneracy pressure suddenly vanishes, and gravity has free rein.

In a fraction of a second, an iron core with a mass comparable to that of our Sun and a size larger than that of Earth collapses into a ball of neutrons just a few kilometers across. The collapse halts only because the neutrons have a degeneracy pressure of their own. The entire core then resembles a giant atomic nucleus. If you recall that ordinary atoms are made almost entirely of empty space [Section 4.3]

FIGURE 14.16 Overall, the average mass per nuclear particle declines from hydrogen to iron and then increases. Selected nuclei are labeled to provide reference points. (This graph shows the most general trends only; a more detailed graph would show numerous up-and-down bumps superimposed on the general trend.)

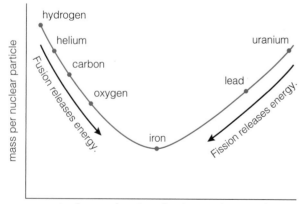

FIGURE 14.17 During the final, catastrophic collapse of a high-mass stellar core, electrons and protons combine to form neutrons, accompanied by the release of neutrinos.

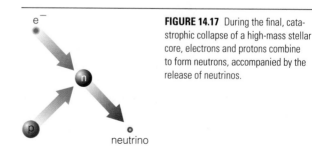

and that almost all their mass is in their nuclei, you'll realize that a giant atomic nucleus must have an astoundingly high density.

The gravitational collapse of the core releases an enormous amount of energy—more than a hundred times what the Sun will radiate over its entire 10-billion-year lifetime! Where does this energy go? It drives the outer layers off into space in a titanic explosion—a **supernova**. The ball of neutrons left behind is called a **neutron star**. In some cases, the remaining mass may be so large that gravity also overcomes neutron degeneracy pressure, and the core continues to collapse until it becomes a *black hole*.

Theoretical models of supernovae successfully reproduce the observed energy outputs of real supernovae, but the precise mechanism of the explosion is not yet clear. Two general processes could contribute to the explosion. In the first process, neutron degeneracy pressure halts the gravitational collapse, causing the core to rebound slightly and ram into overlying material that is still falling inward. Until recently, most astronomers thought that this *core-bounce* process ejected the star's outer layers. But current models of supernovae suggest that the more important process involves the neutrinos formed when electrons and protons combine to make neutrons. Although these ghostly particles rarely interact with anything [Section 12.3], so many are produced when the core implodes that they drive a shock wave that propels the star's upper layers into space.

The shock wave sends the star's former surface zooming outward at a speed of 10,000 km/s, heating it so that it shines with dazzling brilliance. For about a week, a supernova blazes as powerfully as 10 billion Suns, rivaling the luminosity of a moderate-size galaxy. The ejected gases slowly cool and fade in brightness over the next several months, but they continue to expand outward until they eventually mix with other gas in interstellar space. The scattered debris from the supernova carries with it the variety of elements produced in the star's nuclear furnace, as well as additional elements created when some of the neutrons produced during the core collapse slam into other nuclei. Millions or billions of years later, this debris may be incorporated into a new generation of stars.

The Origin of Elements

Before we leave the subject of massive-star life cycles, it's useful to consider the evidence that indicates we actually understand the origin of the elements. We cannot see inside stars, so we cannot directly observe elements being created in the ways we've discussed. However, the signature of nuclear reactions in massive stars is written in the patterns of elemental abundances across the universe.

For example, if massive stars really produce heavy elements (that is, elements heavier than hydrogen and helium) and scatter these elements into space when they die, the total amount of these heavy elements in interstellar gas should gradually increase with time (because additional massive stars have died). Thus, we should expect stars born recently to contain a greater proportion of heavy elements than stars born in the distant past because they formed from interstellar gas that contained more heavy elements. Stellar spectra confirm this prediction: Older stars do indeed contain smaller amounts of heavy elements than younger stars. For very old stars in globular clusters, elements besides hydrogen and helium typically make up as little as 0.1% of the total mass. In contrast, about 2–3% of the mass of young stars that formed in the recent past is in the form of heavy elements.

We gain even more confidence in our model of elemental creation when we compare the abundances of different elements. For example, because helium-capture reactions add two protons (and two neutrons) at a time, we expect nuclei with even numbers of protons to outnumber those with odd numbers of protons that fall between them. Sure enough, even-numbered nuclei such as carbon, oxygen, and neon are relatively abundant (Figure 14.18). Similarly, because elements heavier than iron are made only by rare fusion reactions shortly before and during a supernova, we expect these elements to be extremely

FIGURE 14.18 This graph shows the observed relative abundances of elements in the galaxy in comparison to the abundance of hydrogen. For example, the abundance of nitrogen is about 10^{-4}, which means that there are about $10^{-4} = 0.0001$ times as many nitrogen atoms in the galaxy as hydrogen atoms.

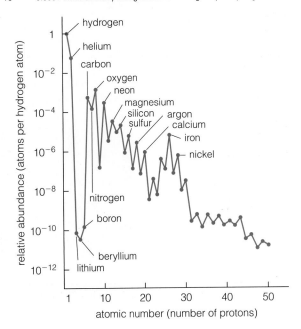

rare. Again, this prediction made by our model of nuclear creation is verified by observations.

Supernova Observations

The study of supernovae owes a great debt to astronomers of many different epochs and cultures. Careful scrutiny of the night skies allowed the ancients to identify several supernovae whose remains still adorn the heavens. The most famous example concerns the Crab Nebula in the constellation Taurus. The Crab Nebula is a **supernova remnant**—an expanding cloud of debris from a supernova explosion (Figure 14.19). A spinning neutron star lies at the center of the Crab Nebula, providing evidence that supernovae really do create neutron stars. Photographs taken years apart show that the nebula is growing larger at a rate of several thousand kilometers per second. Calculating backward from its present size, we can trace the nebula's birth to somewhere near A.D. 1100.

No supernova has been seen in our own galaxy since 1604, but today astronomers routinely discover supernovae in other galaxies. The nearest of these extragalactic supernovae, and the only one near enough to be visible to the naked eye, burst into view in 1987. Because it was the first supernova detected that year, it was given the name **Supernova 1987A**. Supernova 1987A was the explosion of a star in the *Large Magellanic Cloud,* a small galaxy that orbits the Milky Way and is visible only from southern latitudes. The Large Magellanic Cloud is about 150,000 light-years away, so the star really exploded some 150,000 years ago.

As the nearest supernova witnessed in four centuries, Supernova 1987A provided a unique opportunity to study a supernova and its debris in detail. Older photographs of the Large Magellanic Cloud allowed astronomers to determine precisely which star exploded (Figure 14.20). It turned out to be a blue star, not the red supergiant expected when core fusion has ceased. The most likely explanation is that the star's outer layers were unusually thin and warm near the end of its life (perhaps because some of its matter spilled onto a close binary companion), changing its appearance from that of a red supergiant to a blue one. The surprising color of the pre-explosion star demonstrates that we still have much to learn about supernovae. Reassuringly, most other theoretical predictions of stellar life cycles were well matched by observations of Supernova 1987A.

One of the most remarkable findings from Supernova 1987A was that bursts of neutrinos coming from its direction were recorded by two neutrino detectors, one in Japan and one in Ohio. These data confirmed that the explosion released most of its energy in the form of neutrinos, suggesting that we are correct in believing that the stellar core undergoes sudden collapse to a ball of neutrons.

FIGURE 14.19 The Crab Nebula is the remnant of the supernova observed in A.D. 1054. The speed at which the Crab Nebula is growing indicates that the original explosion occured around A.D. 1100. Chinese observers recorded a "guest star" near this location on July 4, 1054, undoubtedly the supernova that created this nebula. This photograph was taken with the new Very Large Telescope at the European Southern Observatory in Chile.

before

after

FIGURE 14.20 Before-and-after photos of the location of Supernova 1987A. In the before picture, the arrow indicates the star that exploded. The supernova actually appeared as a bright *point* of light; it appears larger than a point in the after photograph only because of overexposure.

TIME OUT TO THINK *When Betelgeuse explodes as a supernova, it will be more than 10 times brighter than the full moon in our sky. If Betelgeuse had exploded a few hundred or a few thousand years ago, do you think it could have had any effect on human history? How do you think our modern society would react if we saw Betelgeuse explode tomorrow?*

Summary of Stellar Lives

We have seen that the primary factor determining how a star lives its life is its mass. Low-mass stars live long lives and die in planetary nebulae, leaving behind white dwarfs. High-mass stars live short lives and die in supernovae, leaving behind neutron stars and black holes. Both types of stars are crucial to life. Near the ends of their lives, low-mass stars can become carbon stars, which are the source of most

of the carbon in our bodies. High-mass stars produce the vast array of other chemical elements on which life depends. Figure 14.21 summarizes the life cycles of stars of different masses.

14.5 The Lives of Close Binary Stars

For the most part, stars in binary systems proceed from birth to death as if they were isolated and alone. The exceptions are close binary stars. Algol, the "demon star" in the constellation Perseus, consists of two stars that orbit each other closely: a $3.7M_{Sun}$ main-sequence star and a $0.8M_{Sun}$ subgiant. A moment's thought reveals that something quite strange is going on. The stars of a binary system are born at the same time and therefore must both be the same age. We know that more massive stars live shorter

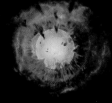

Protostars: A star system forms when a cloud of interstellar gas collapses under gravity. The central protostar is surrounded by a protostellar disk in which planets may eventually form.

Blue main-sequence star: Star is fueled by hydrogen fusion in its core. In high-mass stars, hydrogen fusion proceeds by the series of reactions known as the CNO cycle.

Red supergiant: After core hydrogen is exhausted, the core shrinks and heats. Hydrogen shell burning begins around the inert helium core, causing the star to expand into a red supergiant.

Helium core-burning supergiant: Helium fusion begins when enough helium has collected in the core. The core then expands, slowing the fusion rate and allowing the star's outer layers to shrink somewhat. Hydrogen shell burning continues at a reduced rate.

Life of a 20M_{Sun} Star.
Main-sequence lifetime: 8 million years
Duration of later stages: 1 million years

Multiple shell-burning supergiant: After core helium is exhausted, the core shrinks until carbon fusion begins, while helium and hydrogen continue to burn in shells surrounding the core. Late in its life, the star fuses heavier elements like carbon and oxygen in shells while iron collects in the inert core.

Supernova: Iron cannot provide fusion energy, so it accumulates in the core until the mass of the iron core approaches the 1.4M_{Sun} limit. Then it can no longer support its own weight and collapses, leading to the catastrophic explosion of the star.

Neutron star: During the core collapse of the supernova, electrons combine with protons to make neutrons. The leftover core is therefore made almost entirely of neutrons.

Yellow main-sequence star: Star is fueled by hydrogen fusion in its core, which converts four hydrogen nuclei into one helium nucleus. In low-mass stars, hydrogen fusion proceeds by the series of reactions known as the proton-proton chain.

Helium core-burning star: Helium fusion, in which three helium nuclei fuse to form a single carbon nucleus, begins when enough helium has collected in the core. The core then expands, slowing the fusion rate and allowing the star's outer layers to shrink somewhat. Hydrogen shell burning continues at a reduced rate.

Red giant star: After core hydrogen is exhausted, the core shrinks and heats. Hydrogen shell burning begins around the inert helium core, causing the star to expand into a red giant.

Life of a 1M$_{Sun}$ Star.
Main-sequence lifetime: 10 billion years
Duration of later stages: 1 billion years

Double shell-burning red giant: After core helium is exhausted, the core again shrinks and heats. Helium shell burning begins around the inert carbon core and the star enters its second red giant phase. Hydrogen shell burning continues.

Planetary nebula: The dying star expels its outer layers in a planetary nebula, leaving behind the exposed inert core.

White dwarf: The remaining white dwarf is made primarily of carbon and oxygen because the core never grew hot enough to fuse these elements into anything heavier.

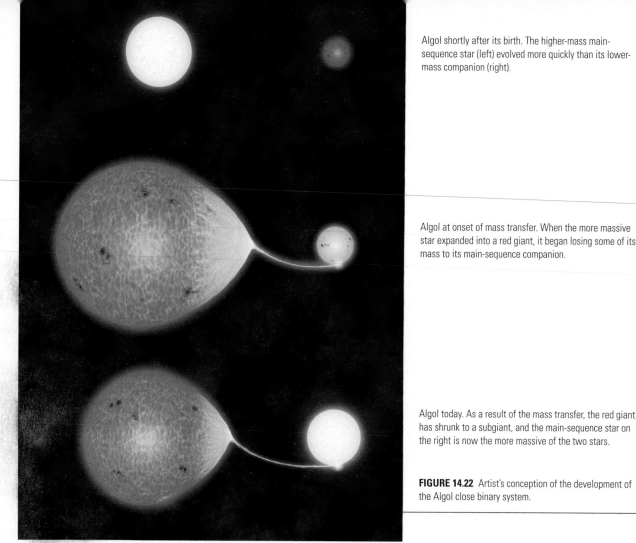

Algol shortly after its birth. The higher-mass main-sequence star (left) evolved more quickly than its lower-mass companion (right).

Algol at onset of mass transfer. When the more massive star expanded into a red giant, it began losing some of its mass to its main-sequence companion.

Algol today. As a result of the mass transfer, the red giant has shrunk to a subgiant, and the main-sequence star on the right is now the more massive of the two stars.

FIGURE 14.22 Artist's conception of the development of the Algol close binary system.

lives, and therefore the more massive star must exhaust its core hydrogen and become a subgiant before the less massive star does. How, then, can Algol's less massive star be a subgiant while the more massive star is still burning hydrogen as a main-sequence star?

This so-called *Algol paradox* reveals some of the complications in ordinary stellar life cycles that can arise in close binary systems. The two stars in close binaries are near enough to exert significant tidal forces on each other [Section 5.4]. The gravity of each star attracts the near side of the other star more strongly than it attracts the far side. The stars therefore stretch into football-like shapes rather than remaining spherical. In addition, the stars become *tidally locked* so that they always show the same face to each other, much as the Moon always shows the same face to Earth.

During the time that both stars are main-sequence stars, the tidal forces have little effect on their lives. But when the more massive star (which exhausts its core hydrogen sooner) begins to expand into a red giant, gas from its outer layers can spill over onto its companion. This **mass exchange** occurs when the giant grows so large that its tidally distorted outer layers succumb to the gravitational attraction of the smaller companion star. The companion then begins to gain mass at the expense of the giant.

The solution to the Algol paradox should now be clear (Figure 14.22). The $0.8M_{Sun}$ subgiant *used to be* much more massive. As the more massive star, it was the first to begin expanding into a red giant. But as it expanded, so much of its matter spilled over onto its companion that it is now the less massive star.

The future may hold even more interesting events for Algol. The $3.7M_{Sun}$ star is still gaining mass from its subgiant companion. Thus, its life cycle is actually accelerating as its increasing gravity raises its core hydrogen fusion rate. Millions of years from now, it will exhaust its hydrogen and begin to expand into a red giant itself. At that point, it can begin to transfer mass *back* to its companion. Even stranger things can happen in other mass-exchange systems, particularly when one of the stars is a white dwarf or a neutron star. But that is a topic for the next chapter.

THE BIG PICTURE

In this chapter, we answered the question of the origin of elements that we first discussed in Chapter 1. As you look back over this chapter, remember these "big picture" ideas:

- Virtually all elements besides hydrogen and helium were forged in the nuclear furnaces of stars. Carbon can be released from low-mass stars near the ends of their lives (carbon stars), and many other elements are released into space by massive stars in supernova explosions.

- The tug-of-war between gravity and pressure determines how stars behave from the time of their birth in a cloud of molecular gas to their sometimes violent death.

- Low-mass stars like our Sun live long lives and die with the ejection of a planetary nebula, leaving behind a white dwarf.

- High-mass stars live fast and die young, exploding dramatically as supernovae.

- Close binary stars can exchange mass, altering the usual course of stellar evolution.

Review Questions

1. What is *thermal pressure?* What are the two energy sources that can help a star maintain its internal thermal pressure?

2. What is a *molecular cloud?* Briefly describe the process by which a *protostar* and *protostellar disk* form from gas in a molecular cloud.

3. Under what conditions does a *close binary* form?

4. Describe some of the activity seen in protostars, such as strong *protostellar winds* and *jets*.

5. What do we mean by a star's *life track* on an H–R diagram? How does an H–R diagram that shows life tracks differ from a standard H–R diagram?

6. Describe the observed trends in the number of stars that are born with different masses. What types of stars are most common? What types are least common?

7. What is *degeneracy pressure?* How does it differ from thermal pressure? Explain why degeneracy pressure can support a stellar core against gravity even when the core becomes very cold.

8. What is a *brown dwarf?*

9. What happens to the core of a star when it exhausts its hydrogen supply? Why does *hydrogen shell burning* begin around the inert core?

10. Why does *helium fusion* require much higher temperatures than hydrogen fusion? Briefly describe the overall reaction by which helium fuses into carbon.

11. Explain why a star's overall radius shrinks from its peak size as a red giant after helium fusion begins.

12. What is a *planetary nebula?* What happens to the core of a star after a planetary nebula occurs?

13. Briefly describe how the Sun will change, and how Earth will be affected by these changes, over the next several billion years.

14. What is *radiation pressure?* Why is it more important in high-mass stars than in low-mass stars? What effects does radiation pressure have on the structure of a high-mass star?

15. Why does continued nuclear burning occur in high-mass stars but not in low-mass stars?

16. What special feature of iron nuclei determines how massive stars end their lives? What happens during a *supernova?*

17. Summarize some of the observational evidence supporting our ideas about how the elements formed and showing that supernovae really occur.

18. What is the Algol paradox? What is its resolution? Use your answer to summarize how the lives of stars in close binaries can differ from the lives of single stars.

Discussion Questions

1. *Connections to the Stars.* In ancient times, many people believed that our lives were somehow influenced by the patterns of the stars in the sky, a belief that survives to this day in astrology. Modern science has not found any evidence to support this belief but instead has found that we have a connection to the stars on a much deeper level: In the words of Carl Sagan, we are "star stuff." Discuss in some detail our real connections to the stars as established by modern astronomy. Do you think these connections have any philosophical implications in terms of how we view our lives and our civilization? Explain.

2. *Humanity in A.D. 5,000,000,000.* Do you think it is likely that humanity will survive until the Sun begins to expand into a red giant 5 billion years from now? Why or why not? If the human race does survive, how do you think people in A.D. 5,000,000,000 will differ from people today? What do you think they will do when faced with the impending death of the Sun? Debate these questions, and see if you and your friends can come to any agreement on possible answers.

Problems

Sensible Statements? For **problems 1–6**, decide whether the statement is sensible and explain why it is or is not.

1. The iron in my blood came from a star that blew up over 4 billion years ago.

2. A protostellar cloud spins faster as it contracts because it is gaining angular momentum.

3. When helium fusion begins in the core of a low-mass star, the extra energy generated causes the star's luminosity to rise.

4. Humanity will eventually have to find another planet to live on, because one day the Sun will blow up as a supernova.

5. I sure am glad hydrogen has a higher mass per nuclear particle than many other elements. If it had the lowest mass per nuclear particle, none of us would be here.

6. I just discovered a $3.5 M_{Sun}$ main-sequence star orbiting a $2.5 M_{Sun}$ red giant. I'll bet that red giant was more massive than $3 M_{Sun}$ when it was a main-sequence star.

Homes to Civilizations. We do not yet know how many stars have Earth-like planets, nor do we know the likelihood that such planets might harbor advanced civilizations like our own. However, some stars can probably be ruled out as candidates for advanced civilizations. For example, given that it took a few billion years for humans to evolve on Earth, it seems unlikely that advanced life would have had time to evolve around a star that is only a few million years old. For each of the stars in **problems 7–11**, decide whether you think it is possible that it could harbor an advanced civilization, and explain your reasoning in one or two paragraphs.

7. A $10 M_{Sun}$ main-sequence star.

8. A carbon star.

9. A $1.5 M_{Sun}$ red giant.

10. A $1 M_{Sun}$ helium-burning star.

11. A red supergiant.

12. *Rare Elements.* Lithium, beryllium, and boron are elements with atomic numbers 3, 4, and 5, respectively. But despite their being three of the five simplest elements, Figure 14.18 shows that they are rare compared to many heavier elements. Suggest a reason for their rarity. (*Hint:* Consider the process by which helium fuses into carbon.)

13. *Future Skies.* As a red giant, the Sun's angular size in the Earth's sky will be about 30°. What will sunset and sunrise be like? About how long will they take? Do you think the color of the sky will be different from what it is today? Explain.

14. *Research: Historical Supernovae.* Historical accounts exist for supernovae in the years 1006, 1054, 1572, and 1604. Choose one of these supernovae and learn more about historical records of the event. Did the supernova influence human history in any way? Write a two- to three-page summary of your research findings.

Web Projects

Find useful links for Web projects on the text Web site.

1. *Coming Fireworks in Supernova 1987A.* Astronomers believe that the show from Supernova 1987A is not yet over. In particular, sometime between now and about 2010, the expanding cloud of gas from the supernova is expected to ram into surrounding material, and the heat generated by the impact is expected to create a new light show. Learn more about how Supernova 1987A is changing and what we might expect to see from it in the future. Summarize your findings in a one- to two-page report.

2. *Picturing Star Birth and Death.* Photographs of stellar birthplaces (i.e., molecular clouds) and death places (e.g., planetary nebulae and supernova remnants) can be strikingly beautiful, but only a few such photographs are included in this chapter. Search the Web for additional photographs of these types; look not only for photos taken in visible light, but also for false-color photographs made from observations in other wavelengths of light. Put each photograph you find into a personal on-line journal, along with a one-paragraph description of what the photograph shows. Try to compile a journal of at least 20 such photographs.

CHAPTER 15

The Bizarre Stellar Graveyard

Welcome to the afterworld of stars, the fascinating domain of white dwarfs, neutron stars, and black holes. To scientists, these dead stars are ideal laboratories for testing the most extreme predictions of general relativity and quantum theory. To most other people, the eccentric behavior of stellar corpses demonstrates that the universe is stranger than they had ever imagined.

Dead stars behave in unusual and unexpected ways that challenge our minds and stretch the boundaries of what we believe is possible. Stars that have finished nuclear burning have only one hope of staving off the crushing power of gravity: the strange quantum mechanical effect of degeneracy pressure. But even this strange pressure cannot save the most massive stellar cores, which collapse into oblivion as black holes. Prepare to be amazed by the eerie inhabitants of the stellar graveyard!

15.1 A Star's Final Battle

In the previous chapter, we saw that an ongoing "battle" between the inward crush of gravity and the outward push of pressure governs a star's life from the time when it first begins to form in an interstellar cloud to the time when it finally exhausts its nuclear fuel. Throughout most of this time, the pressure that holds gravity at bay is *thermal pressure,* which results from the heat produced as the star fuses light elements into heavier ones in its core. But the nuclear fuel eventually runs out. In the end, after the star dies in a planetary nebula or supernova, the fate of the stellar corpse lies in the outcome of a final battle between gravity and pressure—but this time the source of the pressure is the quantum mechanical effect called *degeneracy pressure.*

We have already discussed how degeneracy pressure successfully resists the crush of gravity in the stellar corpses known as white dwarfs and neutron stars, as well as in several other cases (e.g., brown dwarfs and inert stellar cores). Because these objects are supported by degeneracy pressure, they are known collectively as *degenerate objects,* and the matter within them is called *degenerate matter.* Recall that degeneracy pressure arises when subatomic particles are packed as closely as the laws of quantum mechanics allow [Section 14.2]. More specifically, white dwarfs are supported against the crush of gravity by **electron degeneracy pressure**, in which the pressure arises from densely packed electrons. In neutron stars, it is *neutrons* that are packed tightly together, thereby generating **neutron degeneracy pressure**.

White dwarfs and neutron stars would be strange enough if the story ended here, but it does not. Sometimes, gravity wins the battle with degeneracy pressure, and the stellar corpse collapses without end, crushing itself out of existence and forming a *black hole.* A black hole is truly a hole in the universe. If you enter a black hole, you leave our observable universe and can never return. In this chapter, we will study the bizarre properties and occasional catastrophes of the stellar corpses known as white dwarfs, neutron stars, and black holes.

15.2 White Dwarfs

A **white dwarf** is the inert core left over after a star has ceased nuclear burning, so its composition reflects the products of the star's final nuclear-burning stage. The white dwarf left behind by a $1M_{Sun}$ star like our Sun will be made mostly of carbon, the product of the star's final helium-burning stage (Figure 15.1). The cores of very low mass stars never become hot enough to fuse helium and thus end up as helium white dwarfs. Some intermediate-mass stars progress

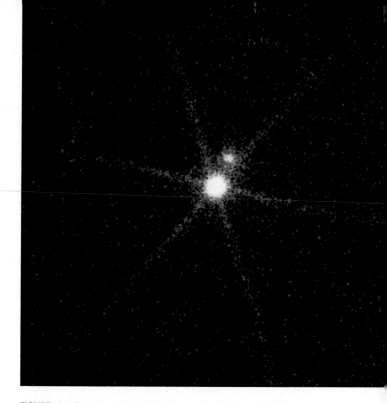

FIGURE 15.1 The binary star system Sirius as seen in X-ray light by the Chandra Telescope. Sirius A, the brightest star in the night sky to human eyes, is actually the dimmer of the two stars in this picture. Sirius B, its white dwarf companion, is much hotter and therefore much brighter in X-ray light. (The spikes emanting from Sirius B are not real; they are artifacts created by the telescope's optics.)

to carbon burning but do not create any iron (and hence do not explode as supernovae). These stars leave behind white dwarfs containing large amounts of oxygen or even heavier elements.

Despite the ordinary-sounding compositions of white dwarfs, their degenerate matter is unlike anything ever seen on Earth. Gravity and electron degeneracy pressure have battled to a draw in white dwarfs. Electron degeneracy pressure finally halts the collapse of a $1M_{Sun}$ stellar corpse when it shrinks to about the size of the Earth. If you recall that our Earth is smaller than a typical sunspot, it should be clear that it's no small feat to pack the entire mass of the Sun into the volume of the Earth. The density of such a white dwarf is so high that a pair of standard dice made from its material would weigh about 5 tons.

The White Dwarf Limit

Theoretical calculations show that the mass of a white dwarf cannot exceed a **white dwarf limit** of about $1.4M_{Sun}$, commonly called the *Chandrasekhar limit* after its discoverer. The limit comes about because the electron speeds are higher in more massive white dwarfs, and these speeds approach the speed of light in white dwarfs with masses approaching

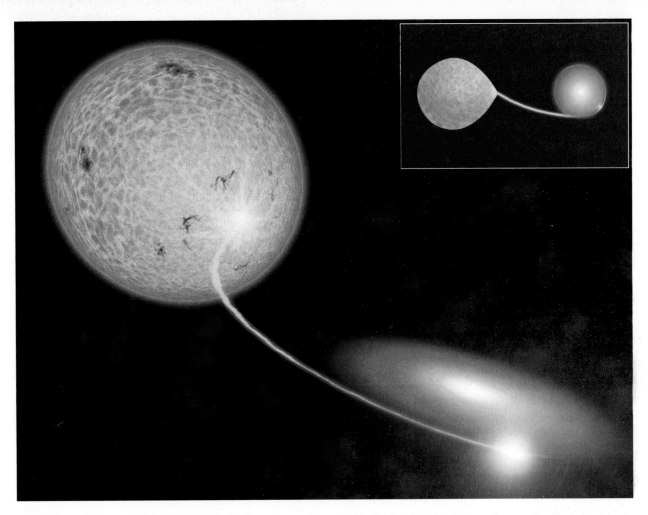

FIGURE 15.2 This artist's conception shows how matter spilling from a companion star (left) toward a white dwarf (right) forms an accretion disk around the white dwarf. The white dwarf itself is in the center of the accretion disk—too small to see on this scale. Matter streaming onto the disk creates a hot spot at the point of impact. The inset shows how the system looks from above the pole, which you can relate to the perspective shown in the painting.

$1.4M_{Sun}$. Neither electrons nor anything else can travel faster than the speed of light, so electrons can do nothing to halt the crush of gravity in a stellar corpse with a mass above the white dwarf limit. Such an object must inevitably collapse into a compact ball of neutrons, at which point *neutron* degeneracy pressure can stop the crush of gravity.

Strong observational evidence supports this theoretical limit on the mass of a white dwarf. Many known white dwarfs are members of binary systems, and hence their masses can be measured [Section 13.4]. In every observed case, the white dwarfs have masses below $1.4M_{Sun}$, just as we would expect from theory.

White Dwarfs in Close Binary Systems

Left to itself, a white dwarf will never succumb to the crush of gravity because its electron degeneracy pressure does not lessen even as the white dwarf cools into a cold black dwarf. However, white dwarfs in close binary systems do not necessarily rest in peace.

A white dwarf in a close binary system can gain substantial quantities of mass if its companion is a main-sequence or giant star. When a clump of matter first spills over from the companion to the white dwarf, it has some small orbital velocity. The law of conservation of angular momentum dictates that the clump must orbit faster and faster as it falls toward the white dwarf's surface. The infalling matter therefore forms a whirlpool-like disk around the white dwarf (Figure 15.2). The process in which material falls onto another body is called *accretion* [Section 7.4], so this rapidly rotating disk is called an **accretion disk**.

TIME OUT TO THINK *Explain why the infalling matter forms a disk around the white dwarf (as opposed to making, say, a spherical distribution of matter). (Hint: See Sections 7.3 and 14.2.) How is an accretion disk similar to a protoplanetary disk? How is it different?*

Accreting gas gradually spirals inward through the accretion disk and eventually falls onto the surface of the white dwarf. This happens because the individual gas particles in the accretion disk, like anything else that orbits a massive body, obey Kepler's laws [Section 3.4]. Gas in the inner parts of the accretion disk moves faster than gas in the outer parts; because of these differences in orbital speed, gas in any particular part of the accretion disk "rubs" against slower-moving gas just outside of it. This "rubbing" generates friction and heat in the same way that rubbing your palms together makes them warm. The friction slowly causes the orbits of individual gas particles to decay, making the orbits smaller and smaller until they fall onto the white dwarf surface. Meanwhile, additional gas spilling over from the companion constantly replenishes the accretion disk.

Accretion can provide the "dead" white dwarf with a new energy source. Theory predicts that the heat generated by friction should make the accretion disk hot enough to radiate optical and ultraviolet light, and sometimes even X rays. Thus, although accretion disks are far too small to be seen directly, we should be able to detect their intense ultraviolet or X-ray radiation. Searches for this radiation have turned up strong evidence for accretion disks around many white dwarfs. In some cases, the brightness of these systems is highly variable—sudden increases in brightness by a factor of 10 or more may persist for a few days and then fade away, only to repeat a few weeks or months later. Such brightening probably arises when instabilities in the accretion disk cause some of the matter to fall suddenly onto the white dwarf surface, with an accompanying release of gravitational potential energy. (This type of brightening is sometimes called a *dwarf nova*.)

Accreting white dwarfs occasionally flare up even more dramatically. Remember that a white dwarf is "dead" because it has no hydrogen left to fuse. This situation changes in an accreting white dwarf. The gas spilling onto an accreting white dwarf comes from the upper layers of its companion star and thus is composed mostly of hydrogen. A thin shell of fresh hydrogen builds up on the surface of the white dwarf as more and more material rains down from the accretion disk. The pressure and temperature at the bottom of this shell increase as the shell grows, and hydrogen fusion ignites when the temperature exceeds 10 million K. The white dwarf suddenly blazes back to life as its hydrogen shell burns. This thermonuclear flash causes the binary system to shine for a few glorious weeks as a **nova**, which can radiate as brightly as 100,000 Suns (Figure 15.3a). The nova generates heat and pressure, ejecting most of the

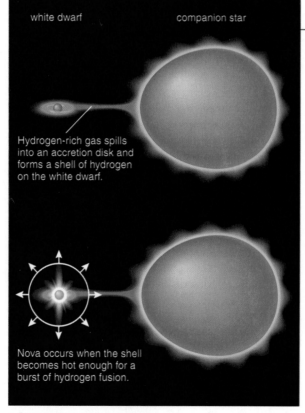

white dwarf   companion star

Hydrogen-rich gas spills
into an accretion disk and
forms a shell of hydrogen
on the white dwarf.

Nova occurs when the shell
becomes hot enough for a
burst of hydrogen fusion.

a Diagram of the nova process.

FIGURE 15.3 A nova occurs when hydrogen fusion ignites on the surface of a white dwarf in a binary star system.

b Hubble Space Telescope image showing blobs of gas ejected from the nova T Pyxidis. The bright spot at the center of the blobs is the binary star system that generated the nova.

material that has accreted onto the white dwarf. This material expands outward, creating a *nova remnant* that sometimes remains visible years after the nova explosion (Figure 15.3b).

Accretion resumes after a nova explosion subsides, so the entire process can repeat itself. The time between successive novae in a particular system depends on the rate at which hydrogen accretes on the white dwarf surface and on how highly compressed this hydrogen becomes. The compression of hydrogen is greatest for the most massive white dwarfs, which have the strongest surface gravities. In some cases, novae have been observed to repeat after just a few decades. More commonly, accreting white dwarfs may have 10,000 years between nova outbursts.

Note that, according to our modern definitions, a nova and a supernova are quite different events: A nova is a relatively minor detonation of hydrogen fusion on the surface of a white dwarf in a close binary, while a supernova is the total explosion of a star. However, the word *nova* simply means "new," and historically a nova referred to any star that appeared to the naked eye where none was visible before. Because supernovae generate far more power than novae—the light of 10 billion Suns in a supernova versus 100,000 Suns in a nova—a very distant supernova can appear as bright in our sky as a nova that is relatively close. Thus, people could not distinguish between novae and supernovae prior to modern times.

White Dwarf Supernovae

Each time a nova occurs, the white dwarf ejects some of its mass. Each time a nova subsides, the white dwarf begins to accrete matter again. Theoretical models cannot yet tell us whether the net result is a gradual increase or decrease in the white dwarf's mass. Nevertheless, in at least some cases, accreting white dwarfs in binary systems continue to gain mass as time passes. If such a white dwarf gains enough mass, it can one day reach the $1.4M_{Sun}$ white dwarf limit. This day is the white dwarf's last.

The electron degeneracy pressure that has been supporting the white dwarf gives out as soon as it reaches the $1.4M_{Sun}$ limit. Gravity is suddenly unopposed, and the white dwarf begins to collapse. The gravitational potential energy released in the collapse quickly heats the white dwarf, which soon reaches the temperature needed for carbon fusion. When carbon fusion begins, it ignites almost instantly throughout the star, and the white

dwarf explodes completely in what we will call a **white dwarf supernova** (also known as a Type Ia supernova).

TIME OUT TO THINK *Why do we expect that most collapsing white dwarfs undergo carbon fusion, as opposed to fusion of hydrogen, helium, or some other element?*

A white dwarf supernova shines as brilliantly as a supernova that occurs at the end of a high-mass star's life [Section 14.4]. To distinguish the two types, we'll refer to the latter as a **massive star supernova** (also known as a Type Ib or Type II supernova). Both white dwarf supernovae and massive star supernovae result in the destruction of a star. A massive star supernova is thought inevitably to leave behind either a neutron star or a black hole, but nothing remains after the explosion of a white dwarf supernova.

Astronomers can distinguish between white dwarf and massive star supernovae by studying their light. Because white dwarfs contain almost no hydrogen, the spectra of white dwarf supernovae show no spectral lines of hydrogen. In contrast, massive stars usually have plenty of hydrogen in their outer layers at the time they explode, so hydrogen lines are prominent in the spectra of most massive star supernovae. A second way to distinguish the two types of supernovae is to plot *light curves* that show how their luminosity fades with time. Figure 15.4 contrasts typical light curves for white dwarf and massive star supernovae. Note that both types of supernovae reach a peak luminosity of about 10 billion Suns ($10^{10}L_{Sun}$), but white dwarf supernovae fade steadily, while massive star supernovae fade in two distinct stages. Not only are white dwarf supernovae dramatic, but they also provide one of the primary means by which we measure large distances in the universe. Massive star

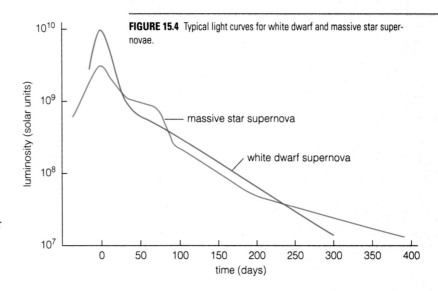

FIGURE 15.4 Typical light curves for white dwarf and massive star supernovae.

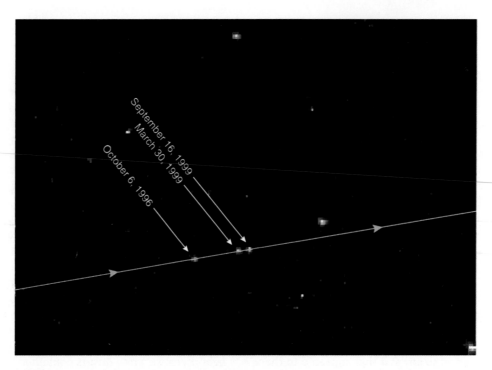

FIGURE 15.5 This composite image from the Hubble Space Telescope shows a neutron star streaking through our galaxy. The neutron star, created in a supernova explosion about a million years ago, glows in optical light because it is still very hot—about 600,000 K.

supernovae differ in intrinsic brightness, but white dwarf supernovae always occur in white dwarfs that have just reached the $1.4M_{Sun}$ limit. The light curves of all white dwarf supernovae therefore look amazingly similar, and their maximum luminosities are nearly identical. This fact is extremely useful: Once we know the true luminosity of one white dwarf supernova, we essentially know the luminosities of them all. Thus, whenever we discover a white dwarf supernova in a distant galaxy, we can determine the galaxy's distance by using the luminosity–distance formula [Section 13.2].

15.3 Neutron Stars

White dwarfs with densities of 5 tons per teaspoon may seem incredible, but neutron stars are stranger still. A **neutron star** is the ball of neutrons created by the collapse of the iron core in a massive star supernova (Figure 15.5). Typically just 10 kilometers across yet more massive than the Sun, neutron stars are essentially giant atomic nuclei, with two important differences: (1) They are made almost entirely of neutrons, and (2) gravity, not the strong force, is what binds them together.

The force of gravity at the surface of a neutron star is truly awe-inspiring. Escape velocity is about half

the speed of light. The strong gravity causes photons to emerge with a *gravitational redshift* that increases their wavelengths by about 30%, just as predicted by Einstein's general theory of relativity. If the Earth were visited by a neutron star, it would be squashed into a shell no thicker than your thumb on the surface of the neutron star.

Pulsars

Theorists first speculated about neutron stars in the 1930s. But until observational proof of their existence arrived in 1967, most astronomers assumed that neutron stars were too strange to exist. The proof came largely through the efforts of Jocelyn Bell. In October 1967 she detected *pulses* of radio waves coming from somewhere near the direction of the constellation Cygnus at precise 1.337301-second intervals (Figure 15.6). No known astronomical object pulsated so regularly. In fact, the pulsations came at such precise intervals that they were nearly as reliable for measuring time as the most precise human-made clocks. For a while, the mysterious source of the radio waves was dubbed "LGM" for Little Green Men, only half-jokingly. Today we refer to such rapidly pulsing radio sources as **pulsars**.

The mystery of pulsars was soon resolved. By the end of 1968, astronomers had found two smoking guns: Pulsars sat at the centers of both the Crab

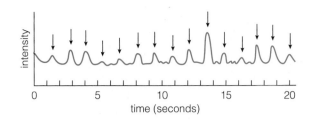

FIGURE 15.6 About 20 seconds of data from the first pulsar discovered by Jocelyn Bell in 1967. Arrows mark the pulses, which come precisely 1.337301 seconds apart.

Nebula and the Vela Nebula, the gaseous remnants of supernovae (Figure 15.7). The pulsars are neutron stars left behind by the supernova explosions.

The pulsations arise because the neutron star is spinning rapidly as a result of the conservation of angular momentum: As an iron core collapses into a neutron star, its rotation rate must increase as it shrinks in size. The collapse also bunches the magnetic field lines running through the core far more tightly, greatly amplifying the strength of the magnetic field. Shortly after the supernova event, the magnetic field of the remaining neutron star is a trillion times stronger than the Earth's. These intense magnetic fields somehow direct beams of radiation out along the magnetic poles, although we do not yet know exactly how the radiation is generated. If a neutron star's magnetic poles do not align with its rotation axis (just as the Earth's magnetic poles do not coincide with its geographic poles), the beams of radiation sweep round and round (Figure 15.8). Like lighthouses, these neutron stars actually emit a fairly steady beam of light, which we see as pulses of light each time the beam sweeps past Earth.

Pulsars are not quite perfect clocks, because each revolution of a pulsar takes slightly longer. This gradual slowing of the neutron star's spin occurs because the continual twirling of the magnetic field generates electromagnetic radiation that carries away energy and angular momentum. The pulsar in the Crab Nebula, for example, currently spins about 30 times per second. Two thousand years from now, it will spin less than half as fast. Eventually, a pulsar's spin slows so much and its magnetic field weakens so much that we can no longer detect it. In addition, some spinning neutron stars may be oriented so that their beams do not sweep past our location. Thus, we have the following rule: All pulsars are neutron stars, but not all neutron stars are pulsars.

TIME OUT TO THINK *Suppose we do not see pulses from a particular neutron star and hence do not call it a pulsar. Is it possible that a civilization living in some other star system would see this neutron star as a pulsar? Explain.*

We know that pulsars must be neutron stars, because no other massive object could spin so fast. A white dwarf, for example, can spin no faster than about once per second; an increase in spin would tear it apart because its surface would be rotating faster than the escape velocity. Yet pulsars have been discovered that rotate as fast as 625 times per second. Only an object as extremely small and dense as a neutron star could spin so fast without breaking apart.

Neutron Stars in Close Binary Systems

Like their white dwarf brethren, neutron stars in close binary systems can brilliantly burst back to life. As is the case with white dwarfs, gas overflowing from a companion star can create a hot, swirling accretion disk around the neutron star. However, in the neutron star's mighty gravitational field, infalling matter releases an amazing amount of gravitational potential energy. Dropping a brick onto a neutron star would liberate as much energy as an atomic bomb. Because the gravitational energies of accretion disks around neutron stars are so tremendous, they are much hotter and much more luminous than those around white dwarfs. The high temperatures in the inner regions of the accretion disk make it radiate

FIGURE 15.7 This time-lapse image of the pulsar at the center of the Crab Nebula shows its main pulse recurring every 0.033 second. The fainter pulses are thought to come from the pulsar's other lighthouselike beam. (Photo from the Very Large Telescope of the European Southern Observatory.)

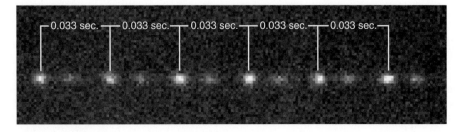

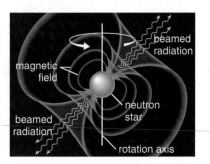

a A pulsar is a rapidly rotating neutron star that beams radiation along its magnetic axis.

FIGURE 15.8 Radiation from a neutron star can appear to pulse if the neutron star is rotating.

b This artwork likens a pulsar (top) to a lighthouse (bottom). If a pulsar's radiation beams are not aligned with its rotation axis, they will sweep through space. Each time one of these beams sweeps across Earth, we see a pulse of radiation.

powerfully in X rays. Some close binaries with neutron stars emit 100,000 times more energy in X rays than our Sun emits in all wavelengths of light combined. Due to this intense X-ray emission, close binaries that contain accreting neutron stars are often called **X-ray binaries**. Their existence confirms many of the strange properties of neutron stars. Today we know of hundreds of X-ray binaries in the Milky Way and they are concentrated in the disk, just like most of our galaxy's stars, gas, and dust.

Like accreting white dwarfs that occasionally erupt into novae, accreting neutron stars sporadically erupt with enormous luminosities. Hydrogen-rich material from the companion star builds up on the surface of the neutron star. Pressures at the bottom of this accreted hydrogen shell, only a meter thick, maintain steady fusion. Helium accumulates beneath the hydrogen-burning shell, and helium fusion suddenly ignites when the temperature builds to 100 million K. The helium burns rapidly to carbon and heavier elements, generating a burst of energy that flows from the neutron star in the form of X rays. These **X-ray bursters** typically flare every few hours to every few days. Each burst lasts only a few seconds, but during that short time the system radiates power equivalent to 100,000 Suns, all in X rays. Within a minute after a burst, the X-ray burster cools back down and resumes accreting.

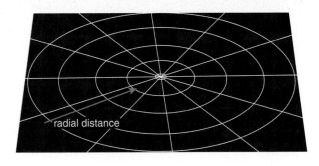

a A two-dimensional representation of "flat" spacetime. The circumference of each circle is 2π times its radius.

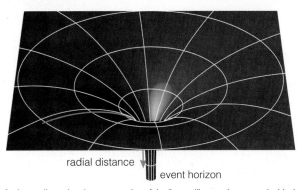

b A two-dimensional representation of the "curved" spacetime around a black hole. The black hole's mass distorts spacetime, making the radial distance between two circles larger than it would be in "flat" spacetime.

FIGURE 15.9 Spacetime is strongly curved near a black hole.

15.4 Black Holes: Gravity's Ultimate Victory

We know that white dwarfs cannot exceed $1.4 M_{Sun}$ because gravity overcomes electron degeneracy pressure above that mass. The mass of a neutron star has a similar limit. The precise *neutron star limit* is not known, but it certainly lies below $3 M_{Sun}$. A collapsing stellar core that weighs more than $3 M_{Sun}$ faces the ultimate oblivion: becoming a **black hole**.

A black hole originates in the collapse of the iron core that forms just prior to the supernova of a very high mass star. Any star born with more than about $8 M_{Sun}$ dies in a supernova, but most of the star's mass is blown into space by the explosion. As a result, the collapsed cores left behind by most supernovae are neutron stars. However, theoretical models show that the most massive stars might not succeed in blowing away all their upper layers. If enough matter falls back onto the core, raising its mass above the neutron star limit, neutron degeneracy pressure will not be able to fend off gravity.

A core whose mass exceeds the neutron star limit will continue to collapse catastrophically. Then, when the core has only a fraction of a second left, gravity plays its cruelest trick. Usually, the gravitational potential energy released as a star collapses boosts its temperature and pressure enough to fight off gravity. However, in a star destined to become a black hole, the enhanced temperature and pressure just make gravity stronger. According to Einstein's theory of relativity, energy is equivalent to mass ($E = mc^2$) and thus must exert some gravitational attraction. The gravity of pure energy usually is negligible, but not in a stellar core collapsing beyond the neutron star limit. Here the energy associated with the temperature and pressure concentrated in the tiny core acts like additional mass, hastening the collapse. To the best of our current understanding, *nothing* can halt the crush of gravity. The core collapses without end, forming a black hole. Gravity has achieved its ultimate triumph.

A Hole in the Universe

The idea of a black hole was first suggested in the late 1700s by British philosopher John Mitchell and French physicist Pierre Laplace. It was already known from Newton's laws that the escape velocity from any object depends only on its mass and size: More compact objects of a particular mass have higher escape velocities [Section 5.5]. Mitchell and Laplace speculated about objects so compact that their escape velocity exceeded the speed of light. Because they worked in a time before it was known that light always travels at the same speed, they assumed that light emitted from such an object would behave like a rock thrown upward, slowing to a stop and falling back down.

Einstein's work showed that black holes are considerably more bizarre. He found that space and time are not distinct, as we usually think of them, but instead are bound up together as four-dimensional **spacetime**. Moreover, in his general theory of relativity, Einstein showed that what we perceive as gravity arises from *curvature of spacetime*. The concept "curvature of spacetime" is quite strange, because we cannot even visualize four dimensions, let alone visualize their curvature. However, we can draw an analogy to spacetime curvature with a "rubber sheet" diagram in which we show how different masses would affect a stretched rubber sheet. In this analogy, a black hole is a region in which spacetime is stretched so far that it becomes a bottomless pit (Figure 15.9). Keep in mind that a black hole is not really shaped like a funnel; the illustration is only a two-dimensional analogy. Black holes are actually spherical, and they are black because not even light can escape from them.

A black hole really is a *hole* in the observable universe, and the boundary between the inside of the black hole and the universe outside is called the **event horizon**. Within the event horizon, the escape velocity exceeds the speed of light, so nothing—not even light—can ever get out. The event horizon gets its name because information can never reach us from events that occur within it.

The radius of the event horizon is known as the **Schwarzschild radius**, named for Karl Schwarzschild (1873–1916), who computed it from Einstein's general theory of relativity in 1916. As the rubber-sheet analogy in Figure 15.9b shows, it is possible to draw a series of concentric circles around a black hole. However, it is not really possible to measure a *radius* for these circles. Because of the extreme curvature of spacetime near a black hole, their centers are within the event horizon and hence are not part of our observable universe. Thus, we define the radius of a circle around a black hole as the radius it *would* have if geometry were flat, as shown in Figure 15.9a. (That is, radius = circumference ÷ 2π.) The Schwarzschild radius of a black hole depends only on its mass: Black holes of greater mass have a larger event horizon and thus a larger Schwarzschild radius.

There is no way to tell what has fallen into a black hole in the past. A black hole that forms from the collapse of a stellar core has the mass of that core, but no recognizable material remains. Because stellar cores normally rotate, conservation of angular momentum dictates that black holes should be rotating rapidly when they form. Aside from mass and angular momentum, the only other measurable property of a black hole is its electrical charge, and most black holes are probably electrically neutral. Any other information carried by objects that plunge into a black hole is irrevocably lost from the universe.

Voyage to a Black Hole

Imagine that you are a pioneer of the future, making the first visit to a black hole. You've selected a black hole with a mass of $10M_{Sun}$ and a Schwarzschild radius of 30 km. As your spaceship approaches the black hole, you fire its engines to put the ship on a circular orbit a few thousand kilometers above the event horizon. Note that this orbit will be perfectly stable—there is no need to worry about getting "sucked in."

Your first task is to test Einstein's general theory of relativity. This theory predicts that time should run more slowly as the force of gravity grows stronger. It also predicts that light coming out of a strong gravitational field should show a redshift, called a *gravitational redshift,* that is due to gravity rather than to the Doppler effect. You test these predictions with the aid of two identical clocks whose numerals glow with blue light. You keep one clock aboard the ship and push the other one, with a small rocket attached, directly toward the black hole (Figure 15.10). The small rocket automatically fires its engines just enough so that the clock falls only gradually toward the event horizon. Sure enough, the clock on the rocket ticks slower as it heads toward the black hole, and its light becomes increasingly redshifted. When the clock reaches a distance of about 10 km above the event horizon, you see it ticking only half as fast as the clock on your spaceship, and its numerals are red instead of blue.

The rocket has to expend fuel rapidly to keep the clock hovering in the strong gravitational field, and it soon runs out of fuel. The clock plunges toward the black hole. From your safe vantage point inside the spaceship, you see the clock ticking slower and slower as it falls. However, you soon need a radio telescope to "see" it, as the light from the clock face shifts from the red part of the visible spectrum, through the infrared, and on into the radio. Finally, its light is so far redshifted that no conceivable telescope could detect it. As the clock vanishes from view, you see that the time on its face has frozen to a stop.

Curiosity overwhelms the better judgment of one of your colleagues. He hurriedly climbs into a space suit, grabs the other clock, resets it, and jumps out of the airlock on a trajectory aimed straight for the black hole. Down he falls, clock in hand. He watches the clock, but because he and the clock are traveling together, its time seems to run normally and its numerals stay blue. From his point of view, time seems to neither speed up nor slow down. In fact, he'd say that *you* were the one with the strange time, as he would see your time running increasingly fast and your light becoming increasingly blueshifted. When the clock reads, say, 00:30, he and his clock pass

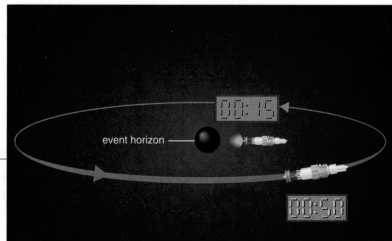

FIGURE 15.10 Time runs slower on the clock nearer to the black hole, and gravitational redshift makes its glowing blue numerals appear red from your spaceship.

through the event horizon. There is no barrier, no wall, no hard surface. The event horizon is a mathematical boundary, not a physical one. From his point of view, the clock keeps ticking. He is inside the event horizon, the first human being ever to leave our universe.

Back on the spaceship, you watch in horror as your overly curious friend plunges to his death. Yet, from your point of view, he will *never* cross the event horizon. You'll see time come to a stop for him and his clock as he vanishes from view due to the huge gravitational redshift of light. When you return home, you can play a videotape for the judges at your trial, proving that your friend is still a part of our universe. Strange as it may seem, all of this is true according to Einstein's theory. From your point of view, your friend takes *forever* to cross the event horizon; from his point of view, it is but a moment's plunge before he leaves the universe.

The truly sad part of this story is that your friend did not live to experience the crossing of the event horizon. The force of gravity he felt as he approached the black hole grew so quickly that it actually pulled much harder on his feet than on his head, simultaneously stretching him lengthwise and squeezing him from side to side (Figure 15.11). In essence, your friend was stretched in the same way the oceans are stretched by the tides, except that the *tidal force* near the black hole is trillions of times stronger than the tidal force of the Moon on the Earth [Section 5.4]. No human could survive it.

If he had thought ahead, your friend might have waited to make his jump until you visited a much larger black hole, like one of the *supermassive black holes* thought to reside in the center of many galaxies [Section 17.6]. A 1 billion M_{Sun} black hole has a Schwarzschild radius of 3 billion kilometers—about the distance from our Sun to Uranus. Although the gravitational forces at the event horizon of all black holes are equally great, the larger size of the supermassive black hole makes its tidal forces much weaker and hence nonlethal. Your friend could safely plunge through the event horizon. Again, from your point of view, the crossing would take forever, and you would see time come to a stop for him just as he vanished from sight because of the gravitational redshift. Again, he would experience time running normally, and he would see time in the outside universe running increasingly fast as he approached the event horizon. Unfortunately, anything he learned as he watched the future of the universe unfold would do him little good as he plunged to oblivion inside the black hole.

Singularity and the Limits to Knowledge

The center of a black hole is thought to be a place where gravity crushes all matter to an infinitely tiny and infinitely dense point called a **singularity**. This

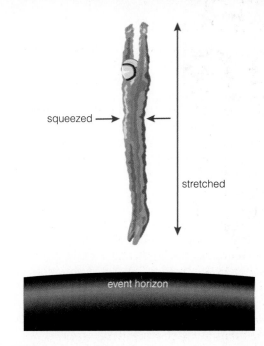

FIGURE 15.11 Tidal forces would be lethal near a black hole formed by the collapse of a star. The black hole would pull more strongly on the astronaut's feet than on his head, stretching him lengthwise and squeezing him from side to side.

singularity is the point at which all the mass that created the black hole resides.

We can never know what really happens inside a black hole, because no information can ever emerge from within the event horizon. Nevertheless, we can use Einstein's theory of relativity to predict conditions inside the black hole, as long as we don't try to describe conditions too close to the singularity. The singularity itself is more puzzling, because the equations of modern physics yield conflicting predictions when we try to apply them to an infinitely compressed mass. General relativity predicts that spacetime should grow infinitely curved as it enters the pointlike singularity. Quantum physics predicts that spacetime should fluctuate chaotically in regions smaller than 10^{-35} meter across. No current theory adequately accommodates these divergent claims. We will not fully understand how a singularity behaves until we have developed a quantum theory of gravity that encompasses both general relativity and quantum mechanics. This uncertainty in our current knowledge is a gold mine for science fiction writers, who speculate about using black holes for exotic forms of travel through spacetime.

Evidence for Black Holes

Have we discovered any black holes yet? We think so, but the evidence is indirect. Black holes emit no light. They are impossible to see, so we must look for their effects on surrounding matter. Black holes in close binaries should be among the easiest to iden-

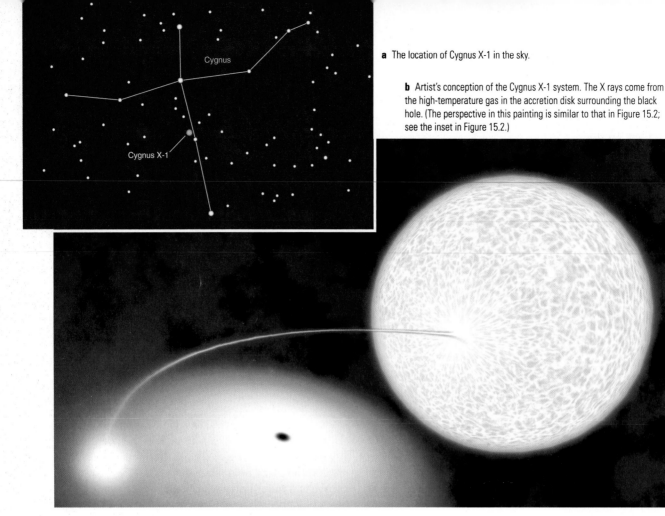

a The location of Cygnus X-1 in the sky.

b Artist's conception of the Cygnus X-1 system. The X rays come from the high-temperature gas in the accretion disk surrounding the black hole. (The perspective in this painting is similar to that in Figure 15.2; see the inset in Figure 15.2.)

FIGURE 15.12 Cygnus X-1, a binary system containing a black hole candidate.

tify. Gas overflowing a black hole's stellar companion will form a hot, X ray–emitting accretion disk similar to the disks that circle accreting neutron stars. The X rays can escape because the disk emits them from well outside the event horizon.

We strongly suspect that a few X-ray binaries contain black holes rather than neutron stars. The trick is to tell the difference, because the accretion disk is likely to be just as hot and to emit about as many X rays whether it circles a neutron star or a black hole. However, we can distinguish between the two possibilities if we can measure the mass of the accreting object. One of the most promising *black hole candidates* is in an X-ray binary called Cygnus X-1 (Figure 15.12). This system contains an extremely bright star with an estimated mass of $18M_{Sun}$; based on Doppler shifts of its spectral lines, this star orbits an unseen companion with a mass of about $10M_{Sun}$. Although there is some uncertainty in these mass estimates, the mass of the invisible accreting object clearly exceeds the $3M_{Sun}$ neutron star limit. More-over, careful studies of variations in the system's X-ray emission indicate that this massive accreting object must be very small in size—far too small to

be an ordinary star [Section 17.6]. Thus, based on our current knowledge, the accreting object in Cygnus X-1 cannot be anything other than a black hole.

TIME OUT TO THINK *As we discussed earlier, some X-ray binaries that contain neutron stars emit frequent X-ray bursts and are called X-ray bursters. Could an X-ray binary that contains a black hole exhibit the same type of X-ray bursts? Why or why not? (Hint: What is the source of the X-ray bursts from an X-ray binary with a neutron star, and where is it located? Does a similar location exist for a system containing a black hole?)*

Of course, confirming that black holes are real with 100% certainty is very difficult. But our current theories successfully explain neutron stars, and the general theory of relativity that leads to the idea of black holes is also on solid ground. Unless something is dramatically wrong in our current theories about the mass limit of neutron stars or some other, un-known, type of compact object can have a huge mass, black holes must be real.

What would happen if our Sun suddenly became a black hole? For some reason, it has become part of our popular culture for most people to believe that Earth and the other planets would inevitably be "sucked in" by the black hole. This is not true. Although the sudden disappearance of light and heat from the Sun would be bad news for life, the Earth's orbit would not change. Newton's law of gravity tells us that the allowed orbits in a gravitational field are ellipses, hyperbolas, and parabolas [Section 5.3]; note that "sucking" is not on the list! A spaceship would get into trouble only if it came so close to a black hole—within about three times its Schwarzschild radius—that gravity would deviate significantly from Newton's law. Otherwise, a spaceship passing near a black hole would simply swing around it on an ordinary orbit (ellipse, parabola, or hyperbola). In fact, because most black holes are so small—typical Schwarzschild radii are smaller than any star or planet, and smaller even than most asteroids—a black hole is actually one of the most difficult things in the universe to fall into by accident.

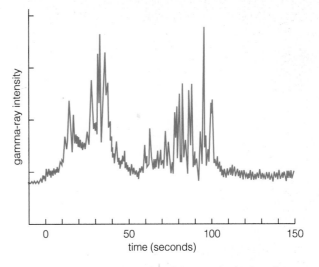

FIGURE 15.13 A typical gamma-ray burst light curve, showing dramatic fluctuations in gamma-ray intensity over a period of two minutes.

15.5 The Mystery of Gamma-Ray Bursts

In the early 1960s, the United States began launching a series of top-secret satellites designed to look for gamma rays emitted by nuclear bomb tests. The satellites soon began detecting occasional bursts of gamma rays, typically lasting a few seconds (Figure 15.13). It took several years for military scientists to become convinced that these **gamma-ray bursts** were coming from space, not from some sinister human activity. They publicized the discovery in 1973.

An increasingly large armada of satellites launched over the next two decades detected hundreds of other gamma-ray bursts. However, not only was their origin unknown, but it was very difficult to tell what direction the bursts came from. The problem is that gamma rays are very difficult to focus. A detector can record that it has been hit by a gamma ray but can provide little information on the direction of the gamma ray. With no specific proof, most astronomers assumed that gamma-ray bursts, like X-ray bursts, came from explosive events associated with neutron stars in X-ray binaries.

In 1991, NASA launched the *Compton Gamma Ray Observatory,* or *Compton* for short, which carried an array of eight detectors designed expressly to study gamma-ray bursts. By comparing the data recorded by all eight detectors, scientists could determine the direction of a gamma-ray burst within about 1°. The results were stunning. Compton detected gamma-ray bursts at a rate of about one per day and compiled a catalog of more than a thousand gamma-ray bursts within a few years. However, these bursts were *not* concentrated in the disk of the Milky Way like X-ray binaries and thus must not be associated with neutron stars in the disk of our galaxy.

So where do gamma-ray bursts come from? The very even distribution across the sky of gamma-ray bursts discovered by Compton ruled out the possibility that they come from anywhere in the Milky Way Galaxy. If the bursts came from objects distributed spherically about the Milky Way Galaxy, we would see a concentration of them in the direction of the galactic center. (Remember that we are located more than halfway out from the center of our galaxy.) Gamma-ray bursts must therefore originate far outside our own galaxy. Additional evidence for this conclusion came in 1997, when astronomers first observed the afterglow of a gamma-ray burst in other wavelengths. The higher resolution possible in these other wavelengths allowed astronomers to pinpoint the burst's origin in a distant galaxy. Since then, several other bursts have also been traced to explosions in distant galaxies (Figure 15.14).

We now know that gamma-ray bursts come from very distant explosions, but we still face a great mystery: How can something so distant be so bright? The afterglows of some gamma-ray bursts can be seen with binoculars, even though they are coming from galaxies billions of light-years away—making them by far the most powerful bursts of energy that ever occur in the universe. If these bursts shine their light equally in all directions, like a light bulb, then the total luminosity of a burst can briefly exceed the combined luminosity of a million galaxies like our Milky

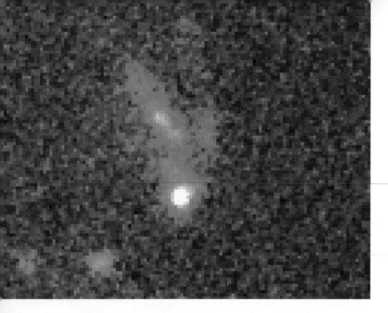

FIGURE 15.14 The bright dot near the center of this image is the visible-light afterglow of a gamma-ray burst, as seen by the Hubble Space Telescope. The elongated blob extending above the dot is the distant galaxy in which the burst occurred.

Way! Because such a high luminosity is very difficult to explain, some scientists speculate that gamma-ray bursts channel their energy into narrow searchlight beams, like pulsars. A burst whose beam is pointed directly at Earth would look unusually bright, but even in this arrangement the burst's luminosity would still surpass that of many thousands of galaxies like our Milky Way.

Astronomers still do not know what causes such massive outbursts of energy. One hypothesis suggests that the bursts come from the collision of *two* neutron stars in a binary system. However, why a neutron star collision would produce the peculiar spectra and light curves of gamma-ray bursts is not clear, and many scientists have raised theoretical objections to this model.

Another hypothesis, currently favored by many astronomers, is that gamma-ray bursts come from unusually powerful supernovae. An ordinary supernova that forms a neutron star does not release enough energy to power the luminosity of the brightest gamma-ray bursts. However, a supernova that forms a black hole crushes even more matter into an even smaller radius, releasing many times more gravitational potential energy than one that forms a neutron star. This kind of event, sometimes called a *hypernova,* might be powerful enough to explain the most extreme gamma-ray bursts. Some circumstantial evidence supports this idea. The extremely massive stars that lead to hypernovae are very short-lived and should be found only in places where stars are actively forming. Indeed, all the distant galaxies known to have produced gamma-ray bursts also appear to be forming new stars. In addition, at least one rela-

tively nearby gamma-ray burst appears to be linked to a supernova explosion. These observations have inspired many theoretical models that attempt to explain how hypernovae might generate gamma-ray bursts, but none has met with much success. Thus, we are left with one of the greatest mysteries in all of science: Gamma-ray bursts are the most powerful events in the universe, but we don't know what causes them.

THE BIG PICTURE

We have now seen what happens to stars after they die. What a mind-bending experience! Nevertheless, try to keep these "big picture" ideas straight in your head:

- Despite the strange nature of stellar corpses, clear evidence exists for white dwarfs and neutron stars, and the case for black holes is very strong.

- White dwarfs, neutron stars, and black holes can all have close stellar companions from which they accrete matter. These binary systems produce some of the most spectacular events in the universe, including novae, white dwarf supernovae, and X-ray bursters.

- Black holes are truly holes in the observable universe that strongly warp space and time around them. The nature of black hole singularities remains beyond the frontier of current scientific understanding.

- Gamma-ray bursts were once thought to be related to neutron stars in our galaxy, but recent evidence indicates that this idea is wrong. Their origin is one of the greatest mysteries in the universe.

Review Questions

1. What is the difference between *electron degeneracy pressure* and *neutron degeneracy pressure?* Which type supports a white dwarf? Why are neutron stars so much smaller in size than white dwarfs?

2. With degeneracy pressure, how are the speeds of electrons (or neutrons) related to the mass of the degenerate object? Explain why this relationship between speed and mass implies upper limits on the possible masses of white dwarfs and neutron stars. What is the *white dwarf limit?* What is the neutron star limit?

3. What is an *accretion disk?* Under what conditions does an accretion disk form? Explain how the accretion disk provides a white dwarf with a new source of energy that we can detect from Earth.

4. Describe the process of a *nova*. Why do novae only occur in close binary star systems? Can the same star system undergo a nova event more than once? Explain.

5. Contrast the process of a *white dwarf supernova* with that of a *massive star supernova*. Observationally, how can we distinguish between these two types of supernovae?

6. What is a *pulsar*? Briefly describe how pulsars were discovered and how they get their name. How do we know that pulsars must be neutron stars?

7. Explain why neutron stars in close binary systems are often called *X-ray binaries*.

8. What is an *X-ray burster*? How is the process of an X-ray burst similar to that of a nova? How do X-ray bursts differ from novae observationally?

9. Briefly explain the process by which the core of a very high mass star can collapse to form a *black hole*.

10. What is the *event horizon* of a black hole? How does it get its name? How is it related to the *Schwarzschild radius*?

11. Suppose you are orbiting a black hole. Describe how you will see the passage of time on an object approaching the black hole. What will you notice about light from the object? Explain why an object falling toward the black hole will eventually fade from view yet never cross the event horizon from your point of view.

12. Suppose you are falling into a black hole. How will you perceive the passage of your own time? How will you perceive the passage of time in the universe around you? Briefly explain why your trip will be lethal if the black hole is relatively small in mass, and why you may survive to cross the event horizon of a supermassive black hole.

13. What do we mean by the *singularity* of a black hole? How do we know that our current theories are inadequate to explain what happens at the singularity?

14. Briefly describe the observational evidence supporting the idea that Cygnus X-1 contains a black hole.

15. What are *gamma-ray bursts*? Why are they mysterious?

Discussion Questions

1. *Too Strange to Be True?* Despite strong theoretical arguments for their existence, many scientists rejected the possibility that neutron stars or black holes could really exist until they were confronted with very strong observational evidence. Some people claim that this type of scientific skepticism demonstrates that scientists are too unwilling to give up their deeply held scientific beliefs. Others claim that this type of skepticism is necessary for scientific advancement. What do you think? Defend your opinion.

2. *Black Holes in Popular Culture.* Phrases such as "it disappeared into a black hole" are now common in popular culture. Give a few examples in which the term *black hole* is used in popular culture but is not meant to be taken literally. In what ways are these uses correct in their analogies to real black holes? In what ways are they incorrect? Why do you think such an esoteric scientific idea as that of a black hole has so captured the public imagination?

Problems

Sensible Statements? For **problems 1–7**, decide whether the statement is sensible and explain why it is or is not.

1. Most white dwarf stars have masses close to that of our Sun, but a few white dwarf stars are up to three times more massive than the Sun.

2. White dwarf supernovae are useful distance indicators.

3. Before pulsars were discovered, no one knew for sure whether neutron stars existed.

4. If you want to find a pulsar, you might want to look near the remnant of a supernova described by ancient Chinese astronomers.

5. If a black hole 10 times more massive than our Sun were lurking just beyond Pluto's orbit, we'd have no way of knowing it was there.

6. If the Sun suddenly became a 1 M_{Sun} black hole, the orbits of the nine planets would not change at all.

7. We can detect black holes with X-ray telescopes because matter falling into a black hole emits X rays after it smashes into the event horizon.

Life Stories of Stars. Write a one- to two-page short story for the scenarios in **problems 8 and 9**. Each story should be detailed and scientifically correct, but also creative. That is, it should be entertaining while at the same time proving that you understand stellar evolution. Be sure to state whether "you" are a member of a binary system.

8. You are a white dwarf of 0.8M_{Sun}. Tell your life story.

9. You are a neutron star of 1.5M_{Sun}. Tell your life story.

Web Projects

Find useful links for Web projects on the text Web site.

1. *Gamma-Ray Bursts.* Go to the Web site for NASA's High Energy Transient Experiment (HETE) or another mission studying gamma-ray bursts and find the latest information about gamma-ray bursts. Write a one- to two-page essay on recent discoveries and how they may shed light on the mystery of gamma-ray bursts.

2. *White Dwarf Supernovae.* Learn more about how astronomers are using white dwarf supernovae to determine distances for distant galaxies. What have these observations taught us about the universe? What have they taught us about white dwarf supernovae?

PART V
Galaxies and Beyond

CHAPTER 16

Our Galaxy

In previous chapters, we saw how stars forge new elements and expel them into space. We also studied how interstellar gas clouds enriched with these stellar by-products form new stars and planetary systems. These processes do not occur in isolation. Instead, they are part of a dynamic system that acts throughout our Milky Way Galaxy.

You are probably familiar with the idea that all living species on Earth interact with one another and with the land, water, and air to form a large, interconnected ecosystem. In a similar way, but on a much larger scale, our galaxy is a nearly self-contained system that cycles matter from stars into interstellar space and back into stars again. The birth of our solar system and the evolution of life on Earth would not have been possible without this "galactic ecosystem."

In this chapter, we will study our Milky Way Galaxy. We will investigate the galactic processes that maintain an ongoing cycle of stellar life and death, examine the structure and motion of the galaxy, and explore the mysteries of the galactic center. Through it all, we will see that we are not only "star stuff" but "galaxy stuff"—the product of eons of complex recycling and reprocessing of matter and energy in the Milky Way Galaxy.

16.1 The Milky Way Revealed

On a dark night, you can see a faint band of light slicing across the sky through several constellations, including Sagittarius, Cygnus, Perseus, and Orion. This band of light looked like a flowing ribbon of milk to the ancient Greeks, and we now call it the *Milky Way*. In the early 1600s, Galileo used his telescope to prove that the light of the Milky Way comes from a myriad of individual stars. Together these stars make up the kind of stellar system we call a *galaxy*, echoing the Greek word for "milk," *galactos*.

Today we know that our Milky Way Galaxy holds over 100 billion stars and is just one among tens of billions of galaxies in the observable universe. If we could stand outside our galaxy, we would see it as a flat **disk** of stars with a bright central **bulge**, spectacular **spiral arms**, and a dimmer, rounder **halo** surrounding everything (Figure 16.1). A few hundred **globular clusters** of stars [Section 13.6] circle our galaxy's center in orbits extending tens of thousands of light-years into the halo. The Milky Way is a relatively large galaxy, so its gravity strongly influences smaller galaxies in its vicinity. Two small galaxies, known as the Large and Small Magellanic Clouds, orbit the Milky Way at distances of some 150,000 light-years (50 kpc). (A third small galaxy, called the Sagittarius Dwarf Elliptical, lies even closer but is obscured from view by the Milky Way's galactic plane.) Although the Magellanic Clouds are relatively small for galaxies (a few billion stars each), they are far larger than globular clusters (which typically contain a few hundred thousand stars). Both Magellanic Clouds are visible to the naked eye from the Southern Hemisphere.

Our knowledge of the Milky Way's true size and shape was long in coming. Clouds of interstellar gas and dust known collectively as the **interstellar medium** fill the galactic disk, obscuring our view when we try to peer directly through it. The smoggy nature of the interstellar medium hides most of our galaxy from us and long fooled astronomers into believing that we lived near our galaxy's center. Astronomer Harlow Shapley finally proved otherwise in the 1920s, when he demonstrated that the Milky Way's globular clusters orbit a point tens of thousands of light-years from our Sun. He concluded that this point, not our Sun, must be the center of the galaxy. We now know that our Sun lies near the outskirts of the galactic disk, about 28,000 light-years (8.5 kpc) from its center.

Our own galaxy can be difficult to study not only because it is dusty, but also because we see it from the inside. The Milky Way's dustiness is no longer such a hindrance, because technologies developed in the past few decades allow us to observe the Milky Way's radio and infrared light. These wavelengths penetrate the enshroud-

FIGURE 16.1 The Milky Way Galaxy

a Artist's conception of the Milky Way viewed from its outskirts.

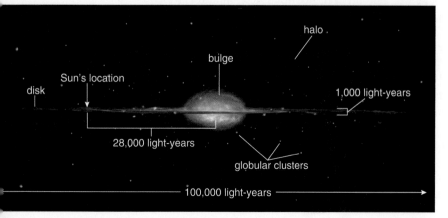

b Edge-on schematic view of the Milky Way.

ing interstellar medium, enabling us to see into regions of the galaxy previously obscured from view. Still, determining our galaxy's structure from our location within the galactic disk is somewhat like trying to draw a picture of your house without ever leaving your bedroom. Just as it is easier to draw pictures of other houses that you can see out the window, it is easier to measure the sizes and shapes of other galaxies than to measure our own. Nevertheless, the Milky Way is still the only galaxy whose inner workings we can examine up close.

We once thought of the Milky Way as a band of light, and later we saw it as a collection of stars. But now we see it revealed as a dynamic and complex system. In the rest of this chapter, we will see that our galaxy in many ways is like a forest of stars whose ecology is shaped by the cycle of stellar life and death and whose structure is determined by gravity and the majestic rotation of the galactic disk.

16.2 The Star–Gas–Star Cycle

Stars have formed, fused atomic nuclei, and exploded throughout the history of our galaxy. Generations of stars continually recycle the same galactic matter through their cores, gradually raising the overall abundance of elements made by fusion. Elements heavier than helium, usually called *heavy elements* by astronomers, now constitute about 2% of the galaxy's gaseous content; the overall composition of the galaxy is about 70% hydrogen, 28% helium, and 2% heavy elements by mass. The process of adding to the abundance of heavy elements, called **chemical enrichment**, is an inevitable by-product of continual star formation. However, the synthesis of elements is only one part of the galactic ecocycle. If the galaxy did not reincorporate these elements into new stars and their accompanying planetary systems, the mineral riches fused in the bellies of stars would be wasted.

TIME OUT TO THINK *Based on the idea of chemical enrichment, which types of stars must contain a higher proportion of heavy elements: stars in globular clusters or stars in open clusters? (Hint: Recall from Chapter 13 that stars in globular clusters are all very old, while stars in open clusters are relatively young.)*

Capturing the newly made elements released by stars is not an easy task. When a star explodes as a supernova, the ejected matter flies out at speeds of several thousand kilometers per second—far exceeding the escape velocity for the galaxy. Were it not for the interstellar medium, the new heavy elements released in the supernova would fly straight out of the Milky Way into intergalactic space. Instead, the blobs of matter expelled from the supernova collide with the interstellar medium, slow down, and eventually stop.

Supernovae of one star after another stir and heat the interstellar medium while feeding it new heavy elements. These elements eventually blend with the older, less chemically enriched hydrogen gas in the vicinity. However, before new stars can form, this gas must cool and form clouds. The hot gas cools first into clouds of atomic hydrogen (that is, neutral hydrogen atoms as opposed to ionized hydrogen or hydrogen molecules) and then into clouds of molecular hydrogen (H_2). The cooling of interstellar gas into clouds of molecular hydrogen takes millions of years. These clouds subsequently give birth to new stars more highly enriched in heavy elements, thus completing the star–gas–star cycle (Figure 16.2). Let's look at the stages of this cycle more closely.

Gas from Stars

All stars return much of their original mass to interstellar space in two basic ways: through stellar winds that blow throughout their lives, and through "death events" of planetary nebulae (for low-mass stars) or supernovae (for high-mass stars). Low-mass stars generally have weak stellar winds while they are on the main sequence, but their winds grow stronger and carry more material into space when they become red giants. By the time a low-mass star like the Sun ends its life with the ejection of a planetary nebula [Section 14.3], it has returned almost half its original mass to the interstellar medium (Figure 16.3).

High-mass stars lose mass much more dynamically and explosively. The powerful winds from supergiants and massive O and B stars recycle large amounts of matter into the galaxy. At the ends of their lives, these stars explode as supernovae. The high-speed gas ejected into space by these winds and supernovae sweeps up surrounding interstellar material, excavating a **bubble** of hot, ionized gas around the exploding star. Although the bubble in Figure 16.4 looks much like a soap bubble, it is actually an expanding shell of hot gas. The glowing surface is the edge of the expanding bubble, where gas piles up as the bubble sweeps outward through the interstellar medium. These hot, tenuous bubbles are quite common, filling roughly 20–50% of the Milky Way's disk. However, they are not always easy to detect. While some are hot enough to emit profuse amounts of X rays and others emit strongly in visible light, many others are evident only through radio emission from the shells of atomic hydrogen gas that surround them.

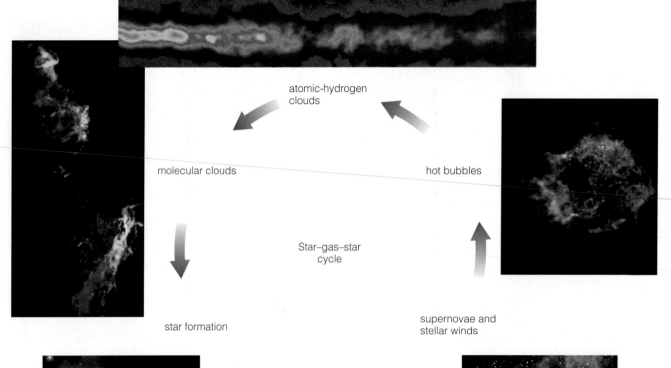

atomic-hydrogen
clouds

molecular clouds

hot bubbles

Star–gas–star
cycle

star formation

supernovae and
stellar winds

stellar burning/
heavy-element
formation

FIGURE 16.2 A pictorial representation of the star–gas–star cycle. (For descriptions of individual photos, proceeding counterclockwise from lower right, see Figures 16.4, 16.5, 16.11a, 16.8, 16.10, and 16.12.)

FIGURE 16.3 A dying low-mass star returns gas to the interstellar medium in a planetary nebula, as photographed by the Hubble Space Telescope.

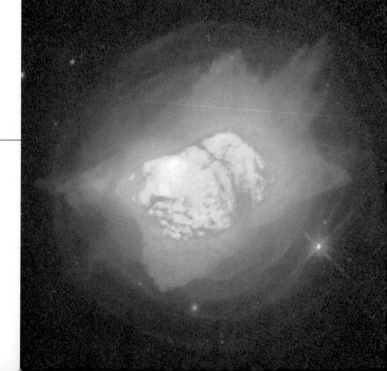

FIGURE 16.4 This photo shows a bubble of hot, ionized gas blown by a wind from the hot star at its center.

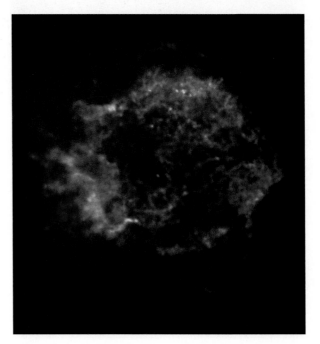

FIGURE 16.5 False-color photograph of X-ray emission from the hot gas in the young supernova remnant Cassiopeia A, as seen by the Chandra X-ray Observatory. Red indicates the lowest-energy X rays; blue indicates the highest-energy X rays.

Shock Waves and Supernova Remnants

Supernovae generate **shock waves**—waves of pressure created by gas moving faster than the speed of sound. A shock wave sweeps up surrounding gas as it travels, creating a "wall" of hot, fast-moving gas on its leading edge. When we observe a *supernova remnant,* we are seeing the aftermath of its shock wave [Section 14.4]. Figure 16.5 shows a young supernova remnant whose shocked gas is hot enough to emit X rays. In contrast, older supernova remnants are cooler because their shock waves have swept up more material and must share their energy among more particles (Figure 16.6). Eventually, the shocked gas radiates away most of its original energy, and the expanding wall of gas slows to subsonic speeds. As the energy dissipates and the gas cools, the supernova's cargo of new elements merges with the surrounding interstellar medium.

Multiple supernovae from a single star cluster can combine forces to create a huge cavity of hot gas over a thousand light-years (300 pc) wide called a **superbubble**. In many places in our galactic disk, we see what appear to be elongated bubbles extending from young clusters of stars to distances of 3,000 light-years (1 kpc) or more above the disk. These probably are places where superbubbles have grown so large that they cannot be contained within the disk of the Milky Way. Once the superbubble breaks out of the disk, where nearly all of the Milky Way's gas resides, nothing remains to slow its expansion except gravity. Such a *blowout* is in some ways similar to a volcanic eruption, but on a galactic scale: Hot plasma erupts from the disk and shoots high into the galactic halo (Figure 16.7).

Movie Madness: The Sound of Space

In many science fiction movies, a thunderous sound accompanies the demolition of a spaceship. If the moviemakers wanted to be more realistic, they would silence the explosion. On Earth, we perceive sound when sound waves—which are waves of alternately rising and falling pressure—cause trillions of gas atoms to push our eardrums back and forth. Although sound waves can and do travel through interstellar gas, the extremely low density of this gas means that only a handful of atoms per second would collide with something the size of a human eardrum. As a result, it would be impossible for a human ear (or a similar-size microphone) to register any sound. Despite the presence of sound waves and shock waves in space, the sound of space is silence.

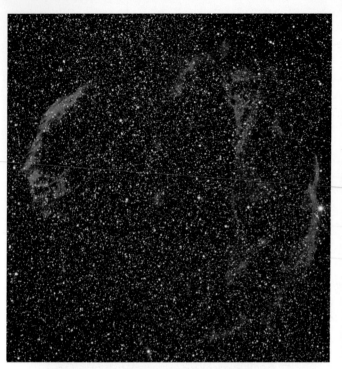

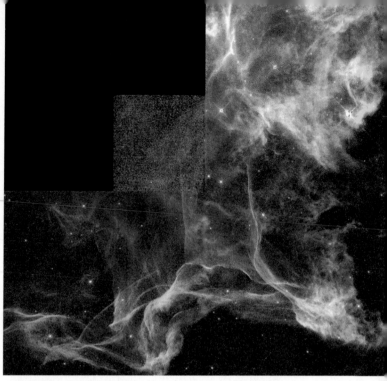

a This large-scale view shows the entire Cygnus Loop supernova remnant glowing in optical light. (The angular size of this remnant in our sky is six times that of the Moon.)

b The close-up view, from the Hubble Space Telescope, displays the fine filamentary structure of the remnant.

FIGURE 16.6 The optical emission from a supernova remnant, the Cygnus Loop.

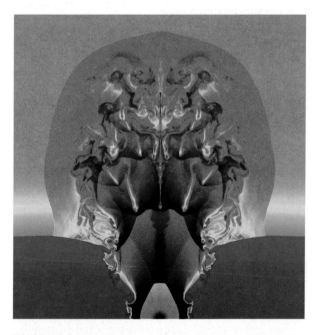

FIGURE 16.7 This image shows a supercomputer simulation of a superbubble blowout (blue = coldest gas; red = hottest gas). The cool gaseous disk of the simulated galaxy runs along the bottom of the picture. Supernova explosions at the center of this disk have ejected a plume of hot gas that rises high above the disk and into the hot halo.

Atomic Hydrogen Gas

The hot, ionized gas in bubbles and superbubbles is dynamic and widespread, but it is a relatively small fraction of the gas in the Milky Way. Atomic hydrogen gas is much more common and emits a spectral line with a wavelength of 21 cm in the radio portion of the electromagnetic spectrum. We see the radio emission from this **21-cm line** coming from all directions, telling us that about 5 billion solar masses of atomic hydrogen gas is distributed throughout the galactic disk.

Matter remains in the warm atomic hydrogen stage of the star–gas–star cycle for millions of years. Gravity slowly draws blobs of this gas together into tighter clumps, which radiate energy more efficiently as they grow denser. The blobs therefore cool and contract, forming the smaller clouds of cool atomic hydrogen. This process of shrinkage and coagulation takes a much longer time than the other steps in the journey from star death to star birth. The slowness of the transition from warm gas to cool clouds accounts for the large amount of gaseous matter in the atomic hydrogen stage of the star–gas–star cycle.

Remember that, although we speak of clouds of *hydrogen,* all interstellar material actually has a composition of about 70% hydrogen, 28% helium, and 2% heavy elements. Some of the heavy elements in regions of atomic hydrogen are in the form of tiny, solid **dust grains**: flecks of carbon and silicon minerals that resemble particles of smoke and form in the winds of red giant stars [Section 14.3]. Dust grains remain in the interstellar medium unless they are heated and destroyed by a passing shock wave or incorporated into a protostar. Although dust grains make up only about 1% of the mass of the atomic hydrogen clouds, they are responsible for the absorption of visible light that prevents us from seeing through the disk of the galaxy.

Molecular Clouds

As the temperature drops further in the center of a cool cloud of atomic hydrogen, hydrogen atoms combine into molecules, making a **molecular cloud**. Molecular clouds are the coldest, densest collections of gas in the interstellar medium, and they often congregate into *giant molecular clouds* that hold up to a million solar masses of gas. The total mass of molecular clouds in the Milky Way is somewhat uncertain, but it is probably about the same as the total mass of atomic hydrogen gas—about 5 billion solar masses. Throughout much of this molecular gas, temperatures hover only a few degrees above absolute zero, and gas densities are a few hundred molecules per cubic centimeter.

Molecular hydrogen (H_2) is by far the most abundant molecule in molecular clouds, but it is difficult to detect. As a result, most of what we know about molecular clouds comes from observing spectral lines of molecules that make up only a tiny fraction of a cloud's mass. The most abundant of these molecules is carbon monoxide (CO, also a common ingredient in car exhaust), which produces strong emission lines in the radio portion of the spectrum at the 10–30 K temperatures of molecular clouds (Figure 16.8). Among the more familiar are water (H_2O), ammonia (NH_3), and alcohol (CH_3OH).

Gravitational forces in molecular clouds gather molecules into the compact *cores* that eventually become protostars, and the final stages of star forma-

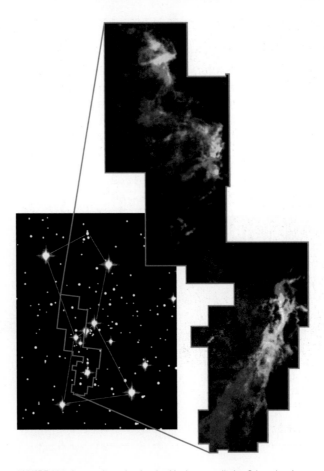

FIGURE 16.8 Image of a molecular cloud in the constellation Orion, showing its complex structure. The picture was made by measuring Doppler shifts of emission lines from carbon monoxide molecules in different locations. The colors indicate gas motions: Relative to the cloud as a whole, bluer parts are moving toward us and redder parts are moving away from us.

tion can be quite disruptive. Recall that protostars often have violent jets of gas spurting outward [Section 14.2]. The turbulence of these jets stirs up nearby regions of the molecular cloud, probably preventing other stars from forming in the vicinity of the growing protostar, at least for a while (Figure 16.9).

FIGURE 16.9 Jets from protostars disrupt nearby regions of molecular clouds. This photo shows a jet emanating from the protostar on the left. All along the jet in the center and on the far right of the photo, gas shocked by the jet is glowing in optical light.

Completing the Cycle

Once a few stars form in a cluster, their radiation begins to erode the surrounding gas in the molecular cloud. Ultraviolet photons from high-mass stars heat and ionize the gas, and winds and radiation pressure push the ionized gas away. This kind of feedback prevents much of the gas in a molecular cloud from turning into stars.

The process of molecular cloud erosion is sometimes spectacular. Figure 16.10 shows the *Eagle Nebula,* a complex of clouds where new stars are currently forming. The dark, lumpy columns are mo-

FIGURE 16.10 A portion of the Eagle Nebula as seen by the Hubble Space Telescope.

Table 16.1 Typical States of Gas in the Interstellar Medium

State of Gas	Primary Constituent	Approximate Temperature	Approximate Density (atoms per cm³)	Description
Hot bubbles	Ionized hydrogen	1,000,000 K	0.01	Pockets of gas heated by supernova shock waves
Warm atomic gas	Atomic hydrogen	10,000 K	1	Fills much of galactic disk
Cool atomic clouds	Atomic hydrogen	100 K	100	Intermediate stage of star–gas–star cycle
Molecular clouds	Molecular hydrogen	30 K	300	Regions of star formation
Molecular cloud cores	Molecular hydrogen	60 K	10,000	Star-forming clouds

lecular clouds. Off to the upper right (outside the picture), newly formed massive stars glow with ultraviolet radiation. This radiation sears the surface of the molecular clouds, destroying molecules and stripping electrons from atoms. As a result, matter "evaporates" from the molecular clouds and joins the hotter ionized gas encircling them. Only the densest knots of gas resist evaporation. Stars are forming in some of these dense knots, which remain compact while the rest of the cloud erodes. These star-forming knots are the tips of the dark "whiskers" protruding from the columns of molecular gas in Figure 16.10.

We have arrived back where we started in the star–gas–star cycle. The most massive stars now forming in the Eagle Nebula will explode within a few million years, filling the region with hot gas and newly formed heavy elements. Farther in the future, this gas will once again cool and coalesce into molecular clouds, forming new stars, new planets, and maybe even new civilizations.

Despite the recycling of matter from one generation of stars to the next, the star–gas–star cycle cannot go on forever. With each new generation, some of the galaxy's gas becomes permanently locked away in brown dwarfs that never return material to space and in stellar corpses left behind when stars die (white dwarfs, neutron stars, and black holes). The interstellar medium therefore is slowly running out of gas, and the rate of star formation will gradually taper off over the next 50 billion years or so. Eventually, star formation will cease.

Putting It All Together:
The Distribution of Gas in the Milky Way

As we look at different regions of the galaxy, we see various stages of the star–gas–star cycle playing themselves out. Because the cycle proceeds over such a long period of time compared to a human lifetime, each stage appears to us as a snapshot. We therefore see the interstellar medium in a wide variety of manifestations, ranging from the tenuous million-degree gas of bubbles to the cold, dense gas of molecular clouds. Table 16.1 summarizes the different states in which we see interstellar gas in the galactic disk.

Figure 16.11 shows seven views of the disk of the Milky Way Galaxy. Each view represents a panorama made by photographing the Milky Way's disk in every direction from Earth. You can visualize how one of these views corresponds to the sky by imagining cutting it out, bringing its ends together to form a circular band, and then lining up the band with the Milky Way on a model of the celestial sphere. Each view shows the Milky Way as it appears in a different set of wavelengths of light and thus reveals different features of the galactic disk.

■ Figure 16.11a shows variations in the intensity of radio emission from the 21-cm line of atomic hydrogen. Thus, it maps the distribution of atomic hydrogen gas, demonstrating that this gas fills much of the galactic disk.

■ Figure 16.11b shows variations in the intensity of radio emission lines from carbon monoxide (CO) and therefore maps the distribution of molecular clouds. Note that these cold, dense clouds are concentrated in a narrow layer near the midplane of the galactic disk.

■ Figure 16.11c shows variations in the intensity of infrared emission (at wavelengths of 60–100 micrometers) from interstellar dust grains. Note that the regions of strongest emission from dust correspond to the locations of molecular clouds in Figure 16.11b.

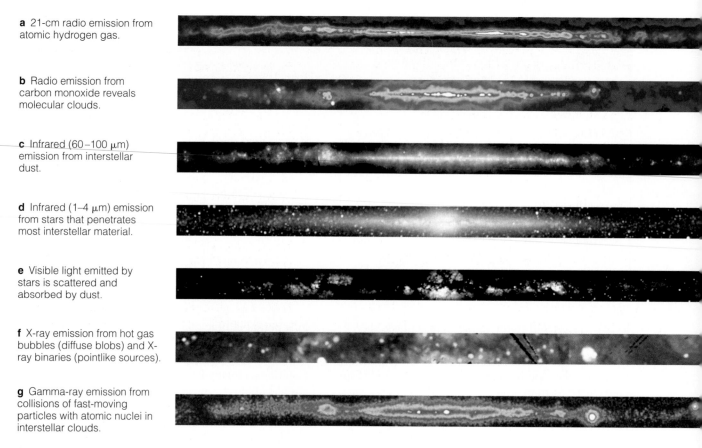

a 21-cm radio emission from atomic hydrogen gas.

b Radio emission from carbon monoxide reveals molecular clouds.

c Infrared (60–100 μm) emission from interstellar dust.

d Infrared (1–4 μm) emission from stars that penetrates most interstellar material.

e Visible light emitted by stars is scattered and absorbed by dust.

f X-ray emission from hot gas bubbles (diffuse blobs) and X-ray binaries (pointlike sources).

g Gamma-ray emission from collisions of fast-moving particles with atomic nuclei in interstellar clouds.

FIGURE 16.11 Panoramic views of the Milky Way in different bands of the spectrum. The center of the galaxy, which lies in the direction of the constellation Sagittarius, is in the center of each strip. The rest of each strip shows all other directions in the Milky Way disk as seen from Earth. (Imagine attaching the left and right ends of each strip to form a circular band that corresponds to the 360° band of the Milky Way in our sky.)

- Figure 16.11d shows infrared light from stars at wavelengths that penetrate clouds of gas and dust (1–4 micrometers). Thus, this image shows how our galaxy would look if there were no dust blocking our view. The galactic bulge is clearly evident at the center.

- Figure 16.11e shows the galactic disk in visible light, just as it appears in the night sky. (Of course, only part of the Milky Way is above the horizon at any one time.) Because visible light cannot penetrate interstellar dust, the dark blotches correspond closely to the bright patches of molecular radio emission and infrared dust emission in Figure 16.11b and c.

- Figure 16.11f shows the distribution of X-ray light from the galactic disk. The pointlike blotches in this view are mostly X-ray binaries [Section 15.3], but the rest of the X-ray emission comes primarily from hot gas bubbles. Note that hot gas tends to rise into the halo, so it is less concentrated toward the midplane than the atomic and molecular gas. (The prominent yellow blob on the lower right is the Vela supernova remnant.)

- Figure 16.11g shows gamma-ray emission from the Milky Way. Most of the gamma-ray emission is produced by collisions between fast-moving particles and atomic nuclei in interstellar clouds. Such collisions happen most frequently where gas densities are highest, so the gamma-ray emission corresponds closely to the locations of molecular and atomic gas. (Gamma rays from the pulsar at the center of the Vela supernova remnant are prominent on the lower right.)

TIME OUT TO THINK *Carefully compare and contrast the different views of the Milky Way's disk in Figure 16.11. Why do regions that appear dark in some views appear bright in others? What kinds of general patterns do you notice?*

16.3 Galactic Environments

The star–gas–star cycle has operated continuously since the Milky Way's birth, yet new stars are not spread evenly across the galaxy. Some regions seem much more fertile than others. Galactic environments rich in molecular clouds tend to spawn new stars easily, while gas-poor environments do not. A quick tour of some characteristic galactic environments will help you spot where the action is.

Out in the Halo

A census of stars in the Milky Way's *halo* would turn up many senior citizens and very few newborns. Most of the halo stars are old, red, and dim and much smaller in mass than our Sun. Halo stars also contain far fewer heavy elements than our Sun, sometimes having heavy-element proportions as low as 0.02% (in contrast to about 2% in the Sun). The relative lack of chemical enrichment in the halo indicates that its stars formed early in the galaxy's history—before many supernovae had exploded, adding heavy elements to star-forming clouds.

The halo's gas content corroborates this view. The halo is virtually gas-free compared to the disk, with very few detectable molecular clouds. Apparently, the bulk of the Milky Way's gas settled into the disk long ago. The halo environment is a place where lack of gas caused star formation to cease early in our galaxy's life. Now only very old stars still survive, and new stars are rarely born.

TIME OUT TO THINK *How does the halo of our galaxy resemble the distant future fate of the galactic disk? Why?*

Our Neighborhood

Our own stellar neighborhood is more active than the halo and typifies much of the galactic disk. Within about 33 light-years (10 pc), we know of over 300 stars. Most are dim, red, spectral type M stars. A few, including Sirius, Vega, Altair, and Fomalhaut, are bright, white stars younger than our Sun. While stars of many different ages and different proportions of heavy elements are in our neighborhood, no very massive, short-lived stars (i.e., spectral type O or B) are present. We therefore infer that stars form periodically in our neighborhood but that no star clusters have formed here recently.

Our quiet suburb of the galaxy has not always been as calm as it is today. X-ray telescopes in space reveal hot, X ray–emitting gas coming from nearby in every direction. Surrounding this hot gas, at distances

ranging up to a few hundred light-years (100 pc), lies a region of much cooler gas. Apparently, we and all our stellar neighbors live inside a hot bubble. The existence of this *Local Bubble* means that a number of supernovae must have detonated within our stellar neighborhood over the past several million years.

Hot-Star Hangouts

The hot spots in our galaxy are the neighborhoods of high-mass stars. Because hot, massive stars live fast and die young, they never get a chance to move very far from their birthmates. Thus, we find them in star clusters close to the molecular clouds from which they formed. These environments are highly active and extraordinarily picturesque.

We find colorful, wispy blobs of glowing gas known as **ionization nebulae** throughout the galactic disk, particularly in the spiral arms. (Ionization nebulae are sometimes called *emission nebulae* or *H II regions*; H II is an abbreviation astronomers use for ionized hydrogen.) The energy that powers ionization nebulae comes from neighboring hot stars that irradiate them with ultraviolet photons. These photons ionize and excite the atoms in the nebulae, causing them to emit light. The Orion Nebula, about 1,500 light-years away in the "sword" of the constellation Orion, is among the most famous. Few astronomical objects can match its spectacular beauty (Figure 16.12).

Most of the red light in an ionization nebula comes from excited hydrogen atoms, but the blue and black tints in some nebulae have a different origin.

FIGURE 16.12 A photo of the Orion Nebula, an ionization nebula energized by ultraviolet photons from hot stars.

FIGURE 16.13 The blue tints in this nebula in the constellation Orion are produced by reflected light.

FIGURE 16.14 A photo of the Horsehead Nebula and its surroundings.

of gravity, can tell us how mass is distributed within the Milky Way. When we decipher these movements, we find that the matter binding stars and gas to our galaxy extends far beyond its visible reaches and that the galaxy's spiral arms are propagating waves of new star formation. Let's investigate the motion of the Milky Way in more detail.

Orbits in the Disk and Halo

If you could stand outside the Milky Way and watch it for a few billion years, the disk would resemble a huge merry-go-round. All the stars in the disk, including our Sun, orbit the center in the same direction. The disk stars are spread over a thickness of about 1,000 light-years—which is quite thin in comparison to the 100,000-light-year diameter of the disk (Figure 16.15). (The disk has some thickness because disk stars bob up and down in response to the disk's gravity as they orbit, rather like horses on a merry-go-round.)

The orbits of stars in the halo and bulge are much less organized than the orbits of stars in the disk. Individual bulge and halo stars travel around the galactic center on more or less elliptical paths, but the orientations of these paths are relatively random. Neighboring halo stars can circle the galactic center in opposite directions. They swoop from high above the disk to far below it and back again, plunging through the disk at velocities so high that the disk's gravity hardly alters their trajectories. Near the Sun, we see several fast-moving halo stars in the midst of hasty excursions through the local region of the disk. One such star is Arcturus, the fourth-brightest star in the night sky.

Starlight reflected from dust grains produces the blue colors, because interstellar dust grains scatter blue light much more readily than red light (Figure 16.13). These so-called *reflection nebulae* are always bluer in color than the stars supplying the light. (The effect is similar to the scattering of sunlight in our atmosphere that makes the sky blue [Section 8.4].) The black regions of nebulae are dark, dusty gas clouds that block our view of the stars beyond them. Figure 16.14 shows a multicolored nebula characteristic of a hot-star neighborhood.

TIME OUT TO THINK *In Figure 16.14, identify the red ionized regions, the blue reflecting regions, and the dark obscuring regions. Briefly explain the origin of the colors in each region.*

16.4 The Milky Way in Motion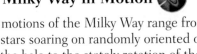

The internal motions of the Milky Way range from the chaos of stars soaring on randomly oriented orbits through the halo to the stately rotation of the galactic disk. These motions, combined with the law

TIME OUT TO THINK *Is there much danger that the Sun or Earth will someday be hit by a halo star swooping through the disk of the galaxy? Why or why not? (Hint: Think about the typical distances between stars, as illustrated by use of the 1-to-10-billion scale in Chapter 1.)*

The Sun's orbital path around the galaxy is called the **solar circle,** and its radius is our 28,000-light-year distance from the galactic center. By measuring the speeds of globular clusters relative to the Sun, we've determined that the Sun and its neighbors orbit the center of the Milky Way at a speed of about

FIGURE 16.15 This painting shows typical orbits of stars in the Milky Way Galaxy. Stars in the disk all orbit the galactic center in the same plane and the same direction. (They also bob slightly up and down in response to the disk's gravity; here these bobs have been exaggerated to make them visible.) In contrast, stars in the halo may orbit in any direction around the galactic center. The overall result is that disk star orbits look very organized, while halo star orbits appear random.

220 km/s. Even at this speed, one circuit of the Sun around the solar circle takes 230 million years. Early dinosaurs ruled the Earth when it last visited this side of the galaxy.

Orbits and Galactic Mass

Recall that Newton's law of gravity determines how quickly objects orbit one another. This fact, embodied in Newton's version of Kepler's third law, allows us to determine the mass of a relatively large object when we know the period and average distance of a much smaller object in orbit around it [Section 5.3]. A closely related law, which we will call the *orbital velocity law,* allows us to "weigh" the galaxy using the Sun's orbital velocity and its distance from the galactic center. The orbital velocity law is an equation that allows us to calculate the mass of the galaxy within the Sun's orbit by knowing only the Sun's orbital velocity and its distance from the galactic center.

Note that the orbital velocity law tells us only how much mass lies *within* the Sun's orbit. It does not tell us the mass of the entire galaxy, because matter lying outside the Sun's orbit has very little effect on the Sun's orbital velocity. Every part of the galaxy exerts gravitational forces on the Sun as it orbits, but the net force from matter outside the Sun's orbit is relatively small because the pulls from opposite sides of the galaxy virtually cancel one another. In contrast, the net gravitational forces from mass within the Sun's orbit all pull the Sun in the same direction—toward the galactic center. Thus, the Sun's orbital velocity responds almost exclusively to the gravitational pull of matter inside its orbit. Substituting the Sun's 28,000-light-year distance and 220-km/s orbital velocity into the orbital velocity law, we find that the total amount of mass within the solar circle is about 2×10^{41} kilograms, or about 100 billion solar masses.

The Distribution of Mass in the Milky Way

Just as we can use the orbit of the Sun to determine the mass of the galaxy within the solar circle, we can use the orbital motion of any other star to measure the mass of the Milky Way within the star's own orbital circle. In principle, we could determine the complete distribution of mass in the Milky Way by applying the orbital velocity law to the orbits of stars at every different distance from the galactic center. In practice, interstellar dust obscures our view of disk stars beyond a few thousand light-years, making it very difficult to measure stellar velocities. However, radio waves penetrate this dust, so we can see the 21-cm line from atomic hydrogen gas no matter where the gas is located in the galaxy. Such studies allow us to build a map of atomic hydrogen clouds. It is such maps that reveal our galaxy's large-scale spiral structure and also its overall pattern of rotation.

We can get a sense of the distribution of mass in the Milky Way by making a diagram called a **rotation curve**, which plots *rotational velocity* against *distance from the center*. As a simple example of the concept, let's construct a rotation curve for a merry-go-round. Every object on a merry-go-round goes around the center with the same rotational period, but objects farther from the center move in larger circles. Thus, objects farther from the center move at faster speeds, and the rotation curve for a merry-go-round is a straight line that rises steadily outward (Figure 16.16a). In contrast, the rotation curve for our solar system drops off with distance from the Sun because inner planets orbit at faster speeds than outer planets (Figure 16.16b). This drop-off in speed with distance occurs because virtually all the mass of the solar system is concentrated in the Sun. The gravitational force holding a planet in its orbit decreases with distance from the Sun, and a smaller force means a lower orbital speed. The rotation curve of any astronomical system whose mass is concentrated toward the center therefore drops steeply.

Figure 16.16c shows the rotation curve for the Milky Way Galaxy. Each individual dot represents the distance from the galactic center and the orbital speed of a particular star or cloud of atomic hydrogen. The curve running through the dots represents a "best fit" to the data. Note that, beyond the inner few thousand light-years, the orbital velocities remain approximately flat. This behavior contrasts sharply with the steeply declining rotation curve of the solar system. Thus, unlike the solar system, most of the mass of the Milky Way must *not* be concentrated at its center. Instead, the orbits of progressively more distant hydrogen clouds must encircle more and more mass. The solar circle encompasses about 100 billion solar masses, but a circle twice as large surrounds twice as much mass, and a larger circle surrounds even more mass. Because of the difficulty involved in finding clouds to measure on the outskirts of the galaxy, we have not yet found the "edge" of this mass distribution.

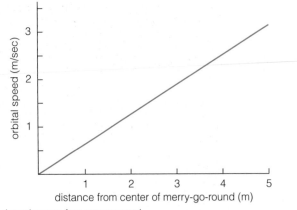

a A rotation curve for a merry-go-round.

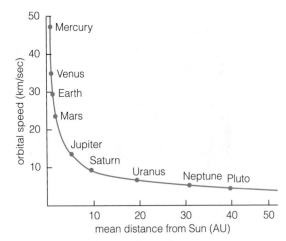

b The rotation curve for the planets in our solar system.

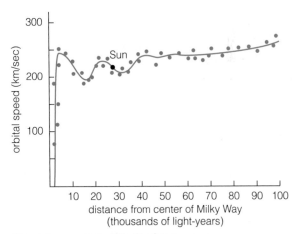

c The rotation curve for the Milky Way Galaxy.

FIGURE 16.16 Rotation curves show how the orbital speed of a system depends on distance from its center. The solar system's rotation curve declines with radius because its mass is concentrated at the center. The Milky Way's rotation curve is flat, indicating that the Milky Way's mass extends well beyond the Sun's orbit.

FIGURE 16.17 Galaxy M51, a spiral galaxy with two prominent spiral arms, as photographed by the Hubble Space Telescope. Note how blue the spiral arms are compared with the yellower tones of the bulge and the regions of the disk between the arms. Because blue stars only live for a few million years, the relative blueness of these spiral arms tells us that stars must be forming more actively here than elsewhere in the galaxy. The red blobs within the spiral arms are ionization nebulae associated with the young, blue stars.

The flatness of the Milky Way's rotation curve came as an immense surprise to astronomers because it implies that most of our galaxy's mass must lie well beyond our Sun, tens of thousands of light-years from the galactic center. In fact, a more detailed analysis suggests that most of this mass is distributed in the galactic halo and that the halo might outweigh all the disk stars *combined* by a factor of 10. However, aside from the light of the relatively small number of halo stars, we have detected very little radiation coming from this enormous amount of mass. Because we see so little light coming from outside the Sun's orbit, most of the halo's huge mass cannot be in the form of orbiting stars. The nature of this mass remains unknown—which means that we do not yet know the nature of the vast majority of matter in our own galaxy. We call this mysterious mass **dark matter** because it does not emit any light that we have yet detected. The nature of this dark matter is one of the greatest mysteries in astronomy today. We will investigate it in much more depth in Chapter 18.

Spiral Arms

The disks of spiral galaxies like the Milky Way display sweeping spiral arms that can stretch tens of thousands of light-years from their bulges. At a glance, spiral arms look as if they ought to rotate with the stars, like fins of a giant pinwheel. However, because stars near the center of the galaxy take less time to complete an orbit than more distant stars, the arms would gradually wind themselves up if they really did rotate with the stars in the disk. By the time a spiral galaxy was a few billion years old, the spiral arms would look like a tightly wound coil. Because we do not see such tightly coiled spiral arms, the explanation for spiral structure must be more complex.

Detailed images show the spiral arms to be home to most of the young, bright, blue stars in a galaxy (Figure 16.17). Clusters of these hot, short-lived massive stars populate the spiral arms. Between the arms, we find many fewer hot stars. We also see enhanced amounts of molecular and atomic gas in the spiral arms, and streaks of interstellar dust often obscure the inner sides of the arms themselves (Figure 16.18). All these clues suggest that spiral arms result from waves of star formation that propagate through the disks of spiral galaxies.

Theoretical models suggest that disturbances called **spiral density waves** are the source of our galaxy's spiral arms. These waves are generated by some sort of disturbance such as a close encounter with another galaxy, and they do not carry stars or interstellar gas with them as they move through the disk. The spiral arms themselves are actually waves of enhanced density that move through the disk. Stars and gas at the location of any wave crest are packed slightly more densely than normal, making the wave crest a location of enhanced gravitational attraction within the disk. This gravitational attraction pulls the stars ahead of the crest more closely together; after the crest passes, the stars move apart into looser groups.

Interstellar clouds react even more strongly to the gravitational attraction of the wave crest. Gas clouds collide with one another most frequently in the densest part of the wave, collecting into groups. Because gas clouds crowd so thickly into the crest of the wave, stars form more profusely in spiral arms than elsewhere in the disk. A froth of young stars and ionization nebulae thus marks the crests of spiral density waves.

TIME OUT TO THINK *Perhaps you have noticed what happens when the scene of an accident blocks one side of a highway. Curious drivers on the other side of the highway slow down to check out the accident, and traffic backs up behind them. After the drivers have passed by, they speed back up, and traffic thins out. How is this disturbance in the flow of traffic similar to a spiral density wave?*

The bluish hue of the spiral arms comes from their prolific star formation. Massive, short-lived, blue stars die out quickly as the wave crest passes by, long before the next spiral wave sweeps through. These luminous hot stars are therefore found close to the spiral arms in which they formed. Longer-lived yellow and red stars live through the passage of many spiral density waves and therefore are distributed relatively evenly throughout the galactic disk.

16.5 The Mysterious Galactic Center

The center of the Milky Way Galaxy lies in the direction of the constellation Sagittarius. This region of the sky does not look particularly special to our unaided eyes; however, if we could remove the interstellar dust that obscures our view, the elliptical shape of the galaxy's central bulge would fill Sagittarius, and its brilliance would be one of the night sky's most spectacular sights.

Because the Milky Way is relatively transparent to long-wavelength radiation, we can use radio and infrared telescopes to peer into its heart (Figure 16.19). Deep in the galactic center, we find swirling clouds of gas and a cluster of several million stars. Bright radio emission traces out the magnetic fields that thread this turbulent region. At the core of it all sits a source of bright radio emission named Sagittarius A* (pronounced "Sagittarius A-star"), or Sgr A* for short, that is quite unlike any other radio source in our galaxy.

The motions of the gas and stars in Sgr A* indicate that it contains a few million solar masses within a region no larger than about 3 light-years across (Figure 16.20). The star cluster in Sgr A* cannot account for all that mass. Many astronomers therefore suspect that Sgr A* contains a black hole weighing around 2.5 million solar masses.

However, the behavior of this suspected black hole is somewhat puzzling. The most plausible black-hole candidates in our galaxy—those in binary systems like Cygnus X-1 [Section 15.4]—accrete matter from their companions. The resulting accretion disks

FIGURE 16.18 The relationship of dust, gas, and new star clusters in a spiral arm.

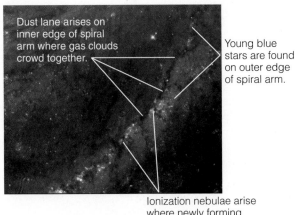

Dust lane arises on inner edge of spiral arm where gas clouds crowd together.

Young blue stars are found on outer edge of spiral arm.

Ionization nebulae arise where newly forming blue stars are ionizing gas clouds.

a In visible light, dusty gas obscures our view of the galactic center, marked by the 0.65° green square.

b In this infrared image depicting a 2° by 4° region around the galactic center, we can see through the dust to the central star cluster.

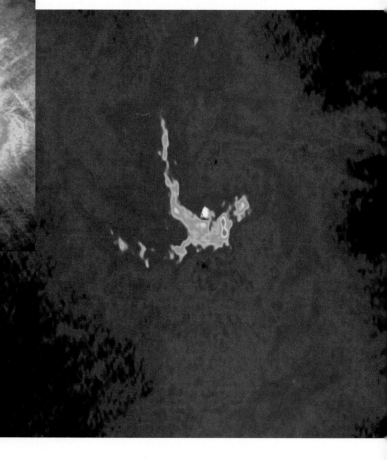

d This false-color radio image of the galactic center shows ionized gas swirling around Sgr A*, which is the white dot near the center of the image.

c This radio image of the central square degree shows the strange environment of Sgr A*; the bright filaments trace magnetic field lines full of charged particles presumably energized by the central black hole.

FIGURE 16.19 The galactic center at various wavelengths.

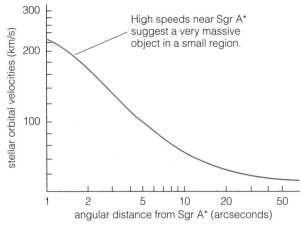

High speeds near Sgr A* suggest a very massive object in a small region.

FIGURE 16.20 Star velocities near the galactic center.

radiate brightly in X rays. If Sgr A* truly is a giant black hole, we might expect it to be accreting some of the surrounding gas from the galactic center. Because of its large mass, it would have a large accretion disk and would shine brightly in X rays. (Many other galaxies have tremendously luminous X-ray sources at their centers that most astronomers believe are accretion disks circling enormous black holes [Section 17.6].) These X rays would shine through the dusty gas of our galaxy, and we would be able to detect them easily. Yet we detect only relatively faint X-ray emission from Sgr A*.

It is possible that Sgr A* contains a huge black hole that simply has run out of gas to accrete. Even more intriguingly, the black hole might be consuming the energy contained in the accreting matter before it has a chance to escape as radiation. In any case, if Sgr A* is indeed a black hole, it does not behave like other black holes whose accretion disks are more apparent. Because the black hole at the center of our galaxy is so well hidden from us, the nature of Sgr A* remains mysterious.

THE BIG PICTURE

In this chapter, we have explored our galaxy, focusing on the recycling of gas that continually produces new stars and on the motions of the Milky Way that reveal its mass. When you review this chapter, pay attention to these "big picture" ideas:

- The inability of visible light to penetrate deeply through interstellar gas and dust concealed the true nature of our galaxy until recent times. Modern astronomical instruments reveal the Milky Way to be a dynamic system of stars and gas that continually gives birth to new stars and planetary systems.

- Stellar winds and explosions make interstellar space a violent place. Hot gas tears through the pervasive atomic hydrogen gas in the galactic disk, leaving expanding pockets of hot plasma and fast-moving clouds in its wake. All this violence might seem quite dangerous, but it performs the great service of mixing essential new elements throughout the Milky Way.

- We can use the orbital speeds of stars and gas clouds around the center of the galaxy to determine the distribution of mass in the Milky Way. To our great surprise, we've discovered that most of the mass of the galaxy lies in the halo, not in the galactic disk where most of the stars are located. The nature of this *dark matter* is one of the greatest mysteries in astronomy today.

- Although the elements from which we are made were forged in stars, we could not exist if stars were not organized into galaxies. The Milky Way Galaxy acts as a giant recycling plant, converting gas expelled from each generation of stars into the next and allowing some of the heavy elements to solidify into planets like our own.

Review Questions

1. Draw simple sketches of our galaxy as it would appear face-on and edge-on, identifying the *disk, bulge, halo,* and *spiral arms.*

2. What do we mean by the *interstellar medium?* How does it affect our view of the galaxy?

3. What do we mean by *heavy elements?* How much of the Milky Way's gas is in the form of heavy elements? Where are heavy elements made?

4. What is the star–gas–star cycle, and how does it lead to *chemical enrichment?* Use the idea of chemical enrichment to explain why stars that formed early in the history of the galaxy contain a smaller proportion of heavy elements than stars that formed more recently.

5. Distinguish between the following forms of gas: ionized hydrogen, atomic hydrogen, and molecular hydrogen.

6. What is a *superbubble?* How is it made? Describe what happens in a blowout in which a superbubble breaks out of the galactic disk.

7. What is the most common form of gas in the interstellar medium? What is the wavelength of the radio emission line characteristic of this gas?

8. What are *dust grains?* Where do they form?

9. How do *molecular clouds* form? How do we observe them?

10. Will the star–gas–star cycle continue forever? Why or why not?

11. Describe the stars we find in the *halo* in terms of mass, luminosity, color, and proportion of heavy elements. Why do halo stars have these characteristics?

12. What is an *ionization nebula?* Why are such nebulae found near massive young stars? What produces their striking colors?

13. Contrast the general patterns of the orbits of stars in the disk with the orbits of stars in the halo.

14. At what speed does the Sun travel around the *solar circle?* About how long does one orbit take? Briefly explain how we can use the characteristics of the Sun's orbit to determine the mass of the Milky Way contained within the solar circle. What is the result?

15. What is a *rotation curve?* Describe the rotation curve of the Milky Way, and explain how it indicates that most of the Milky Way's mass lies outside the solar circle. What do we mean by *dark matter?*

16. How do we know that spiral arms are not simple "pinwheels" as they appear at first glance? What features do we see in images of spiral arms that suggest that they result from waves of star formation propagating through galaxies?

17. What is Sgr A*? What evidence suggests that it contains a massive black hole?

Discussion Questions

1. *Galactic Ecosystem.* The introduction to this chapter likened the star–gas–star cycle in our Milky Way to the ecosystem that sustains life on Earth. Here on our planet, water molecules cycle from the sea to the sky to the ground and back to the sea. Our bodies convert atmospheric oxygen molecules into carbon dioxide, and plants convert the carbon dioxide back into oxygen molecules. How are the cycles of matter on Earth similar to the cycles of matter in the galaxy? How do they differ? Do you think the term *ecosystem* is appropriate to discussions of the galaxy?

2. *Galaxy Stuff.* In the chapters on stars, we learned why we are "star stuff." Based on what you've learned in this chapter, explain why we are also "galaxy stuff." Does the fact that the entire galaxy was involved in bringing forth life on Earth change your perspective on the Earth or on life in any way? If so, how? If not, why not?

Problems

Sensible Statements? For **problems 1–7**, decide whether the statement is sensible and explain why it is or is not.

1. We did not understand the true size and shape of our galaxy until NASA satellites were launched into the galactic halo, enabling us to see what the Milky Way looks like from the outside.

2. Planets like Earth probably didn't form around the very first stars because there were so few heavy elements back then.

3. If I could see infrared light, the galactic center would look much more impressive.

4. Many spectacular ionization nebulae are seen throughout the Milky Way's halo.

5. The carbon in my diamond ring was once part of an interstellar dust grain.

6. The Sun's velocity around the Milky Way tells us that most of our galaxy's dark matter lies within the solar circle.

7. We know that a black hole lies at our galaxy's center because a few of the stars it is sucking in have vanished over the last several years.

8. *Unenriched Stars.* Suppose you discovered a star made purely of hydrogen and helium. How old do you think it would be? Explain your reasoning.

9. *Enrichment of Star Clusters.* The gravitational pull of an isolated globular cluster is rather weak—a single supernova explosion can blow all the interstellar gas out of a globular cluster. How might this fact be related to observations indicating that stars ceased to form in globular clusters long ago? How might it be related to the fact that globular clusters are deficient in elements heavier than hydrogen and helium? Summarize your answers in one or two paragraphs.

10. *High-Velocity Star.* The average speed of stars *relative* to the Sun in the solar neighborhood is about 20 km/s (i.e., the speed at which we see stars moving toward or away from the Sun—*not* their orbital speed around the galaxy). Suppose you discover a star in the solar neighborhood that is moving relative to the Sun at a much higher speed, say 200 km/s. What kind of orbit does this star probably have around the Milky Way? In what part of the galaxy does it spend most of its time? Explain.

11. *Research: Discovering the Milky Way.* Humans have been looking at the Milky Way since long before recorded history, but only in the past century did we verify the true shape of the galaxy and our location within it. Learn more about how conceptions of the Milky Way developed through history. What names did different cultures give the band of light they saw? What stories did they tell about it? How have ideas about the galaxy changed in the past few centuries? Try to locate diagrams that illustrate these changes. Write a two- to three-page summary of your findings.

Web Projects

Find useful links to Web projects on the text Web site.

1. *Images of the Star–Gas–Star Cycle.* Explore the Web to find pictures of nebulae and other forms of interstellar gas in different stages of the star–gas–star cycle. Assemble the pictures into a sequence that tells the story of interstellar recycling, with a one-paragraph explanation accompanying each image.

2. *The Galactic Center.* Search the Web for recent images of the galactic center, along with information about whether the center hides a massive black hole. Present a two- to three-page report, with pictures, giving an update on current knowledge about the center of the Milky Way Galaxy.

CHAPTER 17

A Universe of Galaxies

Far beyond the Milky Way, we see many other galaxies—some similar to our own and some very different—scattered throughout space to the very limits of the observable universe. This fabulous sight inspires some fundamental questions: How far away are the galaxies? How old is the universe? How big is it? Such questions might have seemed ridiculously speculative a century ago. Today, we believe we know the answers to all three questions with respectable accuracy.

Edwin Hubble, the man for whom the Hubble Space Telescope is named, provided the key discovery when he proved conclusively that galaxies exist beyond the Milky Way. The distances he measured revealed an astonishing fact: The more distant a galaxy is, the faster it moves away from us. Hubble's discovery dealt a mortal blow to the traditional belief in a static, eternal, and unchanging universe. The motions of the galaxies away from one another imply instead that the entire universe is expanding and that its age is finite.

In this chapter, we will get acquainted with these islands in space, the galaxies of stars. We will discuss how we know just how far away they are. Because some of these galaxies are billions of light-years away, we are seeing them as they were billions of years ago. Thus, by studying their light, we are gleaning clues about the history of the universe as it is written in the stars and the gas of the distant galaxies.

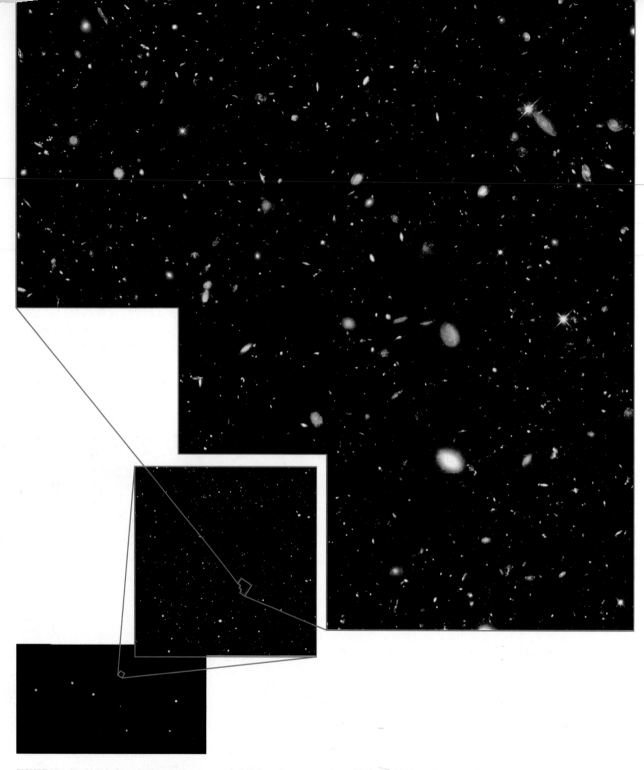

FIGURE 17.1 The Hubble Deep Field, an image composed of 10 days of exposures taken with the Hubble Space Telescope. Some of the galaxies pictured are three-quarters of the way across the observable universe. The field itself is located in the Big Dipper.

17.1 Islands of Stars

Figure 17.1 shows an amazing image of a tiny slice of the sky taken by the Hubble Space Telescope. The telescope pointed toward a single direction in the sky, collecting all the light it could for 10 days. If you held a grain of sand at arm's length, the angular size

of the grain would match the angular size of everything in this picture. This slice of the sky is jam-packed with galaxies of many sizes, colors, and shapes. Some look large, some small. Some are reddish, some whitish. Some appear round, and some appear flat. Small galaxies greatly outnumber large ones, yet the large ones produce most of the light in the universe. Counting the galaxies in this slice of the sky and multiplying by the number of such slices it would take to cover the entire sky, we find that the observable universe contains over 80 billion galaxies.

We would like to tell the life stories of these galaxies as completely as we told the life stories of stars (Chapter 14), but we cannot. Too many aspects of **galaxy evolution**—the formation and development of galaxies—remain mysterious. Galaxies are more challenging to study than stars both because they are far more complex and because most of them are so distant. Nevertheless, galaxy evolution is one of the most active research areas in astronomy, and each year brings new discoveries.

In broad terms, the story we seek should tell us how the hydrogen and helium gas in the early universe evolved into the galaxies we see today. Even this general story still presents puzzles, such as why some galaxies turned out flattened and gas-rich like our own while others did not. But we also need to explain some of the stranger aspects of galaxies. For example, some galaxies have extremely bright centers, known as **active galactic nuclei**, that sometimes outshine all the stars in the galaxy combined. In some cases, these active galactic nuclei appear to be firing powerful jets of material millions of light-years into space. We now believe that the "engines" for active galactic nuclei are enormous black holes and that their tremendous power comes from gigantic accretion disks formed by gas spilling into the black holes. However, we do not yet know where these black holes came from or how their accretion disks got started.

In the rest of this chapter, we will discuss what we currently know about galaxies and their origins. We will begin by outlining how astronomers classify galaxies and will go on to investigate how we measure their distances and estimate their ages. Along the way, we will learn how the first measurements of galaxy distances revolutionized the way we now look at the universe—as an expanding realm within which the story of creation is still unfolding. After laying this groundwork, we will examine what is currently known about galaxy evolution and conclude with a discussion of active galactic nuclei and the evidence for giant black holes at the hearts of galaxies.

17.2 Galaxy Types

Astronomers classify galaxies into three major categories. **Spiral galaxies** look like flat, white disks with yellowish bulges at their centers. The disks are filled with cool gas and dust, interspersed with hotter ionized gas as in the Milky Way, and usually display beautiful spiral arms. **Elliptical galaxies** are redder, more rounded, and often longer in one direction than in the other, like a football. Compared with spiral

galaxies, elliptical galaxies contain very little cool gas and dust, though they often contain very hot, ionized gas. Galaxies that appear neither disklike nor rounded are classified as **irregular galaxies**. The sizes of all three types of galaxy span a wide range, from dwarf galaxies containing as few as 100 million (10^8) stars to giant galaxies with more than 1 trillion (10^{12}) stars.

TIME OUT TO THINK *Take a moment and try to classify the larger galaxies in Figure 17.1. How many appear spiral? Elliptical? Irregular? Do the colors of galaxies seem related to their shapes?*

Spiral Galaxies

Like the Milky Way, other spiral galaxies also have a thin disk extending outward from a central bulge (Figure 17.2). The bulge of a spiral galaxy merges

FIGURE 17.2 Spiral galaxies.

a NGC 1232, a spiral galaxy similar to our Milky Way.

b NGC 891, an edge-on spiral considered to be very similar to the Milky Way—note the dusty disk.

FIGURE 17.3 NGC 1365, a barred spiral galaxy.

FIGURE 17.4 NGC 5078, a lenticular galaxy. Most lenticular galaxies have less cool gas than this one, which displays a prominent black streak of interstellar dust.

smoothly into a halo that can extend to a radius of over 100,000 light-years (30 kpc). Together, the bulge and halo of a spiral galaxy make up its **spheroidal component**, so named because of its rounded shape. Although no clear boundary divides the two parts of the spheroidal component, astronomers usually consider stars within about 10,000 light-years (3 kpc) of the center to be members of the bulge and those outside this radius to be members of the halo.

The **disk component** of a spiral galaxy slices directly through the halo and bulge. The disk of a large spiral galaxy like the Milky Way extends 50,000 light-years (15 kpc) or more from the center. The disks of all spiral galaxies contain an interstellar medium of gas and dust, but the amounts and proportions of the interstellar medium in molecular, atomic, and ionized forms differ from one spiral galaxy to the next. Spiral galaxies with large bulges generally have less interstellar gas and dust than those with small bulges.

Not all galaxies with disks are standard spiral galaxies. Some spiral galaxies appear to have a straight bar of stars cutting across the center, with spiral arms curling away from the ends of the bar. Such galaxies are known as *barred spiral galaxies* (Figure 17.3). Other galaxies have disks but do not appear to have spiral arms (Figure 17.4). These are called *lenticular galaxies* because they look lens-shaped when seen edge-on (*lenticular* means "lens-shaped"). Because they have disks, lenticular galaxies are somewhat like spiral galaxies without arms. But they might more appropriately be considered an intermediate class between spirals and ellipticals—they tend to have less cool gas than normal spirals, but more than ellipticals.

About 75–85% of large galaxies in the universe are spiral or lenticular. Spiral galaxies are often found in loose collections of several galaxies, called **groups**, that extend over a few million light-years (Figure 17.5). Our Local Group is one example, with two large spirals: the Milky Way and the Great Galaxy in Andromeda (M31) [Section 1.1]. Lenticular galaxies are particularly common in **clusters** of galaxies, which can contain hundreds and sometimes thousands of galaxies extending over more than 10 million light-years (Figure 17.6).

Elliptical Galaxies

The major difference between elliptical and spiral galaxies is that ellipticals lack a significant disk component (Figure 17.7). Thus, an elliptical galaxy has only a spheroidal component and looks much like the bulge and halo of a spiral galaxy. (In fact, elliptical galaxies are sometimes called *spheroidal galaxies*.) Most of the interstellar medium in large elliptical galaxies consists of low-density, hot, X ray–emitting gas—rather like the gas in bubbles and superbubbles in the Milky Way [Section 16.2]. Elliptical galaxies usually contain very little dust or cool gas, although they are not completely devoid of either. Some have cold, relatively small gaseous disks rotating at their centers that might be the remnants of a collision with a spiral galaxy.

FIGURE 17.5 Hickson Compact Group 87, a small group of galaxies consisting of a large edge-on spiral galaxy at the bottom of this photo, two smaller spiral galaxies at the center and upper left, and an elliptical galaxy to the right. (The other objects in this photograph are foreground stars in our own galaxy.)

FIGURE 17.6 A portion of the Virgo cluster of galaxies, the nearest large galaxy cluster to the Milky Way.

Elliptical galaxies appear to be more social than spiral galaxies: They are much more common in large clusters of galaxies than outside clusters. Elliptical galaxies make up about half the large galaxies in the cores of clusters, while they represent only about 15% of the large galaxies found outside clusters. Ellipticals are also more common among small galaxies. Small elliptical galaxies with less than a billion stars, called **dwarf elliptical galaxies**, are often found near larger spiral galaxies. At least 10 dwarf elliptical galaxies belong to the Local Group.

FIGURE 17.7 Elliptical galaxies.

a M 87, a giant elliptical galaxy in the Virgo Cluster.

b Leo I, a dwarf elliptical galaxy in the Local Group.

FIGURE 17.8 The Large Magellanic Cloud, a small irregular galaxy that orbits the Milky Way.

Irregular Galaxies

A small percentage of the large galaxies we see nearby fall into neither of the two major categories. This irregular class of galaxies is a miscellaneous class, encompassing small galaxies such as the Magellanic Clouds and "peculiar" galaxies that appear to be in disarray (Figure 17.8). These blobby star systems are usually white and dusty, like the disks of spirals. Telescopic observations probing deep into the universe show that distant galaxies are more likely to be irregu-

lar in shape than those nearby. Because the light of more distant galaxies was emitted longer ago in the past, these observations tell us that irregular galaxies were more common when the universe was younger.

Classifying galaxies helps us identify the patterns underlying their enormous variety, but classification alone tells us little about the size and luminosity of a galaxy. No equivalent to the Hertzsprung–Russell diagram exists for galaxies—they are much more complex than stars. Thus, in order to know how large and luminous a particular galaxy is, we must first measure its distance.

17.3 Measuring Cosmic Distances

Measuring cosmic distances is one of the most fundamental and challenging tasks we face when trying to understand galaxies and the universe as a whole. Our determinations of astronomical distances depend on a chain of methods that begins with knowing distances in our own solar system. At each link in this chain, we establish standards that help us forge the next link. Because each link depends on preceding links, any inaccuracies accumulate as we move up the chain. Nevertheless, we now can measure the distances to the farthest galaxies to an accuracy within better than 20%. The distance chain begins with radar measurements of the solar system's size. The second link involves parallax measurements of the distances to nearby stars [Section 13.2]. We will now follow the rest of this chain, link by link, to the outermost reaches of the observable universe.

THINKING ABOUT . . .

Hubble's Galaxy Classes

Edwin Hubble invented a system for classifying galaxies that remains widely used. It assigns the letter *E*, followed by a number, to each elliptical galaxy. The larger the number, the flatter the galaxy. An E0 galaxy is round, and an E7 galaxy is highly elongated. A spiral galaxy is assigned an uppercase S, or *SB* if it has a bar, and a lowercase *a, b,* or *c*. The lowercase letter indicates the size of the bulge and the dustiness of the disk. An Sa galaxy has a large bulge and a modest amount of dusty gas; an SBc galaxy has a bar, a small bulge, and lots of dusty gas. The lenticular galaxies are designated S0, signifying their intermediate spot between spirals and ellipticals. Irregular galaxies are designated Irr.

Astronomers had once hoped that the classification of galaxies might yield deep insights, just as the classification of stars did in the early twentieth century. The Hubble classification scheme itself was suspected for a time to be an evolutionary sequence in which galaxies flattened and

spread out as they aged. Unfortunately for astronomers, galaxies turn out to be far more complex than stars, and classification schemes like this one have not led to easy answers about the nature of galaxies.

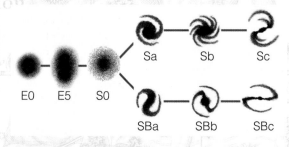

The Hubble "tuning fork" diagram shows Hubble's galaxy classes. The arrangement was once suspected by astronomers to represent an evolutionary sequence, but that is no longer thought to be the case.

Standard Candles

We can always measure an object's apparent brightness, so we can use the luminosity–distance formula to calculate luminosity if we know the distance, or to calculate distance if we know the luminosity [Section 13.2]. For example, suppose we see a distant street lamp and know that all street lamps of its type put out 1,000 watts of light. Then, once we measure the apparent brightness of the street lamp (in units of watts/m²), we can calculate its distance with the luminosity–distance formula.

An object such as a street lamp, for which we are likely to know the true luminosity, represents what astronomers call a **standard candle**. The term *standard candle* is meant to suggest a light source of a known, standard luminosity. Unlike light bulbs, astronomical objects do not come marked with wattage, so an astronomical object can serve as a standard candle only if we have some way of knowing its true luminosity without first measuring its apparent brightness and distance. Fortunately, many astronomical objects meet this requirement. For example, any star that is a twin of our Sun—that is, a main-sequence star with spectral type G2—should have about the same luminosity as the Sun. Thus, if we measure the apparent brightness of a Sun-like star, we can use the luminosity–distance formula to calculate its distance by assuming that it has the same luminosity we have measured for the Sun (3.8×10^{26} watts).

There is always some uncertainty in distances calculated by applying the luminosity–distance formula, because no astronomical object is a perfect standard candle. The challenge of measuring great astronomical distances comes down to the challenge of finding the objects that make the best standard candles. The more confidently we know an object's true luminosity, the less uncertain the distance we calculate with the luminosity–distance formula.

Main-Sequence Fitting

We can use Sun-like stars as standard candles because we know that they are similar to the Sun and because we can measure the Sun's luminosity quite easily. However, Sun-like stars are relatively dim, and we cannot detect them at great distances. To measure distances beyond 1,000 light-years or so, we need brighter standard candles. An obvious first choice is to use brighter main-sequence stars. However, before we can use any main-sequence star as a standard candle, we must first have some way of knowing its true luminosity. The key to understanding this process is to remember that we can use the luminosity–distance formula in two ways:

1. First, we identify a star cluster that is close enough for us to determine its distance by parallax and plot its H–R diagram. Because we know the distances to the cluster stars, we can use the luminosity–distance formula to establish their true luminosities from their apparent brightnesses.

2. Then we can look at stars in other clusters that are too far away for parallax measurements and measure their apparent brightnesses. If we assume that main-sequence stars in other clusters have the same true luminosities as their counterparts in the nearby cluster, we can calculate their distances with the luminosity–distance formula.

The nearest star cluster with a well-populated main sequence, the *Hyades Cluster* in the constellation Taurus, has long been crucial to this technique. We can measure the true distance of the Hyades Cluster through its parallax. Knowing this distance allows us to determine the true luminosities of all its stars with the luminosity–distance formula. We can find the distances to other star clusters by comparing the apparent brightnesses of their main-sequence stars with those in the Hyades Cluster and assuming that all main-sequence stars of the same color have the same luminosity (Figure 17.9). This technique of determining distances by comparing main-sequence stars in different star clusters is called **main-sequence fitting**.

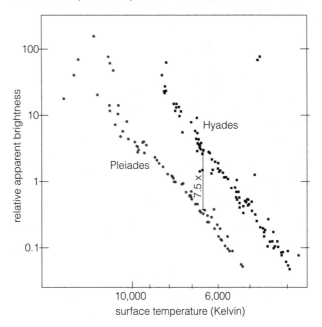

FIGURE 17.9 Comparison of the apparent brightness of stars in the Hyades Cluster with those in the Pleiades Cluster shows that the Pleiades are 2.75 times farther away because they are $2.75^2 = 7.5$ times dimmer.

Cepheid Variables

Main-sequence fitting works well for measuring distances to star clusters throughout the Milky Way, but not for measuring distances to other galaxies. Most main-sequence stars are too faint to be seen in other galaxies, even with our largest telescopes. Instead, we need very bright stars to serve as standard candles for distance measurements beyond the Milky Way, and the most useful bright stars are the Cepheid variables [Section 13.5]. Recall that Cepheids are pulsating variable stars that follow a simple period–luminosity relation: The longer the time period between peaks in brightness, the greater the luminosity of the Cepheid variable star (Figure 17.10). All Cepheids that vary with a particular period have very nearly the same luminosity (within about 10%), so Cepheids effectively scream out their true luminosities when we measure their periods. "Hey, everybody, my luminosity is 10,000 solar luminosities!" is how we would translate the message of a Cepheid variable whose brightness peaks every 30 days. Thus, Cepheid variables are the primary standard candles used to determine distances to nearby galaxies.

Cepheids played a key role in the discovery that the Milky Way is only one of billions of galaxies in the universe. Before the 1920s we had no way of measuring the distances to other galaxies, and many astronomers suspected that these swirling collections of stars were located within the Milky Way. In 1924, Edwin Hubble used the 100-inch telescope atop southern California's Mount Wilson (Figure 17.11)—the largest telescope in the world at the time—to discover Cepheid variables in the Andromeda Galaxy by comparing photographs of the galaxy taken days apart. He used the period–luminosity relation to determine the luminosities of these stars and then used these luminosities in the luminosity–distance formula to compute their distances. His distance measurements proved that the Andromeda Galaxy sat far beyond the outer reaches of stars in the Milky Way, demonstrating that it is a separate galaxy. This single stroke of scientific discovery dramatically changed our view of the universe. Rather than inhabiting a universe that ended with the Milky Way, we suddenly knew that vast numbers of galaxies lay beyond the Milky Way's borders.

Hubble's Law

Hubble's determination of the distance to the Andromeda Galaxy assured him a permanent place in the history of astronomy, but he didn't stop there. Hubble proceeded to measure the distances to many more galaxies. Within just a few years, Hubble made one of the most astonishing discoveries in the history of science: that the universe is expanding.

Astronomers had known since the 1910s that the spectra of most of the faint smudges on the sky known as the "spiral nebulae" tended to be redshifted. In other words, each of the emission and absorption lines in their spectra appeared at longer (redder) wavelengths than expected. Recall that redshifts occur when the object emitting the radiation is moving away from us [Section 6.4]. But, since Hubble had not yet proved that the "spiral nebulae" were distant galaxies, no one understood the true significance of their motions.

In 1929, after measuring distances to a number of "spiral nebulae," Hubble announced his conclusion: The more distant a galaxy, the greater its redshift and hence the faster it is moving away from us. Hubble's original assertion was based on an amazingly small sample of galaxies. Even more incredibly, he had grossly underestimated the luminosities of his standard candles. The "brightest stars" he had been using as standard candles were really entire clusters of bright stars. Fortunately, Hubble was both bold and lucky. Subsequent studies of much larger samples of galaxies showed that they are indeed receding from us, but they are even farther away than Hubble thought.

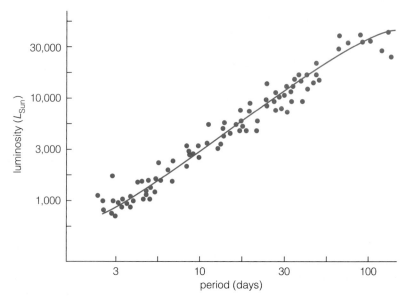

FIGURE 17.10 Cepheid period–luminosity relation. Note that all Cepheids of a particular period have very nearly the same luminosity. (Cepheids actually come in two types with two different period–luminosity relations; the relation here is for Cepheids with heavy-element content similar to that of our Sun, or "Type I Cepheids.")

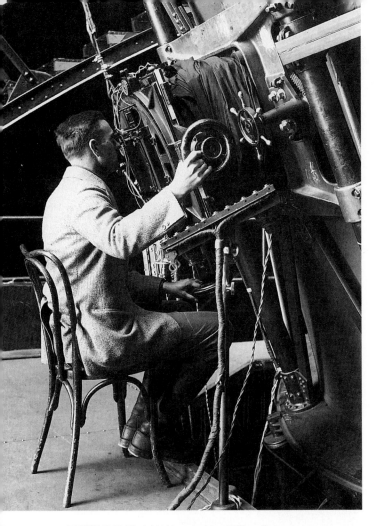

FIGURE 17.11 Edwin Hubble at the Mount Wilson Observatory.

We can express the idea that more distant galaxies move away from us faster with a very simple formula, now known as **Hubble's law**:

$$v = H_0 \times d$$

where v stands for velocity (sometimes called a *recession velocity*), d stands for distance, and H_0 (pronounced "H-naught") is a number called **Hubble's constant**. Astronomers generally quote the Hubble constant in strange-sounding units of *kilometers per second per megaparsec* (km/s/Mpc). The easiest way to explain these units is by example. Suppose Hubble's constant has a value of $H_0 = 100$ km/s/Mpc. Then, according to Hubble's law, we would expect a galaxy located at distance $d = 10$ Mpc (32.5 million light-years) to be moving away from us at a speed of:

$$v = H_0 \times d = 100 \, \frac{\text{km/s}}{\cancel{\text{Mpc}}} \times 10 \, \cancel{\text{Mpc}} = 1{,}000 \text{ km/s}$$

Dividing both sides of Hubble's law by H_0 puts it into a form in which we can use a galaxy's velocity to determine its distance:

$$d = v / H_0$$

In principle, applying this law is one of the best ways to determine distances to galaxies. We find the velocity of a distant galaxy from its redshift; then we divide this velocity by H_0 to find its distance. However, we encounter two major practical difficulties in finding distances in this way:

1. Galaxies do not obey Hubble's law perfectly because they can have velocities relative to one another caused by gravitational tugs that Hubble's law does not take into account.

2. Even when galaxies do obey Hubble's law well, the distances we find with it are only as accurate as our best measurement of Hubble's constant.

The first problem makes Hubble's law difficult to use for nearby galaxies. Within the Local Group, for example, Hubble's law does not work at all: The galaxies in the Local Group are gravitationally bound together with the Milky Way and therefore are not moving away from us in accord with Hubble's law. However, Hubble's law works fairly well for more distant galaxies: The recession speeds of galaxies at large distances are so great that any motions caused by the gravitational tugs of neighboring galaxies are tiny in comparison.

The second problem means that, even for distant galaxies, we can know only relative distances until we pin down the true value of H_0. For example, Hubble's law tells us that a galaxy moving away from us at 20,000 km/s is twice as far away as one moving at 10,000 km/s, but we can determine the actual distances of the two galaxies only if we know H_0. Because of its importance in measuring cosmic distances, the quest to measure H_0 accurately has been one of the main missions of the Hubble Space Telescope.

TIME OUT TO THINK *Suppose a galaxy is moving away from us at 10,000 km/s. Use Hubble's law in the form $d = v/H_0$ to calculate its distance (in Mpc) if $H_0 = 100$ km/s/Mpc. What is its distance if $H_0 = 50$ km/s/Mpc?*

Seeking Distant Standards

The process of measuring H_0 is rather like the process of calibrating a scale. We can calibrate a scale by, for example, making sure that it reads 10 pounds when we place a 10-pound weight on it, 20 pounds when we place a 20-pound weight on it, and so on. Once it is properly calibrated, we can use it to weigh objects whose weights are not known in advance. In a similar way, astronomers calibrate Hubble's law by making sure that it gives correct distances for galaxies whose distances we already know from applying some other method. The superior resolving power of the Hubble Space Telescope enables us to measure

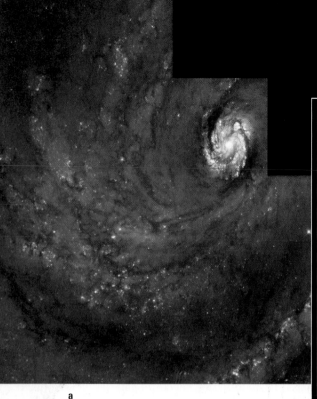

a

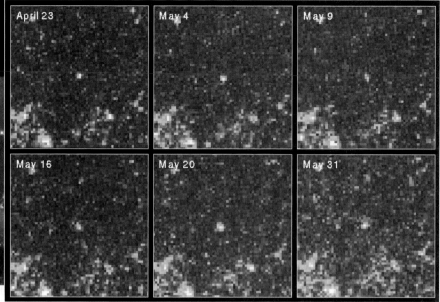

FIGURE 17.12 Hubble Space Telescope observations of (**a**) galaxy M 100 in the Virgo Cluster and (**b**) one of its Cepheid variable stars varying in brightness over several weeks.

April 23   May 4   May 9

May 16   May 20   May 31

b

the periods and apparent brightnesses of Cepheid variable stars in galaxies up to 100 million light-years (30 Mpc) away, but this is not quite far enough (Figure 17.12). At this distance, random gravitational tugs can still cause noticeable deviations from Hubble's law. Nevertheless, these Cepheids can help us by enabling us to calibrate even brighter standard candles, such as supernovae.

White dwarf supernovae are thought to be exploding white dwarf stars that have reached the 1.4-solar-mass limit and should all have nearly the same luminosity [Section 15.2]. Thus, white dwarf supernovae should be good standard candles. We can use Cepheids to find the distances to galaxies in which white dwarf supernovae have occurred, and once we know the galaxy's distance, the luminosity–distance formula provides the supernova's true luminosity. This technique has allowed researchers using the Hubble Space Telescope to determine the true luminosities of several historical white dwarf supernovae. As expected, they are all about the same, and we can now use them as reliable standard candles. Because white dwarf supernovae are so bright—about 10 billion solar luminosities at their peak—we can use them to measure distances to galaxies billions of light-years away (Figure 17.13).

Galaxies themselves can also be used as standard candles. Astronomers studying spiral galaxies have discovered a close relationship between their total luminosities and the rotation speeds of their disks: The faster a spiral galaxy's rotation speed, the more

FIGURE 17.13 Photo of a white dwarf supernova in a galaxy halfway across the observable universe, recorded in March 1996 by the Hubble Space Telescope.

luminous it is (Figure 17.14). This relationship, called the **Tully–Fisher relation** (after its discoverers), holds because both luminosity and rotation speed depend on the galaxy's mass. A galaxy's luminosity depends on the number of stars it contains, which is related to the total amount of matter within it, and the total amount of matter determines a galaxy's rotation speed [Section 16.4]. Once we have measured the rotation speed of a spiral galaxy, the Tully–Fisher relation tells us its true luminosity, which makes the galaxy itself a standard candle that we can use to determine its distance.

Despite the new calibrations of Hubble's law made possible by the Hubble Space Telescope, some uncertainties remain. These uncertainties should decrease as we measure accurate distances to more and more galaxies. As of 2002, the true value of

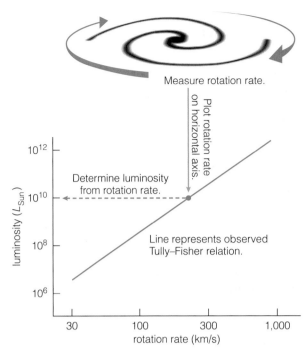

Measure rotation rate.

Plot rotation rate on horizontal axis.

Determine luminosity from rotation rate.

Line represents observed Tully–Fisher relation.

FIGURE 17.14 A schematic diagram of the Tully–Fisher relation. (The precise relation differs among different subclasses of spiral galaxies.)

H_0 appears to be somewhere between 55 and 75 km/s/Mpc.

Summary: The Distance Chain

Distance measurements are fundamental to our understanding of galaxies and the universe. Here we briefly summarize the chain of measurements that allows us to determine ever-greater distances (Figure 17.15). Note that with each link in the distance chain uncertainties become somewhat greater. Thus, although we know the Earth–Sun distance at the base of the chain extremely accurately, distances to the farthest reaches of the universe remain uncertain by about 20%.

- **Radar ranging:** We measure distances within the solar system by bouncing radio waves off planets. Then, with the aid of some geometry, we use these radar measurements to determine the Earth–Sun distance.

- **Parallax:** We measure the distances to nearby stars by observing how their positions change relative to the background stars as the Earth moves around the Sun. These distances thus rely on our knowledge of the Earth–Sun distance.

- **Main-sequence fitting:** We know the distance to the Hyades Cluster in our Milky Way Galaxy through parallax. Comparing the apparent brightness of its main-sequence stars to the brightness of those in other clusters gives us the distances to these other star clusters in our galaxy.

- **Cepheid variables:** By studying Cepheids in star clusters with distances measured by main-sequence fitting, we learn the precise period–luminosity relation for Cepheids. When we find a Cepheid in a more distant star cluster or galaxy, we can determine its true luminosity by measuring the period between its peaks in brightness and then use this true luminosity to determine the distance.

- **Distant standards:** By measuring distances to relatively nearby galaxies with Cepheids, we learn the true luminosities of white dwarf supernovae and the Tully–Fisher relation between

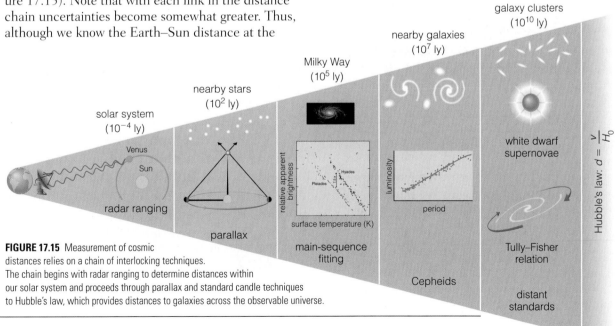

FIGURE 17.15 Measurement of cosmic distances relies on a chain of interlocking techniques. The chain begins with radar ranging to determine distances within our solar system and proceeds through parallax and standard candle techniques to Hubble's law, which provides distances to galaxies across the observable universe.

the luminosities and rotation speeds of spiral galaxies. Thus, supernovae and entire galaxies become useful standard candles for measuring great distances in the universe.

- **Hubble's law:** Distances measured to galaxies with white dwarf supernovae and the Tully–Fisher relation allow us to calibrate Hubble's law and measure Hubble's constant, H_0. Once we know H_0, we can determine a galaxy's distance directly from its redshift.

17.4 Measuring Cosmic Ages

The ages of the galaxies we see in deep images of the universe, such as Figure 17.1, are directly related to their distances. Remember that light travels at a finite speed. Light from the most distant galaxies we can see began its journey to Earth when the universe was a small fraction of its current age. We are therefore viewing these most distant galaxies as they were when they were young, but the light we are detecting is ancient. Because a deep image like this records galaxies at many different distances, it contains information about galaxies of many different ages. For example, we see galaxies 1 billion light-years away as they were 1 billion years ago and galaxies 2 billion light-years away as they were 2 billion years ago. The history of galaxies across the universe is therefore encoded in such images, but to decipher the clues they hold we need to know how the universe's history is related to its expansion. The intimate relationship between the universe's expansion rate and its age helps us age-date galaxies throughout the observable universe.

Universal Expansion

The fact that galaxies all across the universe are moving away from one another in accordance with Hubble's law implies that galaxies must have been closer together in the past. Tracing this convergence back in time, we reason that all the matter in the observable universe started very close together and that the entire universe came into being at a single moment. Ever since that moment, the universe has been expanding, and all the while the gravitational pull of each galaxy on every other galaxy has been working to slow the expansion rate. In the densest regions of the universe, gravity has won out over expansion. The Local Group no longer expands, nor do other clusters of galaxies, but the expansion continues on all larger scales.

Visualizing this expansion requires a little thought. Upon first hearing about the expansion of the universe, most people are tempted to think of the universe as a ball of galaxies expanding into a void. This impression is mistaken. The universe expands, but it is not expanding into anything. Each point moves away from every other point, and the universe looks about the same no matter where you stand. (In this book, whenever we speak about the universe as a whole, we are assuming that matter in the universe is evenly distributed on scales much larger than superclusters of galaxies. This assumption is known as the *Cosmological Principle*. It is impossible to prove that the Cosmological Principle holds outside the observable regions of the universe, but within the observable universe matter does appear to be distributed more evenly as we look at increasingly larger scales [Section 18.5].)

Back in Chapter 1, we likened the expanding universe to a raisin cake rising as it bakes [Section 1.3], but a cake has edges and a center. A better analogy would be something that can expand but that has no center and no edges—like the surface of a balloon. Just as you'll never encounter an edge if you sail around the world, there is no edge to a balloon's surface. And, while a balloon has a center inside it, the center is not part of the surface. Thus, like the universe, the surface of a balloon has no center and no edges. The only significant problem with this analogy is that, because we want to represent the entire universe with the two-dimensional surface of a balloon, we must imagine flattening the universe into only two dimensions. We attach plastic polka dots to the surface of the balloon to represent flattened clusters of galaxies and blow into the balloon to represent the expansion of the universe (Figure 17.16).

As the real universe expands, the distances between clusters of galaxies grow. However, the clusters themselves do not grow because gravity binds each cluster's galaxies together. In a similar way, as the surface of the balloon expands, the polka dots

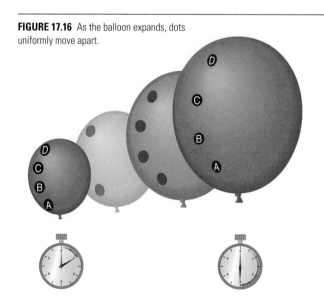

FIGURE 17.16 As the balloon expands, dots uniformly move apart.

move apart, but the dots themselves don't grow. Suppose that, 1 second after you begin blowing into the balloon, dots A, B, C, and D are spaced 1 cm apart along a line. After 2 seconds, the distances between neighboring dots grow to 2 cm, and after 3 seconds to 3 cm. If you recorded the distances of the other dots from Dot B once each second, your observations would look like this:

Measuring Distances from Dot B

Time	Dot A Distance	Dot B Distance	Dot C Distance	Dot D Distance
1 s	1 cm	0 cm	1 cm	2 cm
2 s	2 cm	0 cm	2 cm	4 cm
3 s	3 cm	0 cm	3 cm	6 cm

Miniature scientists on dot B might be tempted to think they are at the center of some explosion. After all, they see every other dot moving away from them, and the farthest dots move fastest. However, miniature scientists on Dot C would record a very similar set of observations:

Measuring Distances from Dot C

Time	Dot A Distance	Dot B Distance	Dot C Distance	Dot D Distance
1 s	2 cm	1 cm	0 cm	1 cm
2 s	4 cm	2 cm	0 cm	2 cm
3 s	6 cm	3 cm	0 cm	3 cm

The miniature scientists on Dot C would also be tempted to think they are at the center of some explosion. Eventually, the miniature scientists would realize that the balloon itself is expanding. Just as we cannot point to a single spot on a balloon's surface and say "the balloon is expanding from here," we cannot identify a single point in space from which the universe is expanding. The universe has no center. Likewise, neither the balloon nor the universe has any edge.

The balloon analogy, while helpful, has its limitations. For example, the balloon has a finite total surface area, but the universe might be infinite in extent. In addition, you might be tempted to say that the balloon is expanding away from a point inside the balloon or that the surface of the balloon has an inside edge and an outside edge. We notice these features of the balloon because the two-dimensional surface of the balloon exists within a three-dimensional space. But the third dimension that we perceive so easily would make no sense to a two-dimensional miniature scientist restricted to the surface of the balloon.

Hubble's Constant and Age

To see how the universe's expansion rate is related to its age, let's return to the miniature scientists living on dot B. Three seconds after the balloon began to expand, they would measure the following:

Dot A is 3 cm away and moving at 1 cm/s.

Dot C is 3 cm away and moving at 1 cm/s.

Dot D is 6 cm away and moving at 2 cm/s.

They could summarize these observations as follows: *Every dot is moving away from our home with a speed that is 1 cm/s for each 3 cm of distance.* And because the expansion of the balloon is uniform, scientists living on any other dot would come to the same conclusion. Each scientist living on the balloon would determine that the following formula relates the distances and velocities of other dots on the balloon:

$$v = \left(\frac{1 \text{ cm/s}}{3 \text{ cm}}\right) \times d \quad \text{or} \quad v = \left(\frac{1}{3 \text{ s}}\right) \times d$$

where v and d are the velocity and distance of any dot, respectively.

TIME OUT TO THINK *Confirm that this formula gives the correct values for the speeds of dots C and D, as seen from dot B, 3 seconds after the balloon began expanding. How fast would a dot located 9 cm from dot B move, according to the scientists on dot B?*

If the miniature scientists think of their balloon as a bubble, they might call the number relating distance to velocity—the term $\frac{1}{3 \text{ s}}$ in the above formula—the "bubble constant." An especially insightful miniature scientist might flip over the "bubble constant" and find that it is exactly equal to the time since the balloon started expanding. That is, the "bubble constant" of $\frac{1}{3 \text{ s}}$ tells them that the balloon has been expanding for 3 seconds. Perhaps you see where we are heading.

Just as the inverse of the "bubble constant" tells the miniature scientists that their balloon has been expanding for 3 seconds, the inverse of the Hubble constant, or $1/H_0$, tells us something about how long our universe has been expanding. The "bubble constant" for the balloon depends on when it is measured, but it is always equal to 1/(time since balloon started expanding). Similarly, the Hubble constant actually changes with time but stays roughly equal to 1/(age of the universe). We call it a constant because it is the same at all locations in the universe.

Moreover, its value does not change noticeably on the time scale of human civilization.

Current estimates based on the value of Hubble's constant put the age of the universe between about 12 and 18 billion years. To derive a more precise value for the universe's age, we need to know whether the expansion has been speeding up or slowing down over time, a question we will examine more closely in Chapter 18. If the gravitational pull of each galaxy on every other galaxy has significantly slowed the expansion rate, then the universe's age is somewhat less than $1/H_0$. If some mysterious force has accelerated the expansion rate, then the universe's age is somewhat more than $1/H_0$. Intriguingly, the oldest stars in globular clusters appear to be somewhere between 12 and 16 billion years old [Section 13.6]. These stars could not possibly have existed before the birth of the universe, so our age estimates based on Hubble's constant probably aren't far off the mark. The best available evidence as of 2002 suggests that the universe is indeed somewhere between 12 and 16 billion years old. If you are reading this book at a much later date, it is likely that the age of the universe is known within more tightly constrained limits.

Lookback Time

All the galaxies we observe must be younger than the universe itself. Pictures of the farthest galaxies show us how galaxies looked when they were only a few billion years old; pictures of the closest show galaxies nearly as old as the universe. However, when we try to specify exactly how the distances of galaxies relate to their ages, we run into complications.

Imagine photons of light from a supernova in a distant galaxy, traveling toward us at the speed of light. If the supernova occurred in a galaxy 400 million light-years away, we are seeing a supernova that occurred 400 million years ago. But what do we mean by the distance of 400 million light-years? Because the universe is expanding, the distance between Earth and the supernova is greater today than it was at the time of the supernova event. Are we talking about the distance from the supernova to Earth when the supernova exploded, or are we referring to the distance between the supernova and Earth when the light arrived here? Do we mean that photons traveled through 400 million light-years of space?

Because the distances between galaxies are always changing, it is easier to speak about faraway galaxies in terms of the time it has taken for their light to reach us—400 million years in the case of the galaxy with the supernova. We call this time the **lookback time** to the supernova. In other words, a distant object's lookback time is the difference between the current age of the universe and the age of the universe when the light left the object. This quantity is much more meaningful than an actual distance when we discuss very distant objects that are moving away from us with the expansion of the universe.

TIME OUT TO THINK *Explain why there is no question about meaning when we talk about distances to objects within the Local Group, but distances become difficult to define when we talk about objects much farther away.*

An object's lookback time is directly related to its redshift. We have seen that the redshifts of galaxies tell us how quickly they are moving away from us. In the context of an expanding universe, redshifts have an additional, more fundamental, interpretation. Let's return one final time to the balloon analogy. Suppose you drew wavy lines on the balloon's surface to represent light waves. As the balloon inflates, these wavy lines stretch out, and their wavelengths increase (Figure 17.17). This stretching closely resembles what happens to photons in an expanding universe. The expansion of the universe stretches out all the photons within it, shifting them to longer, redder wavelengths. We call this effect a **cosmological redshift**.

In a sense, we have a choice when we interpret the redshift of a distant galaxy: We can think of the redshift as being caused either by the Doppler effect as the galaxy moves away from us or by a photon-stretching, cosmological redshift. However, as we look to very distant galaxies, the ambiguity in the meaning of distance also makes it difficult to specify precisely what we mean by a galaxy's speed. Thus, it becomes preferable to interpret the redshift as being due to photon stretching in an expanding universe. From this perspective, it is better to think of space itself as expanding, carrying the galaxies along for the ride, than to think of the galaxies as projectiles flying through a static universe. The cosmological redshift of a galaxy thus tells us how much space has expanded during the lookback time to that galaxy.

FIGURE 17.17 As the universe expands, photon wavelengths stretch like the wavy lines on this expanding balloon.

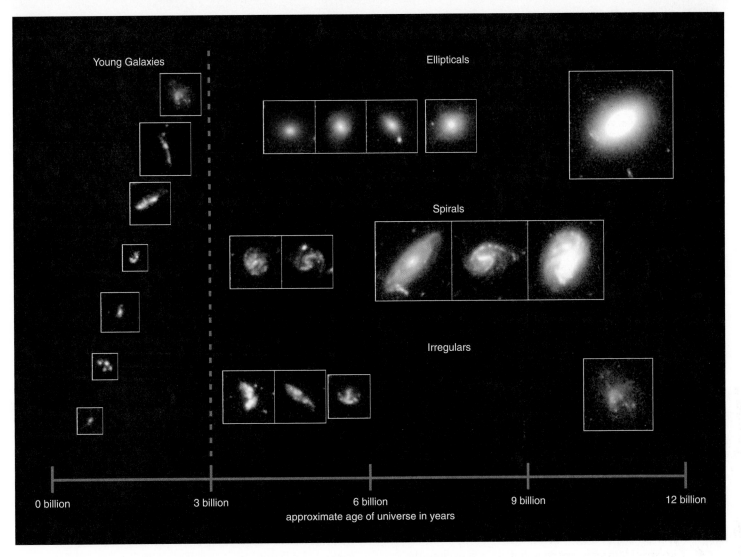

FIGURE 17.18 Family albums for elliptical, spiral, and irregular galaxies at different ages along with some very young, distant galaxies. All of these galaxy photos have been drawn from the Hubble Space Telescope image in Figure 17.1. (Younger galaxies appear smaller because they are more distant.)

Capitalizing on this relationship between redshift and lookback time, astronomers can use the redshifts of galaxies to assemble "family albums" showing galaxies of many different ages (Figure 17.18). The photos in such an album are pictures of galaxies with lookback times ranging from a few million light-years to billions of light-years. Each picture shows one galaxy at a single stage in its life, and we can arrange these pictures to make a family album for galaxies of each type. Subtracting a galaxy's lookback time from the age of the universe tells us about how old the universe was at the time the light left the galaxy. We cannot state the precise ages of these galaxies because we do not know exactly when they formed. However, galaxies whose light reaches us from when the universe was less than 3 billion years old are certainly in their childhood, and those whose light reaches us from when the universe was more than 6 billion years old are considerably more mature.

TIME OUT TO THINK *When speaking of galaxies, astronomers sometimes use the term* today *in a very broad sense. For example, if we look at a relatively nearby galaxy—one located, say, 20 million light-years away—we see it as it was 20 million years ago. In what sense is this "today"? (Hint: How does 20 million years compare to the age of the universe?)*

The Horizon of the Universe

What lies beyond the youngest, most distant galaxies? When we began our discussion of the expanding universe, we stressed that the universe does not have an edge. Yet the universe does have a horizon, a place beyond which we cannot see. This **cosmological horizon** is a boundary in time, not in space. It exists because we cannot see back to a time before the universe began. The lookback time to the cosmological horizon is equal to the age of the universe. For reasons discussed in Chapter 19, we cannot see matter at the cosmological horizon. But if we could, we would see it as it looked at the very beginning of time.

Our quest to measure the distances and ages of galaxies has brought us to the very limits of the observable universe. Now it is time to examine what these galaxies have been doing for the past 10 billion or more years.

17.5 Galaxy Evolution

We pieced together the life stories of stars by observing relatively nearby stars of many different ages, but we cannot do the same for galaxies. Because the youngest galaxies are so distant, our current instruments are not sensitive enough to reveal the whole history of galaxy evolution. The galaxies pictured in Figure 17.18 extend back to fairly young ages, but not all the way back to the infancy of galaxies. Even the superior resolving power and sensitivity of the Hubble Space Telescope cannot reveal what young galaxies look like in detail. Thus, we are left having to weave several different lines of less direct evidence into a single coherent story of galaxy evolution. In this section, we will review the story of galaxy evolution as we currently understand it, keeping in mind that this story remains incomplete.

Modeling Galaxy Birth

Because our telescopes cannot yet see back to the time when galaxies formed their first stars, we must use theoretical modeling to study the earliest stages in galaxy evolution. The most successful models for galaxy formation assume the following:

- Hydrogen and helium gas filled all of space more or less uniformly when the universe was very young—say, in the first million years after its birth.

- This uniformity was not quite perfect, and certain regions of the universe were ever so slightly denser than others.

Beginning from these assumptions, which are supported by a mounting body of evidence (discussed in Chapter 19), we can model galaxy formation using well-established laws of physics to trace how the denser regions in the early universe grew into galaxies. The models show that the regions of enhanced density originally expanded along with the rest of the universe. However, the slightly greater pull of gravity in these regions gradually slowed their expansion. Within about a billion years, the expansion of these denser regions halted and reversed, and the material within them began to contract into **protogalactic**

clouds, the clouds of matter that eventually formed galaxies.

According to the models, protogalactic clouds initially cooled as they contracted, radiating away their thermal energy, and the first generation of stars grew from the densest, coldest clumps of gas. The most massive of these stars lived and died within just a few million years, a short time compared to the time required for the collapse of a protogalactic cloud into a mature galaxy. The supernovae of these massive stars generated shock waves that heated the surrounding interstellar gas, slowing the collapse of young galaxies and the rate at which new stars formed within them.

Old Spheroids, Young Disks

The colors of galaxies provide a crucial clue to what happened next. Recall that the disks of spiral galaxies appear whitish with flecks of blue. This indicates that the stars in the disks of spiral galaxies, sometimes referred to as the *disk population* (or *Population I*), include hot, blue stars as well as cool, red ones. Because only short-lived, massive stars look blue, we conclude that star formation must be an ongoing process in galactic disks—just as it is in the disk of the Milky Way. In contrast, the *spheroidal population* of stars (or *Population II*) found in elliptical galaxies and in the bulges and halos of spiral galaxies looks reddish in color. The absence of blue stars among the spheroidal population indicates that most of the stars in these spheroids are old and that their star formation must have ceased long ago—just as is the case in the halo of the Milky Way [Section 16.3].

The characteristic motions of stars in disks and spheroids provide another important clue to their nature (Figure 17.19). Disk stars generally orbit in orderly circles in the same direction, like the rotation of the Milky Way's disk [Section 16.4]. Stars of the spheroidal population in elliptical galaxies and in the bulges and halos of spiral galaxies tend to orbit in random directions. Astronomers reason that, because the motions of their stars are so different, disks and spheroids must have formed in very different ways.

According to the most basic model for galaxy formation from a protogalactic cloud, the stars of the spheroidal population formed first, while the cloud remained blobby in shape with little or no measurable rotation. The orbits of these stars around the center of the galaxy could have had any orientation, accounting for the randomly oriented orbits of spheroidal population stars that we see today. Later, at least in a spiral galaxy, conservation of angular momentum caused the remaining gas to flatten into a spinning disk as it contracted under the force of gravity (Figure 17.20). Stars that form within this spinning gas disk are born on orbits moving at the same speed and in the same direction as their neighbors and thus become the disk population stars.

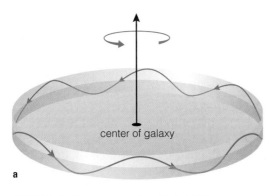

a

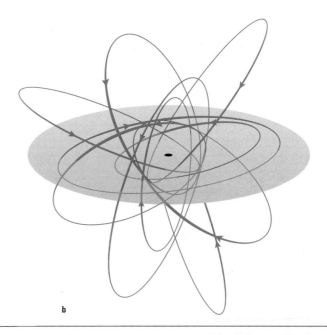

b

FIGURE 17.19 Characteristic stellar orbits in (**a**) disk component and (**b**) spheroidal component. Stars in the disk tend to orbit in the same direction around the center of the galaxy. (The bit of up and down motion is caused by the gravity of the disk itself.) Stars in the bulge and halo also orbit the center of the galaxy, but the orientations of their orbits are more random.

FIGURE 17.20 Schematic model of galaxy formation.

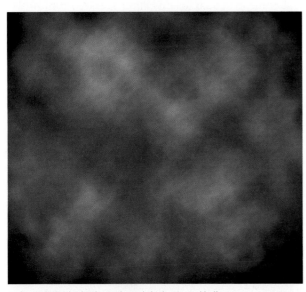

a A protogalactic cloud contains only hydrogen and helium gas.

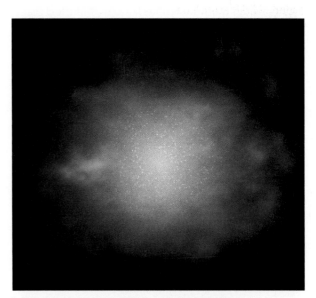

b Halo stars begin to form as the protogalactic cloud collapses.

c Conservation of angular momentum ensures that the remaining gas flattens tion within the disk.

d Billions of years later, the star–gas–star cycle supports ongoing star forma-into a spinning disk. The lack of gas in the halo precludes further star formation outside the disk.

How does the disk of a spiral galaxy resemble a protoplanetary disk? How are the processes that form galactic disks and protoplanetary disks similar?

Why Do Galaxies Differ?

On the whole, the idea that galaxies grow from contracting protogalactic clouds and form their stars over billions of years is consistent with what we know from studying the Milky Way and from our limited observations of very young galaxies. However, this simple model does not explain why galaxies come in such a great variety of shapes and sizes. We now believe that both the initial conditions of the protogalactic cloud and later interactions with other galaxies affected the development of all types of galaxies, leading to the variety we see today. To help differentiate the roles of initial conditions and subsequent interactions, we will consider a question about galaxy evolution that is still unanswered: Why do spiral galaxies have gas-rich disks, while elliptical galaxies do not?

Two potential explanations for the differences between spirals and ellipticals trace a galaxy's current appearance back to the properties of the protogalactic cloud from which it formed. The first explanation concerns angular momentum differences among protogalactic clouds. If a protogalactic cloud has enough angular momentum, it will rotate quickly as it collapses, and the galaxy it produces will tend to form a disk. The resulting galaxy would therefore be a spiral galaxy. If a protogalactic cloud has little or no net angular momentum, its gas might not form a disk at all, and the resulting galaxy would be elliptical.

The second scenario hinges on the ability of the protogalactic gas to cool and form stars. Remember that elliptical galaxies are most common in clusters of galaxies. The density of matter in these clusters was probably higher than elsewhere in the universe when galaxies first formed. Thus, the preponderance of elliptical galaxies in clusters suggests that elliptical galaxies formed from protogalactic clouds with higher densities than those of the clouds from which spirals formed. The higher gas densities in these proto-elliptical clouds would have enabled them to radiate energy more efficiently and to cool more quickly. This cooling would have allowed gravity to collapse clumps of gas into stars more quickly, using up all the gas before the cloud had time to collapse into a disk. The resulting galaxy would look elliptical. In contrast, lower-density, proto-spiral clouds would have had a slower rate of star formation, leaving plenty of gas to form a disk as the cloud collapsed. The resulting spiral galaxy would have a gas-filled disk in which the star–gas–star cycle could maintain ongoing star formation.

Some evidence for this latter scenario comes from a few giant elliptical galaxies at very large distances (Figure 17.21). These galaxies look very red even after we have accounted for their large redshifts. They apparently have no blue or white stars at all, indicating that new stars no longer form within these galaxies—even though we are seeing them as they were when the universe was only a few billion years old. This finding supports the idea that all the stars in these galaxies formed almost simultaneously, leaving no time for a disk to develop.

These two scenarios, in which the formation of a gas-rich disk depends on the angular momentum or density of the protogalactic cloud, probably describe important parts of the overall story. However, they ignore one key fact: Galaxies rarely evolve in perfect isolation. Recall that, in our scale-model solar system in Chapter 1, we made the Sun the size of a grape-

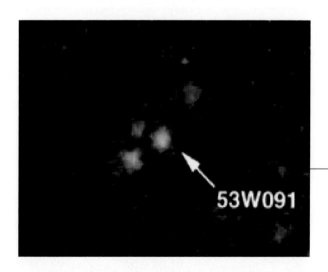

FIGURE 17.21 The redness of this distant elliptical galaxy, LBDS 53W091, indicates that its stars all formed during the first billion years or so after the birth of the universe. Its two red companions to the left are probably also elliptical galaxies with similar star-formation histories.

a

FIGURE 17.22 Galaxy collisions. (**a**) The Antennae, also known as NGC 4038/4039, are a pair of colliding spiral galaxies. The wide-field image taken from the ground reveals vast streamers of stars torn from the galaxies by tidal forces. The close-up from the Hubble Space Telescope shows the fine details of the starburst at the center of the collision. (**b**) NGC 6240, a galaxy collision farther along than the Antennae, is still highly disturbed, but the two galaxies have almost merged into one.

b

fruit. On this scale, the nearest star was like another grapefruit a few thousand kilometers away. Because the average distances between stars are so huge compared to the sizes of stars, direct star–star collisions are extremely rare. However, if we rescale the universe so that our entire galaxy is the size of a grapefruit, the Andromeda Galaxy is like another grapefruit only about a meter away, and a few smaller galaxies lie considerably closer. Thus, the average distances between galaxies are not tremendously larger than the sizes of galaxies, and collisions between galaxies are inevitable. Even the Milky Way galaxy is not immune. About 80,000 light-years away, directly behind the galactic bulge, a small elliptical galaxy (called the Sagittarius dwarf elliptical) is currently crashing through the Milky Way's disk.

Collisions between galaxies are spectacular events that unfold over hundreds of millions of years (Figure 17.22). In our short lifetime, we can at best see an instantaneous snapshot of a collision in progress, as evidenced by the distorted shape of the colliding galaxies. Galactic collisions must have been even more frequent in the past, when the universe was smaller and galaxies were even closer together. Photographs of galaxies at a variety of distances confirm that distorted-looking galaxies—probably galaxy collisions in progress—were more common in the early universe than they are today (Figure 17.23).

We can learn much more about collisions between galaxies with the aid of computer models that simulate interactions between galaxies. Such models allow us to "watch" a collision that takes hundreds of millions of years in nature unfold in a few minutes

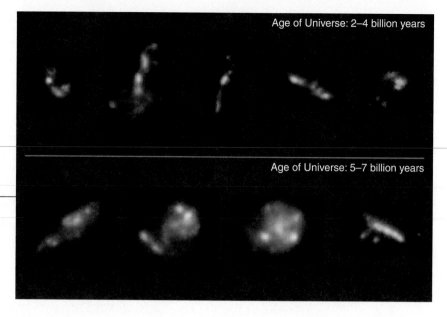

Age of Universe: 2–4 billion years

Age of Universe: 5–7 billion years

FIGURE 17.23 Hubble Space Telescope photos of distorted young galaxies. The larger number of distorted galaxies in the past suggests that collisions between galaxies were more common during the universe's first few billion years.

on a computer screen. These computer models show that a collision between two spiral galaxies can create an elliptical galaxy (Figure 17.24). Tremendous tidal forces between the colliding galaxies tear apart the two disks, randomizing the orbits of their stars. Meanwhile, a large fraction of their gas sinks to the center of the collision and rapidly forms new stars. When the cataclysm finally settles down, the merger of the two spirals has produced a single elliptical galaxy; little gas is left for a disk, and the orbits of the stars have random orientations.

Observational evidence confirms the idea that at least some elliptical galaxies are the result of collisions and subsequent mergers. The fact that elliptical galaxies dominate the galaxy populations at the cores of dense clusters of galaxies, where collisions should be most frequent, may mean that some of the spirals once present became ellipticals through collisions. Stronger evidence comes from structural details of elliptical galaxies, which often attest to a violent past. Some elliptical galaxies have stars and gas clouds in their core that orbit differently than the other stars in the galaxy, suggesting that they are leftover pieces of galaxies that merged in a past collision. Other elliptical galaxies are surrounded by shells of stars that probably formed from stars stripped out of smaller galaxies that once strayed too close and were destroyed by tidal forces (Figure 17.25).

The most decisive evidence that collisions affect the evolution of some elliptical galaxies comes from observations of the **central dominant galaxies** found at the centers of many dense clusters. These giant elliptical galaxies apparently grew to huge sizes by consuming other galaxies through collisions. Central dominant galaxies frequently contain several tightly bound clumps of stars that probably

once were the centers of individual galaxies later swallowed by the giant central galaxy (Figure 17.26). This process of galactic cannibalism can create galaxies over 10 times more massive than the Milky Way, making them the largest galaxies in the universe.

In summary, we have identified several scenarios that may determine whether a galaxy becomes an elliptical or a spiral galaxy. Two of these scenarios depend on initial conditions: One suggests that elliptical galaxies form when protogalactic clouds have low angular momentum, and the other suggests that elliptical galaxies form when high density leads to rapid star formation. But we have also found that interactions such as collisions between galaxies can drastically influence the lives of galaxies and very likely cause some spiral galaxies to become elliptical. Each of these mechanisms probably plays a role in galaxy evolution, and we do not yet know whether any one is more important than the others. Irregular galaxies, which make up only a small percentage of large galaxies, probably arise from a variety of different and unusual circumstances, but they probably are affected by the same processes that affect spirals and ellipticals.

Starburst Galaxies

Most of the galaxies in the present-day universe now appear comfortably settled, steadily forming new stars at modest rates. The Milky Way itself produces an average of about one new star per year. At this rate, the Milky Way won't exhaust the interstellar gas in its disk until long after the Sun has died. Certain galaxies are not so thrifty. Some form new stars at rates exceeding 100 stars per year, more than 100 times the star-formation rate in the Milky Way. These **starburst galaxies** cannot possibly sustain such a

Two simulated spiral galaxies approach each other on a collision course.

The first encounter begins to disrupt the two galaxies and sends them into orbit around each other.

As the collision continues, much of the gas in the disk of each galaxy collapses toward the center.

Gravitational forces between the two galaxies tear out long streamers of stars called tidal tails.

The centers of the two galaxies approach each other and begin to merge.

The single galaxy resulting from the collision and merger is an elliptical galaxy surrounded by debris.

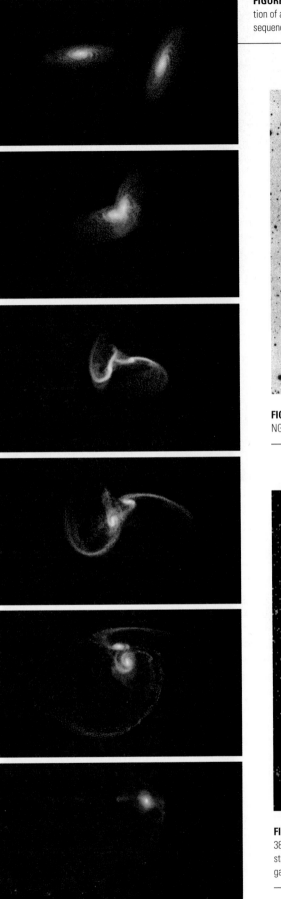

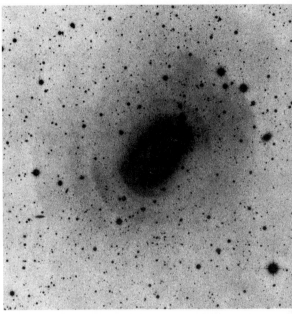

FIGURE 17.25 Shells of stars around the elliptical galaxy NGC 3923.

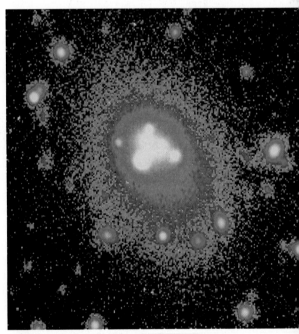

FIGURE 17.26 The central galaxy of the cluster Abell 3827, shown here in false color, has multiple clumps of stars that probably once were the centers of individual galaxies.

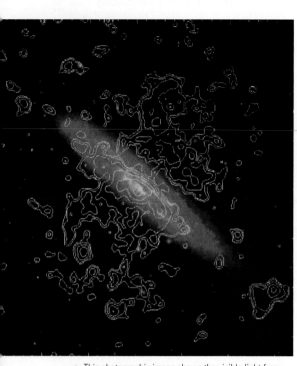

a This photographic image shows the visible light from the starburst galaxy NGC 253; the contour lines represent regions of strong X-ray emission. Note that the X rays appear on either side of the disk, suggesting that they come from hot gas blown out of the disk.

FIGURE 17.27 Galactic winds can blow out the top and bottom of a starburst galaxy's disk.

b This photograph of the starburst galaxy M 82 shows violently disturbed gas (red) poking out from above and below the disk where hot gas is escaping.

torrid pace of star formation for long: At their current rates of star formation, they would consume all their interstellar gas in just a few hundred million years—a relatively short time compared to the 10 billion or more years since galaxies first formed. Starburst galaxies are thus passing through an important but temporary phase in the evolution of at least some galaxies.

A star-formation rate 100 times that of the Milky Way also means an occurrence of supernovae at 100 times the Milky Way's rate. Just as in the Milky Way, each supernova in a starburst galaxy generates a shock wave that creates a bubble of hot gas [Section 16.2]. The shock waves from several nearby supernovae quickly overlap and blend into a much larger superbubble. This is where the story usually ends here in the Milky Way, but in a starburst galaxy the drama is just beginning. Supernovae continue to explode inside the superbubble, adding to its thermal and kinetic energy. When the superbubble starts to break through the disrupted gaseous disk, it expands

even faster. Hot gas erupts into intergalactic space, creating a *galactic wind*.

Galactic winds consist of low-density but extremely hot gas, typically with temperatures of 10–100 million Kelvin. They do not emit much visible light, but they do generate X rays. X-ray telescopes in orbit have detected pockets of X-ray emission surrounding the disks of some starburst galaxies, presumably coming from the outflowing galactic wind (Figure 17.27a). Sometimes we also see the glowing remnants of a punctured superbubble extending out into space (Figure 17.27b).

Many of the most luminous starburst galaxies look violently disturbed. They are neither flat disks with symmetrical spiral arms nor smoothly rounded balls of stars like elliptical galaxies. Streamers of stars are strewn everywhere. Such galaxies are filled with dusty molecular clouds, and deep inside them we see two distinct clumps of stars that look like two different galactic centers. Their appearance suggests that such starbursts result from a collision between two gas-rich spiral galaxies. The collision apparently compresses the gas inside the colliding galaxies, leading to the burst of star formation.

At least some large starbursts represent a short-lived phase in the life of galaxies that occurs when two spiral galaxies collide. The causes of smaller-scale starbursts are less clear: Some may result from direct collisions, while others may be caused by close encounters with other galaxies. The Large Magellanic Cloud is currently undergoing a period of rapid star formation, perhaps because of the tidal influence of the Milky Way. But no matter what their exact causes turn out to be, starburst galaxies clearly represent an important piece in the overall puzzle of galaxy evolution.

Loose Ends

We seem to be well on our way to understanding how and why protogalactic clouds evolve into galaxies. Nevertheless, a few puzzles remain. For example, while the oldest halo stars in our Milky Way have very low fractions of heavy elements, these fractions are not zero. Where did the heavy elements in these stars come from? They probably came from an even earlier generation of stars, but we have not yet discovered any examples of first-generation stars with no heavy elements. An even more fundamental question relates to the formation of the protogalactic clouds themselves. Our models assume that these clouds formed in regions of slightly enhanced density in the early universe, but where did these density enhancements come from? The question of the nature and origin of density enhancements in the early universe is one of the major puzzles in astronomy; we'll revisit it in Chapters 18 and 19.

17.6 Quasars and Other Active Galactic Nuclei

Collisions between galaxies may be spectacular, but some galaxies display even more incredible phenomena: extreme amounts of radiation, and sometimes powerful jets of material, emanating from deep in their centers (Figure 17.28). We generally refer to these unusually bright galactic centers as **active galactic nuclei** and reserve the term **quasar** for the very brightest of them. The brightest quasars shine more powerfully than 1,000 galaxies the size of the Milky Way.

The glory days of quasars are long past. The fact that we find quasars primarily at great distances tells us that these blazingly luminous objects arose during an early stage in the evolution of at least some galaxies. Because we find no nearby quasars and relatively few nearby galaxies with less luminous active galactic nuclei, we conclude that the objects that shine as

FIGURE 17.28 The active galactic nucleus in the elliptical galaxy M 87. The bright yellow spot is the active nucleus, and the blue streak is a jet of particles shooting outward from the nucleus at nearly the speed of light.

quasars in young galaxies must become dormant as the galaxies age. Thus, many nearby galaxies—we don't yet know which ones—that now look quite normal must have centers that once shone brilliantly as quasars.

What could possibly drive the incredible luminosities of quasars, and why did quasars fade away? A growing body of evidence points to a single answer: The energy output of quasars comes from gigantic accretion disks surrounding **supermassive black holes**—black holes with masses millions to billions of times that of our Sun. But before we study how these incredible powerhouses work, let's investigate the evidence that points to their existence.

The Discovery of Quasars

In the early 1960s, a young professor at the California Institute of Technology named Maarten Schmidt was busy identifying cosmic sources of radio-wave emission. Usually, the radio sources turned out to be normal-looking galaxies, but one day he discovered a major mystery. A radio source called 3C 273 looked like a blue star through a telescope but had strong emission lines at wavelengths that did not appear to correspond to any known chemical element. After months of puzzlement, Schmidt suddenly realized that the emission lines were not coming from an unfamiliar element at all. Instead, they were emission lines of hydrogen hugely redshifted from their normal wavelengths.

Schmidt computed the distance to 3C 273 using Hubble's law; then he plugged this distance into the luminosity–distance formula [Section 13.2]. What he found was astonishing: 3C 273 has a luminosity of about 10^{39} watts, well over a trillion (10^{12}) times that of our Sun—making it hundreds of times more powerful than the entire Milky Way Galaxy. Discoveries of similar but even more distant objects soon followed. Because the first few of these objects were strong sources of radio emission that looked like stars through visible-light telescopes, they were named "quasi-stellar radio sources," or quasars for short. Later, astronomers learned that most quasars are not such powerful radio emitters, but the name has stuck.

For many years, a debate raged among astronomers over whether Hubble's law could really be used to determine quasar distances. Some argued that quasars might have high redshifts for other reasons and therefore might be much nearer to us than Hubble's law would suggest. The vast majority of astronomers now consider this debate settled. Improved images show that quasars are indeed the centers of extremely distant galaxies and often are members of very distant galaxy clusters.

Most quasars lie more than halfway to the cosmological horizon. The lines in typical quasar spectra are shifted to more than three times their rest wavelengths, which tells us that the light from these quasars emerged when the universe was less than a third of its present age. Light from the farthest known quasars began its journey when the universe was only about a tenth of its present age.

The extraordinary energy output of quasars emerges across an unusually wide swath of the electromagnetic spectrum. Quasars radiate approximately equal amounts of power from infrared wavelengths all the way through to gamma rays; they also produce strong emission lines. By comparison, most stars and galaxies emit primarily visible light. The wide spread of photon energies coming from quasars implies that they contain matter with a wide range of temperatures. How does this matter manage to radiate such a large luminosity?

Evidence from Nearby Active Galactic Nuclei

Quasars are difficult to study in detail because they are so far away. Luckily, some quasarlike objects are much closer to home. About 1% of present-day galaxies—that is, galaxies we see nearby—have less powerful active galactic nuclei that look very much like weak quasars (such galaxies are called *Seyfert galaxies*). The spectra of these active nuclei range from infrared to gamma rays, just like those of quasars. The only real distinction between nearby active galactic nuclei and quasars is their luminosity. Generally, the power outputs of quasars swamp those of the galaxies that contain them, making the surrounding galaxy hard to detect (Figure 17.29). That is why quasars look "quasi-stellar." The galaxies surrounding dimmer active galactic nuclei are much easier to see because their centers are not so overwhelmingly bright.

The light-emitting regions of active galactic nuclei are so small that even the sharpest images do not resolve them. Our best visible-light images show only that active galactic nuclei must be smaller than 100 light-years (30 parsecs) across. Radio-wave images made with the aid of interferometry show that these nuclei are even smaller: less than 3 light-years (1 parsec) across. But rapid changes in the luminosities of some active galactic nuclei point to an even smaller size.

To understand how variations in luminosity give us clues about an object's size, imagine that you are a master of the universe and you want to signal one of your fellow masters a billion light-years away. An active galactic nucleus would make an excellent signal beacon, because it is so bright. However, suppose the smallest nucleus you can find is 1 light-year across. Each time you flash it on, the photons from the front end of the source reach your fellow master a full year before the photons from the back end. Thus, if you flash it on and off more than once a

FIGURE 17.29 The bright quasar in the nucleus of the galaxy just left of center outshines the rest of the galaxy, which appears disturbed, as if it has recently undergone a collision.

year, your signal will be smeared out. Similarly, if you find a source that is 1 light-day across, you can transmit signals that flash on and off no more than once a day. If you want to send signals just a few hours apart, you need a source no more than a few light-hours across.

Occasionally, the luminosity of an active galactic nucleus doubles in a matter of hours. The fact that we see a clear signal indicates that the source must be less than a few light-hours across. In other words, the incredible luminosities of active galactic nuclei and quasars are apparently being generated in a volume of space not much bigger than our solar system.

Radio Galaxies and Jets

In the early 1950s, a decade before the discovery of quasars, radio astronomers noticed that certain galaxies emit unusually large quantities of radio waves. Today we believe that these **radio galaxies** are another class of celestial powerhouse that is closely related to quasars. Upon close inspection, we find that much of the radio emission comes not from the galaxies themselves, but rather from pairs of huge **radio lobes**, one on either side of the galaxy (Figure 17.30).

The radio waves from the lobes are produced by electrons and protons spiraling around magnetic field lines at nearly the speed of light. Astronomers of the 1950s guessed that this might be the source of the radio emission, but the implied amounts of energy in the lobes seemed implausibly large until the discovery of quasars. British theoretician Martin Rees led the way in showing that radio galaxies contain active galactic nuclei. He argued in 1971 that radio galaxies must have powerful **jets** spurting from their nuclei that transport energy to their lobes. That is, although the lobes lie far outside the visible galaxy, he asserted that the ultimate source of their energy would be found in the galaxy's tiny nucleus, which must therefore be as powerful as a quasar. A few years later, detailed radio images began to support his claims.

Today, radio telescopes resolve the structure of radio galaxies in vivid detail. At the center of a typical radio galaxy sits a tiny radio core—the active galactic nucleus of the radio galaxy—less than a few light-years across. Two jets of plasma shoot out of the core in opposite directions. Frequently, only the jet tilted in our direction is visible. Using time-lapse radio images taken several years apart, we can track the motions of various plasma blobs in the jets. Some of these blobs move at close to the speed of light. The lobes lie at the ends of the jets, sometimes as much as a million light-years from the core.

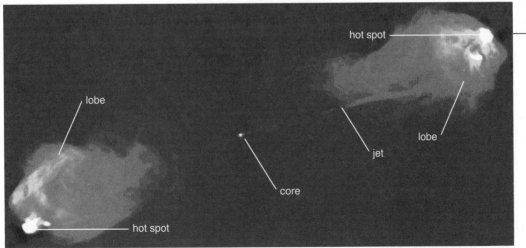

FIGURE 17.30 Radio image of the radio galaxy Cygnus A taken with the Very Large Array in New Mexico. The distance between the lobes is about 400,000 light-years, several times larger than the extent of the galaxy in visible light.

Quasars and radio galaxies may be much more similar than they appear. Indeed, many quasars, including 3C 273, have core–jet–lobe radio structures reminiscent of radio galaxies. For example, Figure 17.31 shows a series of pictures of a jet in which a blob of plasma is moving outward from a quasar at close to the speed of light. Moreover, the active galactic nuclei of many radio galaxies seem to be concealed beneath donut-shaped rings of dark molecular clouds (Figure 17.32). It may be that such structures look like quasars when they are oriented so that we can see the active galactic nucleus at the center, and like the cores of radio galaxies when the ring of dusty gas dims our view of the central object.

Supermassive Black Holes

Astronomers have worked hard to envision physical processes that might explain how radio galaxies, quasars, and other active galactic nuclei release so much energy within such small central volumes. Only one explanation seems to fit: The energy comes from matter falling into a supermassive black hole (Figure 17.33). Gravity converts the potential energy of the infalling matter into kinetic energy. Collisions between infalling particles convert the kinetic energy into thermal energy, and photons carry this thermal

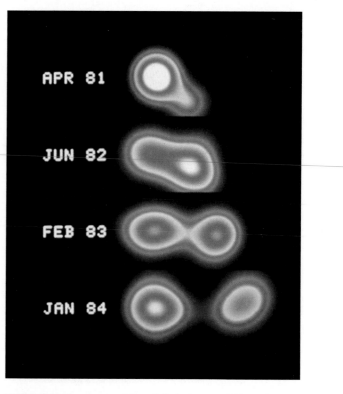

APR 81

JUN 82

FEB 83

JAN 84

FIGURE 17.31 These images of the radio jet in the quasar 3C 345, taken over a period of several years, show a blob of plasma (right) moving away from the core (left) at close to the speed of light.

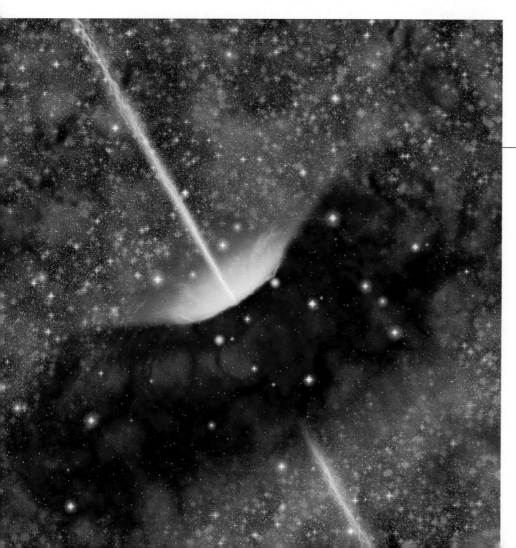

FIGURE 17.32 Artist's conception of the central region of a radio galaxy. The active galactic nucleus, obscured by a ring of dusty clouds, lies at the point from which the jets emerge. If viewed along a direction closer to the jet axis, the active nucleus would not be obscured and would look more like a quasar.

FIGURE 17.33 Artist's conception of an accretion disk surrounding a supermassive black hole. Note that this picture represents only the very center of an object like that shown in Figure 17.32.

energy away. As in X-ray binaries, we expect that the infalling matter swirls through an accretion disk before it disappears beneath the event horizon of the black hole [Section 15.4].

This method of energy generation can be awesomely efficient. The gravitational potential energy lost by a chunk of matter falling into a black hole is equivalent to its mass-energy, $E = mc^2$. As much as 10–40% of this energy can emerge as radiation before the matter crosses the event horizon. Thus, accretion by black holes is far more efficient at producing light than is nuclear fusion, which converts less than 1% of mass-energy into photons. Note that, as with accretion into the black holes formed in supernovae, the light is coming not from the black hole itself but rather from the hot gas surrounding it.

A variety of mechanisms explain why quasars and other active galactic nuclei radiate energy across the electromagnetic spectrum. The hot gas in and above the accretion disk produces copious amounts of ultraviolet and X-ray photons. This radiation ionizes surrounding interstellar gas, energizing intense ionization nebulae that emit visible light. The emission lines produced by these nebulae are the same ones Maarten Schmidt used to measure the first quasar redshifts. The infrared light may come from rings of molecular clouds that encircle the active galactic nucleus (see Figure 17.32). Dust grains in these molecular clouds can absorb the high-energy light and reemit it as infra-

red light. This ring of dense clouds could well be the reservoir for the accretion disk, supplying the matter that the giant black hole eventually consumes. Finally, the radio emission from active galactic nuclei comes from the fast-moving electrons that we sometimes see jetting from these nuclei at nearly the speed of light.

Although it's easy to understand why matter swirls inward through the accretion disk toward the black hole, explaining the powerful jets emerging from active galactic nuclei is more challenging. One plausible model for jet production relies on the twisted magnetic fields thought to accompany accretion disks. As an accretion disk spins, it pulls the magnetic field lines that thread it around in circles. Charged particles start to fly outward along the field lines like beads on a twirling string. The particles careening along the field lines accelerate to speeds near the speed of light, forming a jet that shoots out into space. Some jets blast all the way through the galaxy's interstellar medium and penetrate into the much less dense intergalactic gas. The hot spots at the ends of the lobes in radio galaxies (see Figure 17.30) are the places where the jets are currently ramming into the intergalactic gas. When particles traveling down a jet hit the hot spot, they are deflected into the surrounding radio lobe like water from a firehose hitting a wall. The particles then fill the lobe with energy, generating powerful radio emission.

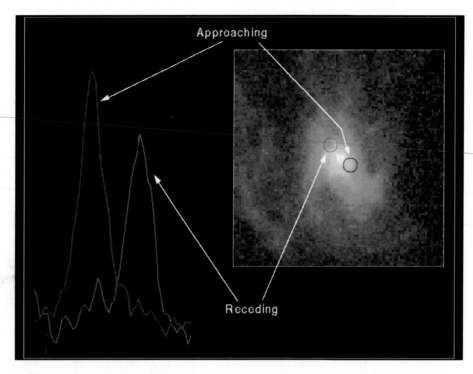

FIGURE 17.34 Doppler shifts of the emission lines in the nucleus of the elliptical galaxy M 87 indicate a 2–3-billion-solar-mass black hole.

The supermassive black hole theory explains many of the observed features of quasars and other active galactic nuclei, but it is incomplete. We do not yet know what would create such giant black holes, nor do we know why quasars eventually run out of gas and stop shining. The preponderance of quasars during the first several billion years of the universe suggests that the formation of supermassive black holes is somehow linked to galaxy formation, a suggestion supported by recent observations showing that the mass of a galaxy's central black hole is closely related to the mass of its spheroidal component. Some scientists have suggested that clusters of neutron stars resulting from extremely dense starbursts at the centers of galaxies might somehow coalesce to form an enormous black hole, but these speculations are still unverified. The origins of supermassive black holes remain mysterious.

Do supermassive black holes really drive the tremendous activity of radio galaxies, quasars, and other active galactic nuclei? The idea that such monsters even exist has been hard for some astronomers to swallow. Proving that we have found a black hole is tricky. Black holes themselves do not emit any light, so we need to infer their existence from the ways in which they alter their surroundings. In the vicinity of a black hole, matter should be orbiting at high speed around something invisible.

The case for supermassive black holes grew much stronger during the 1990s. The relatively nearby galaxy M 87 features a bright nucleus and a jet that emits both radio and visible light. Thus, it was already a prime black-hole suspect when astronomers pointed the Hubble Space Telescope at its core in 1994 (Figure 17.34). The spectra they gathered showed blueshifted emission lines on one side of the nucleus and redshifted emission lines on the other. This pattern of Doppler shifts is the characteristic signature of orbiting gas: On one side of the orbit the gas is coming toward us and hence is blueshifted, while on the other side it is moving away from us and is redshifted. The magnitude of these Doppler shifts shows that the gas, located up to 60 light-years (18 parsecs) from the center, is orbiting something invisible at a speed of hundreds of kilometers per second. This high-speed orbital motion indicates that the central object has a mass some 2–3 billion times that of our Sun. A supermassive black hole is the only thing we know of that could be so massive while remaining unseen.

Observations of NGC 4258, another galaxy with a visible jet, delivered even more persuasive evidence just 1 year later. A ring of molecular clouds orbits the nucleus of this galaxy in a circle less than 1 light-year in radius. We can pinpoint these clouds because they amplify the microwave emission lines of water molecules, generating beams of microwaves very similar to laser beams. The Doppler shifts of these emission lines allow us to determine the orbits of the clouds very precisely. Their orbital motion tells us that the clouds are circling a single, invisible object—presumably a black hole—with a mass of 36 million solar masses.

We may never be 100% certain that we have discovered black holes in other galaxies. The best we can do is rule out all other possibilities. However, the hypothesis that gigantic black holes lie at the cores of quasars, nearby active galactic nuclei, and radio galaxies is withstanding the tests of time and thousands of observations.

THE BIG PICTURE

The picture could hardly get any bigger than it has in this chapter. Looking through both space and time, we have seen nearly to the limits of the observable universe and have found that it is filled with a wide variety of galaxies. As you look back, keep sight of these "big picture" ideas:

- Our measurements of the distances to faraway galaxies rely on a chain of distance measurements that begins with the Earth–Sun distance measured with the aid of radar ranging and continues with parallax and various standard-candle techniques. The distance chain allowed Hubble to measure distances to galaxies and thus to discover the relationship known as Hubble's law and the expansion of the universe.

- As the universe expands, it carries the matter within it along for the ride. From the current expansion rate and from estimates of the overall strength of gravity in the universe and the ages of the oldest stars, we infer that the universe is somewhere between about 12 and 16 billion years old.

- Although we do not yet know the complete story of galaxy evolution, we are rapidly learning more. We know that galaxies grow from protogalactic clouds of gas, but collisions with neighboring galaxies have probably affected many galaxies.

- The tremendous energy outputs of quasars and other active galactic nuclei, including those of radio galaxies, are probably powered by gas accreting onto supermassive black holes. The centers of many present-day galaxies must still contain the supermassive black holes that once enabled them to shine as quasars.

Review Questions

1. Distinguish between *spiral galaxies* and *elliptical galaxies* in terms of their shapes and colors. What are *irregular galaxies*?

2. Distinguish between the *disk component* and the *spheroidal component* of a spiral galaxy. Which component includes the galaxy's spiral arms? Which includes its bulge? Which includes its halo?

3. What is a *group* of galaxies? What is a *cluster* of galaxies?

4. What is the major difference between an elliptical galaxy and a spiral galaxy? How is an elliptical galaxy similar to the halo of the Milky Way?

5. What do we mean by a *standard candle*? Briefly explain how, once we identify an object as a standard candle, we can use the luminosity–distance formula to find its distance.

6. Describe the technique of *main-sequence fitting*. Why isn't this technique useful for measuring distances to other galaxies?

7. Why are Cepheid variables good standard candles? Briefly explain how Edwin Hubble used his discovery of a Cepheid in Andromeda to prove that the "spiral nebulae" were actually entire galaxies.

8. What is *Hubble's law*? How did Hubble discover it? Explain the meaning of Hubble's law when it is expressed mathematically as $v = H_0 \times d$. What are the units of *Hubble's constant, H_0*?

9. Describe how we can use Hubble's law to determine the distance to a distant galaxy. What practical difficulties limit the use of Hubble's law for measuring distances?

10. What makes white dwarf supernovae good standard candles? Briefly describe how we have learned the true luminosities of white dwarf supernovae.

11. What is the *Tully–Fisher relation*? How does it enable us to use entire spiral galaxies as standard candles?

12. What is our current best estimate of Hubble's constant? Summarize all the links in the distance chain that allow us to estimate distances to the farthest reaches of the universe.

13. In what ways is the surface of an expanding balloon a good analogy to the universe? In what ways is this analogy limited? Explain why a miniature scientist living in a polka dot on the balloon would observe all other dots to be moving away, with more distant dots moving away faster.

14. What is a *lookback time*? Why does it make more sense to talk about a distant object in terms of its lookback time than in terms of its distance?

15. What do we mean by a *cosmological redshift*? How does our interpretation of a distant galaxy's redshift differ if we think of it as a cosmological redshift rather than a Doppler shift?

16. What two assumptions underlie theoretical models of galaxy formation? According to these models, what processes eventually slow the collapse of *protogalactic clouds?*

17. Distinguish between disk-population stars and spheroidal-population stars, in terms of their ages, masses, colors, and orbits. Briefly describe how the gravitational contraction of a protogalactic cloud explains the different characteristics of these populations.

18. Briefly explain why we expect that collisions between galaxies should be relatively common, while collisions between stars are extremely rare. Why should galaxy collisions have been more common in the past than they are today?

19. Briefly describe how a collision between two spiral galaxies might lead to the creation of a single elliptical galaxy. What evidence supports this scenario for the formation of elliptical galaxies?

20. Briefly describe the discovery of *quasars.* What evidence convinced astronomers that the high redshifts of quasars really do imply great distances? Why can we learn more about quasars by studying nearby *active galactic nuclei?*

21. Briefly explain how we can use variations in luminosity to set limits on the size of an object's emitting region. For example, if an object doubles its luminosity in 1 hour, how big can it be?

22. What is a *radio galaxy?* Describe *jets* and *radio lobes.* Why do we think that their ultimate energy sources lie in quasarlike galactic nuclei?

23. Briefly explain the general picture of how *supermassive black holes* enable quasars to produce their huge luminosities. Summarize the evidence supporting the idea that supermassive black holes lie at the center of radio galaxies, quasars, and other active galactic nuclei.

Discussion Questions

1. *Cosmology and Philosophy.* One hundred years ago, many scientists believed that the universe was infinite and eternal, with no beginning and no end. When Einstein first developed his general theory of relativity, he found it predicted that the universe should be either expanding or contracting. He believed so strongly in an eternal and unchanging universe that he modified the theory, a modification he would later call his "greatest blunder." Why do you think Einstein and others assumed that the universe had no beginning? Do you think that a universe with a definite beginning in time, some 15 billion or so years ago, has any important philosophical implications? Explain.

2. *The Case for Supermassive Black Holes.* The evidence for supermassive black holes at the center of galaxies is strong. However, it is very difficult to prove absolutely that they exist because the black holes themselves emit no light. We can only infer their existence from their powerful gravitational influences on surrounding matter. How compelling do you find the evidence? Do you think astronomers have proved the case for black holes beyond a reasonable doubt? Defend your opinion.

Problems

Sensible Statements? For **problems 1–9**, decide whether the statement is sensible and explain why it is or is not.

1. If you want to find elliptical galaxies, you'll have better luck looking in clusters of galaxies than elsewhere in the universe.

2. Cepheid variables make good standard candles because they all have exactly the same luminosity.

3. If the standard candles you are using are less luminous than you think they are, then the distances you determine from them will be too small.

4. Galaxy A is moving away from me twice as fast as Galaxy B. That probably means it's twice as far away.

5. The lookback time to the Andromeda Galaxy is about 2.5 million years.

6. Galaxies that are more than 10 billion years old are too far away to see even with our most powerful telescopes.

7. Heavy elements ought to be much more common near the Milky Way's center than at its outskirts.

8. We know that some galaxies contain supermassive black holes because their centers are completely dark.

9. The black hole at the center of our own galaxy may once have powered an active galactic nucleus.

10. *Classifying Galaxies.* Figure 17.1 shows a wide variety of galaxies. Devise your own scheme for classifying them by color, and describe it. Estimate the fraction of galaxies that fall into each color class. Devise and describe your own scheme for classifying galaxies by shape. Estimate the fraction that fall into each shape class. Describe any relationships you see between the shapes and the colors of these galaxies.

11. *Distance Measurements.* The techniques astronomers use to measure distances are not so different from the ones you use every day. Describe how parallax measurements are similar to your visual depth perception. Describe how standard candle measurements are similar to the way you estimate the distance to an oncoming car at night.

12. *Standard Candles.* List at least three qualities that would tend to make a type of astronomical object useful as a standard candle. Explain why each quality is important if we hope to use objects of this type to measure distances in the universe.

13. *Life Story of a Spiral.* Imagine that you are a spiral galaxy. Describe your life history from birth to the present day. Your story should be detailed and scientifically consistent, but also creative. That is, it should be entertaining while at the same time incorporating current scientific ideas about the formation of spiral galaxies.

14. *Life Story of an Elliptical.* Imagine that you are an elliptical galaxy. Describe your life history from birth to the present. There are several possible scenarios for the formation of elliptical galaxies, so choose one and stick to it. Be creative while also incorporating scientific ideas that demonstrate your understanding.

15. *Research Project: Hubble's Constant.* Measurements of Hubble's constant are rapidly growing more reliable. Find at least two measurements of H_0 determined in the past 2 years. Describe how these values were measured. What is the quoted uncertainty of each? By how much do the values differ?

16. *Research: The Quasar Controversy.* For many years, some astronomers argued that quasars were not really as distant as Hubble's law implies. Research the history of the discovery of quasars and the debates that followed. What evidence led some astronomers to think that quasars might be nearer than Hubble's law suggested? Why did most astronomers eventually conclude that quasars really are far away? Write a one- to two-page report summarizing your findings.

Web Project

Find useful links for Web projects on the text Web site.

1. *Future Missions.* The subject of galaxy evolution is a very active area of research. Using the Web, look for information on current and future NASA missions involved in investigating galaxy evolution. How big are the planned telescopes? What wavelengths will they look at? When will they be launched? Assemble a journal from your Web research, with pictures and a paragraph on each mission you find. If possible, make your journal a Web page and include links to the relevant Web sites.

CHAPTER 18

Dark Matter and the Fate of the Universe

The majority of the matter in the universe, the stuff that binds galaxies together with the force of its own gravity, is too dark to see. We know that this matter exists because we can detect its gravitational influence on other things, but what is this so-called *dark matter*? Do armies of Jupiter-size bodies, too dim for us to see at a distance, populate the voids between the stars? Could black holes be responsible? Or is dark matter an entirely new form of matter, still undiscovered here on Earth? We don't yet know the answer. Incredibly, we still haven't identified the most common form of matter in the universe, making dark matter one of the greatest cosmic mysteries.

In this chapter, we will investigate why most astronomers believe that dark matter exists. We'll see how it affects structures the size of galaxies and larger. We'll investigate current hypotheses about the nature of dark matter. We'll see how the fate of the universe relates to the question of just how much dark matter there is. And we'll explore recent evidence suggesting that the universe's expansion may be accelerating.

18.1 The Mystery of Dark Matter

We first encountered dark matter when we studied the Milky Way's rotation in Chapter 16. We saw that atomic hydrogen clouds lying farther from the galactic center than our Sun orbit the galaxy at unexpectedly high speeds. We concluded that much of our galaxy's mass must lie beyond the distance of the Sun's orbit around the galactic center, distributed throughout the galaxy's spherical halo [Section 16.4]. Yet most of the Milky Way's *light* comes from stars lying closer to the galaxy's center than does our Sun. Together, these facts imply that the halo contains large amounts of matter but this matter emits so little light that we cannot see it. Thus, we call it *dark matter*.

Does dark matter really exist? To claim that the galaxy is filled with a form of matter that we cannot yet identify may seem strange, but there are only two possible explanations for the high orbital velocities of the atomic hydrogen clouds: Either their velocities are caused by the gravitational attraction of unseen matter, or we are doing something wrong when we apply the law of gravity to understand their orbits. The latter possibility would imply that we do not understand how gravity operates on galaxy-size scales, but our theories of gravity successfully account for many cosmic phenomena. Thus, the vast majority of astronomers believe that we correctly understand gravity and that dark matter must really exist.

The evidence that the Milky Way contains dark matter qualifies as an interesting surprise on its own, but it is also a glimpse into a more profound mystery. If our suspicions about dark matter are correct, then the luminous part of the Milky Way's disk must be rather like the tip of an iceberg, marking only the center of a much larger clump of mass (Figure 18.1). A more detailed analysis involving the motions of nearby dwarf galaxies shows that the total mass of dark matter in the Milky Way might be 10 times greater than the mass of visible stars. Each galaxy in the universe is similar to our own in this respect, meaning that we probably cannot see most of the matter in the universe. This matter's darkness makes it challenging to study, but its overwhelming dominance of the universe makes it extremely important. Dark matter, by virtue of its mighty gravitational pull, seems to be the glue that holds galaxies and clusters of galaxies together.

Dark matter has even more profound implications for the fate of the universe. The universe ultimately has two possible fates: It might continue to expand forever, or it might someday stop expanding and begin to collapse. We do not yet know which fate awaits our universe because the answer depends on the overall density of matter in the universe—and we cannot determine the overall density until we first determine how much dark matter is out there. Thus, the very fate of the universe hinges on the total amount of dark matter.

In this chapter, we will presume that we understand gravity correctly and consequently that dark matter pervades the universe. If our understanding of gravity is correct, then we already know quite a lot about dark matter, despite the fact that its composition remains a mystery. We do not know the total amount of dark matter in the universe, but we can measure the amounts in individual galaxies and clusters of galaxies from its gravitational effects. We can even make some reasonable guesses as to what it will turn out to be. We'll begin by discussing how we "discover" dark matter and conclude by discussing what our current knowledge predicts about the ultimate fate of the universe.

FIGURE 18.1 Dark matter distribution relative to luminous matter in a spiral galaxy.

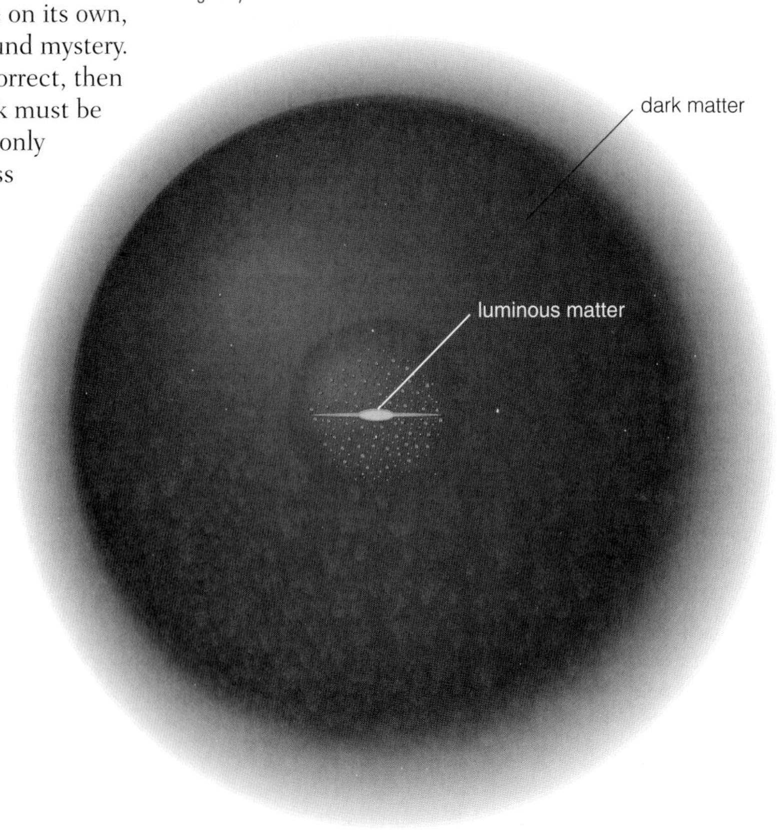

dark matter

luminous matter

18.2 Dark Matter in Galaxies

The claim that dark matter far outweighs the visible matter in the universe might seem farfetched, but it rests on fundamental physical laws. Newton's laws of motion and gravity are among the most trustworthy tools in science. We have used them time and again to measure masses of celestial objects. We found the masses of the Earth and the Sun by applying Newton's version of Kepler's third law [Section 5.3]. We used this same law to calculate the masses of stars in binary star systems, revealing the general relationships between the masses of stars and their outward appearances. Newton's laws have also told us the masses of things we can't see directly, such as the masses of neutron stars in X-ray binaries and of black holes in active galactic nuclei. These laws have proved extremely reliable in many different applications. Thus, when Newton's laws tell us that dark matter exists, we are not inclined to ignore them.

We determine the amount of dark matter in a galaxy from the galaxy's mass and luminosity: We can infer the total mass in stars from the galaxy's luminosity, so whatever additional mass remains unaccounted for must be dark. Measuring a galaxy's luminosity is relatively easy. We simply point a telescope at the galaxy in question, measure its apparent brightness, and calculate its luminosity from the luminosity–distance formula [Section 13.2]. Measuring the mass is more complicated. We would like to apply Newton's laws in the form of the orbital velocity law from Chapter 16 to matter orbiting as far from the galaxy's center as possible. However, the matter farthest from the center of a galaxy is extremely dim. For all we know, the matter at the edge of a galaxy might be completely dark.

Weighing Galaxies

We can weigh a spiral galaxy by measuring the gravitational effects of the galaxy's mass on the orbits of objects in its disk. Even beyond the point in the disk at which starlight fades into the blackness of intergalactic space, we can still see radio waves from atomic hydrogen gas. We can therefore use Doppler shifts of the 21-cm emission line of atomic hydrogen [Section 16.2] to determine how quickly this gas moves toward us or away from us (Figure 18.2). Galaxies beyond the Local Group have cosmological redshifts affecting all their spectral lines. However, on one side of a spiral galaxy the gas is rotating away from us, so its 21-cm line is redshifted a little more than the redshift of the galaxy as a whole. On the other side, the 21-cm line is blueshifted relative to the redshift of the galaxy as a whole because the gas is rotating toward us. From the Doppler shifts of these clouds, we can construct a *rotation curve*—a plot

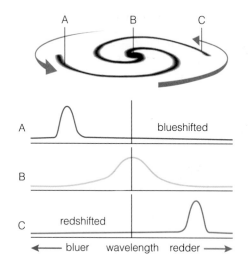

FIGURE 18.2 Measuring the rotation of a spiral galaxy with the 21-cm line of atomic hydrogen. Blueshifted lines on the left side of the disk show how fast that side is rotating toward us. Redshifted lines on the right side show how fast that side is rotating away from us.

showing orbital velocities of gas clouds and stars—just as we did for the Milky Way (Figure 18.3) [Section 16.4]. A rotation curve contains all the information we need to measure the mass contained within the orbits of the outermost gas clouds. (Because the Doppler effect tells us only about the velocity of material directly toward or away from us, we must also take into account the tilt of the galaxy before we construct the rotation curve.)

TIME OUT TO THINK *As a brief review, draw a rotation curve for our solar system. How does the solar system rotation curve differ from the rotation curve of the Milky Way? What does that tell us about the distribution of matter in the Milky Way? Why? (Hint: See Chapter 16 for review.)*

The rotation curves of most spiral galaxies turn out to be remarkably flat as far out as we can see, which means that the orbital speeds of gas clouds remain roughly constant with increasing distance from the galactic center (Figure 18.4). We saw in Chapter 16 that our Milky Way's rotation curve remains more or less flat from well within the radius of the Sun's orbit (28,000 light-years) to beyond twice that radius. The rotation curves of some other spiral galaxies are flat to well beyond 150,000 light-years.

Because the orbital speeds of gas clouds tell us the amount of mass contained within their orbital paths, the flat rotation curves imply that a great deal of matter lies far from the galactic center [Section 16.4]. In particular, if we imagine drawing bigger and bigger circles around a galaxy, the flat rotation curves imply that we must keep encircling more and more matter.

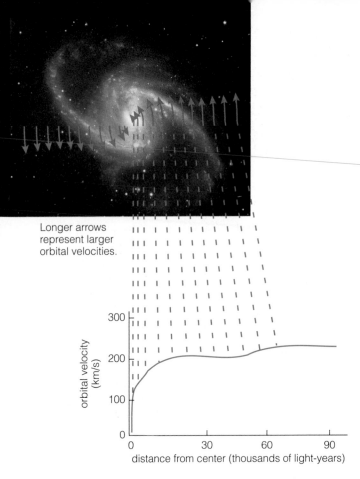

Longer arrows represent larger orbital velocities.

FIGURE 18.3 A rotation curve shows the orbital velocities of stars or gas clouds at different distances from a galaxy's center.

to its luminosity. For example, the Milky Way contains about 90 billion (9×10^{10}) solar masses of material within the Sun's orbit. However, the total luminosity of stars within this region is only about 15 billion (1.5×10^{10}) solar luminosities. Thus, on average, it takes 6 solar masses of matter to produce 1 solar luminosity of light in this region of the galaxy (15 billion × 6 = 90 billion). We therefore say that the Milky Way's **mass-to-light ratio** within the Sun's orbit is about 6 solar masses per solar luminosity. This fact tells us that most matter is dimmer than our Sun, which is not surprising since we know that most stars are smaller and dimmer than our Sun. We find similar mass-to-light ratios for the inner regions of most other spiral galaxies.

TIME OUT TO THINK *Because the Sun's mass is $1M_{Sun}$ and its luminosity is $1L_{Sun}$, its mass-to-light ratio is $1M_{Sun}/1L_{Sun}$ = 1 solar mass per solar luminosity. What is the mass-to-light ratio of a $1M_{Sun}$ red giant with luminosity $100L_{Sun}$? What is the mass-to-light ratio of a $1M_{Sun}$ white dwarf with luminosity $0.001L_{Sun}$?*

Elliptical galaxies contain virtually no luminous high-mass stars, so their stars are less bright on average than the stars in a spiral galaxy. In the inner regions of elliptical galaxies, the orbits of stars indicate a mass-to-light ratio of about 10 solar masses per solar luminosity—nearly double the mass-to-light ratio in the central region of the Milky Way. This is not surprising given the fact that elliptical galaxies have dimmer stars.

The surprise in mass-to-light ratios—and one of the key pieces of evidence for the existence of dark matter—comes when we look to the outer reaches of galaxies. As we look farther from a galaxy's center, we find a lot more mass but not much more light. The mass-to-light ratios for entire spiral galaxies can be as high as 50 solar masses per solar luminosity, and the overall ratios in some dwarf galaxies can be even higher. We are forced to conclude that stars alone cannot account for the amount of mass present in galaxies. For example, if a galaxy's overall mass-

The mass of a spiral galaxy with gas clouds orbiting at 200 km/s at a distance of 150,000 light-years from its center must be at least 500 billion (5×10^{11}) solar masses.

In elliptical galaxies, we can study only the motions of stars, but the basic result remains the same—a great deal of matter exists outside the galaxies' cores.

Mass-to-Light Ratio

We can determine how much dark matter a galaxy contains by comparing the galaxy's measured mass

FIGURE 18.4 Actual rotation curves of four spiral galaxies.

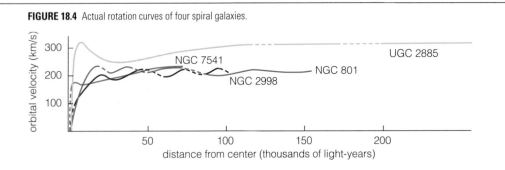

to-light ratio is 60 solar masses per solar luminosity and its stars account for only 6 solar masses per solar luminosity, then the remaining 90% of the galaxy's mass must be dark.

18.3 Dark Matter in Clusters

The problem of dark matter in astronomy is not particularly new. In the 1930s, astronomer Fritz Zwicky was already arguing that clusters of galaxies held enormous amounts of this mysterious stuff (Figure 18.5). Few of his colleagues paid attention, but later observations supported Zwicky's claims. We can now weigh clusters of galaxies in three different ways: by measuring the speeds of galaxies orbiting the center of the cluster, by studying the X-ray emission from hot gas between the cluster galaxies, and by observing how the cluster's gravity bends light. All three techniques indicate that clusters contain huge amounts of dark matter. Let's investigate each of these techniques more closely.

Orbiting Galaxies

Zwicky was one of the first astronomers to think of galaxy clusters as huge swarms of galaxies bound together by gravity. It seemed natural to him that galaxies clumped closely in space should all be orbiting one another, just like the stars in a star cluster. He therefore assumed that he could measure cluster masses by observing galaxy motions and applying the orbital velocity law.

Armed with a spectrograph, Zwicky measured the redshifts of the galaxies in a particular cluster and used these redshifts to calculate the speeds at which the individual galaxies are moving away from us. He determined the velocity of the cluster as a whole by averaging the velocities of its individual galaxies. He then estimated the speed of a galaxy around the cluster—that is, its orbital speed—by subtracting this average velocity from the individual galaxy's velocity. Finally, he plugged these orbital speeds into an appropriate form of the orbital velocity law [Section 16.4] to estimate the cluster's mass and compared this mass to the luminosity of the cluster. To his surprise, Zwicky found that clusters of galaxies have huge mass-to-light ratios: They contain hundreds of solar masses for each solar luminosity of radiation they emit. He concluded that most of the matter within these clusters must be almost entirely dark. Many astronomers disregarded Zwicky's result, believing that he must have done something wrong to arrive at such a strange answer.

Today, far more sophisticated measurements of galaxy orbits in clusters confirm Zwicky's original

FIGURE 18.5 Fritz Zwicky.

finding. If we measure the radial velocities of many galaxies in a cluster, we can estimate the average orbital velocity of all the cluster's galaxies. Cluster masses found in this way imply lots of dark matter. The visible portions of a cluster's galaxies represent less than 10% of the cluster's mass.

Intracluster Medium

A second method for weighing a cluster of galaxies relies on X-ray observations of the hot gas that fills the space between the galaxies in the cluster (Figure 18.6). This gas, also known as the **intracluster medium** (*intra* means "within"), is so hot that it emits primarily X rays and therefore went undetected until the 1960s, when X-ray telescopes were finally launched above the Earth's atmosphere. The temperature of this gas is tens of millions of degrees in many clusters and can exceed 100 million degrees in the largest clusters. Even though the intracluster medium is invisible in optical light, its mass often exceeds the mass of all the visible stars in all the cluster's galaxies combined. The largest clusters of galaxies contain up to five times more matter in the form of hot gas than in the form of stars.

The temperature of the hot intracluster gas depends on the mass of the cluster itself, enabling us to use X-ray telescopes to measure the masses of galaxy clusters. The intracluster medium in most clusters is

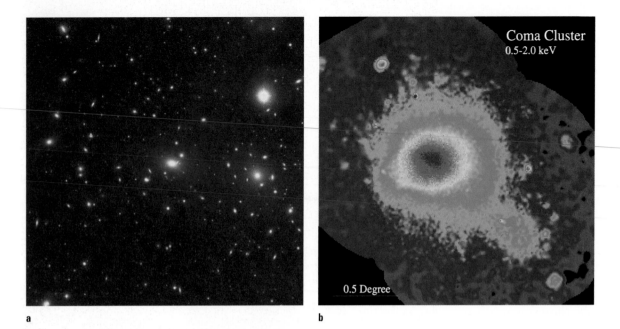

a   b

FIGURE 18.6 (**a**) The Coma Cluster of galaxies seen in visible light. Virtually every pictured object here is a galaxy. (**b**) A map of X-ray emission from the Coma Cluster. Hot, X ray–emitting gas (the intracluster medium) bound to the cluster by gravity fills the spaces between the galaxies.

nearly in a state of *gravitational equilibrium*—that is, the outward gas pressure balances gravity's inward pull [Section 12.1]. In this state of balance, the average kinetic energies of the gas particles are determined primarily by the strength of gravity and hence by the amount of mass within the cluster. Because the temperature of a gas reflects the average kinetic energies of its particles, the gas temperatures we measure with X-ray telescopes tell us the average speeds of the X ray–emitting particles. We can then use these speeds and the orbital velocity law to weigh the cluster.

The results obtained with this method agree well with the results found by studying the orbital motions of the cluster's galaxies. Again, we find that mass-to-light ratios in clusters of galaxies generally exceed 100 solar masses per solar luminosity. Even after accounting for the hot intracluster gas, it is clear that clusters must contain huge quantities of dark matter binding all the galaxies together.

Gravitational Lensing

We have so far relied exclusively on methods derived from Newton's laws, such as the orbital velocity law, to measure galaxy and cluster masses. These laws keep telling us that the universe holds far more matter than we can see. Can we trust these laws? Today, we have another tool for measuring masses: *gravitational lensing.*

Gravitational lensing occurs because masses distort spacetime—the "fabric" of the universe [Section 15.4]. Massive objects can therefore act as **gravitational lenses** that bend light beams passing nearby. This prediction of Einstein's theory of general relativity was first verified in 1919 during an eclipse of the Sun. Because the light-bending angle of a gravitational lens depends on the mass of the object doing the bending, we can measure the masses of objects by observing how strongly they distort light paths.

Figure 18.7 shows a striking example of how a cluster of galaxies can act as a gravitational lens. Many of the yellow elliptical galaxies concentrated toward the center of the picture belong to the cluster, but at least one of the galaxies pictured does not. At several positions on various sides of the central clump of yellow galaxies you will notice multiple images of the same blue galaxy. Each one of these images, whose sizes differ, looks like a distorted oval with an off-center smudge.

The blue galaxy seen in these multiple images lies almost directly behind the center of the cluster, at a much greater distance. Multiple images arise because photons traveling from this blue galaxy to Earth do not follow straight paths. Instead, the cluster's gravity bends their paths as they pass through the cluster, enabling light from this galaxy to arrive at Earth from a few slightly different directions. Each alternative path produces a separate, distorted image of the blue galaxy, making this one galaxy look like several galaxies (Figure 18.8).

FIGURE 18.7 Hubble Space Telescope picture of a galaxy cluster acting as a gravitational lens. The yellow elliptical galaxies are cluster members, and the odd, blue ovals are multiple images of a single galaxy that lies almost directly behind the cluster's center.

Multiple images of a gravitationally lensed galaxy are rare, occurring only when a distant galaxy lies directly behind the lensing cluster, but single images of gravitationally lensed galaxies behind clusters are quite common. Figure 18.9 shows a more typical example. This picture shows numerous normal-looking galaxies and several arc-shaped galaxies. The oddly curved galaxies are not members of the cluster, nor are they really curved. They are normal galaxies lying far beyond the cluster whose images have been distorted by the cluster's gravity.

Careful analyses of the distorted images in pictures of clusters like these enable us to weigh the clusters without resorting to the orbital velocity law. Instead, Einstein's theory of general relativity tells us how massive these clusters must be to generate the observed distortions. It is reassuring that cluster masses derived in this way generally agree with those derived from galaxy velocities and X-ray temperatures. The three different methods all indicate that clusters of galaxies hold very substantial amounts of dark matter.

TIME OUT TO THINK *Should the fact that we have three different ways of measuring cluster masses give us greater confidence that we really do understand gravity and that dark matter really does exist? Why or why not?*

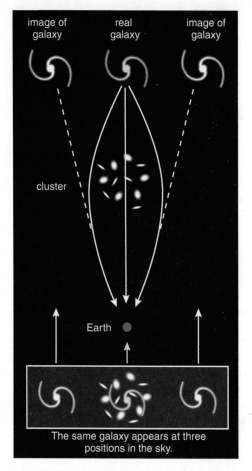

FIGURE 18.8 A cluster's powerful gravity bends light paths from background galaxies to Earth. If light can arrive from several different directions, we see multiple images of the same galaxy.

FIGURE 18.9
Hubble Space Telescope picture of the cluster Abell 2218. The thin, elongated galaxies are the images of background galaxies distorted by the cluster's gravity.

18.4 Dark Matter: Ordinary or Extraordinary?

We do not yet know what the dark matter in galaxies and clusters of galaxies is composed of, but our educated guesses fit into two basic categories. First, some or all of the dark matter could be *ordinary,* made of protons, neutrons, and electrons. In that case, the only unusual thing about dark matter is that it is dim; otherwise it is made of the same stuff as all the "bright matter" that we can see. The second possibility is that some or all of the dark matter is *extraordinary,* made of particles that we have yet to discover. A bit more terminology will be useful. Protons and neutrons belong to a category of particles called *baryons,* so ordinary matter is sometimes called *baryonic matter.* By extension, extraordinary matter is called **nonbaryonic matter**.

Ordinary Dark Matter: MACHOs

Matter need not be extraordinary to be dark. In astronomy, "dark" merely means not as bright as a normal star and therefore not visible across vast distances of space. Your body is dark matter. Everything you own is dark matter. Earth and the planets are also dark matter. The process of star formation itself leaves behind dark matter in the form of *brown dwarfs,* those dim "failed stars" that are not quite massive enough to sustain nuclear fusion [Section 14.2]. Some scientists believe that trillions of faint red stars, brown dwarfs, and Jupiter-size objects left over from the Milky Way's formation still roam our galaxy's halo, providing much of its mass. They fancifully term these objects **MACHOs**, for *massive compact halo objects.*

Because MACHOs are too faint for us to see directly, astronomers who wish to detect them must resort to clever techniques. One innovative way to detect faint starlike objects in the Milky Way's halo takes advantage of gravitational lensing. If trillions of these MACHOs really exist, every once in a while a MACHO should drift across our line of sight to a more distant star. When a MACHO lies almost directly between us and the farther star, the MACHO's gravity will focus the star's light directly toward the Earth. The distant star will appear much brighter than usual for several days or weeks as the MACHO passes in front of it (Figure 18.10). We cannot see the MACHO itself, but the duration of the lensing event reveals its mass.

Gravitational lensing events such as these are rare; they happen to about one star in a million each year. To detect MACHO lensing events, we therefore must monitor huge numbers of stars. Current large-scale monitoring projects now record numerous lensing events annually. These events demonstrate that MACHOs do indeed populate our galaxy's halo, but probably not in large enough numbers to account for all the Milky Way's dark matter. Something else lurks unseen in the outer reaches of our galaxy. Maybe it is also normal matter made from baryons, but maybe not.

Extraordinary Dark Matter: WIMPs

A more exotic possibility is that most of the dark matter in galaxies and clusters of galaxies is not made of ordinary, baryonic matter at all. Let's begin to explore this possibility by taking another look at those nonbaryonic particles we discussed in Section 12.3: neutrinos. These unusual particles are dark by their very nature, because they have no electrical charge and hence cannot emit electromagnetic radiation of any kind. Moreover, they are never bound together with charged particles in the way that neutrons are bound

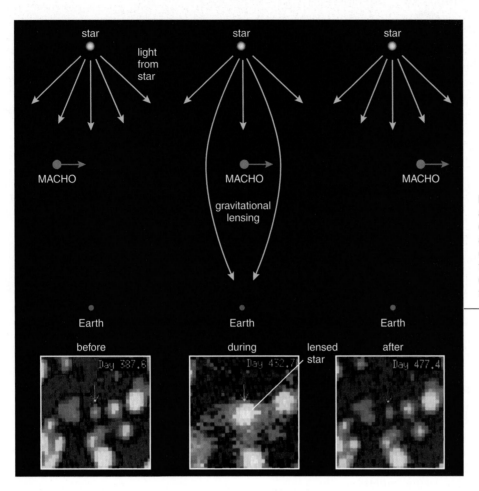

FIGURE 18.10 A lensing event in which a MACHO passing in front of a more distant star temporarily makes the star appear brighter, revealing the presence of the dark MACHO. Multiple images of the star are possible, but they are too close together to resolve with current telescopes.

in atomic nuclei, so their presence cannot be revealed by associated light-emitting particles. In fact, particles like neutrinos interact with other forms of matter through only two of the four forces: gravity and a force called the *weak force*. For this reason, they are said to be *weakly interacting particles*. If you recall that trillions of neutrinos from the Sun are passing through your body at this very moment without doing any damage, you'll see why the name *weakly interacting* fits well.

The dark matter in galaxies cannot be made of neutrinos, because these tiny particles travel through the universe at enormous speeds and can easily escape a galaxy's gravitational pull. (However, as we'll discuss shortly, neutrinos might contribute to the amount of dark matter outside galaxies.) But what if other weakly interacting particles exist that are similar to neutrinos but considerably heavier? They too would evade direct detection, but they would move more slowly and could collect into galaxies, adding mass without adding light. Such hypothetical particles are called **WIMPs**, for *weakly interacting massive particles*. (Weakly interacting particles that are slow-moving enough to collect into galaxies are sometimes called *cold dark matter* to set them apart from faster-moving *hot dark matter* particles such as neutrinos.)

WIMPs could make up most of our galaxy's mass, but they would be completely invisible in all wavelengths of light.

It might surprise you that scientists would suspect the universe to be filled with particles they haven't yet discovered. However, WIMPs could also explain why dark matter doesn't behave like the visible matter in galaxies. During the early stages of the Milky Way's formation, the ordinary (baryonic) matter settled toward our galaxy's center and then flattened into a disk. Meanwhile, the dark matter stayed where it was, out in the galaxy's halo. This resistance of the dark matter to settling is exactly what we would expect from weakly interacting particles. Because WIMPs do not emit electromagnetic radiation, they cannot radiate away their energy. They are therefore stuck orbiting out at large distances and cannot collapse with the rest of the protogalactic cloud.

TIME OUT TO THINK *What do you think of the idea that much of the universe is made of as-yet-undiscovered particles? Can you think of other instances in the history of science in which the existence of something was predicted before it was discovered?*

18.5 Structure Formation

Dark matter remains enigmatic, but every year we are learning more about its role in the universe. Because galaxies and clusters of galaxies seem to contain much more dark matter than luminous matter, we believe that dark matter's gravitational pull must be the primary force holding these structures together. Thus, we strongly suspect that the gravitational attraction of dark matter is what pulled galaxies and clusters together in the first place.

Growth of Structure

Stars, galaxies, and clusters of galaxies are all *gravitationally bound systems,* meaning that their gravity is strong enough to hold them together. In most of the gravitationally bound systems we have discussed so far, gravity has completely overwhelmed the expansion of the universe. That is, while the universe as a whole is expanding, space is *not* expanding within our solar system, our galaxy, or our Local Group of galaxies.

Our best guess at how galaxies formed, briefly outlined in Section 17.5, envisions them growing from slight density enhancements that were present in the very early universe. During the first few million years after the Big Bang, the universe expanded everywhere. But the stronger gravity in regions of enhanced density gradually pulled in matter until these regions stopped expanding and became protogalactic clouds—even as the universe as a whole continued (and still continues) to expand. If dark matter is indeed the most common form of mass in galaxies, it must have provided most of the gravitational attraction responsible for creating the protogalactic clouds. The hydrogen and helium gas in the protogalactic clouds collapsed inward and gave birth to stars, while weakly interacting dark matter remained in the outskirts because of its inability to radiate away its orbital energy. According to this model, the luminous matter in each galaxy must still be nestled inside the larger cocoon of dark matter that initiated the galaxy's formation, just as observational evidence seems to suggest.

The formation of galaxy clusters probably echoes the formation of galaxies. Early on, all the galaxies that will eventually constitute a cluster are flying apart with the expansion of the universe. But the gravity of the dark matter associated with the cluster eventually reverses the trajectories of these galaxies. The galaxies ultimately fall back inward and start orbiting each other randomly, like the stars in the halo of our galaxy.

Some clusters of galaxies apparently have not yet finished forming: Their immense gravity continues to draw in new members. For example, the nearby Virgo cluster of galaxies appears to be tugging on our own Local Group. Right now the Local Group is still moving away from the Virgo Cluster, but not as quickly as a simple application of Hubble's law would predict. Hubble's law lets us calculate the speed at which universal expansion carries us away from the Virgo Cluster [Section 17.3], but it does not account for any gravitational effects. The discrepancy between a straight application of Hubble's law and our actual velocity away from the Virgo Cluster is about 400 km/s. That is, we are moving away from the Virgo Cluster 400 km/s more slowly than we would be as a result of the expansion of the universe alone. Astronomers call this 400-km/s deviation from Hubble's law a **peculiar velocity**, but it's really not so peculiar. It is simply the effect of the gravitational attraction pulling us back toward Virgo against the flow of universal expansion. The overall effect is rather like that of swimming upstream against a strong current: You move "upstream" relative to other objects floating in the current, but you're still headed "downstream" relative to the shore because of the strong current. The speed of the current is like the speed of expansion of the universe, and your swimming speed "upstream" is your peculiar velocity.

Just as the Earth's gravity slows a rising baseball and eventually turns it around and pulls it back toward the ground, the gravitational tug of the Virgo Cluster may eventually turn the galaxies of our Local Group around and pull them into the cluster. Similar processes are taking place on the outskirts of other large clusters of galaxies, where we see many galaxies with large peculiar velocities pointing in the direction of the cluster. Their peculiar velocities indicate that the cluster's gravity is pulling on them. Eventually, some or perhaps all of these galaxies will fall into the cluster. Thus, many clusters are still attracting galaxies, adding to the hundreds they already contain. On even larger scales, clusters themselves have surprisingly large peculiar velocities, hinting that they are parts of even bigger gravitationally bound systems, called **superclusters**, that are just beginning to form (Figure 18.11).

Large-Scale Structures

Beyond about 300 million light-years, we can no longer measure distances accurately enough to determine peculiar velocities. We can estimate the distances to faraway galaxies only by measuring their redshifts and applying Hubble's law. Today we have redshift measurements for many thousands of distant galaxies. When we convert the redshifts to distance estimates, we can create three-dimensional maps of the universe. Such maps have been made for only a few "slices" of the sky so far, but they are already revealing **large-scale structures** much vaster than clusters of galaxies.

Among the most famous depictions of large-scale structure are the slices of the universe pictured in Figure 18.12a. Many astronomers collaborated to measure the redshifts of all the galaxies they could find in these particular parts of the sky. This project dramatically revealed the complex structure of our corner of the universe. Each dot in such pictures represents an entire galaxy of stars. The arrangement of the dots reveals huge sheets of galaxies spanning many millions of light-years. Clusters of galaxies are located at the intersections of these sheets. Between

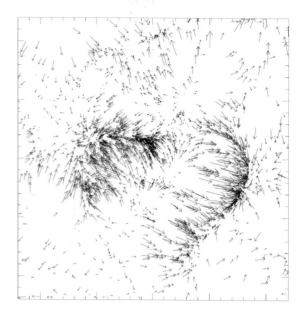

FIGURE 18.11 Peculiar velocities of galaxies flowing into superclusters. Each arrow shows the peculiar velocity of a galaxy inferred from a combination of observations and modeling. The Milky Way is at the center of the picture, and not all galaxies are shown. The area pictured is about 600 million light-years from side to side. Note how the galaxies tend to flow into regions where the density of galaxies is already high. These vast, high-density regions are probably superclusters in the process of formation.

FIGURE 18.12 These "slices" show galaxies in very thin, fanlike swaths extending outward from the Earth. Each dot represents a galaxy. Note that galaxies are not scattered randomly but instead seem to lie along sheets and strings interspersed with voids that contain very few galaxies.

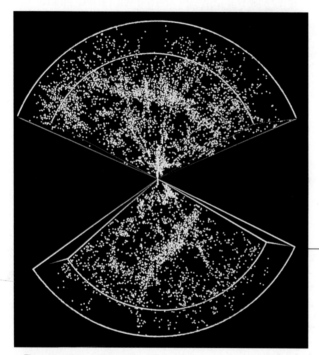

a These two slices, which point in different directions, extend to a distance of about 600 million light-years.

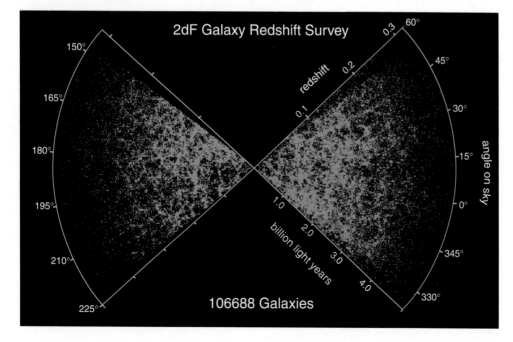

b The galaxies in these slices of the sky extend much deeper into space. Many voids, walls, and filaments of galaxies are evident, and the distribution of galaxies on scales larger than a billion light-years is nearly uniform.

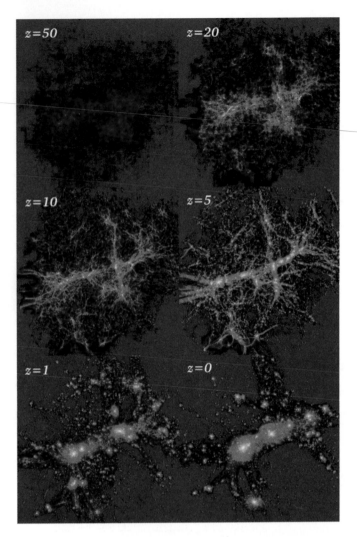

z=50 z=20
z=10 z=5
z=1 z=0

FIGURE 18.13 These six frames from a supercomputer simulation of structure formation depict the development of a region the size of our Local Group. As the age of the universe progresses from about 60 million years (cosmological redshift z = 50) to the present (z = 0), the high-density portions of the universe, shown in orange, continue to attract more and more matter.

density regions of different sizes. The voids in the distribution of galaxies probably started as mildly low-density regions.

If this picture of structure formation is correct, then the structures we see in today's universe mirror the original distribution of dark matter very early in time. Supercomputer models of structure formation in the universe can now simulate the growth of galaxies, clusters, and larger structures from tiny density enhancements as the universe evolves. The results of these models look remarkably similar to the slices of the universe in Figure 18.12, bolstering our confidence in this scenario. However, the models do not tell us *why* the universe started with these slight density enhancements—that is a topic for the next chapter. Nevertheless, it seems increasingly clear that these "lumps" in the early universe were the seeds of all the marvelous structures we see today.

18.6 Fire or Ice?

Some say the world will end in fire,
Some say in ice.
From what I've tasted of desire
I hold with those who favor fire.
But if it had to perish twice,
I think I know enough of hate
To say that for destruction ice
Is also great
And would suffice.
ROBERT FROST

We now arrive at one of the ultimate questions in astronomy: How will the universe end? Edwin Hubble's work established that the galaxies in the universe are rapidly flying away from one another [Section 17.3], but the gravitational pull of each galaxy on every other galaxy acts to slow the expansion. These facts suggest two possible fates for the universe. If gravity is strong enough, the expansion will someday halt and the universe will begin collapsing, eventually ending in a cataclysmic crunch. Alternatively, if the expansion can overcome the pull of gravity, the universe will continue to expand forever, growing ever colder as its galaxies grow ever farther apart. The fate of the universe thus boils down to a simple question: Does the expanding universe have enough kinetic energy to escape its own gravitational pull?

Density: Not Quite Critical

Hubble's constant essentially tells us the current kinetic energy of the universe, but we do not yet know the overall strength of the universe's gravita-

the sheets of galaxies lie giant empty regions called **voids** (Figure 18.12b).

Some of the structures we see in the universe are amazingly large. The so-called *Great Wall* of galaxies stretches across an expanse measuring some 180 million light-years side to side. Immense structures such as these apparently have not yet collapsed into randomly orbiting, gravitationally bound systems. They haven't had enough time. The universe may still be growing structures on ever larger scales even though it is billions of years old.

Most astronomers believe that all these large-scale structures grew from slight density enhancements in the early universe, just as galaxies did (Figure 18.13). Galaxies, clusters, superclusters, and the Great Wall probably all started as mildly high-

tional pull. The strength of this pull depends on the density of matter in the universe: The greater the density, the greater the overall strength of gravity and the higher the likelihood that gravity will someday halt the expansion. Precise calculations show that gravity can win out over expansion if the current density of the universe exceeds a seemingly minuscule 10^{-29} gram per cubic centimeter, which is roughly equivalent to a few hydrogen atoms in a volume the size of a closet. The precise density marking the dividing line between eternal expansion and eventual collapse is called the **critical density**.

Observations of the luminous matter in galaxies show that the mass contained in stars falls far short of the critical density. The visible parts of galaxies contribute less than 1% of the matter density needed to halt the universe's expansion. The fate of the universe thus rests with the dark matter. Is there enough dark matter to halt the expansion of the universe?

Recall that we can estimate the amount of dark matter by looking at the mass-to-light ratio. For dark matter to contribute enough mass for the universe to have the critical density, the average mass-to-light ratio throughout the universe would have to be approximately 1,000 solar masses per solar luminosity. Clusters of galaxies have mass-to-light ratios of a few hundred solar masses per solar luminosity, still a few times less than the ratio needed to halt the expansion. Thus, if the proportion of dark matter in the universe at large is similar to that in clusters, the universe will expand forever. For gravity to reverse the expansion and pull the universe back together, even more dark matter would have to lie beyond the boundaries of clusters.

If large-scale structures really do contain a higher proportion of dark matter than do clusters, the influence of that extra dark matter should show up in the velocities of galaxies near those large-scale structures: Larger amounts of dark matter should cause greater deviations from Hubble's law. To date, studies of these galaxy velocities remain inconclusive. Some researchers claim to find evidence that large-scale structures have significantly higher mass-to-light ratios than do clusters of galaxies, in which case the universe's matter density could be close to the critical value. As of 2001, however, most studies hold the line near the value we infer from clusters, which is about 25% of the critical density required to reverse the expansion. If that is the case, the universe will continue to expand forever.

Mysterious Acceleration

In the past few years, observations of distant white dwarf supernovae have enabled us to probe the fate of the universe in an entirely new way. Because white dwarf supernovae are such good standard candles [Section 17.3], we can use them to determine whether gravity has been slowing the universe's expansion, as it must if the universe is destined to end in a cataclysmic crunch. However, the astronomers who set out to measure gravity's influence over the universe by observing supernovae discovered something completely unexpected. Instead of slowing because of gravity, the expansion of the universe appears to be speeding up, suggesting that some mysterious repulsive force is pushing all the universe's galaxies apart. This discovery, if it holds up to further scrutiny, has far-reaching implications for both the fate of the universe and our understanding of the forces that govern its behavior on large scales. Before examining the evidence for an accelerating expansion, we must further explore the possible fates of the universe.

Astronomers subdivide the two general possibilities for the fate of the universe, expanding forever or someday collapsing, into four broad categories, each representing a particular pattern of change in the future expansion rate. We will call these four possible expansion patterns *recollapsing, critical, coasting,* and *accelerating.* The first three possibilities assume that gravity is the only force that affects the expansion rate of the universe:

- If the matter density of the universe is *larger* than the critical density, the collective gravity of all its matter will eventually halt the universe's expansion and reverse it. The galaxies will come crashing back together, and the entire universe will end in a fiery "Big Crunch." If this is the fate of our universe, then we live in a **recollapsing universe**. (The overall geometry of a recollapsing universe closes upon itself like the surface of a sphere, which is why this type of universe is sometimes called a *closed universe.*)

- If the matter density of the universe *equals* the critical density, the collective gravity of all its matter is exactly the amount needed to balance the expansion. The universe will never collapse but will expand more and more slowly as time progresses. If this is the fate of our universe, then we live in a **critical universe**. (Mathematically speaking, a critical universe stops expanding after infinite time, and its overall geometry is flat like the surface of a table; thus, a critical universe is one example of what astronomers call a *flat universe.*)

- If the matter density of the universe is *smaller* than the critical density, the collective gravity of all its matter cannot halt the expansion. The universe will keep expanding forever, with little change in its rate of expansion. If this is the fate of our universe, then we live in a **coasting universe**. (A coasting universe is sometimes called an *open universe,* because its overall geometry is more like the open surface of a saddle than like the closed surface of a sphere.)

Because observations of distant supernovae suggest that a repulsive force opposes gravity on very large scales, astronomers are now seriously considering a fourth possibility:

■ If a repulsive force causes the expansion of the universe to accelerate with time, then we live in an **accelerating universe**. Its galaxies will recede from one another increasingly faster, and it will become cold and dark more quickly than a coasting universe. (Depending on the strength of gravity relative to the repulsive force, the overall geometry of an accelerating universe could be flat, open, or closed; however, most astronomers favor a flat geometry for this case.)

TIME OUT TO THINK *Do you think that one of the potential fates of the universe is preferable to the others? If so, why? If not, why not?*

Figure 18.14 illustrates how the average distance between galaxies changes with time for each possibility. The lines for the accelerating, coasting, and critical universes always continue upward as time increases, because in these cases the universe is always expanding; the steeper the slope, the faster the expansion. In the recollapsing case, the line begins on an upward slope but eventually turns around and de-

clines as the universe contracts. All the lines pass through the same point and have the same slope at the moment labeled "now," because the current separation between galaxies and the current expansion rate in each case must agree with observations of the present-day universe.

Note that the age for the universe that we infer from its expansion rate differs in each case. A recollapsing universe requires the least amount of time to arrive at the current separation between galaxies— the example in Figure 18.14 goes from zero separation to the current separation in only about 5 billion years. The cases for which gravity is less important require more time to achieve the current separation between galaxies. The ages we would infer from the examples in Figure 18.14 are 10 billion years for a critical universe, 15 billion years for a coasting universe, and around 16 billion years for an accelerating universe.

Exactly why the universe might be accelerating remains a deep mystery, because no known force would act to push the universe's galaxies apart. Some speculative ideas for what might generate such a force go by the names *dark energy, quintessence,* and *cosmological constant.* No matter what the mechanism turns out to be, it now seems likely that the universe is indeed doomed to expand forever, its galaxies receding ever more quickly into an icy, empty future.

This is the way the world ends
This is the way the world ends
This is the way the world ends
Not with a bang but a whimper.
FROM *THE HOLLOW MEN* BY T. S. ELIOT

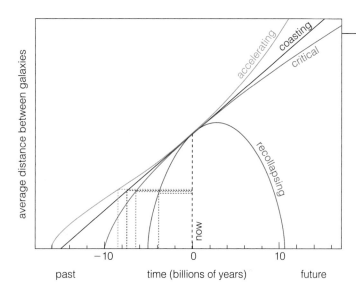

FIGURE 18.14 Four different models for the expansion of the universe. Each curve shows how the average distance between galaxies changes with time. A rising curve means that the universe is expanding, and a falling curve means that the universe is contracting. All the curves are aligned at the time labeled "now," where the expansion rate and average distance between galaxies are the same for all the models. Note that the lookback time to the birth of the universe, which is the moment at which the average distance between galaxies is essentially zero, depends on how strongly gravity affects the expansion rate. If gravity is strong enough to slow the universe's expansion, as in the critical and recollapsing universe models, the lookback time is relatively short—10 billion years or less. In the coasting and accelerating universe models, the pull of gravity is relatively weak, and the lookback time to the birth of the universe is closer to 16 billion years. Observations of distant white dwarf supernovae suggest that the universe is indeed accelerating and is destined to expand forever.

THE BIG PICTURE

In this chapter, we have found that there may be much more to the universe than meets the eye. Dark matter too dim for us to see seems to far outweigh the stars. Here are some key "big picture" points to remember about this mysterious matter:

- Either dark matter exists, or we do not understand how gravity operates across galaxy-size distances. We have many reasons to have confidence in our understanding of gravity, so the majority of astronomers believe that dark matter is real.

- Measurements of the mass and luminosity of galaxies and galaxy clusters indicate that they contain far more mass in dark matter than in stars.

- Despite the fact that dark matter is by far the most abundant form of mass in the universe, we still have little idea what it is.

- Superclusters, walls, and voids much larger than clusters of galaxies extend many millions of light-years across the universe. Each of these structures probably began as a very slight enhancement in the density of dark matter early in time, and these enormous structures are still in the process of forming.

- Dark matter holds the key to the fate of the universe. If there is enough of it, its gravity will cause the expansion of the universe to halt and reverse someday. Current indications are that there is not enough dark matter to halt the expansion, in which case the universe will continue to expand forever. Recent observations of an accelerating expansion only reinforce the suggestion that the expansion will never cease.

Review Questions

1. In what sense is dark matter "dark"?

2. What evidence suggests that the Milky Way contains dark matter?

3. Briefly describe how we construct rotation curves for spiral galaxies and how these curves lead us to conclude that spiral galaxies contain dark matter.

4. Briefly describe how we can infer the average orbital speeds of stars in an elliptical galaxy from the widths of its spectral lines and how this allows us to weigh elliptical galaxies.

5. What is a *mass-to-light ratio?* Explain why higher mass-to-light ratios imply more dark matter.

6. What are typical mass-to-light ratios for inner regions of spiral galaxies? For inner regions of elliptical galaxies? How do these mass-to-light ratios compare to the mass-to-light ratios we find when we look farther from a galaxy's center? What does this tell us about dark matter in galaxies?

7. Briefly describe the three different ways of measuring the mass of a cluster of galaxies. Do the results from the different methods agree? What do they tell us about dark matter in galaxy clusters?

8. What is gravitational lensing? Why does it occur? How can we use it to estimate the masses of lensing objects?

9. Briefly explain what we mean by baryonic and nonbaryonic matter. Which type are you made of?

10. What do we mean by *MACHOs?* Describe several possibilities for the nature of MACHOs. Based on evidence from gravitational lensing events, can MACHOs explain all the dark matter in the Milky Way? Why or why not?

11. What do we mean by *WIMPs?* Why do many astronomers believe that dark matter consists largely of WIMPs?

12. What do we mean by gravitationally bound systems? How did such structures come to exist in an expanding universe?

13. What are *peculiar velocities?* What causes them?

14. Explain how the total amount of dark matter in the universe holds the key to its final fate. According to current evidence about the amount of dark matter, what is the fate of the universe?

15. Briefly describe the four possible patterns for the expansion of the universe. What is the fate of the universe in each case? How does the current age of the universe differ among the four cases?

Discussion Questions

1. *Dark Matter or Revised Gravity.* One possible explanation for the large mass-to-light ratios we measure in galaxies and clusters is that we are currently using the wrong law of gravity to measure the masses of very large objects. If we really do misunderstand gravity, then many fundamental theories of physics, including Einstein's theory of general relativity, will need to be revised. Which explanation for these mass-to-light ratios do you find more appealing, dark matter or revised gravity? Explain why. Why do you suppose most astronomers find dark matter more appealing?

2. *Our Fate.* Scientists, philosophers, and poets alike have speculated on the fate of the universe. How would you prefer the universe as we know it to end, in a big crunch or through eternal expansion? Explain the reasons behind your preference.

Problems

True Statements? For **problems 1–6**, decide whether the statement is true and explain why it is or is not.

1. Astronomers now believe that most of any galaxy's mass lies beyond the portions of the galaxy that we can see.

2. If our galaxy had less dark matter, its mass-to-light ratio would be lower.

3. A cluster of galaxies is held together by the mutual gravitational attraction of all the stars in the cluster's galaxies.

4. We can estimate the total mass of a cluster of galaxies by studying the distorted images of galaxies whose light passes through the cluster.

5. Clusters of galaxies are the largest structures that we have so far detected in the universe.

6. There is no doubt remaining among astronomers that the fate of the universe is to expand forever.

7. *Strange Repulsive Force.* The idea that a strange repulsive force may be accelerating the universe's expansion is quite recent. What do we mean by an accelerating expansion, and what evidence suggests that such an acceleration may be occurring? If the acceleration is real, what is the fate of the universe? Explain.

8. *Mass-to-Light Ratio.* Suppose you discovered a galaxy with a mass-to-light ratio of 0.1 solar mass per solar luminosity. Would you be surprised? Explain why or why not. What would this measurement say about the nature of the stars in this galaxy?

Web Projects

Find useful links for Web projects on the text Web site.

1. *Gravitational Lenses.* Gravitational lensing occurs in numerous astronomical situations. Compile a catalog of examples from the Web. Try to find pictures of lensed stars, quasars, and galaxies. Supply a one-paragraph explanation of what is going on in each picture.

2. *Accelerating Universe.* Search for the most recent information about possible acceleration of the expansion of the universe. Write a one- to three-page report on your findings.

3. *The Nature of Dark Matter.* Find and study recent reports on new ideas about the possible nature of dark matter. Write a one- to three-page report that summarizes the latest ideas about what dark matter is made of.

Somewhere, something incredible is waiting to be known.

CARL SAGAN

CHAPTER 19

The Beginning of Time

The universe has been expanding for over 10 billion years. During that time, matter collected into galaxies. Stars formed in those galaxies, producing heavy elements that were recycled into later generations of stars. One of these late-coming stars formed about 4.6 billion years ago, in a remote corner of a galaxy called the Milky Way. This star was born with a host of planets that formed in a flattened disk surrounding it. One of these planets soon became covered with life that gradually evolved into more and more complex forms. Today, the most advanced life-form on this planet, human beings, can look back on this series of events and marvel at how the universe created conditions suitable for life.

To this point in the book, we have discussed how the matter in the early universe collected into galaxies, stars, planets, and ultimately people. We have seen that we are star stuff, and galaxy stuff as well. More to the point, we now recognize that we are the product of more than 10 billion years of cosmic evolution. We have one question left to address: Where did the *matter itself* come from?

To answer this ultimate question, we must go beyond the most distant galaxies and even beyond what we can see near the horizon of the universe. We must go back not only to the origins of matter and energy but to the beginning of time itself.

19.1 Running the Expansion Backward

Is it really possible to study the origin of the entire universe? Not long ago, questions about creation were considered unfit for scientific study. But that attitude began to change with Hubble's discovery that the universe is expanding. This discovery led to the insight that all things very likely sprang into being at a single moment in time, in an event that we have come to call the *Big Bang*. Today, powerful telescopes allow us to view how galaxies have changed over the last 10 billion years, and at great distances we see young galaxies still in the process of forming [Section 17.5]. These observations confirm that the universe is gradually aging, just as we should expect if the entire universe really was born some 12–16 billion years ago.

Unfortunately, we cannot see back to the very beginning of time. Light from the most distant galaxies shows us what the universe looked like when it was a few billion years old. Beyond these galaxies, we have not yet found any objects shining brightly enough for us to see them. But we face an even more fundamental problem. The universe is filled with a faint glow of radiation that appears to be the remnant heat of the Big Bang. This faint glow comes from a time when the universe was about 500,000 years old, and it prevents us from seeing directly to earlier times. Thus, just as we must rely on theoretical modeling to determine what the Earth or the Sun is like on the inside, we must use modeling to investigate what happened in the universe in its first 500,000 years.

Fortunately, calculating conditions in the early universe is straightforward if we consider some fundamental principles. We know that the expanding universe is cooling and becoming less dense as it grows. Therefore, the universe must have been hotter and denser in the past. Calculating exactly how hot and dense the universe must have been when it was more compressed is similar to calculating the temperatures and densities in a car engine as the pistons compress the fuel mix, except that the conditions become much more extreme. Figure 19.1 shows just how hot the universe was during its earliest moments, according to such calculations. To understand these early moments, we need to know how matter and energy behave under such extraordinarily hot conditions.

Recent advances in physics have enabled us to predict the behavior of matter and energy under conditions far more extreme than in the centers of the hottest stars. When we apply our understanding of modern physics to the early universe, we find something so astonishing that it almost sounds preposterous: Today, we may understand the behavior of matter and energy well enough to describe what was

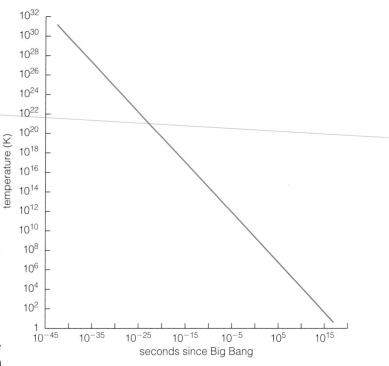

FIGURE 19.1 Temperature of the universe from the Big Bang to the present (10^{10} years $\approx 3 \times 10^{17}$ seconds).

happening in the universe just *one ten-billionth* (10^{-10}) of a second after the Big Bang. And that's not all. While our understanding of physics is less certain under the more extreme conditions that prevailed even earlier, we have some ideas about what the universe was like when it was a mere 10^{-38} second old, and perhaps a glimmer of what it was like at the age of just 10^{-43} second. These tiny fractions of a second are so small that, for all practical purposes, we are studying the very moment of creation.

19.2 A Scientific History of the Universe

The **Big Bang theory**—the scientific theory of the universe's earliest moments—presumes that all we see today, from Earth to the cosmic horizon, began as an incredibly tiny, hot, and dense collection of matter and radiation. It describes how expansion and cooling of this unimaginably intense mixture of particles and photons could have led to the present universe of stars and galaxies, and it explains several aspects of today's universe with impressive accuracy. Our ideas about the earliest moments of the universe are quite speculative because they depend on aspects of physics that we do not yet fully comprehend. However, the evidence supporting this theory grows stronger and stronger as we focus on progressively

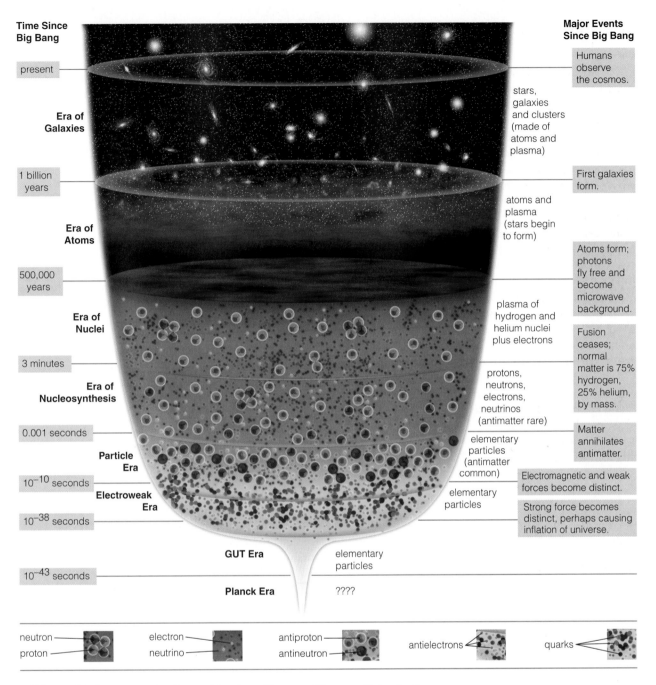

Time Since Big Bang			Major Events Since Big Bang

Time Since Big Bang (left labels):

- present
- Era of Galaxies
- 1 billion years
- Era of Atoms
- 500,000 years
- Era of Nuclei
- 3 minutes
- Era of Nucleosynthesis
- 0.001 seconds
- Particle Era
- 10^{-10} seconds
- Electroweak Era
- 10^{-38} seconds
- GUT Era
- 10^{-43} seconds
- Planck Era

State of matter (center labels):

- stars, galaxies and clusters (made of atoms and plasma)
- atoms and plasma (stars begin to form)
- plasma of hydrogen and helium nuclei plus electrons
- protons, neutrons, electrons, neutrinos (antimatter rare)
- elementary particles (antimatter common)
- elementary particles
- elementary particles
- ????

Major Events Since Big Bang (right labels):

- Humans observe the cosmos.
- First galaxies form.
- Atoms form; photons fly free and become microwave background.
- Fusion ceases; normal matter is 75% hydrogen, 25% helium, by mass.
- Matter annihilates antimatter.
- Electromagnetic and weak forces become distinct.
- Strong force becomes distinct, perhaps causing inflation of universe.

Bottom key: neutron, proton, electron, neutrino, antiproton, antineutron, antielectrons, quarks

FIGURE 19.2 This diagram summarizes the eras of the universe. The names of the eras and their ending times are indicated on the left, and the state of matter during each era and the events marking the end of each era are indicated on the right.

later moments in time. We will discuss the evidence supporting the Big Bang theory later in this chapter. First, in order to help you understand the significance of the evidence, let's examine the story of creation according to this theory.

Figure 19.2 summarizes the story by dividing the overall history of the universe into a series of *eras,* or time periods. Each era is distinguished from the next by some major change in the conditions of the uni-

verse. You'll find it easiest to keep track of the various eras if you refer back to this figure as we discuss each era in detail. (In more specialized books on cosmology, these eras are subdivided into many more eras, providing additional details about the early universe not covered here.)

Before we delve into the story of creation, it's important to be aware of a key aspect of the early universe: it was so hot that photons could transform themselves into matter, and vice versa, in accordance with Einstein's formula $E = mc^2$ [Section 4.2]. Today, physicists can study many such reactions by reproducing them in their laboratories, but reactions that create and destroy matter are now relatively rare in the universe at large.

One example of such a reaction is the creation or destruction of an *electron–antielectron pair*. When two photons collide with a total energy greater than twice the mass-energy of an electron (i.e., the electron's mass times c^2), they can create two brand-new particles: a negatively charged electron and its positively charged twin, the *antielectron*, also known as a *positron* (Figure 19.3). The electron is a particle of **matter**, and the antielectron is a particle of **antimatter**. The reaction that creates an electron–antielectron pair also runs in reverse: When an electron and an antielectron meet, they *annihilate* each other totally, transforming all their mass-energy back into photon energy.

FIGURE 19.3 Electron–antielectron creation and annihilation.

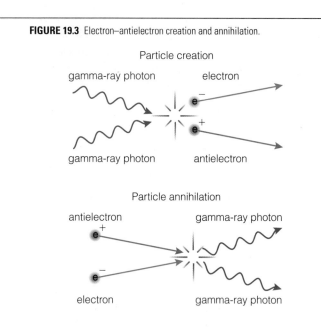

Similar reactions can produce or destroy any particle–antiparticle pair, such as a proton and antiproton or neutron and antineutron. The early universe was therefore filled with an extremely dynamic blend of photons, matter, and antimatter, converting furiously back and forth. Despite all these vigorous reactions, describing conditions in the early universe is straightforward, because we can calculate the proportions of the various forms of radiation and matter from the universe's temperature and density at any time.

The First Instant

The first 10^{-43} second of the universe's history is called the **Planck era** after physicist Max Planck, one of the founders of the science of quantum mechanics. According to the laws of quantum mechanics and general relativity, random mass and energy fluctuations were so large during the Planck era that our current theories are powerless to describe what might have been happening. The problem is that we do not yet have a theory that links quantum mechanics and general relativity. In a sense, quantum mechanics is our successful theory of the very small (subatomic particles), and general relativity is our successful theory of the very big (the large-scale structure of spacetime). Perhaps someday we will be able to merge these theories of the very small and the very big into a single "theory of everything." Until that happens, science cannot describe the universe's first instant.

The GUT Era

Understanding the transition that marked the beginning of the next era requires thinking in terms of the *forces* that operate in the universe. Everything that happens in the universe today is governed by four distinct forces: *gravity, electromagnetism,* the *strong force,* and the *weak force.* Gravity dominates large-scale action, and electromagnetism dominates chemical and biological reactions. The strong and weak forces determine what happens within atomic nuclei: The strong force binds nuclei together, and the weak force mediates nuclear reactions such as fission or fusion.

These forces were not so distinct in the early universe. The four forces that act so differently today turn out to be separate aspects of a smaller number of more fundamental forces, probably only one or two (Figure 19.4). Physicists have shown that the electromagnetic and weak forces lose their separate identities under conditions of very high temperature or energy and merge together into a single *electroweak force.* Physicists also believe that the electroweak force and the strong force ultimately lose their separate identities at even higher energies. The theories that predict this merger of the electroweak and strong forces are called **grand unified theories**, or **GUTs** for short; their merger is usually referred to as the *GUT force.* According to these theories, when the universe entered the GUT era at an age of 10^{-43} second, only two forces operated in the universe: gravity and the GUT force.

The GUT era lasted but a tiny fraction of a second, coming to an end when the universe had cooled to 10^{29} K at an age of 10^{-38} second. Grand unified theories predict that the strong force separated from

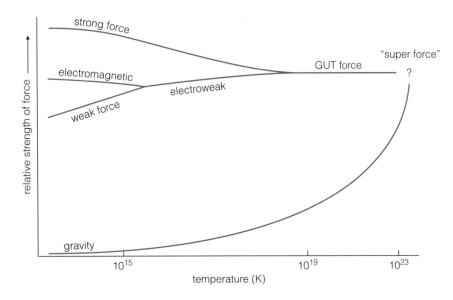

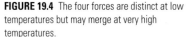

FIGURE 19.4 The four forces are distinct at low temperatures but may merge at very high temperatures.

the GUT force at this point, leaving three forces operating in the universe: gravity, the strong force, and the electroweak force. As we will discuss later, it now seems possible that this separation of the strong force released an enormous amount of energy. This energy release would have caused the universe to undergo a sudden and dramatic expansion that we call **inflation**. In the space of a mere 10^{-36} second, pieces of the universe the size of an atomic nucleus may have grown to the size of our solar system. Inflation sounds bizarre, but it might actually explain some puzzling features of today's universe.

The Electroweak Era

The end of the GUT era marks the beginning of the **electroweak era**, so named because the electromagnetic and weak forces were still unified into the electroweak force. Intense radiation filled all of space, as it had since the Planck era, spontaneously producing matter and antimatter particles of many different sorts that immediately reverted back to energy. The universe continued to expand and cool throughout the electroweak era, dropping to a temperature of 10^{15} K when it reached an age of 10^{-10} second. This temperature is still 100 million times hotter than the temperature in the core of the Sun, but it was low enough for the electromagnetic and weak forces to separate from the electroweak force. After this instant (10^{-10} s), all four forces were forever distinct in the universe.

The end of the electroweak era marks an important transition, not only in the universe, but also in human understanding of the universe. The theory that unified the weak and electromagnetic forces, developed in the 1970s, predicted the emergence of new types of particles (called the W and Z bosons, or

weak bosons) when temperatures rose above the 10^{15} K that pervaded the universe when it was 10^{-10} second old. In 1983, experiments performed in a huge particle accelerator near the French/Swiss border reached energies equivalent to such high temperatures for the first time. The new particles showed up just as predicted, produced from the extremely high energy in accord with $E = mc^2$. Thus, we have direct experimental evidence concerning the conditions in the universe at the end of the electroweak era. Prior to that time, we do not have any direct experimental evidence. Our theories concerning the earlier parts of the electroweak era and the GUT era consequently are much more speculative than our theories describing the universe from the end of the electroweak era to the present.

The Particle Era

Spontaneous creation and annihilation of particles continued throughout the next era, which we call the **particle era** because tiny particles of matter were as numerous as photons. These particles included electrons, neutrinos, and *quarks*—the building blocks of heavier particles such as protons, antiprotons, neutrons, and antineutrons. Near the end of the particle era, when the universe was about 0.0001 second old, the temperature grew too cool for quarks to exist on their own, and quarks combined in groups of three to form protons and neutrons. The particle era came to an end when the universe reached an age of 1 millisecond (0.001 s) and the temperature had fallen to 10^{12} K. At this point, it was no longer hot enough to spontaneously produce protons and antiprotons (or neutrons and antineutrons) from pure energy.

Einstein's Greatest Blunder

Shortly after Einstein completed his general theory of relativity in 1915, he found that it predicted that the universe could not be standing still: The mutual gravitational attraction of all the matter would make the universe collapse. Because Einstein thought at the time that the universe should be eternal and static, he decided to alter his equations. In essence, he inserted a "fudge factor" called the *cosmological constant* that acted as a repulsive force to counteract the attractive force of gravity. Had he not been so convinced that the universe should be standing still, he might instead have come up with the correct explanation for why the universe is not collapsing: because it is still expanding from the event of its birth. After Hubble discovered that the universe is expanding, Einstein called his invention of the cosmological constant the greatest blunder of his career.

Despite Einstein's disavowal, variations both on the idea of an eternal universe and on the cosmological constant continued to crop up over the years. One of the cleverest ideas, developed in the late 1940s, was called the *steady state theory* of the universe. This theory accepted the fact that the universe is expanding but rejected the idea of a Big Bang and instead postulated that the universe is infinitely old. This idea may seem paradoxical at first: If the universe has been expanding forever, shouldn't every galaxy be infinitely far away from every other galaxy? The steady state theory suggested that new galaxies continually form in the gaps that open up as the universe expands, thereby keeping the same average distance between galaxies at all times. In a sense, the steady state theory said that the creation of the universe is an ongoing and eternal process rather than having happened all at once with a Big Bang.

Two key discoveries caused the steady state theory to lose favor. First, the 1965 discovery of the cosmic microwave background matched a prediction of the Big Bang theory but could not be adequately explained by the steady state theory. Second, the steady state theory predicts that the universe looks about the same at all times, which is inconsistent with observations showing that galaxies at great distances look younger than nearby galaxies. As a result of these predictive failures, the steady state theory is no longer taken seriously by most astronomers.

More recently, astronomers have begun to take the idea of a cosmological constant more seriously. In the mid-1990s, a few observations suggested that the ages of the oldest stars are slightly older than the age of the universe derived from Hubble's constant under the assumption that gravity is the only force affecting the universe's expansion. Clearly, stars cannot be older than the universe. If these observations were being interpreted correctly, the universe had to be older than Hubble's constant implies. Recall that we estimate the age of the universe by running the expansion backward while accounting for how gravity and other forces have modified the expansion rate [Section 18.6]. If the expansion rate has accelerated, so that the universe is expanding faster today than it was in the past, then the age of the universe would be greater than that ordinarily found from Hubble's constant. What could cause the expansion of the universe to accelerate over time? The repulsive force represented by a cosmological constant, of course.

Further study of the troubling observations has shown that the oldest stars probably are *not* older than the age of the universe derived from Hubble's constant. However, other observations still suggest that a cosmological constant is necessary. As we discussed in Chapter 18, measurements of distances to high-redshift galaxies using white dwarf supernovae as standard candles now suggest that the expansion *is* accelerating. A cosmological constant could account for this startling finding, but we'll need many more observations before we'll know whether it is correct. Einstein's greatest blunder, it seems, just won't go away.

If the universe had contained equal numbers of protons and antiprotons (or neutrons and antineutrons) at the end of the particle era, all of the pairs would have annihilated each other, creating photons and leaving essentially no matter in the universe. In today's universe, photons outnumber protons by over a billion to one. We conclude that protons must have slightly outnumbered antiprotons at the end of the particle era: For every billion antiprotons, there were about a billion and one protons. Thus, for each 1 billion protons and antiprotons that annihilated each other, creating photons, a single proton was left over. This seemingly slight excess of ordinary matter over antimatter makes up all the matter in the present-day universe. That is, the protons (and neutrons) left over from when the universe was 0.001 second old are the very ones that make up our bodies.

The Era of Nucleosynthesis

So far, everything we have discussed occurred within the first 0.001 second of the universe's existence—a time span shorter than the time it takes you to blink an eye. At this point, the protons and neutrons left over after the annihilation of antimatter attempted to fuse into heavier nuclei. However, the heat of the universe remained so high that most nuclei broke apart as fast as they formed. This dance of fusion and breakup marked the **era of nucleosynthesis**.

The era of nucleosynthesis ended when the universe was about 3 minutes old. By this time, the density in the expanding universe had dropped so much that fusion no longer occurred, even though the temperature was still about a billion Kelvin (10^9 K)—much hotter than the temperature at the center of the Sun today. When fusion ceased, about 75% of the mass of the ordinary (baryonic) matter in the universe remained as individual protons, or hydrogen nuclei. The other 25% of this mass had fused into helium nuclei, with trace amounts of deuterium (hydrogen with a neutron) and lithium. Except for the relatively small amount of matter that stars later forged into heavier elements, the chemical composition of the universe remains unchanged today.

The Era of Nuclei

At the end of the era of nucleosynthesis, the universe consisted of a very hot plasma of hydrogen nuclei, helium nuclei, and free electrons. This basic picture held for about the next 500,000 years as the universe continued to expand and cool. The fully ionized nuclei moved independently of electrons during this period (rather than being bound with electrons in neutral atoms), which we call the **era of nuclei**. Throughout this era, photons bounced rapidly from one electron to the next, just as they do deep inside the Sun today, never managing to travel far between collisions. Any time a nucleus managed to capture an electron to form a complete atom, one of the photons quickly ionized it.

The era of nuclei came to an end when the expanding universe was about 500,000 years old, at which point the temperature had fallen to about 3,000 K—roughly half the temperature of the Sun's surface today. Hydrogen and helium nuclei finally captured electrons for good, forming stable, neutral atoms for the first time. With electrons now bound into atoms, the universe suddenly became transparent. Photons, formerly trapped among the electrons, began to stream freely across the universe. We still see these photons today as the *cosmic microwave background*, which we will discuss shortly.

The Era of Atoms and the Era of Galaxies

We've already discussed the rest of the story in earlier chapters. The end of the era of nuclei marked the beginning of the **era of atoms**, when the universe consisted of a mixture of neutral atoms and plasma. Thanks to the slight density enhancements present in the universe at this time and the gravitational attraction of dark matter, the atoms and plasma slowly assembled into protogalactic clouds [Section 18.5]. Stars formed in these clouds, transforming the gas clouds into galaxies. The first full-fledged galaxies had

formed by the time the universe was about 1 billion years old, beginning what we call the **era of galaxies**.

The era of galaxies continues to this day. Generation after generation of star formation in galaxies steadily builds elements heavier than helium and incorporates them into new star systems. Some of these star systems develop planets, and on at least one of these planets life burst into being a few billion years ago. And here we are, thinking about it all. Describing both the follies and the achievements of the human race, Carl Sagan once said, "These are the things that hydrogen atoms do—given 15 billion years of cosmic evolution."

19.3 Evidence for the Big Bang

Like any scientific theory, the Big Bang theory is a model designed to explain a set of facts. If it is close to the truth, it should be able to make predictions about the real universe that we can verify through observations or experiments. The Big Bang model has gained wide scientific acceptance for two key reasons:

■ The Big Bang model predicts that the radiation that began to stream across the universe at the end of the era of nuclei should still be present today. Sure enough, we find that the universe is filled with what we call the **cosmic microwave background**. Its characteristics precisely match what we expect according to the Big Bang model.

■ The Big Bang model predicts that some of the original hydrogen in the universe should have fused into helium during the era of nucleosynthesis. Observations of the actual helium content of the universe closely match the amount of helium predicted by the Big Bang model. Fusion of hydrogen to helium in stars could have produced only about 10% of the observed helium.

The Cosmic Microwave Background

The first major piece of evidence supporting the Big Bang theory arrived in 1965. Arno Penzias and Robert Wilson, two physicists working at Bell Laboratories in New Jersey, were calibrating a sensitive microwave antenna designed for satellite communications (Figure 19.5). (*Microwaves* fall within the radio portion of the electromagnetic spectrum.) Much to their chagrin, they kept finding unexpected "noise" in every measurement they made with the antenna. The noise was the same no matter where they pointed their antenna, indicating that it came from all directions in the sky and ruling out the possibility that it came from any particular astronomical object or from

an astronomer who told him of the Princeton calculations. The Princeton group soon met with Penzias and Wilson to compare notes. The "noise" in the Bell Labs antenna was not an embarrassment after all. Instead, it was the cosmic microwave background—and the first strong evidence that the Big Bang had really happened. Penzias and Wilson received the 1978 Nobel Prize in physics for their discovery of the cosmic microwave background.

The cosmic microwave background radiation consists of photons arriving at Earth directly from the end of the era of nuclei, when the universe was about 500,000 years old. Because neutral atoms could remain stable for the first time, they captured most of the electrons in the universe. With no more free electrons to block them, the photons from that epoch have flown unobstructed through the universe ever since (Figure 19.6). When we observe the cosmic microwave background, we essentially are seeing back to a time when the universe was only 500,000 years old. In that sense, we are seeing light from the most distant regions of the observable universe— only 500,000 light-years from our cosmological horizon [Section 17.4]. Surprisingly, it does not take a particularly powerful telescope to "see" this radiation. In fact, you can pick it up with an ordinary television antenna. If you set an antenna-fed television (i.e., *not* cable or satellite TV) to a channel for which there is no local station, you will see a screen full of static "snow." About 1% of this static is due to photons in the cosmic microwave background.

The cosmic microwave background came from the heat of the universe itself and therefore should have an essentially perfect thermal radiation spectrum [Section 6.3]. When this radiation first broke free

any place on the Earth. Embarrassed by their inability to explain the noise, Penzias and Wilson prepared to "bury" their discovery at the end of a long scientific paper about their antenna.

Meanwhile, physicists at nearby Princeton University were busy calculating the expected characteristics of the radiation left over from the heat of the Big Bang. They concluded that, if the Big Bang had really occurred, this radiation should be permeating the entire universe and should be detectable with a microwave antenna. On a fateful airline trip home from an astronomical meeting, Penzias sat next to

FIGURE 19.6 Photons frequently collided with free electrons during the era of nuclei and thus could travel freely only after electrons became bound into atoms. The photons released at the end of the era of nuclei make up the cosmic microwave background.

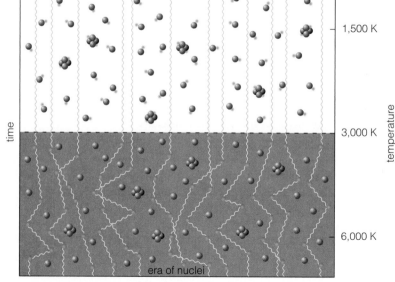

500,000 years after the Big Bang, the temperature of the universe was about 3,000 K, not too different from that of a red giant star's surface. Thus, the spectrum of the cosmic microwave background originally peaked in visible light, just like the thermal radiation from a red star, with wavelengths of a few hundred nanometers. However, the universe has expanded by a factor of about 1,000 since that time, stretching the wavelengths of these photons by the same amount [Section 17.4]. Their wavelengths should therefore have shifted to about a millimeter, squarely in the microwave portion of the spectrum and corresponding to a temperature of a few degrees above absolute zero. In the early 1990s, a NASA satellite called the *Cosmic Background Explorer (COBE)* was launched to test these ideas about the cosmic microwave background. The results were a stunning success for the Big Bang theory. As shown in Figure 19.7, the cosmic microwave background does indeed have a perfect thermal radiation spectrum, with a peak corresponding to a temperature of 2.73 K. In a very real sense, the temperature of the night sky is a frigid 3 degrees above absolute zero.

TIME OUT TO THINK *Suppose the cosmic microwave background did not really come from the heat of the universe itself but instead came from many individual stars and galaxies. Explain why, in that case, we would not expect it to have a perfect thermal radiation spectrum. How does the spectrum of the cosmic microwave background lend support to the Big Bang theory?*

COBE achieved an even greater success mapping the temperature of the cosmic microwave background in all directions. It was already known that the cosmic microwave background is extraordinarily

uniform throughout the universe. Conditions in the early universe must have been extremely uniform to produce such a smooth radiation field. For a time, this uniformity was considered a strike against the Big Bang theory because, as we discussed in Chapters 17 and 18, the universe must have contained some regions of enhanced density in order to explain the formation of galaxies. The COBE measurements restored confidence in the Big Bang theory because they showed that the cosmic microwave background is *not quite* perfectly uniform. Instead, its temperature varies very slightly from one place to another by a few parts in 100,000 (Figure 19.8). These variations in temperature indicate that the density of the early universe really did differ slightly from place to place—the seeds of structure formation were indeed present during the era of nuclei.

COBE's discovery of density enhancements bolstered the idea that some of the dark matter consists of WIMPs (weakly interacting massive particles [Section 18.4]) that we have not yet identified and that the gravity of this dark matter drove the formation of structure in the universe. Regions of enhanced density can grow into galaxies because the extra gravity in these regions draws matter together even while the rest of the universe expands. The greater the density enhancements, the faster matter should have collected into galaxies. Detailed calculations show that, to explain the fact that galaxies formed within a few billion years, the density enhancements at the end of the era of nuclei must have been significantly greater than the few parts in 100,000 suggested by the temperature variations in the cosmic microwave background. Because WIMPs are weakly interacting and do not interact with photons, we do not expect them to influence the temperature of the cosmic microwave background directly. However, the gravity of the WIMPs can collect ordinary baryonic matter into clumps that *do* interact with photons. Thus, the small density enhancements detected by COBE may actually echo the much larger density enhancements made up of WIMPs. Careful modeling of the COBE temperature variations shows that they are consistent with dark-matter density enhancements large enough to account for the structure we see in the universe today.

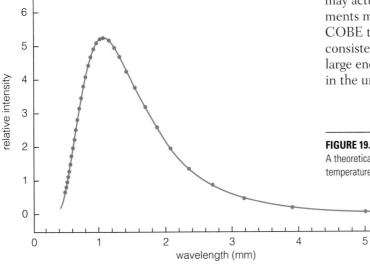

FIGURE 19.7 Spectrum of the cosmic microwave background from COBE. A theoretically calculated thermal radiation spectrum (smooth curve) for a temperature of 2.73 K perfectly fits the data (dots).

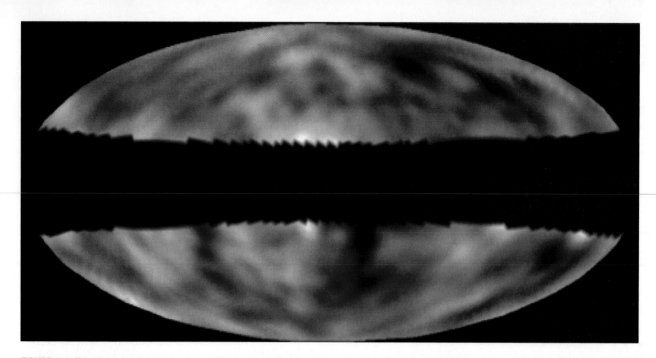

FIGURE 19.8 This all-sky map shows temperature differences in the cosmic microwave background measured by COBE. The background temperature is about 2.73 K everywhere, but the brighter regions of this picture are slightly less than 0.0001 K hotter than the darker regions—indicating that the early universe was very slightly lumpy. We are essentially seeing what the universe was like at the surface marked "500,000 years" in Figure 19.2. (The central strip of this map, which corresponds to the disk of the Milky Way, has been masked out because the brightness differences there stem primarily from radio noise in the Milky Way.

Synthesis of Helium

The discovery of the cosmic microwave background in 1965 quickly solved another long-standing astronomical problem: the origin of cosmic helium. Everywhere in the universe, about one-quarter of the mass of ordinary matter (i.e., not dark matter) is helium. The Milky Way's helium fraction is about 28%, and no galaxy has a helium fraction lower than 25%. A small proportion of this helium comes from hydrogen fusion in stars, but most does not. The majority of the helium in the universe must already have been present in the protogalactic clouds that preceded the formation of galaxies. In other words, the universe itself must once have been hot enough to fuse hydrogen into helium. The current microwave background temperature of 2.73 K tells us precisely how hot the universe was in the distant past and exactly how much helium it should have made. The result—25% helium—is another impressive success of the Big Bang theory.

A helium nucleus contains two protons and two neutrons, so we need to understand what protons and neutrons were doing during the era of nucleosynthesis in order to see why the primordial fraction of helium was 25%. Early in this era, when the universe's temperature was 10^{11} K, nuclear reactions could convert protons into neutrons, and vice versa. As long as the universe remained hotter than 10^{11} K,

these reactions kept the numbers of protons and neutrons nearly equal.

As the universe cooled from 10^{11} K to 10^{10} K, neutron–proton conversion reactions began to favor protons. Neutrons are slightly heavier than protons, and therefore reactions that convert protons to neutrons require energy to proceed (in accordance with $E = mc^2$). By the time the temperature of the universe fell to 10^{10} K, protons began to outnumber neutrons because the conversion reactions ran only in one direction: Neutrons changed into protons, but the protons didn't change back.

At 10^{10} K, the universe was still hot and dense enough for nuclear fusion to operate. Protons and neutrons constantly combined to form *deuterium*—the rare form of hydrogen nuclei that contains a neutron in addition to a proton—and deuterium nuclei fused to form helium. However, during the early parts of the era of nucleosynthesis, the helium nuclei were almost immediately blasted apart by one of the many gamma rays that filled the universe.

Fusion finally created long-lasting helium nuclei when the universe was about 1 minute old, at which time the destructive gamma rays were gone. Calculations show that the proton-to-neutron ratio at this time should have been about 7 to 1. Moreover, virtually all the available neutrons should have become incorporated into nuclei of helium-4. Figure 19.9 shows that, based on the 7-to-1 ratio of protons to neutrons, the universe should have had a composi-

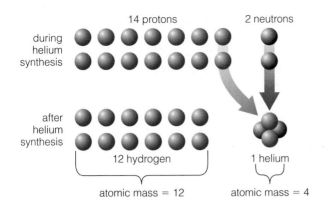

FIGURE 19.9 During helium synthesis, protons outnumbered neutrons 7 to 1, which is the same as 14 to 2. The result was 12 hydrogen nuclei (individual protons) for each helium nucleus. Thus, the hydrogen-to-helium mass ratio is 12 to 4, which is the same as 75% to 25%.

tion of 75% hydrogen and 25% helium by mass at the end of the era of nucleosynthesis.

Thus, the Big Bang theory makes a very concrete prediction about the chemical composition of the universe: It should be 75% hydrogen and 25% helium by mass. The fact that observations confirm this predicted ratio of hydrogen to helium is another striking success of the Big Bang theory.

TIME OUT TO THINK *Briefly explain why it should not be surprising that some galaxies contain a little more than 25% helium, but it would be very surprising if some galaxies contained less. (Hint: Think about how the relative amounts of hydrogen and helium in the universe can be affected by fusion in stars.)*

19.4 Inflation

The Big Bang model relies heavily on our knowledge of particle physics, which has been tested to temperatures of 10^{15} K, corresponding to the end of the electroweak era. Our knowledge of earlier times rests on a weaker foundation because we are less certain of the physical laws at work. In fact, the best laboratory for studying the laws of physics at such high temperatures is the Big Bang itself. Different guesses as to how matter might behave at such high energies predict different outcomes for the universe we see today. If a particular model predicts that our universe should look different from the way it really does, then that model must be wrong. On the other hand, if a new model of particle physics explains some previously unexplained aspects of the universe, then it may be on the right track. The grand unified theories of particle physics discussed in Section 19.2 have not yet been proved, but many scientists believe these theories are on the right track for just this reason.

Prior to the early 1980s, scientists had identified several major aspects of our actual universe that were unexplained by the Big Bang model. Three of the most pressing unanswered questions were the following:

■ *Where does structure come from?* According to the Big Bang theory, the process of galaxy formation required the density of the early universe to differ slightly from place to place. Otherwise, the pull of gravity would have been equally balanced everywhere, prohibiting the collapse of protogalactic clouds. The subtle temperature differences seen in the cosmic microwave background tell us that regions of enhanced density did indeed exist at the end of the era of nuclei, when the universe was 500,000 years old. However, the Big Bang theory cannot explain where these density enhancements came from.

■ *Why is the large-scale universe so smooth?* Not only is the slight lumpiness of the early universe a problem, but its large-scale smoothness is also a problem. Observations of the cosmic microwave background show that, overall, the density of the universe at the end of the era of nuclei varied from place to place by no more than about 0.01%. The standard Big Bang theory does not explain why distant reaches of the universe look so similar.

■ *Why is the density almost critical?* The density of matter in the universe is somewhere around 20–100% of the critical density [Section 18.6]— remarkably close to the critical density considering that the standard Big Bang theory doesn't say anything about what the density should be. Why isn't it 1,000 times the critical density, or 0.0000001 the critical density?

Physicist Alan Guth realized in 1981 that grand unified theories could potentially answer all three questions. These theories predict that the separation of the strong force from the GUT force should have released enormous energy, causing the universe to expand dramatically, perhaps by as much as 10^{30} times in less than 10^{-36} second. This dramatic expansion is what we call *inflation*, and while it sounds outrageous, it may have shaped the way the universe looks today.

Giant Quantum Fluctuations

To understand how inflation solves the structure question, we need to recognize a special feature of energy fields. Laboratory-tested principles of quantum mechanics require that the energy fields at any point in space be always fluctuating. Thus, the distribution of energy through space on very small scales is slightly irregular, even in a complete vacuum. These tiny quantum "ripples" can be characterized by a wavelength that corresponds roughly to their size. In principle, quantum ripples in the very early universe could have been the seeds for density enhancements that later grew into galaxies. However, the wavelengths of the original ripples were far too small to explain density enhancements like those we see imprinted on the cosmic microwave background.

Inflation would have dramatically altered these quantum fluctuations. The fantastic growth of the universe during the period of inflation would have stretched tiny ripples in space to enormous wavelengths (Figure 19.10). Ultimately, inflation could have caused these quantum ripples to grow large enough to become the density enhancements that later formed large structures in the universe. Amazingly, all the structure we see today in the universe may have started as tiny quantum fluctuations just before the period of inflation.

Equalizing Temperatures and Densities

The idea that different parts of the universe should be very similar shortly after the Big Bang may seem quite natural at first, but on further inspection the smoothness of the universe becomes difficult to ex-

plain. Imagine observing the cosmic microwave background in a certain part of the sky. You are seeing that region as it was only 500,000 years after it formed. The microwaves coming from it have taken some 12 billion years or more—almost the entire age of the universe—to travel all the way to Earth. Now imagine turning around and looking at the background radiation coming from the opposite direction. You are also seeing this region at an age of 500,000 years, and it looks virtually identical. The two microwave-emitting regions are billions of light-years apart, but we are seeing them as they were when they were only 500,000 years old. They can't possibly have had communication with each other then. A signal traveling at the speed of light from one to the other would barely have started its journey. So how do they come to be exactly the same temperature?

The idea of inflation solves this problem because it says that the entire observable universe was less than 10^{-38} light-second across before the episode of enormous expansion. Radiation signals traveling at the speed of light would have had time to equalize all the temperatures and densities in this region. Inflation then pushed these equalized regions to much greater distances, far out of contact with one another. Like criminals getting their stories straight before being locked in separate jail cells, all parts of the observable universe came to the same temperature and density before inflation spread them far apart. Thus, inflation accounts for how vastly separated regions can look so similar.

Balancing the Universe

The third question answered by inflation asks why the matter density of the universe is so close to the critical density. Another way to say that the universe's density is close to critical is to say that the overall geometry of the universe is remarkably "flat" [Section 18.6]. If the density of the universe exactly equals the critical density, then the kinetic energy of expansion precisely balances the universe's overall gravitational pull. According to Einstein's theory of general relativity, an imbalance in these energies causes curvature of spacetime, and deviations from precise balance grow more severe as the universe evolves. For example, if the universe had been 10% denser at the end of the era of nuclei, it would have collapsed long ago. On the other hand, if it had been 10% less dense at that time, galaxies would never have formed before expansion spread all the matter too thin. Thus, the universe had to start out remarkably balanced to be even remotely close to flat today.

Inflation balances the universe by ensuring that the mass-energy providing the universe's gravity precisely matches its kinetic energy of expansion. Once this balance is set, it should never waver. In terms of

FIGURE 19.10 During inflation, ripples in spacetime would have stretched by a factor of perhaps 10^{30}.

size of ripple before inflation = size of atomic nucleus

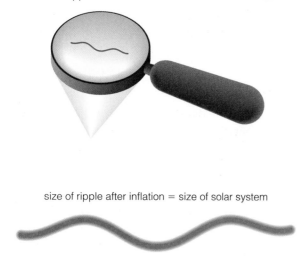

size of ripple after inflation = size of solar system

FIGURE 19.11 As a balloon expands, its surface seems increasingly flat to an ant crawling along it.

Einstein's theory, the effect of inflation on spacetime curvature is similar to the flattening of a balloon's surface when you blow into it (Figure 19.11). The flattening of space during the period of inflation would have been so enormous that it would virtually have eliminated any curvature the universe might have had previously.

The fine balance created by inflation turns out to be both a success and a potential pitfall of the inflation theory. On one hand, it explains how the universe managed to stay balanced long enough to bring forth galaxies. However, our studies of dark matter suggest that the current universe might be slightly out of balance: The best current estimates put the actual density of the universe at only about 25% of the critical density [Section 18.6], in which case the universe has about four times more kinetic energy than gravity can overcome.

Inflation theory has difficulty explaining this imbalance, but a loophole in Einstein's theory may solve the problem: Energy associated with a large-scale repulsive force can compensate for the shortfall in the matter density, making the present-day universe flatter than it would otherwise be. Remarkably, the supernova measurements indicating that our universe's expansion is accelerating [Section 18.6] also show that the strength of the implied repulsive force is about right to render the universe perfectly flat, just as inflation predicts.

Testing for Inflation

Verifying that the universe underwent an episode of inflation is difficult, because this inflationary episode would have happened extremely early in the history of the universe—much earlier than we can observe directly. Nevertheless, we can test the idea of inflation by exploring whether its predictions are consistent with our observations of the universe at later times. Such observations cannot prove the theory of inflation, but if they are inconsistent with the theory, then inflation probably never happened. Scientists are only beginning to make observations that test inflation, but to date the findings are consistent with the idea that an early inflationary episode smoothed and flattened the universe while planting the seeds of structure formation.

The earliest moment we can directly observe is the end of the era of nuclei, when photons began to stream freely across the universe, ultimately becoming the cosmic microwave background. COBE was the first telescope to reveal the density enhancements present at that moment in time, but many telescopes are following in COBE's footsteps and further refining our view of the microwave background. Because physical models of inflation make specific predictions about the nature of the density enhancements that ought to have been present in the early universe, measurements of the cosmic microwave background offer a way to test those models.

In a certain sense, cosmic microwave observations are recording the "genetic code" of the universe, because the characteristics of the density enhancements present in the early universe governed the subsequent growth of galaxies, clusters, and larger-scale structures. To the extent that we have been able to read this code, it is quite similar to what we expect from inflation. These observations also indicate that the universe has a flat geometry, again in accord with inflation. More stringent tests are under way, and others are planned for the near future. Instruments like NASA's Microwave Anisotropy Probe (MAP) will be delivering details about the universe's genetic code over the next several years, so keep listening for the latest news about inflation.

The Bottom Line

All things considered, inflation does a remarkable job of addressing the problems inherent in the standard Big Bang theory. Many astronomers and physicists believe that some process akin to inflation did affect the early universe, but the details of the interaction between high-energy particle physics and the evolving universe remain unclear. If these details can be worked out successfully, we face an amazing prospect—a breakthrough in our understanding of the very smallest particles achieved by studying the universe on the largest observable scales.

19.5 Did the Big Bang Really Happen?

You might occasionally read an article in a newspaper or a magazine questioning whether the Big Bang really happened. We may never be able to prove with absolute certainty that the Big Bang theory is correct. However, no one has come up with any other model of the universe that so successfully explains so much of what we see. As we have discussed, the Big Bang model makes at least two specific predictions that we have observationally verified: the characteristics of the cosmic microwave background and the composition of the universe. It also explains quite naturally many other features of the universe. And, at least so far, we know of nothing that is absolutely inconsistent with the Big Bang model.

The Big Bang theory's very success has also made it a target for respected scientists, skeptical nonscientists, and crackpots alike. The nature of scientific work requires that we test established wisdom to make sure it is valid. A sound scientific disproof of the Big Bang theory would be a discovery of great importance. However, stories touted in the news media as disproofs of the Big Bang usually turn out to be dis-

THINKING ABOUT . . .

The End of Time

According to the Big Bang theory, time and space have a beginning in the Big Bang. Do they also have an end? If we live in a recollapsing universe, the answer seems to be a clear yes: Sometime in the distant future, the universal expansion will cease and reverse, and the universe will eventually come to an end in a fiery *Big Crunch*. But it appears more likely that the universe will continue to expand forever.

In a critical, coasting, or accelerating universe, the end will come much more gradually. The star–gas–star cycle in galaxies cannot continue forever, because not all the material is recycled: With each generation of stars, more mass becomes locked up in planets, brown dwarfs, white dwarfs, neutron stars, and black holes. Eventually, about a trillion years from now, even the longest-lived stars will burn out, and the galaxies will fade into darkness.

At this point, the only new action in the universe will occur on the rare occasions when two objects—such as two brown dwarfs or two white dwarfs—collide within a galaxy. The vast distances separating star systems in galaxies make such collisions extremely rare. For example, the probability of our Sun (or the white dwarf that it will become) colliding with another star is so small that it would be expected to happen only once in a quadrillion (10^{15}) years. But forever is a long time, and even low-probability events will eventually happen many times. If a star system experiences a collision once in a quadrillion years, it will experience about 100 collisions in 100 quadrillion (10^{17}) years. By the time the universe reaches an age of 10^{20} years, star systems will have suffered an average of 100,000 collisions each, making a time-lapse history of any galaxy look like a cosmic game of pinball.

These multiple collisions will severely disrupt galaxies. As in any gravitational encounter, some objects lose energy and some gain energy. Objects that gain enough energy will be flung into intergalactic space, where they will drift ever farther from all other objects with the expansion of the universe. Objects that lose energy will eventu- ally fall to the galactic center, forming a gigantic black hole. The remains of the universe will consist of black holes with masses as great as a trillion solar masses widely separated from a few scattered planets, brown dwarfs, and stellar corpses. If the Earth somehow survives, it will be a frozen chunk of rock in the darkness of the expanding universe, billions of light-years from any other solid object.

If grand unified theories are correct, the Earth still cannot last forever. These theories predict that protons will eventually fall apart. The predicted lifetime of protons is extremely long: a half-life of at least 10^{32} years. However, if protons really do decay, then by the time the universe is 10^{40} years old the Earth and all other atomic matter will have disintegrated into radiation and subatomic particles, such as electrons and neutrinos.

The final phase is predicted to come with the disintegration of the giant, black-hole corpses of galaxies. Black holes are predicted to slowly evaporate through the process of *Hawking radiation,* finally disappearing in brilliant bursts of radiation. The largest black holes last longest, but even trillion-solar-mass black holes will evaporate sometime after the universe reaches an age of 10^{100} years. From then on, the universe will consist only of individual photons and subatomic particles, each separated by enormous distances from the others. Nothing new will ever happen, and no events will ever occur that would allow an omniscient observer to distinguish past from future. In a sense, the universe will finally have reached the end of time.

Lest any of this sound depressing, keep in mind that 10^{100} years is an indescribably long time. As an example, imagine that you wanted to write on a piece of paper a number that consisted of a 1 followed by 10^{100} zeros (i.e., the number $10^{10^{100}}$). It sounds easy, but a piece of paper large enough to fit all those zeros *would not fit in the observable universe* today. And, if that still does not alleviate your concerns, you may be glad to know that a few creative thinkers speculate about ways in which the universe might undergo rebirth, even after the end of time.

agreements over details rather than fundamental problems that threaten to bring down the whole edifice. Nevertheless, scientists must keep refining the theory and tracking down the disagreements, because every once in a while a small disagreement blossoms into a full-blown scientific revolution.

You don't need to believe all you have read without question. The next time you are musing on the universe's origins, try an experiment for yourself. Go outside on a clear night and look at the sky. Notice how dark it is, and then ask yourself *why* it is dark. If the universe were infinite, unchanging, and everywhere the same, then the entire night sky would blaze as brightly as the Sun. Johannes Kepler [Section 3.4] was perhaps the first person to realize that the night sky in such a universe should be bright, but this realization is now called **Olbers' paradox** after Heinrich Olbers, a German astronomer of the 1800s.

To understand how Olbers' paradox comes about, imagine that you are in a dense forest on a flat plain. If you look in any given direction, you'll likely see a

tree. If the forest is small, you might be able to see through some gaps in the trees to the open plains, but larger forests have fewer gaps (Figure 19.12). An infinite forest would have no gaps at all—a tree trunk would block your view along any direction.

The universe is like a forest of stars in this respect. In an unchanging universe with an infinite number of stars, we would see a star in every direction, making every point in the sky as bright as the Sun's surface. Obscuring dust doesn't change this conclusion. The intense starlight would heat the dust over time until it too glowed like the Sun or evaporated away.

There are only two ways out of this dilemma: Either the universe has a finite number of stars, in which case we would not see a star in every direction, or it changes over time in some way that prevents us from seeing an infinite number of stars. For several centuries after Kepler first recognized the dilemma, astronomers leaned toward the first option. Kepler himself preferred to believe that the universe had a finite number of stars because he thought it

FIGURE 19.12 In a large forest (left) you'll see trees no matter where you look. In a small forest (below), you can see open spaces beyond the trees.

had to be finite in space, with some kind of dark wall surrounding everything. Astronomers in the early twentieth century preferred to believe that the universe was infinite in space but that we lived inside a finite collection of stars; they thought of the Milky Way as an island floating in a vast black void. However, subsequent observations showed that galaxies fill all of space more or less uniformly. We are therefore left with the second option: The universe changes in time.

The Big Bang theory solves Olbers' paradox in a particularly simple way: It tells us that we can see only a finite number of stars because the universe began at a particular moment. While the universe may contain an infinite number of stars, we can see only those that lie within our cosmological horizon [Section 17.4]. There are other ways in which the universe could change in time and prevent us from seeing an infinite number of stars, so Olbers' paradox does not *prove* that the universe began with a Big Bang. But we must have some explanation for why the sky is dark at night, and no explanation besides the Big Bang also explains so many other observed properties of the universe so well.

THE BIG PICTURE

Our "big picture" is now about as complete as it gets. We've discussed the universe from the Earth outward, and from the beginning to the end. When you think back on this chapter, keep in mind the following ideas:

- Predicting conditions in the early universe is straightforward. The only real question is how matter and energy behave under such extreme conditions.

- Our current understanding of physics allows us to reconstruct the conditions that prevailed in the universe all the way back to the first 10^{-10} second. Our understanding is less certain back to 10^{-38} second. Beyond 10^{-43} second, we run up against the present limits of human knowledge.

- Although it may sound strange to talk about the universe during its first fraction of a second, our ideas about the Big Bang rest on a solid foundation of observational, experimental, and theoretical evidence. We cannot say with absolute certainty that the Big Bang really happened, but no other model ever proposed has so successfully explained how our universe came to be as it is.

Review Questions

1. Briefly explain why the early universe must have been much hotter and denser than the universe is today.

2. What do we mean by the *Big Bang theory*?

3. What is *antimatter*? How were particle–antiparticle pairs created in the early universe? How were they destroyed?

4. Briefly summarize the characteristics marking each of the various eras shown in Figure 19.2.

5. What are the *Planck era* and the *Planck time*? Why can't our current theories describe the history of the universe during the Planck era?

6. What are *grand unified theories*? According to these theories, how many forces operated in the universe during the *GUT era*? Why?

7. What do we mean by *inflation*? Briefly describe why inflation might have occurred at the end of the GUT era.

8. What characterized the universe during the *electroweak era*? What evidence from particle accelerators supports the idea that the weak and electromagnetic forces were once unified as a single electroweak force?

9. What characterized the universe during the *particle era*? What happened to the quarks that existed freely in the universe early in the particle era?

10. How long did the *era of nucleosynthesis* last? Why was this era so important in determining the chemical composition of the universe forever after?

11. Briefly explain why radiation was trapped for 500,000 years during the *era of nuclei* and why the cosmic microwave background broke free at the end of this era.

12. What do we mean by the *era of atoms* and the *era of galaxies*? What era do we live in? Explain.

13. Briefly describe the two key pieces of evidence that support the Big Bang theory.

14. Briefly describe how the *cosmic microwave background* was discovered.

15. Why does the Big Bang theory predict that the cosmic microwave background should have a perfect thermal radiation spectrum? How did the Cosmic Background Explorer (COBE) support this prediction of the Big Bang theory? What is the temperature of the cosmic microwave background?

16. Briefly describe how theoretical calculations allow us to predict the fraction of ordinary matter in the universe that should consist of helium as opposed to hydrogen.

17. Why do we think that tiny quantum ripples should have been present in the early universe? How might

inflation have caused these tiny ripples to become the seeds that later grew into galaxies?

18. How does inflation explain why the universe appears so uniform?

19. Briefly explain why the theory of inflation predicts that the universe ought to be flat. How is inflation being tested by astronomers?

20. What is *Olbers' paradox?* Explain how this simple paradox provides at least some support for the idea that the universe began in a Big Bang.

Discussion Questions

1. *The Moment of Creation.* You've probably noticed that, in discussing the Big Bang theory, we never quite talk about the first instant. Even our most speculative theories at present take us back only to within 10^{-43} second of creation. Do you think it will *ever* be possible for science to consider the moment of creation itself? Will we ever be able to answer questions such as *why* the Big Bang happened? Defend your opinions.

2. *The Big Bang.* How convincing do you find the evidence for the Big Bang model of the universe's origin? What are the strengths of the theory? What does it fail to explain? Overall, do *you* think the Big Bang really happened? Defend your opinion.

3. *The End of Time.* According to our current understanding, the universe as we know it will eventually come to an end—either in a fiery Big Crunch or in a slow death as all the stars eventually die. However, some people speculate that the universe might then undergo some type of rebirth. In the case of a recollapsing universe, the rebirth might consist of a new Big Bang; some people even speculate that the universe might repeatedly oscillate through cycles of Big Bangs and Big Crunches. Discuss some ideas about how the universe might undergo rebirth if it continues to expand forever. (*Note:* You may wish to read Isaac Asimov's short story *The Last Question,* published in 1956.)

4. *Forever.* If you could live forever, would you choose to do so? Suppose the universe will keep expanding forever, as the bulk of observations currently suggest. How would you satisfy your energy needs as time goes on? How would you grow food? What would you do with all that time?

Problems

True Statements? For **problems 1–8**, decide whether the statement is true and explain why it is or is not.

1. According to the Big Bang theory, the universe's temperature was greater than 10 billion (10^{10}) K when the universe was less than one ten-billionth (10^{-10}) of a second old.

2. Although the universe today appears to be made mostly of matter and not antimatter, the Big Bang theory suggests that the early universe had nearly equal amounts of matter and antimatter.

3. According to the Big Bang theory, the cosmic microwave background was created when energetic photons ionized the neutral hydrogen atoms that originally filled the universe.

4. While the existence of the cosmic microwave background is consistent with the Big Bang theory, it is also easily explained by assuming that it comes from individual stars and galaxies.

5. According to the Big Bang theory, most of the helium in the universe was created by nuclear fusion in the cores of stars.

6. The theory of inflation suggests that the structure in the universe today may have originated as tiny quantum fluctuations.

7. Within the next decade, observations of the cosmic microwave background will prove definitively whether inflation really occurred in the early universe.

8. The fact that the night sky is dark tells us that the universe cannot be infinite, unchanging, and everywhere the same.

9. *Life Story of a Proton.* Tell the life story of a proton from its formation shortly after the Big Bang to its presence in the nucleus of an oxygen atom you have just inhaled. Your story should be creative and imaginative, but it also should demonstrate your scientific understanding of as many stages in the proton's life as possible. You can draw on material from the entire book, and your story should be three to five pages long.

Web Projects

Find useful links for Web projects on the text Web site.

1. *New Tests of the Big Bang Theory.* The Cosmic Background Explorer (COBE) satellite provided striking confirmation of several predictions of the Big Bang theory. But new satellites are already being designed that will test the Big Bang theory further, primarily by observing the subtle variations in the cosmic microwave background with a much higher sensitivity than COBE. Use the Web to gather pictures and information about the COBE mission and its planned successors, such as MAP or Planck. Write a one- to two-page report about the strength of the evidence compiled by COBE and how much more we might learn from the upcoming missions.

2. *Decay of the Proton.* One of the most startling predictions of grand unified theories is that protons will eventually decay, albeit with a half-life of more than 10^{32} years. If this is true, it may be possible to observe an occasional proton decay, despite the

extraordinarily long half-life. Several experiments to look for proton decay are under way or being planned. Find out about one or more of these experiments, and write a one- to two-page summary in which you describe the experiment(s), any results to date, and what these results (or potential results) mean to the grand unified theories.

3. *New Ideas in Inflation.* The idea of inflation solves many of the puzzles associated with the standard Big Bang theory, but we are still a long way from finding strong evidence that inflation really occurred. Find recent articles that discuss some of the latest ideas about inflation and how we might test these ideas. Write a two- to three-page summary of your findings.

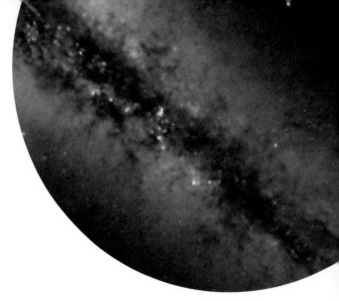

We, this people, on a small and lonely planet
Travelling through casual space
Past aloof stars, across the way of indifferent suns
To a destination where all signs tell us
It is possible and imperative that we learn
A brave and startling truth.

MAYA ANGELOU, EXCERPTED FROM
A BRAVE AND STARTLING TRUTH

CHAPTER 20

Interstellar Travel
And the Search for
Extraterrestrial Civilizations

According to science fiction, our descendants will soon travel among the stars as routinely as we jet about the Earth in airplanes. They'll race around the galaxy in starships of all sizes and shapes, circumventing nature's prohibition on faster-than-light travel by entering hyperspace, wormholes, or warp drive. They'll witness incredible cosmic phenomena firsthand, including stars and planets in all stages of development, accretion disks around white dwarfs and neutron stars, and the distortion of spacetime near black holes. Along the way, they'll encounter numerous alien species, most of which will look a lot like us and share virtually the same level of technological advancement.

Real interstellar travel is likely to be limited by the speed of light. Moreover, given the billions of years over which the galaxy has existed, finding an alien species that shares our level of technological development would be an extraordinary coincidence; more likely, any other species we encounter will be either far ahead of us or far behind us.

Nevertheless, if our civilization survives long enough, it seems almost inevitable that our descendants will leave the confines of our solar system. In this final chapter, we will explore how that might happen. We will also consider whether other species might already have progressed beyond this point and, if so, how we might contact them.

20.1 Starships: Distant Dream or Near-Reality?

Before our ancestors domesticated camels and horses, and before the invention of the wheel, the fastest way to travel was on foot. Traveling on foot, our ancestors could sustain speeds of no more than a few kilometers per hour. Today, our fastest interplanetary spacecraft travel through the solar system at speeds of a few *tens of thousands* of kilometers per hour— a factor of 10,000 faster than our ancestors traveled on foot. But we will need to go much faster to travel among the stars (at least if we are interested in making round-trip journeys within human lifetimes). At the speeds of our current interplanetary spacecraft, the journey to even the nearest stars (besides the Sun) would take more than 50,000 years.

The vast distances between the stars mean that practical starships would have to travel near the speed of light in order to complete a journey within a human lifetime. Today's seemingly fast interplanetary spacecraft travel less than 1/10,000 the speed of light— meaning that starships will need to go 10,000 times faster. From this standpoint, we are as far from interstellar flight as cavemen were from the space age.

From another standpoint, however, starflight might be just around the corner. Nearly all of the increase in speed we have achieved over our ancestors occurred within the past century or so, with the advent of flight and the dawn of the space age. If we can achieve a comparable speed advancement in the next century, our great-grandchildren may be building starships.

Regardless of how far in the future interstellar travel might be, we will have to overcome huge technological and social hurdles to achieve it. On the technological side, we will need entirely new types of engines to reach speeds close to the speed of light. We'll also need new types of shielding to protect crew members from instant death: When a starship travels through interstellar gas at near–light speed, ordinary atoms and ions will hit it like a flood of high-energy particles.

The social hurdles may be even more difficult to overcome. Building starships will almost certainly require construction facilities in space and probably cannot be achieved until we first learn to mine materials from the Moon or nearby asteroids. Thus, at least a few major space stations and outposts on the Moon are probably prerequisites to any era of starship exploration. Given global political realities, such colonization of space will probably not occur without far greater international trust and cooperation than we enjoy today.

Even if we overcome the technological and social hurdles, interstellar travel will pose further difficulties for starship crews. Einstein's theory of relativity offers high-speed travelers a "ticket to the stars":

Time will run slow for the travelers, enabling them to make journeys across many light-years of space in relatively short amounts of time. For example, in a ship traveling at an average speed of 99.9% of the speed of light, the 50-light-year round-trip to Vega would take the travelers only about 2 years. But more than 50 years would pass on Earth while they were gone, which would make their lives very difficult when they returned. Family and friends would be older or gone, new technologies might have made their knowledge and skills obsolete, and it might take them many years just to understand the political and social changes that occurred in their absence.

Despite all these hurdles, we seem bound to become interstellar travelers unless we *choose* otherwise. Such a choice might be deliberate: Many people argue that money for space exploration would be better spent here on Earth, and others argue against space colonization on philosophical grounds. Or it might be the consequence of a catastrophe: Despite the lessening of international tensions since the end of the Cold War, more than 10,000 nuclear weapons remain in the global arsenal, and many nations possess the technology to build chemical and biological weapons of mass destruction. Our civilization might also succumb to disasters brought on by human activity, such as overpopulation, epidemic disease, or global warming.

If an alien civilization were watching us, they might well conclude that we are poised on the brink of the most significant turning point in human history. If we destroy ourselves, all our achievements in science, art, and philosophy will be lost forever. But if we survive and choose to continue the exploration of space, we may be embarking on a path that will take us to the stars.

20.2 Starship Design

So far we have spoken of starships that could carry humans across vast interstellar distances. But if we expand the definition to include robotic spacecraft, then we have already launched our first starships. Four interplanetary probes—*Pioneers 10* and *11,* and *Voyagers 1* and *2*—are traveling fast enough to escape the solar system. It will take them more than 10,000 years to cover each light-year of distance, and their trajectories will not take them on close passes of any nearby stars. Nevertheless, these spacecraft should suffer little damage during their journeys and are likely to look almost as good as new for millions of years. Each carries a greeting from Earth, in case someone comes across one of them someday (Figure 20.1).

Current spacecraft, including both robotic probes like Pioneer and Voyager and crew-carrying spacecraft such as the Space Shuttle, are launched by *chemical rockets.* Their engines use energy from chem-

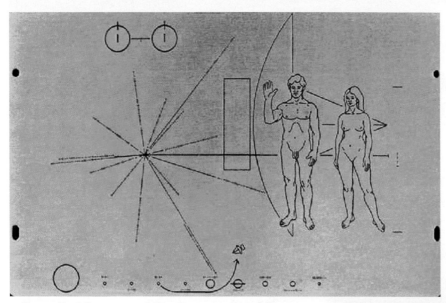

a *Pioneers 10* and *11* carry a copy of this etching as a greeting to anyone who might find them.

b *Voyagers 1* and *2* carry a phonograph record—a 12-inch gold-plated copper disk containing music, greetings, and images from Earth.

FIGURE 20.1 Messages aboard the Pioneer and Voyager spacecraft, which are bound for the stars.

ical reactions to drive hot gas out the back of the rocket, which causes the rocket to accelerate forward. Chemical rockets will never be practical for human interstellar travel because a limiting process prevents them from reaching very high speeds: Achieving higher speeds requires more fuel, but the added weight of additional fuel makes it more difficult to increase the rocket's speed. Detailed calculations of this limiting process show that chemical rockets, even as propulsion systems for relatively small robotic spacecraft, cannot exceed speeds of about 0.001*c*—in which case the journey to Alpha Centauri would take more than 4,000 years. For the much larger starships needed to support human crews, chemical rockets are completely out of the question. Fortunately, other technologies may allow much faster travel.

Nuclear Rocket Propulsion

Rockets would be far more powerful if their engines used nuclear power rather than chemical power. Nuclear reactions generate far more power than chem-

ical reactions because they convert some of the mass in atomic nuclei into energy in accord with $E = mc^2$. Fusing a kilogram of hydrogen, for example, generates more than a million times more energy than chemical reactions in a kilogram of hydrogen and oxygen.

One way to realize the benefits of fusion power would be to power starships with nuclear bombs. This idea, called *nuclear pulse propulsion*, involves generating energy with repeated detonations of relatively small H-bombs. One design, created under the name Project Orion, envisions the explosions taking place a few tens of meters behind the spaceship (Figure 20.2). The vaporized debris from each explosion would impact a "pusher plate" on the back of the spacecraft, propelling the ship forward. In principle, we could build an Orion spacecraft with existing technology, although it would be very expensive.

Another design for a fusion-powered rocket, called Project Daedalus, relies on generation of a continuous stream of energy from an onboard controlled nuclear fusion reactor (Figure 20.3). Because we

FIGURE 20.2 Artist's conception of the Project Orion starship, showing one of the small H-bomb detonations that propel it. Debris from the detonation impacts the flat disk, called the pusher plate, at the back of the spaceship. The central sections (enclosed in lattice) hold the bombs, and the crew lives in the front sections.

FIGURE 20.3 Artist's conception of a robotic Project Daedalus starship. The front section (right) holds the scientific instruments. The large spheres hold the hydrogen for the central fusion reactor.

FIGURE 20.4 Artist's conception of a spaceship propelled by a solar sail, shown as it approaches a forming planet in a young solar system. The sail is many kilometers across; the scientific payload is at the central meeting point of the four scaffoldlike structures.

have not yet achieved controlled nuclear fusion (as opposed to "uncontrolled" bombs), Daedalus remains beyond our current technological capabilities.

Fusion-powered starships such as Orion or Daedalus could probably achieve speeds of about 10% of the speed of light, completing journeys to nearby stars in a few decades. Such speeds are certainly sufficient for sending colonists on one-way voyages, particularly if the crew can be put into some sort of hibernation that prevents them from aging during the trip.

TIME OUT TO THINK *From the sixteenth to the nineteenth century, many people left their homes in Europe on one-way journeys to the "new world" of America. In the future, similar one-way trips to colonies on Earth-like planets around other stars may be possible. If offered the opportunity, would you go? Do you know anyone who would?*

Matter–Antimatter Rocket Engines

In principle, there is an even more powerful energy source for rocket engines than fusion: matter–antimatter annihilation. Whereas fusion converts less than 1% of the mass of atomic nuclei into energy, matter–antimatter annihilation converts *all* the annihilated mass into energy. Starships with matter–antimatter engines could probably reach 90% or more of the speed of light. At these speeds, the slow-

ing of time predicted by relativity becomes noticeable, putting many nearby stars within a few years' journey for the crew members.

However, while matter–antimatter engines may someday be the propulsion systems of choice for starships, this technology appears to be far in the future. The two major technological challenges to developing matter–antimatter engines are finding a way to produce sufficient quantities of antimatter to be used in the engines and finding a way to safely store the antimatter until it is needed. Neither challenge seems likely to be met soon.

Sails and Beamed Energy Propulsion

In principle, spacecraft with large, lightweight sails could be propelled by the solar wind and the radiation pressure of sunlight. Solar sailing may well prove to be a fairly inexpensive way of navigating within the solar system (Figure 20.4). Surprisingly, it may even be useful for interstellar travel. Energy flowing from the Sun would give an optimally designed sailing starship a small but continuous acceleration as it headed out of the solar system. By the time the ship reached the distance at which the solar wind could no longer provide acceleration, it might already be traveling at speeds of a few percent of the speed of light.

We might achieve even higher velocities by shining powerful lasers at the spacecraft sails. Such *beamed energy propulsion* theoretically could accelerate ships

FIGURE 20.5 Artist's conception of a spaceship powered by an interstellar ramjet. The giant scoop in the front (left) collects interstellar hydrogen for use as fusion fuel.

to substantial fractions of the speed of light—but only if we could build enormous lasers. For example, accelerating a ship to half the speed of light within a few years would require a laser that uses 1,000 times more power than all current human power consumption. As a result, starships propelled by beamed energy probably remain far in the future.

Interstellar Ramjets

All the propulsion systems we've discussed so far have limited capabilities: Rockets can carry only a limited amount of fuel no matter what their fuel source, and beamed energy becomes too weak to accelerate starships when they get very far from home. One way around these difficulties would be to build a starship that collects fuel as it goes, using a giant scoop to gather interstellar hydrogen as fuel for a fusion engine. Such ships are known as *interstellar ramjets* (Figure 20.5); in principle, they can accelerate continuously by collecting and using fuel continuously.

Imagine an interstellar ramjet that accelerates at 1g for half the voyage to its destination and then turns around and decelerates at 1g until it arrives. The crew would find the trip quite comfortable, experiencing Earth-like gravity the entire way. During most of the journey, the ship would be traveling relative to Earth (and its destination) at a speed very close to the speed of light, so time on the ship would

pass very slowly compared to time on Earth. Longer trips would mean reaching top speeds closer to the speed of light and therefore more extreme effects on time. For example, such a ship could make a trip to a star 500 light-years away in only about 12 years of ship time. It could travel the 28,000 light-years to the center of the Milky Way Galaxy, where it could observe firsthand the mysterious galactic center [Section 16.5], in only about 21 years of ship time. It could make the trip to a star system in the Andromeda Galaxy in only about 29 years of ship time. Thus, the crew could travel to the Andromeda Galaxy, spend 2 years studying one of its star systems and taking pictures of the Milky Way Galaxy to bring home, and return in just 60 years of ship time. However, because the Andromeda Galaxy is about 2.5 million light-years away, the crew would find that about 5 million years had passed on Earth by the time they returned home.

Science Fiction

If you are a science fiction fan, this discussion of interstellar travel may be depressing. Interstellar tourism and commerce seem out of the question, even with interstellar ramjets that reach speeds very close to the speed of light. If we are ever to travel about the galaxy the way we now travel about the Earth, we will need spacecraft that can get us from here to there much faster than the speed of light.

Einstein's theory of relativity leaves little hope that we can ever find a way to travel *through* space faster than the speed of light, but science fiction writers have imagined all kinds of novel shortcuts that don't necessarily violate relativity or any other known laws of physics. Some of these go by names like hyperspace, wormholes, and warp drive. The bottom line is that our present knowledge does not allow us to say one way or the other whether any of these technologies are possible.

20.3 Radio Communication: Virtual Interstellar Travel

We cannot travel at the speed of light, but we can send information at the speed of light using radio waves, lasers, or other forms of light. Because radio waves penetrate interstellar gas and dust fairly easily, they probably will prove to be the most useful form of interstellar communication. Although we usually think of radio communication as a way of sending and receiving fairly simple messages, there is really no limit to how much information we could transmit or receive. In principle, we could become *virtual* interstellar travelers. Imagine, for example, that we sent a robotic probe to Alpha Centauri at 10% of the speed of light. The probe would take 40 years to get there, but it could send data back to Earth via radio in just 4 years. Nearby stars, at least, could be explored with robotic spacecraft in the same way we now explore the planets of our own solar system.

An even better way to gather data about distant worlds would be to receive signals from civilizations in other star systems. If such civilizations exist, they may already be broadcasting messages to us. The **search for extraterrestrial intelligence**, or **SETI**, is a project in which scientists use large radio telescopes to scan the skies for signals from other civilizations (Figure 20.6). With enough data and the computer technology known as virtual reality, we could actually experience what it would be like to walk among the cities of an alien civilization.

The more immediate hope of SETI efforts is simply to learn whether anyone else is out there—and happens to be broadcasting messages. But how could we interpret these messages? After all, we could hardly expect alien civilizations to share any of our social customs, let alone to speak English. Fortunately, we know that all civilizations must share some things in common: the laws of mathematics and science that are the same everywhere. Imagine that we wanted to send a message to a distant civilization. As popularized in the movie *Contact,* we might begin by broadcasting radio pulses in the order of the prime numbers (1, 2, 3, 5, 7, 11, . . .)—clearly identifying the message as something sent by an intelligent species, since no known natural process can broadcast prime numbers. We could then represent the prime numbers with a binary code and use this same code to begin representing some of the laws of mathematics. Because alien civilizations would know these laws already, they could learn the intricacies of our code. Once they knew the full code, they could receive as much additional information about Earth and our civilization as we wished to send them.

FIGURE 20.6 This 210-foot radio telescope in New South Wales, Australia, is being used for Project Phoenix, a project to search nearby, Sun-like star systems for signals from extraterrestrial civilizations. Project Phoenix is sustained by private funding.

Scientists working on SETI hope that someone out there is already broadcasting similar coded messages to Earth. Even if such messages are being continually broadcast to Earth, finding them may be rather like finding the proverbial needle in a haystack: There are billions of stars to search, and we don't know the radio frequencies at which other civilizations might choose to broadcast their messages. Nevertheless, we can make some guesses. For example, we might focus our attention on Sun-like stars and assume that the aliens would broadcast near some "natural" frequency, such as the frequency of 21-cm radio waves from hydrogen. In addition, new technologies allow us to search many frequencies simultaneously, increasing the odds of success. However, SETI efforts remain controversial because they require extensive time on powerful and expensive radio telescopes when we have no way of knowing whether anyone is even out there.

TIME OUT TO THINK *SETI supporters argue that, despite the costs and the unknowns, the efforts are justified because contact with an extraterrestrial intelligence would be one of the most dramatic discoveries in human history. Do you agree? Defend your opinion.*

20.4 Civilizations

Whether we consider the virtual travel of radio communication or actual ships bound for the stars, a recurrent theme of interstellar travel is contact with other civilizations. Thus, we are led to one of the most exciting questions in science: Is anyone else out there?

Unfortunately, we simply don't yet know. But we can speculate—and perhaps even learn something important in the process. Let's try a systematic approach to guessing the number of alien civilizations that exist. We'll concentrate on the Milky Way Galaxy and then extend our results to the entire observable universe. For simplicity's sake, we'll also assume that alien life, like life on Earth, will need liquid water and therefore that it must evolve on a planet that lies within the *habitable zone* around a star [Section 11.6].

How Many Civilizations Are Out There?

In principle, we could calculate the number of civilizations in the Milky Way Galaxy if we knew just a few basic facts. We would start with the number of planets that lie within habitable zones, which we'll call N_P (for "number of planets"). Then we would determine the fraction of those planets that actually have life; let's use f_{life} to stand for this fraction. Multiplying N_P by f_{life} would tell us the number of life-bearing planets in the Milky Way. Next, we'd multiply this number by the fraction of life-bearing planets on which intelligent beings evolve and develop a civilization, which we'll call f_{civ}. Finally, we'd multiply by the fraction of planets with civilizations that have those civilizations *now* (as opposed to the distant past or future), which we'll call f_{now}. Putting all these ideas together gives us the following simple equation (a variation on what is called the Drake equation) for the number of civilizations now inhabiting the Milky Way Galaxy:

$$\text{Number of civilizations} = N_P \times f_{life} \times f_{civ} \times f_{now}$$

The only difficulty in using this equation is that we don't know the value of any of its terms! In fact,

the only term for which we can even make a reasonable guess is the number of planets, N_P. Our theories of star formation tell us that planetary systems ought to be common, and recent discoveries of extrasolar planets support this idea. Studies of our own solar system suggest that both Earth and Mars lie within the Sun's habitable zone, and Venus is probably close to this zone. If other planetary systems are similar, we might expect the typical number of planets within the habitable zone to be between one and three. In that case, it seems likely that many, and perhaps even most, of the several hundred billion stars in the Milky Way have at least one planet within their habitability zone. The technology for detecting planets beyond our solar system is rapidly advancing, so we may be able to improve or confirm this estimate within the next couple of decades.

For the moment, we have no rational way to guess the fraction f_{life} of planets on which life actually arose. The fact that life arose quickly on Earth [Section 11.4] suggests that life develops fairly easily under the right conditions, but it is by no means conclusive. As far as we know today, f_{life} could be anywhere between about 1 in 100 billion (within the Milky Way Galaxy, it happened only on Earth) and 1 (it happened on all planets suitable for life).

Similarly, we have little basis on which to guess the fraction f_{civ} of life-bearing planets that develop a civilization. However, the fact that life flourished on Earth for some 4 billion years before the rise of humans suggests that developing intelligent life is much more difficult than developing microbial life. The question is *how much* more difficult. Roughly half the stars in the Milky Way are older than our Sun, so if 4 billion years is a typical time for a civilization to develop then it should have happened already on about half the planets with life. On the other hand, if the Earth was "fast" and the typical time needed for intelligent life to develop is much longer, then most life-bearing planets may be covered with nothing more advanced than bacteria.

The huge uncertainties in f_{life} and f_{civ} make it impossible for us to determine the number of civilizations that have arisen in the Milky Way's history. If 1 in 1,000 suitable planets ends up with a civilization, then we are newcomers among a billion civilizations in the Milky Way. If 1 in 1 million develops a civilization, there would still have been some 100,000 civilizations out there in the Milky Way Galaxy alone. But if only 1 in 1 trillion planets ends up with a civilization, then we may be the first in the Milky Way— and the only civilization within millions of light-years.

The final step in our calculation, determining the number of civilizations that exist *now*, requires knowing how long a civilization typically lasts in a planet's history. Here we must define what we mean by *civilization*; let's use a definition that presumes a civilization advanced enough to travel in space. On the nearly 5-billion-year-old planet Earth, the space age began only about 50 years ago. Thus, the fraction of Earth's history during which it has had a space-faring civilization is about 50 in 5 billion, or about 1 in 100 million. If this fraction is typical, there would have to be some 100 million civilization-bearing planets in the Milky Way in order for us to have good odds of finding another civilization out there *now*. However, we'd expect this fraction to be typical only if we are on the brink of self-destruction, since the fraction will grow larger as long as our civilization continues to thrive. Thus, if civilizations are at all common, the key factor in whether any are out there now is their survivability. If most civilizations self-destruct shortly after achieving the technology for space travel, then we probably are alone in the galaxy at present. But if most survive and become interstellar travelers, the Milky Way may be brimming with civilizations—most far more advanced than ours.

TIME OUT TO THINK *What do you think is the likelihood that our civilization will survive to become interstellar travelers? Explain.*

A Paradox: Where Are the Aliens?

Imagine that we survive and become interstellar travelers and that we begin colonizing habitable planets around nearby stars. As the colonies grow at each new location, some of the people may decide to set out for other star systems. Even if our starships traveled at relatively low speeds—say, a few percent of the speed of light—we could have dozens of outposts around nearby stars within a few centuries. In 10,000 years, our descendants would be spread among stars within a few hundred light-years of Earth. In a few million years, we could have outposts scattered throughout the Milky Way and would likely have explored nearly every star system in the galaxy. We would have become a true galactic civilization.

Now, if we take the idea that *we* could develop a galactic civilization within a few million years and combine it with the reasonable (though unproved) idea that civilizations ought to be common, we are led to an astonishing conclusion: Someone else should already have created a galactic civilization.

Thus, we encounter a strange paradox: Plausible arguments suggest that a galactic civilization should already exist, yet we have so far found no evidence of such a civilization. There are many possible solutions to this paradox, but most fall into one of three basic categories:

1. There is no galactic civilization because civilizations are not common. Perhaps we are the first, or perhaps civilizations are so rare that many star systems, including ours, remain unexplored.

2. There is no galactic civilization because civilizations do not leave their home worlds—either because they are uninterested in interstellar travel or because they destroy themselves before achieving it.

3. There *is* a galactic civilization, but it has deliberately avoided revealing its existence to us.

We do not know which, if any, of these three explanations is the correct solution to the paradox, but each leads to intriguing possibilities. If the first is correct, then our civilization is an astonishing achievement—something that few if any other species have ever accomplished. From this point of view, humanity becomes all the more precious, and the collapse of our civilization would be all the more tragic. The second possible explanation is much less satisfying. Unless we "think differently" than other civilizations, it says that we will soon either lose our interest in interstellar travel or destroy ourselves—hardly a comforting thought. The third possibility is perhaps the most intriguing. It says that we are newcomers on the scene of a galactic civilization that has existed for millions or billions of years before us. Perhaps members of this civilization are deliberately leaving us alone for the time being and will invite us to join them when we prove ourselves worthy. If so, our entire species may be on the verge of beginning a journey every bit as incredible as that of a baby emerging from the womb and coming into the world.

THE BIG PICTURE

Throughout our study of astronomy, we have taken the "big picture" view of trying to understand how we fit into the universe. Here, at last, we have returned to Earth and examined the role of our own generation in the big picture of human history. Tens of thousands of past human generations have walked this Earth, but ours is the first generation with the technology to study the far reaches of our universe and to travel beyond the Earth. It is up to us to decide whether we will use this technology to advance our species or to destroy it.

Take a long view, a view down the centuries and millennia from this moment. Imagine our descendants living among the stars, having created or joined a great galactic civilization. They will have the privilege of experiencing ideas, worlds, and discoveries far beyond our wildest imagination. Perhaps, in their history lessons, they will learn of our generation—the generation that history placed at the turning point and that chose life over death and destruction.

Review Questions

1. How do speeds of current spacecraft compare to the speed of light?

2. Describe several hurdles, both technological and social, that we must overcome in order to achieve interstellar flight.

3. Why do we say that *Pioneers 10* and *11* and *Voyagers 1* and *2* are our first interstellar spacecraft? How long will it take them to reach nearby stars?

4. Why are chemical rockets inadequate for interstellar travel?

5. Briefly describe the prospects and limitations for each of the following alternative starship designs: nuclear propulsion, matter–antimatter engines, sails and beamed energy propulsion, and interstellar ramjets.

6. What is *SETI*? Outline how we might recognize a signal from another civilization.

7. Briefly describe and interpret the equation given in the text that allows us to estimate the number of civilizations in the Milky Way Galaxy. Why is the term f_{now} particularly important?

8. Describe several potential solutions to the paradox of "where are the aliens?"

Discussion Questions

1. *Distant Dream or Near-Reality?* Considering all the issues surrounding interstellar flight, when (if ever) do you think we are likely to begin traveling among the stars? Why?

2. *Where Are the Aliens?* Consider the paradox that arises from the fact that we do not yet have evidence of a galactic civilization. What do *you* think is the solution to this paradox? Why?

3. *Aliens in the Movies.* Choose a science fiction movie (or television show) that involves alien species, and discuss what aspects of it are realistic and what aspects are not. Pay particular attention to the depiction of alien species and encounters between different species. Do you think such encounters are portrayed realistically?

Problems

Fantasy or Science Fiction? Each of **problems 1–5** describes some futuristic device or discovery. In each case, decide whether the device or discovery is plausible according to our present understanding of science or whether it is unlikely to be possible. Explain your reasoning.

1. A brilliant teenager discovers a way to build a rocket that burns coal as its fuel and can travel at half the speed of light.

2. Using beamed energy propulsion from a laser powered by energy produced at a windmill farm in the California desert, NASA engineers are able to send a solar sailing ship on a journey to Alpha Centauri that will take only 50 years.

3. Future human colonization of the moons of Saturn occurs using spaceships powered by nuclear pulse propulsion.

4. In the year 2750, we receive a signal from a civilization around a nearby star telling us that the *Voyager 2* spacecraft recently crash-landed on their planet.

5. In the year 2047, we receive a greeting from a galactic civilization that has already existed for more than a billion years.

6. *Research: Pioneers of Flight.* Write a one- to two-page biography of one of the following pioneers of flight: Otto Lilienthal, the Wright brothers, Charles Lindbergh, Amelia Earhart.

7. *Research: Development of Rocketry.* Write a one- to two-page paper about the development of rocketry. Be sure to include the roles of Konstantin Tsiolkovsky and Robert Goddard.

8. *Research: Starship Design.* Find more details about one of the proposals for starship propulsion and design discussed in this chapter. How would the starship actually be built? What new technologies would be needed, and what existing technologies could be used? Summarize your findings in a one- to two-page research report.

Web Projects

Find useful links for Web projects on the text Web site.

1. *SETI.* Visit the Web site for the SETI Institute and learn more about the search for other civilizations. Choose some aspect of this issue that interests you, and write a one- to two-page report on what you learn.

2. *Advance Spaceship Design.* NASA supports many efforts to incorporate new technologies into spaceships. Although few of these technologies reach the level of being suitable for interstellar colonization, most are innovative and fascinating. Learn about one such NASA project and write a short summary of your findings.

APPENDIXES

APPENDIX A

USEFUL NUMBERS

Astronomical Distances

1 AU $\approx 1.496 \times 10^8$ km

1 light-year $\approx 9.46 \times 10^{12}$ km

1 parsec (pc) $\approx 3.09 \times 10^{13}$ km ≈ 3.26 light-years

1 kiloparsec (kpc) $= 1,000$ pc $\approx 3.26 \times 10^3$ light-years

1 megaparsec (Mpc) $= 10^6$ pc $\approx 3.26 \times 10^6$ light-years

Astronomical Times

1 solar day (average) $= 24^{\text{h}}$

1 sidereal day $\approx 23^{\text{h}} \, 56^{\text{m}} \, 4.09^{\text{s}}$

1 synodic month (average) ≈ 29.53 solar days

1 sidereal month (average) ≈ 27.32 solar days

1 tropical year ≈ 365.242 solar days

1 sidereal year ≈ 365.256 solar days

Universal Constants

Speed of light: $\qquad c = 3 \times 10^5$ km/s $= 3 \times 10^8$ m/s

Gravitational constant: $\qquad G = 6.67 \times 10^{-11} \dfrac{\text{m}^3}{\text{kg} \times \text{s}^2}$

Planck's constant: $\qquad h = 6.626 \times 10^{-34}$ joule $\times$ s

Stefan–Boltzmann constant: $\qquad \sigma = 5.7 \times 10^{-8} \dfrac{\text{watt}}{\text{m}^2 \times \text{Kelvin}^4}$

mass of a proton: $\qquad m_{\text{p}} = 1.67 \times 10^{-27}$ kg

mass of an electron: $\qquad m_{\text{e}} = 9.1 \times 10^{-31}$ kg

Useful Sun and Earth Reference Values

Mass of the Sun: $1M_{\text{Sun}} \approx 2 \times 10^{30}$ kg

Radius of the Sun: $1R_{\text{Sun}} \approx 696,000$ km

Luminosity of the Sun: $1L_{\text{Sun}} \approx 3.8 \times 10^{26}$ watts

Mass of the Earth: $1M_{\text{Earth}} \approx 5.97 \times 10^{24}$ kg

Radius (equatorial) of the Earth: $1R_{\text{Earth}} \approx 6,378$ km

Acceleration of gravity on Earth: $g = 9.8$ m/s^2

Escape velocity from surface of Earth: $v_{\text{escape}} = 11$ km/s $= 11,000$ m/s

Energy and Power Units

Basic unit of energy: 1 joule $= 1 \dfrac{\text{kg} \times \text{m}^2}{\text{s}^2}$

Basic unit of power: 1 watt $= 1$ joule/s

Electron-volt: 1 eV $= 1.60 \times 10^{-19}$ joule

APPENDIX B

USEFUL FORMULAS

■ Universal law of gravitation for the force between objects of mass M_1 and M_2, distance d between their centers:

$$F = G\frac{M_1 M_2}{d^2}$$

■ Newton's version of Kepler's third law; p and a are period and semimajor axis, respectively, of either orbiting mass:

$$p^2 = \frac{4\pi^2}{G(M_1 + M_2)}\, a^3$$

■ Escape velocity at distance R from center of object of mass M:

$$v_{\text{escape}} = \sqrt{\frac{2GM}{R}}$$

■ Relationship between a photon's wavelength (λ), frequency (f), and the speed of light (c):

$$\lambda \times f = c$$

■ Energy of a photon of wavelength λ or frequency f:

$$E = hf = \frac{hc}{\lambda}$$

■ Stefan–Boltzmann law for thermal radiation at temperature T (in Kelvin):

$$\text{emitted power per unit area} = \sigma T^4$$

■ Wien's law for the peak wavelength (λ_{max}) thermal radiation at temperature T (in Kelvin):

$$\lambda_{\text{max}} = \frac{2{,}900{,}000}{T}\ \text{nm}$$

■ Doppler shift (radial velocity is positive if the object is moving away from us and negative if it is moving toward us):

$$\frac{\text{radial velocity}}{\text{speed of light}} = \frac{\text{shifted wavelength} - \text{rest wavelength}}{\text{rest wavelength}}$$

■ Angular separation (α) of two points with an actual separation s, viewed from a distance d (assuming d is much larger than s):

$$\alpha = \frac{s}{2\pi d} \times 360°$$

■ Luminosity–distance formula:

$$\text{apparent brightness} = \frac{\text{luminosity}}{4\pi d^2}$$

(where d is the distance to the object)

■ Parallax formula (distance d to a star with parallax angle p in arcseconds):

$$d\ (\text{in parsecs}) = \frac{1}{p\ (\text{in arcseconds})}$$

■ The orbital velocity law, to find the mass M_r contained within the circular orbit of radius r for an object moving at speed v:

$$M_r = \frac{r \times v^2}{G}$$

APPENDIX C

A FEW MATHEMATICAL SKILLS

This appendix reviews the following mathematical skills: powers of 10, scientific notation, working with units, the metric system, and finding a ratio. You should refer to this appendix as needed while studying the textbook, particularly if you are having difficulty with the Mathematical Insights.

C.1 Powers of 10

Powers of 10 simply indicate how many times to multiply 10 by itself. For example:

$$10^2 = 10 \times 10 = 100$$

$$10^6 = 10 \times 10 \times 10 \times 10 \times 10 \times 10 = 1,000,000$$

Negative powers are the reciprocals of the corresponding positive powers. For example:

$$10^{-2} = \frac{1}{10^2} = \frac{1}{100} = 0.01$$

$$10^{-6} = \frac{1}{10^6} = \frac{1}{1,000,000} = 0.000001$$

Table C.1 lists powers of 10 from 10^{-12} to 10^{12}. Note that powers of 10 follow two basic rules:

1. A positive exponent tells how many zeros follow the 1. For example, 10^0 is a 1 followed by no zeros, and 10^8 is a 1 followed by eight zeros.

2. A negative exponent tells how many places are to the right of the decimal point, including the 1. For example, $10^{-1} = 0.1$ has one place to the right of the decimal point; $10^{-6} = 0.000001$ has six places to the right of the decimal point.

Table C.1 Powers of 10

Zero and Positive Powers			Negative Powers		
Power	Value	Name	Power	Value	Name
10^0	1	One			
10^1	10	Ten	10^{-1}	0.1	Tenth
10^2	100	Hundred	10^{-2}	0.01	Hundredth
10^3	1,000	Thousand	10^{-3}	0.001	Thousandth
10^4	10,000	Ten thousand	10^{-4}	0.0001	Ten thousandth
10^5	100,000	Hundred thousand	10^{-5}	0.00001	Hundred thousandth
10^6	1,000,000	Million	10^{-6}	0.000001	Millionth
10^7	10,000,000	Ten million	10^{-7}	0.0000001	Ten millionth
10^8	100,000,000	Hundred million	10^{-8}	0.00000001	Hundred millionth
10^9	1,000,000,000	Billion	10^{-9}	0.000000001	Billionth
10^{10}	10,000,000,000	Ten billion	10^{-10}	0.0000000001	Ten billionth
10^{11}	100,000,000,000	Hundred billion	10^{-11}	0.00000000001	Hundred billionth
10^{12}	1,000,000,000,000	Trillion	10^{-12}	0.000000000001	Trillionth

Multiplying and Dividing Powers of 10

Multiplying powers of 10 simply requires adding exponents, as the following examples show:

$$10^4 \times 10^7 = \underbrace{10,000}_{10^4} \times \underbrace{10,000,000}_{10^7} = \underbrace{100,000,000,000}_{10^{4+7}\,=\,10^{11}} = 10^{11}$$

$$10^5 \times 10^{-3} = \underbrace{100,000}_{10^5} \times \underbrace{0.001}_{10^{-3}} = \underbrace{100}_{10^{5+(-3)}\,=\,10^2} = 10^2$$

$$10^{-8} \times 10^{-5} = \underbrace{0.00000001}_{10^{-8}} \times \underbrace{0.00001}_{10^{-5}} = \underbrace{0.0000000000001}_{10^{-8+(-5)}\,=\,10^{-13}} = 10^{-13}$$

Dividing powers of 10 requires subtracting exponents, as in the following examples:

$$\frac{10^5}{10^3} = \underbrace{100,000}_{10^5} \div \underbrace{1,000}_{10^3} = \underbrace{100}_{10^{5-3}\,=\,10^2} = 10^2$$

$$\frac{10^3}{10^7} = \underbrace{1,000}_{10^3} \div \underbrace{10,000,000}_{10^7} = \underbrace{0.0001}_{10^{3-7}\,=\,10^{-4}} = 10^{-4}$$

$$\frac{10^{-4}}{10^{-6}} = \underbrace{0.0001}_{10^{-4}} \div \underbrace{0.000001}_{10^{-6}} = \underbrace{100}_{10^{-4-(-6)}\,=\,10^2} = 10^2$$

Powers of Powers of 10

We can use the multiplication and division rules to raise powers of 10 to other powers or to take roots. For example:

$$(10^4)^3 = 10^4 \times 10^4 \times 10^4 = 10^{4+4+4} = 10^{12}$$

Note that we can get the same end result by simply multiplying the two powers:

$$(10^4)^3 = 10^{4 \times 3} = 10^{12}$$

Because taking a root is the same as raising to a fractional power (e.g., the square root is the same as the 1/2 power, the cube root is the same as the 1/3 power, etc.), we can use the same procedure for roots, as in the following example:

$$\sqrt{10^4} = (10^4)^{1/2} = 10^{4 \times (1/2)} = 10^2$$

Adding and Subtracting Powers of 10

Unlike with multiplication and division, there is no shortcut for adding or subtracting powers of 10. The values must be written in longhand notation. For example:

$$10^6 + 10^2 = 1{,}000{,}000 + 100 = 1{,}000{,}100$$

$$10^8 + 10^{-3} = 100{,}000{,}000 + 0.001 = 100{,}000{,}000.001$$

$$10^7 - 10^3 = 10{,}000{,}000 - 1{,}000 = 9{,}999{,}000$$

Summary

We can summarize our findings using n and m to represent any numbers:

- To *multiply* powers of 10, *add* exponents: $10^n \times 10^m = 10^{n+m}$

- To *divide* powers of 10, *subtract* exponents: $\dfrac{10^n}{10^m} = 10^{n-m}$

- To *raise* powers of 10 to other powers, multiply exponents: $(10^n)^m = 10^{n \times m}$

C.2 Scientific Notation

When we are dealing with large or small numbers, it's generally easier to write them with powers of 10. For example, it's much easier to write the number $6{,}000{,}000{,}000{,}000$ as 6×10^{12}. This format, in which a number *between* 1 and 10 is multiplied by a power of 10, is called **scientific notation**.

Converting a Number to Scientific Notation

We can convert numbers written in ordinary notation to scientific notation with a simple two-step process:

1. Move the decimal point to come after the *first* nonzero digit.

2. The number of places the decimal point moves tells you the power of 10; the power is *positive* if the decimal point moves to the left and *negative* if it moves to the right.

   Examples:

   $$3{,}042 \xrightarrow[\text{3 places to left}]{\text{decimal needs to move}} 3.042 \times 10^3$$

   $$0.00012 \xrightarrow[\text{4 places to right}]{\text{decimal needs to move}} 1.2 \times 10^{-4}$$

   $$226 \times 10^2 \xrightarrow[\text{2 places to left}]{\text{decimal needs to move}} (2.26 \times 10^2) \times 10^2 = 2.26 \times 10^4$$

Converting a Number from Scientific Notation

We can convert numbers written in scientific notation to ordinary notation by the reverse process:

1. The power of 10 indicates how many places to move the decimal point; move it to the *right* if the power of 10 is positive and to the *left* if it is negative.

2. If moving the decimal point creates any open places, fill them with zeros.

   Examples:

$$4.01 \times 10^2 \xrightarrow[\text{2 places to right}]{\text{move decimal}} 401$$

$$3.6 \times 10^6 \xrightarrow[\text{6 places to right}]{\text{move decimal}} 3{,}600{,}000$$

$$5.7 \times 10^{-3} \xrightarrow[\text{3 places to left}]{\text{move decimal}} 0.0057$$

Multiplying or Dividing Numbers in Scientific Notation

Multiplying or dividing numbers in scientific notation simply requires operating on the powers of 10 and the other parts of the number separately.

Examples:

$$(6 \times 10^2) \times (4 \times 10^5) = (6 \times 4) \times (10^2 \times 10^5) = 24 \times 10^7 = (2.4 \times 10^1) \times 10^7 = 2.4 \times 10^8$$

$$\frac{4.2 \times 10^{-2}}{8.4 \times 10^{-5}} = \frac{4.2}{8.4} \times \frac{10^{-2}}{10^{-5}} = 0.5 \times 10^{-2-(-5)} = 0.5 \times 10^3 = (5 \times 10^{-1}) \times 10^3 = 5 \times 10^2$$

Note that, in both these examples, we first found an answer in which the number multiplied by a power of 10 was *not* between 1 and 10. We therefore followed the procedure for converting the final answer to scientific notation.

Addition and Subtraction with Scientific Notation

In general, we must write numbers in ordinary notation before adding or subtracting.

Examples:

$$(3 \times 10^6) + (5 \times 10^2) = 3{,}000{,}000 + 500 = 3{,}000{,}500 = 3.0005 \times 10^6$$

$$(4.6 \times 10^9) - (5 \times 10^8) = 4{,}600{,}000{,}000 - 500{,}000{,}000 = 4{,}100{,}000{,}000 = 4.1 \times 10^9$$

When both numbers have the *same* power of 10, we can factor out the power of 10 first.

Examples:

$$(7 \times 10^{10}) + (4 \times 10^{10}) = (7 + 4) \times 10^{10} = 11 \times 10^{10} = 1.1 \times 10^{11}$$

$$(2.3 \times 10^{-22}) - (1.6 \times 10^{-22}) = (2.3 - 1.6) \times 10^{-22} = 0.7 \times 10^{-22} = 7.0 \times 10^{-23}$$

C.3 Working with Units

Showing the units of a problem as you solve it usually makes the work much easier and also provides a useful way of checking your work. If an answer does not come out with the units you expect, you probably did something wrong. In general, working with units is very similar to working with numbers, as the following guidelines and examples show.

Five Guidelines for Working with Units

Before you begin any problem, think ahead and identify the units you expect for the final answer. Then operate on the units along with the numbers as you solve the problem. The following five guidelines may be helpful when you are working with units:

1. Mathematically, it doesn't matter whether a unit is singular (e.g., meter) or plural (e.g., meters); we can use the same abbreviation (e.g., m) for both.

2. You cannot add or subtract numbers unless they have the *same* units. For example, 5 apples + 3 apples = 8 apples, but the expression 5 apples + 3 oranges cannot be simplified further.

3. You *can* multiply units, divide units, or raise units to powers. Look for key words that tell you what to do.

   ■ *Per* suggests division. For example, we write a speed of 100 kilometers per hour as:

   $$100 \, \frac{km}{hr} \quad or \quad 100 \, \frac{km}{1 \, hr}$$

   ■ *Of* suggests multiplication. For example, if you launch a 50-kg space probe at a launch cost *of* $10,000 per kilogram, the total cost is:

   $$50 \, \cancel{kg} \times \frac{\$10,000}{\cancel{kg}} = \$500,000$$

   ■ *Square* suggests raising to the second power. For example, we write an area of 75 square meters as 75 m^2.

   ■ *Cube* suggests raising to the third power. For example, we write a volume of 12 cubic centimeters as 12 cm^3.

4. Often the number you are given is not in the units you wish to work with. For example, you may be given that the speed of light is 300,000 km/s but need it in units of m/s for a particular problem. To convert the units, simply multiply the given number by a *conversion factor*: a fraction in which the numerator (top of the fraction) and denominator (bottom of the fraction) are equal, so that the value of the fraction is 1; the number in the denominator must have the units that you wish to change. In the case of changing the speed of light from units of km/s to m/s, you need a conversion factor for kilometers to meters. Thus, the conversion factor is:

   $$\frac{1,000 \, m}{1 \, km}$$

   Note that this conversion factor is equal to 1, since 1,000 meters and 1 kilometer are equal, and that the units to be changed (km) appear in the denominator. We can now convert the speed of light from units of km/s to m/s simply by multiplying by this conversion factor:

   $$\underbrace{300,000 \frac{\cancel{km}}{s}}_{\substack{\text{speed of light} \\ \text{in km/s}}} \times \underbrace{\frac{1,000 \, m}{1 \, \cancel{km}}}_{\substack{\text{conversion from} \\ \text{km to m}}} = \underbrace{3 \times 10^8 \frac{m}{s}}_{\substack{\text{speed of light} \\ \text{in m/s}}}$$

   Note that the units of km cancel, leaving the answer in units of m/s.

5. It's easier to work with units if you replace division with multiplication by the reciprocal. For example, suppose you want to know how many minutes are represented by 300 seconds. We can find the answer by dividing 300 seconds by 60 seconds per minute:

   $$300 \, s \div 60 \, \frac{s}{min}$$

   However, it is easier to see the unit cancellations if we rewrite this expression by replacing the division with multiplication by the reciprocal (this process is easy to remember as "invert and multiply"):

   $$300 \, s \div 60 \, \frac{s}{min} = 300 \, \cancel{s} \times \underbrace{\frac{1 \, min}{60 \, \cancel{s}}}_{\substack{\text{invert} \\ \text{and multiply}}} = 5 \, min$$

We now see that the units of seconds (s) cancel in the numerator of the first term and the denominator of the second term, leaving the answer in units of minutes.

More Examples of Working with Units

Example 1. How many seconds are there in 1 day?

Solution: We can answer the question by setting up a *chain* of unit conversions in which we start with 1 *day* and end up with *seconds*. We use the facts that there are 24 hours per day (24 hr/day), 60 minutes per hour (60 min/hr), and 60 seconds per minute (60 s/min):

$$1 \text{ day} \times \underbrace{\frac{24 \text{ hr}}{\text{day}}}_{\substack{\text{conversion} \\ \text{from} \\ \text{day to hr}}} \times \underbrace{\frac{60 \text{ min}}{\text{hr}}}_{\substack{\text{conversion} \\ \text{from} \\ \text{hr to min}}} \times \underbrace{\frac{60 \text{ s}}{\text{min}}}_{\substack{\text{conversion} \\ \text{from} \\ \text{min to s}}} = 86{,}400 \text{ s}$$

Note that all the units cancel except *seconds*, which is what we want for the answer. There are 86,400 seconds in 1 day.

Example 2. Convert a distance of 10^8 cm to km.

Solution: The easiest way to make this conversion is in two steps, since we know that there are 100 centimeters per meter (100 cm/m) and 1,000 meters per kilometer (1,000 m/km):

$$\underbrace{10^8 \text{ cm}}_{\substack{\text{starting} \\ \text{value}}} \times \underbrace{\frac{1 \text{ m}}{100 \text{ cm}}}_{\substack{\text{conversion} \\ \text{from} \\ \text{cm to m}}} \times \underbrace{\frac{1 \text{ km}}{1{,}000 \text{ m}}}_{\substack{\text{conversion} \\ \text{from} \\ \text{m to km}}} = 10^8 \text{ cm} \times \frac{1 \text{ m}}{10^2 \text{ cm}} \times \frac{1 \text{ km}}{10^3 \text{ m}} = 10^3 \text{ km}$$

Alternatively, if we recognize that the number of kilometers should be smaller than the number of centimeters (because kilometers are larger), we might decide to do this conversion by dividing as follows:

$$10^8 \text{ cm} \div \frac{100 \text{ cm}}{\text{m}} \div \frac{1{,}000 \text{ m}}{\text{km}}$$

In this case, before carrying out the calculation, we replace each division with multiplication by the reciprocal:

$$10^8 \text{ cm} \div \frac{100 \text{ cm}}{\text{m}} \div \frac{1{,}000 \text{ m}}{\text{km}} = 10^8 \text{ cm} \times \frac{1 \text{ m}}{100 \text{ cm}} \times \frac{1 \text{ km}}{1{,}000 \text{ m}}$$

$$= 10^8 \text{ cm} \times \frac{1 \text{ m}}{10^2 \text{ cm}} \times \frac{1 \text{ km}}{10^3 \text{ m}}$$

$$= 10^3 \text{ km}$$

Note that we again get the answer that 10^8 cm is the same as 10^3 km, or 1,000 km.

Example 3. Suppose you accelerate at 9.8 m/s^2 for 4 seconds, starting from rest. How fast will you be going?

Solution: The question asked "how fast?" so we expect to end up with a speed. Therefore, we multiply the acceleration by the amount of time you accelerated:

$$9.8 \, \frac{\text{m}}{\text{s}^2} \times 4 \text{ s} = (9.8 \times 4) \, \frac{\text{m} \times \text{s}}{\text{s}^2} = 39.2 \, \frac{\text{m}}{\text{s}}$$

Note that the units end up as a speed, showing that you will be traveling 39.2 m/s after 4 seconds of acceleration at 9.8 m/s^2.

Example 4. A reservoir is 2 km long and 3 km wide. Calculate its area, in both square kilometers and square meters.

Solution: We find its area by multiplying its length and width:

$$2 \text{ km} \times 3 \text{ km} = 6 \text{ km}^2$$

Next we need to convert this area of 6 km² to square meters, using the fact that there are 1,000 meters per kilometer (1,000 m/km). Note that we must square the term 1,000 m/km when converting from km² to m²:

$$6 \text{ km}^2 \times \left(1{,}000 \frac{\text{m}}{\text{km}}\right)^2 = 6 \text{ km}^2 \times 1{,}000^2 \frac{\text{m}^2}{\text{km}^2} = 6 \text{ km}^2 \times 1{,}000{,}000 \frac{\text{m}^2}{\text{km}^2}$$

$$= 6{,}000{,}000 \text{ m}^2$$

The reservoir area is 6 km², which is the same as 6 million m².

C.4 The Metric System (SI)

The modern version of the metric system, known as *Système Internationale d'Unites* (French for "International System of Units") or **SI**, was formally established in 1960. Today, it is the primary measurement system in nearly every country in the world with the exception of the United States. Even in the United States, it is the system of choice for science and international commerce.

The basic units of length, mass, and time in the SI are:

■ The **meter** for length, abbreviated m

■ The **kilogram** for mass, abbreviated kg

■ The **second** for time, abbreviated s

Multiples of metric units are formed by powers of 10, using a prefix to indicate the power. For example, *kilo* means 10^3 (1,000), so a kilometer is 1,000 meters; a microgram is 0.000001 gram, because *micro* means 10^{-6}, or one millionth. Some of the more common prefixes are listed in Table C.2.

Metric Conversions

Table C.3 lists conversions between metric units and units used commonly in the United States. Note that the conversions between kilograms and pounds are valid only on Earth, because they depend on the strength of gravity.

Example 1. International athletic competitions generally use metric distances. Compare the length of a 100-meter race to that of a 100-yard race.

Table C.2 SI (Metric) Prefixes

Small Values			Large Values		
Prefix	Abbreviation	Value	Prefix	Abbreviation	Value
Deci	d	10^{-1}	Deca	da	10^1
Centi	c	10^{-2}	Hecto	h	10^2
Milli	m	10^{-3}	Kilo	k	10^3
Micro	μ	10^{-6}	Mega	M	10^6
Nano	n	10^{-9}	Giga	G	10^9
Pico	p	10^{-12}	Tera	T	10^{12}

Table C.3 Metric Conversions

To Metric	From Metric
1 inch = 2.540 cm	1 cm = 0.3937 inch
1 foot = 0.3048 m	1 m = 3.28 feet
1 yard = 0.9144 m	1 m = 1.094 yards
1 mile = 1.6093 km	1 km = 0.6214 mile
1 pound = 0.4536 kg	1 kg = 2.205 pounds

Solution: Table C.3 shows that 1 m = 1.094 yd, so 100 m is 109.4 yd. Note that 100 meters is almost 110 yards; a good "rule of thumb" to remember is that distances in meters are about 10% longer than the corresponding number of yards.

Example 2. How many square kilometers are in 1 square mile?

Solution: We use the square of the miles-to-kilometers conversion factor:

$$(1 \text{ mi}^2) \times \left(\frac{1.6093 \text{ km}}{1 \text{ mi}}\right)^2 = (1 \text{ mi}^2) \times \left(1.6093^2 \frac{\text{km}^2}{\text{mi}^2}\right) = 2.5898 \text{ km}^2$$

Therefore, 1 square mile is 2.5898 square kilometers.

C.5 Finding a Ratio

Suppose you want to compare two quantities, such as the average density of the Earth and the average density of Jupiter. The way we do such a comparison is by dividing, which tells us the *ratio* of the two quantities. In this case, the Earth's average density is 5.52 grams/cm^3 and Jupiter's average density is 1.33 grams/cm^3 (see Table 9.1), so the ratio is:

$$\frac{\text{average density of Earth}}{\text{average density of Jupiter}} = \frac{5.52 \text{ g/cm}^3}{1.33 \text{ g/cm}^3} = 4.15$$

Notice how the units cancel on both the top and bottom of the fraction. We can state our result in two equivalent ways:

■ The ratio of the Earth's average density to Jupiter's average density is 4.15.

■ The Earth's average density is 4.15 times Jupiter's average density.

Sometimes, the quantities that you want to compare may each involve an equation. In such cases, you could, of course, find the ratio by first calculating each of the two quantities individually and then dividing. However, it is much easier if you first express the ratio as a fraction, putting the equation for one quantity on top and the other on the bottom. Some of the terms in the equation may then cancel out, making any calculations much easier.

Example 1. Compare the kinetic energy of a car traveling at 100 km/hr to that of a car traveling at 50 km/hr.

Solution: We do the comparison by finding the ratio of the two kinetic energies, recalling that the formula for kinetic energy is $\frac{1}{2} mv^2$. Since we are not told the mass of the car, you might at first think that we don't have enough information to find the ratio. However, notice what happens when we put the equations for each kinetic energy into the ratio, calling the two speeds v_1 and v_2:

$$\frac{\text{K.E. car at } v_1}{\text{K.E. car at } v_2} = \frac{\frac{1}{2} m_{car} v_1^2}{\frac{1}{2} m_{car} v_2^2} = \frac{v_1^2}{v_2^2} = \left(\frac{v_1}{v_2}\right)^2$$

All the terms cancel except those with the two speeds, leaving us with a very simple formula for the ratio. Now we put in 100 km/hr for v_1 and 50 km/hr for v_2:

$$\frac{\text{K.E. car at } 100 \text{ km/hr}}{\text{K.E. car at } 50 \text{ km/hr}} = \left(\frac{100 \text{ km/hr}}{50 \text{ km/hr}}\right)^2 = 2^2 = 4$$

The ratio of the car's kinetic energies at 100 km/hr and 50 km/hr is 4. That is, the car has four times as much kinetic energy at 100 km/hr as it has at 50 km/hr.

Example 2. Compare the strength of gravity between the Earth and the Sun to the strength of gravity between the Earth and the Moon.

Solution: We do the comparison by taking the ratio of the Earth–Sun gravity to the Earth–Moon gravity. In this case, each quantity is found from the equation of Newton's law of gravity. (See Section 5.3.) Thus, the ratio is:

$$\frac{\text{Earth–Sun gravity}}{\text{Earth–Moon gravity}} = \frac{\cancel{G} \dfrac{\cancel{M_{\text{Earth}}} M_{\text{Sun}}}{(d_{\text{Earth–Sun}})^2}}{\cancel{G} \dfrac{\cancel{M_{\text{Earth}}} M_{\text{Moon}}}{(d_{\text{Earth–Moon}})^2}} = \frac{M_{\text{Sun}}}{(d_{\text{Earth–Sun}})^2} \times \frac{(d_{\text{Earth–Moon}})^2}{M_{\text{Moon}}}$$

Note how all but four of the terms cancel; the last step comes from replacing the division with multiplication by the reciprocal (the "invert and multiply" rule for division). We can simplify the work further by rearranging the terms so that we have the masses and distances together:

$$\frac{\text{Earth–Sun gravity}}{\text{Earth–Moon gravity}} = \frac{M_{\text{Sun}}}{M_{\text{Moon}}} \times \frac{(d_{\text{Earth–Moon}})^2}{(d_{\text{Earth–Sun}})^2}$$

Now it is just a matter of looking up the numbers (see Appendix E) and calculating:

$$\frac{\text{Earth–Sun gravity}}{\text{Earth–Moon gravity}} = \frac{1.99 \times 10^{30} \cancel{\text{kg}}}{7.35 \times 10^{22} \cancel{\text{kg}}} \times \frac{(384.4 \times 10^3 \cancel{\text{km}})^2}{(149.6 \times 10^6 \cancel{\text{km}})^2} = 179$$

In other words, the Earth–Sun gravity is 179 times stronger than the Earth–Moon gravity.

APPENDIX D

THE PERIODIC TABLE OF THE ELEMENTS

Key

12	— Atomic number
Mg	— Element's symbol
Magnesium	— Element's name
24.305	— Atomic mass*

*Atomic masses are fractions because they represent a weighted average of atomic masses of different isotopes—in proportion to the abundance of each isotope on Earth.

1 **H** Hydrogen 1.00794																	2 **He** Helium 4.003
3 **Li** Lithium 6.941	4 **Be** Beryllium 9.01218											5 **B** Boron 10.81	6 **C** Carbon 12.011	7 **N** Nitrogen 14.007	8 **O** Oxygen 15.999	9 **F** Fluorine 18.988	10 **Ne** Neon 20.179
11 **Na** Sodium 22.990	12 **Mg** Magnesium 24.305											13 **Al** Aluminum 26.98	14 **Si** Silicon 28.086	15 **P** Phosphorus 30.974	16 **S** Sulfur 32.06	17 **Cl** Chlorine 35.453	18 **Ar** Argon 39.948
19 **K** Potassium 39.098	20 **Ca** Calcium 40.08	21 **Sc** Scandium 44.956	22 **Ti** Titanium 47.88	23 **V** Vanadium 50.94	24 **Cr** Chromium 51.996	25 **Mn** Manganese 54.938	26 **Fe** Iron 55.847	27 **Co** Cobalt 58.9332	28 **Ni** Nickel 58.69	29 **Cu** Copper 63.546	30 **Zn** Zinc 65.39	31 **Ga** Gallium 69.72	32 **Ge** Germanium 72.59	33 **As** Arsenic 74.922	34 **Se** Selenium 78.96	35 **Br** Bromine 79.904	36 **Fr** Krypton 83.80
37 **Rb** Rubidium 85.468	38 **Sr** Strontium 87.62	39 **Y** Yttrium 88.9059	40 **Zr** Zirconium 91.224	41 **Nb** Niobium 92.91	42 **Mo** Molybdenum 95.94	43 **Tc** Technetium (98)	44 **Ru** Ruthenium 101.07	45 **Rh** Rhodium 102.906	46 **Pd** Palladium 106.42	47 **Ag** Silver 107.868	48 **Cd** Cadmium 112.41	49 **In** Indium 114.82	50 **Sn** Tin 118.71	51 **Sb** Antimony 121.75	52 **Te** Tellurium 127.60	53 **I** Iodine 126.905	54 **Xe** Xenon 131.29
55 **Cs** Cesium 132.91	56 **Ba** Barium 137.34		72 **Hf** Hafnium 178.49	73 **Ta** Tantalum 180.95	74 **W** Tungsten 183.85	75 **Re** Rhenium 186.207	76 **Os** Osmium 190.2	77 **Ir** Iridium 192.22	78 **Pt** Platinum 195.08	79 **Au** Gold 196.967	80 **Hg** Mercury 200.59	81 **Ti** Thallium 204.383	82 **Pb** Lead 207.2	83 **Bi** Bismuth 208.98	84 **Po** Polonium (209)	85 **At** Astatine (210)	86 **Rn** Radon (222)
87 **Fr** Francium (223)	88 **Ra** Radium 226.0254		104 **Rf** Rutherfordium (261)	105 **Db** Dubnium (262)	106 **Sg** Seaborgium (263)	107 **Bh** Bohrium (262)	108 **Hs** Hassium (265)	109 **Mt** Meitnerium (266)	110 **Uun** Ununnilium (269)	111 **Uuu** Unununium (272)	112 **Uub** Ununbium (277)						

Lanthanide Series

57 **La** Lanthanum 138.906	58 **Ce** Cerium 140.12	59 **Pr** Praseodymium 140.908	60 **Nd** Neodymium 144.24	61 **Pm** Promethium (145)	62 **Sm** Samarium 150.36	63 **Eu** Europium 151.96	64 **Gd** Gadolinium 157.25	65 **Tb** Terbium 158.925	66 **Dy** Dysprosium 162.50	67 **Ho** Holmium 164.93	68 **Er** Erbium 167.26	69 **Tm** Thulium 168.934	70 **Yb** Ytterbium 173.04	71 **Lu** Lutetium 174.967

Actinide Series

89 **Ac** Actinium 227.028	90 **Th** Thorium 232.038	91 **Pa** Protactinium 231.036	92 **U** Uranium 238.029	93 **Np** Neptunium 237.048	94 **Pu** Plutonium (244)	95 **Am** Americium (243)	96 **Cm** Curium (247)	97 **Bk** Berkelium (247)	98 **Cf** Californium (251)	99 **Es** Einsteinium (252)	100 **Fm** Fermium (257)	101 **Md** Mendelevium (258)	102 **No** Nobelium (259)	103 **Lr** Lawrencium (260)

APPENDIX E

PLANETARY DATA

Table E.1 Physical Properties of the Sun and Planets

Name	Radius (Eq[a]) (km)	Radius (Eq) (Earth units)	Mass (kg)	Mass (Earth units)	Average Density (g/cm^3)	Surface Gravity (Earth = 1)
Sun	695,000	109	1.99×10^{30}	333,000	1.41	27.5
Mercury	2,440	0.382	3.30×10^{23}	0.055	5.43	0.38
Venus	6,051	0.949	4.87×10^{24}	0.815	5.25	0.91
Earth	6,378	1.00	5.97×10^{24}	1.00	5.52	1.00
Mars	3,397	0.533	6.42×10^{23}	0.107	3.93	0.38
Jupiter	71,492	11.19	1.90×10^{27}	317.9	1.33	2.53
Saturn	60,268	9.46	5.69×10^{26}	95.18	0.71	1.07
Uranus	25,559	3.98	8.66×10^{25}	14.54	1.24	0.91
Neptune	24,764	3.81	1.03×10^{26}	17.13	1.67	1.14
Pluto	1,160	0.181	1.31×10^{22}	0.0022	2.05	0.07

[a]Eq = equatorial.

Table E.2 Orbital Properties of the Sun and Planets

Name	Distance from Sun[a] (AU)	Distance from Sun[a] (10^6 km)	Orbital Period (years)	Orbital Inclination[b] (degrees)	Orbital Eccentricity	Sidereal Rotation Period (Earth days)[c]	Axis Tilt (degrees)
Sun	—	—	—	—	—	25.4	7.25
Mercury	0.387	57.9	0.2409	7.00	0.206	58.6	0.0
Venus	0.723	108.2	0.6152	3.39	0.007	−243.0	177.4
Earth	1.00	149.6	1.0	0.00	0.017	0.9973	23.45
Mars	1.524	227.9	1.881	1.85	0.093	1.026	23.98
Jupiter	5.203	778.3	11.86	1.31	0.048	0.41	3.08
Saturn	9.539	1,427	29.46	2.49	0.056	0.44	26.73
Uranus	19.19	2,870	84.01	0.77	0.046	−0.72	97.92
Neptune	30.06	4,497	164.8	1.77	0.010	0.67	29.6
Pluto	39.54	5,916	248.0	17.15	0.248	−6.39	118

[a]Semimajor axis of the orbit.

[b]With respect to the ecliptic.

[c]A negative sign indicates rotation relative to other planets.

Table E.3 Satellites of the Solar System[a]

Planet Satellite	Radius or Dimensions[b] (km)	Distance from Planet (10³ km)	Orbital Period[c] (Earth days)	Mass[d] (kg)	Density[d] (g/cm³)	Notes About the Satellites
Earth						**Earth**
Moon	1,738	384.4	27.322	7.349×10^{22}	3.34	*Moon:* Probably formed in giant impact.
Mars						**Mars**
Phobos	13×11×9	9.38	0.319	1.3×10^{16}	2.2	*Phobos, Deimos:* Probable captured asteroids.
Deimos	8×6×5	23.5	1.263	1.8×10^{15}	1.7	
Jupiter						**Jupiter**
Small inner moons (5 moons)	10 to 135×82×75	128–222	0.295–0.6745	—	—	*Metis, Adrastea, Amalthea, Thebe, 1999 J1:* Small moonlets within and near Jupiter's ring system.
Io	1,821	421.6	1.769	8.933×10^{22}	3.57	*Io:* Most volcanically active object in the solar system.
Europa	1,565	670.9	3.551	4.797×10^{22}	2.97	*Europa:* Possible oceans under icy crust.
Ganymede	2,634	1,070.0	7.155	1.482×10^{23}	1.94	*Ganymede:* Largest satellite in solar system; unusual ice geology.
Callisto	2,403	1,883.0	16.689	1.076×10^{23}	1.86	*Callisto:* Cratered iceball.
Irregular group 1 (5 moons)	7–22	11,400–17,100	453–829	—	—	*Leda, Himalia, Elara, Lysithea, 2000 J11:* Probable captured moons with inclined orbits.
Irregular group 2 (14 moons)	5–15	17,400–18,000	−854 to −901	—	—	*Ananke, Carme, Pasiphae, Sinope, 2000 J1–J10:* Probable captured moons in inclined backward orbits.
Saturn						**Saturn**
Small inner moons (6)	10 to 97×95×77	134–151	0.574–0.695	—	—	*Pan, Atlas, Prometheus, Pandora, Epimetheus, Janus:* Small moonlets within and near Saturn's ring system.
Mimas	199	185.52	0.942	3.70×10^{19}	1.17	*Mimas, Enceladus, Tethys:* Small and medium-size iceballs, many with interesting geology.
Enceladus	249	238.02	1.370	1.2×10^{20}	1.24	
Tethys	530	294.66	1.888	6.17×10^{20}	1.26	
Calypso	15×8×8	294.66	1.888	4×10^{15}	—	*Calypso, Telesto:* Small moonlets sharing Tethys's orbit.
Telesto	15×13×8	294.67	1.888	6×10^{15}	—	
Dione	559	377.4	2.737	1.08×10^{21}	1.44	*Dione:* Medium-size iceball, with interesting geology.
Helene	18×?×15	377.4	2.737	1.6×10^{16}	—	*Helene:* Small moonlet sharing Dione's orbit.
Rhea	764	527.04	4.518	2.31×10^{21}	1.33	*Rhea:* Medium-size iceball, with interesting geology.
Titan	2,575	1,221.85	15.945	1.3455×10^{23}	1.88	*Titan:* Dense atmosphere shrouds surface; ongoing geological activity possible.
Hyperion	180×140×112	1,481.1	21.277	2.8×10^{19}	—	*Hyperion:* Only satellite known not to rotate synchronously.

Name	Radius or Dimensions (km)[b]	Distance (10³ km)	Orbital Period (days)[c]	Mass (kg)[d]	Density[d]	Notes[a]
Iapetus	718	3,561.3	79.331	1.59×10^{21}	1.21	*Iapetus:* Bright and dark hemispheres show greatest contrast in the solar system.
Phoebe	110	12,952	−550.4	1×10^{19}	—	*Phoebe:* Very dark; material ejected from Phoebe may coat one side of Iapetus.
Irregular group 1 (4 moons)	7–22	11,400–17,100	453–829	—	—	2000 S2, S3, S5, S6: Probable captured moons with highly inclined orbits.
Irregular group 2 (3 moons)	5–15	17,400–18,000	854–901	—	—	2000 S4, S10, S11: Probable captured moons in inclined orbits.
Irregular group 3 (5 moons)	4–10	15,600–23,400	−723 to −1,325	—	—	2000 S1, S7, S8, S9, S12: Probable captured moons in inclined backward orbits.
Uranus						**Uranus**
Small inner moons (11 moons)	10 to 97×95×77	134–151	0.574–0.695	—	—	*Cordelia, Ophelia, Bianca, Cressida, Desdemona, Juliet, Portia, Rosalind, Belinda, Puck, 1986 U10:* Small moonlets within and near Uranus's ring system.
Miranda	236	129.8	1.413	6.6×10^{19}	1.26	*Miranda, Ariel, Umbriel, Titania, Oberon:* Small and medium-size iceballs, with some interesting geology.
Ariel	579	191.2	2.520	1.35×10^{21}	1.65	
Umbriel	584.7	266.0	4.144	1.17×10^{21}	1.44	
Titania	788.9	435.8	8.706	3.52×10^{21}	1.59	
Oberon	761.4	582.6	13.463	3.01×10^{21}	1.50	
Irregular group (5 moons)	???–60	7,170–25,000	580–2,280	—	—	*Caliban, Sycorax, Stephano, Prospero, Setebos:* Too recently discovered for accurate determination of their properties; several in backward orbits.
Neptune						**Neptune**
Small inner moons (5 moons)	29 to 104×?×89	48–74	0.296–0.554	—	—	*Naiad, Thalassa, Despina, Galatea, Larissa:* Small moonlets within and near Neptune's ring system.
Proteus	218×208×201	117.6	1.121	6×10^{19}	—	
Triton	1,352.6	354.59	−5.875	2.14×10^{22}	2.0	*Triton:* Probable captured Kuiper belt object—largest captured object in solar system.
Nereid	170	5,588.6	360.125	3.1×10^{19}	—	*Nereid:* Small, icy moon; very little known.
Pluto						**Pluto**
Charon	635	19.6	6.38718	1.56×10^{21}	1.6	*Charon:* Unusually large compared to its planet; may have formed in giant impact.

[a] *Note:* Authorities differ substantially on many of the values in this table.

[b] a × b × c values for the Dimensions are the approximate lengths of the axes for irregular moons.

[c] Negative sign indicates backward orbit.

[d] Masses and densities are most accurate for those satellites visited by a spacecraft on a flyby. Masses for the smallest moons have not been measured but can be estimated from the radius and an assumed density.

STELLAR DATA

Table F.1 Stars Within 12 Light-Years

Star	Distance (ly)	Spectral Type		RA h	m	Dec °	′	Luminosity (L/L_{Sun})
Sun	0.000016	G2	V	—	—	—	—	1.0
Proxima Centauri	4.2	M5.5	V	14	30	−62	41	0.0006
α Centauri A	4.4	G2	V	14	40	−60	50	1.6
α Centauri B	4.4	K0	V	14	40	−60	50	0.53
Barnard's Star	6.0	M4	V	17	58	+04	42	0.005
Wolf 359	7.8	M6	V	10	56	+07	01	0.0008
Lalande 21185	8.3	M2	V	11	03	+35	58	0.03
Sirius A	8.6	A1	V	06	45	−16	42	26.0
Sirius B	8.6	DA2	—	06	45	−16	42	0.002
Luyten 726-8A	8.7	M5.5	V	01	39	−17	57	0.0009
Luyten 726-8B	8.7	M6	V	01	39	−17	57	0.0006
Ross 154	9.7	M3.5	V	18	50	−23	50	0.004
Ross 248	10.3	M5.5	V	23	42	+44	11	0.001
ε Eridani	10.5	K2	V	03	33	−09	28	0.37
Lacaille 9352	10.7	M1.5	V	23	06	−35	51	0.05
Ross 128	10.9	M4	V	11	48	+00	49	0.003
EZ Aquarii A	11.3	M5	V	22	39	−15	18	0.0006
EZ Aquarii B	11.3	M6	V	22	39	−15	18	0.0004
EZ Aquarii C	11.3	M6.5	V	22	39	−15	18	0.0003
61 Cygni A	11.4	K5	V	21	07	+38	42	0.15
61 Cygni B	11.4	K7	V	21	07	+38	42	0.09
Procyon A	11.4	F5	IV–V	07	39	+05	14	7.4
Procyon B	11.4	DA	—	07	39	+05	14	0.0005
Gliese 725 A	11.4	M3	V	18	43	+59	38	0.02
Gliese 725 B	11.4	M3.5	V	18	43	+59	38	0.01
Gliese 15 A	11.6	M1.5	V	00	18	+44	01	0.03
Gliese 15 B	11.6	M3.5	V	00	18	+44	01	0.003
DX Cancri	11.8	M6.5	V	08	30	+26	47	0.0003
ε Indi	11.8	K5	V	22	03	−56	45	0.26
τ Ceti	11.9	G8	V	01	44	−15	57	0.59
GJ 1061	11.9	M5.5	V	03	36	−44	31	0.0009

Note: These data were provided by the RECONS project, courtesy of Dr. Todd Henry. The luminosities are all total (bolometric) luminosities. The DA stellar types are white dwarfs. The coordinates are for the year 2000.

Table F.2 Twenty Brightest Stars

Star	Constellation	RA h	RA m	Dec °	Dec ′	Distance (ly)	Spectral Type		Apparent Magnitude	Luminosity (L/L_{Sun})
Sirius	Canis Major	6	45	−16	42	8.6	A1	V	−1.46	26
Canopus	Carina	6	24	−52	41	313	F0	Ib–II	−0.72	13,000
α Centauri	Centaurus	14	40	−60	50	4.4	G2	V	−0.01	1.6
							K0	V	1.3	0.53
Arcturus	Boötes	14	16	+19	11	37	K2	III	−0.06	170
Vega	Lyra	18	37	+38	47	25	A0	V	0.04	60
Capella	Auriga	5	17	+46	00	42	G0	III	0.75	70
							G8	III	0.85	77
Rigel	Orion	5	15	−08	12	772	B8	Ia	0.14	70,000
Procyon	Canis Minor	7	39	+05	14	11.4	F5	IV–V	0.37	7.4
Betelgeuse	Orion	5	55	+07	24	427	M2	Iab	0.41	38,000
Achernar	Eridanus	1	38	−57	15	144	B5	V	0.51	3,600
Hadar	Centaurus	14	04	−60	22	525	B1	III	0.63	100,000
Altair	Aquila	19	51	+08	52	17	A7	IV–V	0.77	10.5
Acrux	Crux	12	27	−63	06	321	B1	IV	1.39	22,000
							B3	V	1.9	7,500
Aldebaran	Taurus	4	36	+16	30	65	K5	III	0.86	350
Spica	Virgo	13	25	−11	09	260	B1	V	0.91	23,000
Antares	Scorpio	16	29	−26	26	604	M1	Ib	0.92	38,000
Pollux	Gemini	7	45	+28	01	34	K0	III	1.16	45
Fomalhaut	Piscis Austrinus	22	58	−29	37	25	A3	V	1.19	18
Deneb	Cygnus	20	41	+45	16	2,500	A2	Ia	1.26	170,000
β Crucis	Crux	12	48	−59	40	352	B0.5	IV	1.28	37,000

Note: Three of the stars on this list, Capella, α Centauri, and Acrux, are binary systems with members of comparable brightness. They are counted as single stars because that is how they appear to the naked eye. All the luminosities given are total (bolometric) luminosities. The coordinates are for the year 2000.

GALAXY DATA

Table G.1 Galaxies of the Local Group

Galaxy Name	Distance (millions of ly)	Type[a]	RA h	RA m	Dec °	Dec ′	Luminosity (millions of L_{Sun})
Milky Way	—	Sbc	—	—	—	—	15,000
WLM	3.0	Irr	00	02	−15	30	50
NGC 55	4.8	Irr	00	15	−39	13	1,300
IC 10	2.7	dIrr	00	20	+59	18	160
NGC 147	2.4	dE	00	33	+48	30	131
And III	2.5	dE	00	35	+36	30	1.1
NGC 185	2.0	dE	00	39	+48	20	120
NGC 205	2.7	E	00	40	+41	41	370
M 32	2.6	E	00	43	+40	52	380
M 31	2.5	Sb	00	43	+41	16	21,000
And I	2.6	dE	00	46	+38	00	4.7
SMC	0.19	Irr	00	53	−72	50	230
Sculptor	0.26	dE	01	00	−33	42	2.2
LGS 3	2.6	dIrr	01	04	+21	53	1.3
IC 1613	2.3	Irr	01	05	+02	08	64
And II	1.7	dE	01	16	+33	26	2.4
M 33	2.7	Sc	01	34	+30	40	2,800
Phoenix	1.5	dIrr	01	51	−44	27	0.9
Fornax	0.45	dE	02	40	−34	27	15.5
EGB0427+63	4.3	dIrr	04	32	+63	36	9.1
LMC	0.16	Irr	05	24	−69	45	1,300
Carina	0.33	dE	06	42	−50	58	0.4
Leo A	2.2	dIrr	09	59	+30	45	3.0
Sextans B	4.4	dIrr	10	00	+05	20	41
NGC 3109	4.1	Irr	10	03	−26	09	160
Antlia	4.0	dIrr	10	04	−27	19	1.7
Leo I	0.82	dE	10	08	+12	18	4.8
Sextans A	4.7	dIrr	10	11	−04	42	56
Sextans	0.28	dE	10	13	−01	37	0.5
Leo II	0.67	dE	11	13	+22	09	0.6
GR 8	5.2	dIrr	12	59	+14	13	3.4
Ursa Minor	0.22	dE	15	09	+67	13	0.3
Draco	2.7	dE	17	20	+57	55	0.3
Sagittarius	0.08	dE	18	55	−30	29	18
SagDIG	3.5	dIrr	19	30	−17	41	6.8
NGC 6822	1.6	Irr	19	45	−14	48	94
DDO 210	2.6	dIrr	20	47	−12	51	0.8
IC 5152	5.2	dIrr	22	03	−51	18	70
Tucana	2.9	dE	22	42	−64	25	0.5
UKS2323-326	4.3	dE	23	26	−32	23	5.2
Pegasus	3.1	dIrr	23	29	+14	45	12

[a]Types beginning with S are spiral galaxies classified according to Hubble's system (see Chapter 17). Type E galaxies are elliptical or spheroidal. Type Irr galaxies are irregular. The prefix d denotes a dwarf galaxy.

Table G.2 Nearby Galaxies in the Messier Catalog[a,b]

Galaxy Name (M / NGC)[c]	RA h	RA m	Dec °	Dec ′	RV$_{hel}$[d]	RV$_{gal}$[e]	Type[f]	Nickname
M 31 / NGC 224	00	43	+41	16	−300 ± 4	−122	Spiral	Andromeda
M 32 / NGC 221	00	43	+40	52	−145 ± 2	32	Elliptical	
M 33 / NGC 598	01	34	+30	40	−179 ± 3	−44	Spiral	Triangulum
M 49 / NGC 4472	12	30	+08	00	997 ± 7	929	Elliptical/ Lenticular/Seyfert	
M 51 / NGC 5194	13	30	+47	12	463 ± 3	550	Spiral/Interacting	Whirlpool
M 58 / NGC 4579	12	38	+11	49	1,519 ± 6	1,468	Spiral/Seyfert	
M 59 / NGC 4621	12	42	+11	39	410 ± 6	361	Elliptical	
M 60 / NGC 4649	12	44	+11	33	1,117 ± 6	1,068	Elliptical	
M 61 / NGC 4303	12	22	+04	28	1,566 ± 2	1,483	Spiral/Seyfert	
M 63 / NGC 5055	13	16	+42	02	504 ± 4	570	Spiral	Sunflower
M 64 / NGC 4826	12	57	+21	41	408 ± 4	400	Spiral/Seyfert	Black Eye
M 65 / NGC 3623	11	19	+13	06	807 ± 3	723	Spiral	
M 66 / NGC 3627	11	20	+12	59	727 ± 3	643	Spiral/Seyfert	
M 74 / NGC 628	01	37	+15	47	657 ± 1	754	Spiral	
M 77 / NGC 1068	02	43	−00	01	1,137 ± 3	1,146	Spiral/Seyfert	
M 81 / NGC 3031	09	56	+69	04	−34 ± 4	73	Spiral/Seyfert	
M 82 / NGC 3034	09	56	+69	41	203 ± 4	312	Irregular/Starburst	
M 83 / NGC 5236	13	37	−29	52	516 ± 4	385	Spiral/Starburst	
M 84 / NGC 4374	12	25	+12	53	1,060 ± 6	1,005	Elliptical	
M 85 / NGC 4382	12	25	+18	11	729 ± 2	692	Spiral	
M 86 / NGC 4406	12	26	+12	57	−244 ± 5	−298	Elliptical/ Lenticular	
M 87 / NGC 4486	12	30	+12	23	1,307 ± 7	1,254	Elliptical/Central Dominant/Seyfert	Virgo A
M 88 / NGC 4501	12	32	+14	25	2,281 ± 3	2,235	Spiral/Seyfert	
M 89 / NGC 4552	12	36	+12	33	340 ± 4	290	Elliptical	
M 90 / NGC 4569	12	37	+13	10	−235 ± 4	−282	Spiral/Seyfert	
M 91 / NGC 4548	12	35	+14	30	486 ± 4	442	Spiral/Seyfert	
M 94 / NGC 4736	12	51	+41	07	308 ± 1	360	Spiral	
M 95 / NGC 3351	10	44	+11	42	778 ± 4	677	Spiral/Starburst	
M 96 / NGC 3368	10	47	+11	49	897 ± 4	797	Spiral/Seyfert	
M 98 / NGC 4192	12	14	+14	54	−142 ± 4	−195	Spiral/Seyfert	
M 99 / NGC 4254	12	19	+14	25	2,407 ± 3	2,354	Spiral	
M 100 / NGC 4321	12	23	+15	49	1,571 ± 1	1,525	Spiral	
M 101 / NGC 5457	14	03	+54	21	241 ± 2	360	Spiral	
M 104 / NGC 4594	12	40	−11	37	1,024 ± 5	904	Spiral/Seyfert	Sombrero
M 105 / NGC 3379	10	48	+12	35	911 ± 2	814	Elliptical	
M 106 / NGC 4258	12	19	+47	18	448 ± 3	507	Spiral/Seyfert	
M 108 / NGC 3556	11	09	+55	57	695 ± 3	765	Spiral	
M 109 / NGC 3992	11	55	+53	39	1,048 ± 4	1,121	Spiral	
M 110 / NGC 205	00	38	+41	25	−241 ± 3	−61	Elliptical	

[a]Galaxies identified in the catalog published by Charles Messier in 1781; these galaxies are relatively easy to observe with small telescopes.

[b]Data obtained from NED: NASA/IPAC Extragalactic Database (http://ned.ipac.caltech.edu). The original Messier list of galaxies was obtained from SED, and the list data were updated to 2001 and M 102 was dropped.

[c]The galaxies are identified by the Messier number (M followed by a number) and by their NGC numbers, which come from the *New General Catalog* published in 1888.

[d]Radial velocity in km/s, with respect to the Sun (heliocentric). Positive values mean motion away from the Sun, and negative values are toward the Sun.

[e]Radial velocity in km/s, with respect to the Milky Way Galaxy, calculated from the RV$_{hel}$ values with a correction for the Sun's motion around the galactic center.

[f]Galaxies are first listed by their primary type (spiral, elliptical, or irregular) and then by any other special categories that apply (see Chapter 17).

Table G.3 Nearby, X-ray Bright Clusters of Galaxies

Cluster Name	Redshift	Distance[a] (billions of ly)	Temperature of Intracluster Medium (millions of K)	Average Orbital Velocity of Galaxies[b] (km/sec)	Cluster Mass[c] ($10^{15}\ M_{Sun}$)
Abell 2142	0.0907	1.20	101. ± 2	1,132 ± 110	1.8
Abell 2029	0.0766	1.07	100. ± 3	1,164 ± 98	1.8
Abell 401	0.0737	1.03	95.2 ± 5	1,152 ± 86	1.6
Coma	0.0233	0.34	95.1 ± 1	821 ± 49	1.6
Abell 754	0.0539	0.77	93.3 ± 3	662 ± 77	1.6
Abell 2256	0.0589	0.83	87.0 ± 2	1,348 ± 86	1.4
Abell 399	0.0718	1.01	81.7 ± 7	1,116 ± 89	1.3
Abell 3571	0.0395	0.57	81.1 ± 3	1,045 ± 109	1.3
Abell 478	0.0882	1.22	78.9 ± 2	904 ± 281	1.2
Abell 3667	0.0566	0.80	78.5 ± 6	971 ± 62	1.2
Abell 3266	0.0599	0.85	78.2 ± 5	1,107 ± 82	1.2
Abell 1651a	0.0846	1.17	73.1 ± 6	685 ± 129	1.2
Abell 85	0.0560	0.80	70.9 ± 2	969 ± 95	1.2
Abell 119	0.0438	0.63	65.6 ± 5	679 ± 106	0.94
Abell 3558	0.0480	0.69	65.3 ± 2	977 ± 39	0.94
Abell 1795	0.0632	0.89	62.9 ± 2	834 ± 85	0.88
Abell 2199	0.0314	0.46	52.7 ± 1	801 ± 92	0.68
Abell 2147	0.0353	0.51	51.1 ± 4	821 ± 68	0.65
Abell 3562	0.0478	0.68	45.7 ± 8	736 ± 49	0.55
Abell 496	0.0325	0.47	45.3 ± 1	687 ± 89	0.54
Centaurus	0.0103	0.15	42.2 ± 1	863 ± 34	0.49
Abell 1367	0.0213	0.31	41.3 ± 2	822 ± 69	0.47
Hydra	0.0126	0.19	38.0 ± 1	610 ± 52	0.42
C0336	0.0349	0.50	37.4 ± 1	650 ± 170	0.41
Virgo	0.0038	0.06	25.7 ± 0.5	632 ± 41	0.23

Note: This table lists the 25 brightest clusters of galaxies in the X-ray sky from a catalog by J. P. Henry (2000).

[a]Cluster distances were computed using a value for Hubble's constant of 65 km/sec/Mpc.

[b]The average orbital velocities of galaxies given in this column are the velocity dispersions of the clusters' galaxies.

[c]This column gives each cluster's mass within the largest radius at which the intracluster medium can be in gravitational equilibrium. Because our estimates of that radius depend on Hubble's constant, these masses are inversely proportional to Hubble's constant, which we have assumed to be 65 km/s/Mpc.

APPENDIX H

SELECTED ASTRONOMICAL WEB SITES

The Web contains a vast amount of astronomical information. For all your astronomical Web surfing, the best starting point is the Web site for this textbook:

The Astronomy Place
www.astronomyplace.com

The following are some other sites that may be of particular use. In case any of the links change, you can always find live links to these sites, and many more, on The Astronomy Place.

Key Mission Sites

The following table lists the Web pages for major current astronomy missions.

Site	Description	Web Address
NASA's Office of Space Science Missions Page	Direct links to all past, present, and planned NASA space science missions	http://spacescience.nasa.gov/missions
Cassini/Huygens	Mission scheduled to arrive at Saturn in 2004	http://www.jpl.nasa.gov/cassini
Chandra X-Ray Observatory	Latest discoveries, educational activities, and other information from the Chandra X-Ray Observatory	http://chandra.harvard.edu
Far Ultraviolet Spectroscopic Explorer (FUSE)	Ultraviolet observatory in space	http://fuse.pha.jhu.edu
Galileo	Mission orbiting Jupiter	http://www.jpl.nasa.gov/galileo
Hubble Space Telescope	Latest discoveries, educational activities, and other information from the Hubble Space Telescope	http://hubble.stsci.edu
Mars Exploration Program	Information on current and planned Mars missions	http://mars.jpl.nasa.gov
Microwave Anisotropy Probe (MAP)	Mission to study the cosmic microwave background	http://map.gsfc.nasa.gov
Space Infrared Telescope Facility (SIRTF)	Infrared observatory scheduled for launch in 2002	http://sirtf.jpl.nasa.gov
Stratospheric Observatory for Infrared Astronomy (SOFIA)	Airborne observatory scheduled to begin flights in 2002	http://sofia.arc.nasa.gov

Key Observatory Sites

The following table lists the Web pages leading to major ground-based observatories.

Site	Description	Web Address
World's Largest Optical Telescopes	**Direct links to most of the world's major optical observatories**	http://www.seds.org/billa/bigeyes.html
Arecibo Observatory (Puerto Rico)	World's largest single-dish radio telescope	http://www.naic.edu
Cerro Tololo Inter-American Observatory	Links to major observatories on site in Cerro Tololo, Chile	http://www.ctio.noao.edu
European Southern Observatory	Links to European telescope projects in Chile, including the Very Large Telescope	http://www.eso.org
Mauna Kea Observatories	Links to major observatories in Hawaii, including Keck, Gemini, Subaru, CFHT, and others	http://www.ifa.hawaii.edu/mko
Mt. Palomar Observatory	Powerful telescope near San Diego	http://www.astro.caltech.edu/palomarpublic
National Optical Astronomy Observatory	Home page for United States national observatories in Arizona, Hawaii, and Chile	http://www.noao.edu
National Radio Astronomy Observatory	Home page for United States national radio observatories, including the Very Large Array (VLA)	http://www.nrao.edu

More Astronomical Web Sites

The following Web sites are some of the authors' favorites among many other non-commercial resources for astronomy.

Site	Description	Web Address
The Astronomy Place	**Don't forget to start here for all your astronomical Web surfing.**	**http://www.astronomyplace.com**
American Association of Variable Star Observers (AAVSO)	One of the largest organizations of amateur astronomers in the world. Check this site if you are interested in serious amateur astronomy.	http://www.aavso.org
Astronomical Society of the Pacific	An organization for both professional astronomers and the general public, devoted largely to astronomy education.	http://www.aspsky.org
Astronomy Picture of the Day	An archive of beautiful pictures, updated daily.	http://antwrp.gsfc.nasa.gov/apod
AstroWeb	Listing of major resources for astronomy on the Web.	http://www.cv.nrao.edu/fits/www/astronomy.html
Canadian Space Agency	Home page for Canada's space program.	http://www.space.gc.ca
European Space Agency (ESA)	Home page for this international agency.	http://www.esa.int
The Extrasolar Planets Encyclopedia	Information about the search for and discoveries of extrasolar planets.	http://cfa-www.harvard.edu/planets
NASA Home Page	Learn almost anything you want about NASA.	http://www.nasa.gov
NASA Science News	Read the latest news from NASA; has option to subscribe to e-mail notices of news releases.	http://science.nasa.gov
The Nine Planets (University of Arizona)	A multimedia tour of the solar system.	http://seds.lpl.arizona.edu/nineplanets/nineplanets/nineplanets.html
The Planetary Society	Has more than 100,000 members who are interested in planetary exploration and the search for life in the universe.	http://planetary.org
The SETI Institute	Devoted to the search for other civilizations.	http://www.seti.org
Voyage Scale Model Solar System	Take a virtual tour of the Voyage Scale Model Solar System.	http://www.voyageonline.org

THE 88 CONSTELLATIONS

Constellation Names (English equivalent in parentheses)

Andromeda (The Chained
  Princess)
Antlia (The Air Pump)
Apus (The Bird of Paradise)
Aquarius (The Water Bearer)
Aquila (The Eagle)
Ara (The Altar)
Aries (The Ram)
Auriga (The Charioteer)
Boötes (The Herdsman)
Caelum (The Chisel)
Camelopardalis (The Giraffe)
Cancer (The Crab)
Canes Venatici (The Hunting Dogs)

Canis Major (The Great Dog)
Canis Minor (The Little Dog)
Capricornus (The Sea Goat)
Carina (The Keel)
Cassiopeia (The Queen)
Centaurus (The Centaur)
Cepheus (The King)
Cetus (The Whale)
Chamaeleon (The Chameleon)
Circinus (The Drawing Compass)
Columba (The Dove)
Coma Berenices (Berenice's Hair)
Corona Australis (The Southern
  Crown)

Corona Borealis (The Northern
  Crown)
Corvus (The Crow)
Crater (The Cup)
Crux (The Southern Cross)
Cygnus (The Swan)
Delphinus (The Dolphin)
Dorado (The Goldfish)
Draco (The Dragon)
Equuleus (The Little Horse)
Eridanus (The River)
Fornax (The Furnace)
Gemini (The Twins)
Grus (The Crane)

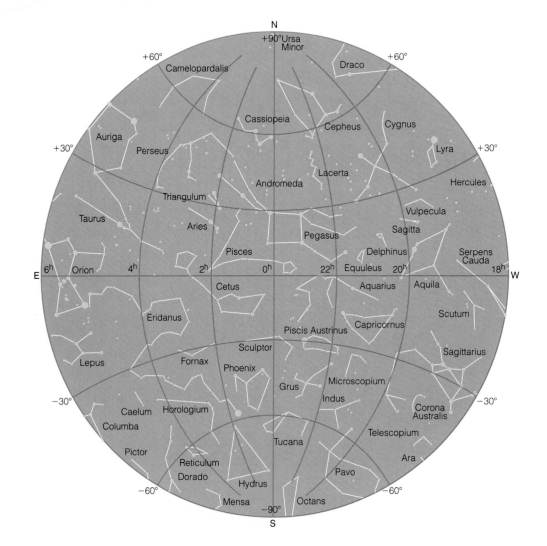

Hercules
Horologium (The Clock)
Hydra (The Sea Serpent)
Hydrus (The Water Snake)
Indus (The Indian)
Lacerta (The Lizard)
Leo (The Lion)
Leo Minor (The Little Lion)
Lepus (The Hare)
Libra (The Scales)
Lupus (The Wolf)
Lynx (The Lynx)
Lyra (The Lyre)
Mensa (The Table)
Microscopium (The Microscope)
Monoceros (The Unicorn)
Musca (The Fly)
Norma (The Level)
Octans (The Octant)
Ophiuchus (The Serpent Bearer)
Orion (The Hunter)
Pavo (The Peacock)

Pegasus (The Winged Horse)
Perseus (The Hero)
Phoenix (The Phoenix)
Pictor (The Painter's Easel)
Pisces (The Fish)
Piscis Austrinus (The Southern Fish)
Puppis (The Stern)
Pyxis (The Compass)
Reticulum (The Reticle)
Sagitta (The Arrow)
Sagittarius (The Archer)
Scorpius (The Scorpion)
Sculptor (The Sculptor)
Scutum (The Shield)
Serpens (The Serpent)
Sextans (The Sextant)
Taurus (The Bull)
Telescopium (The Telescope)
Triangulum (The Triangle)
Triangulum Australe (Southern Triangle)

Tucana (The Toucan)
Ursa Major (The Great Bear)
Ursa Minor (The Little Bear)
Vela (The Sail)
Virgo (The Virgin)
Volans (The Flying Fish)
Vulpecula (The Fox)

Constellation Locations

Each of the charts on these pages shows half of the celestial sphere in projection, so you can use them to learn the approximate locations of the constellations. The grid lines are marked by right ascension and declination.

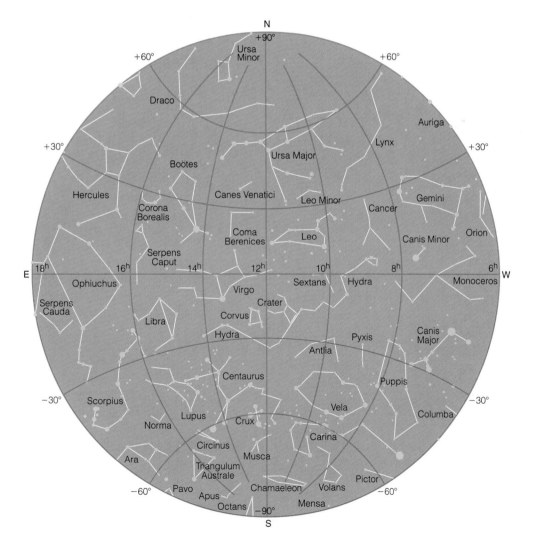

STAR CHARTS

How to use the star charts:

Check the times and dates under each chart to find the best one for you. Take it outdoors within an hour or so of the time listed for your date. Bring a dim flashlight to help you read it.

On each chart, the round outside edge represents the horizon all around you. Compass directions around the horizon are marked in yellow. Turn the chart around so the edge marked with the direction you're facing (for example, north, southeast) is down. The stars above this horizon now match the stars you are facing. Ignore the rest until you turn to look in a different direction.

The center of the chart represents the sky overhead, so a star plotted on the chart halfway from the edge to the center can be found in the sky halfway from the horizon to straight up.

The charts are drawn for 40°N latitude (for example, Denver, New York, Madrid). If you live far south of there, stars in the southern part of your sky will appear higher than on the chart and stars in the north will be lower. If you live far north of there, the reverse is true.

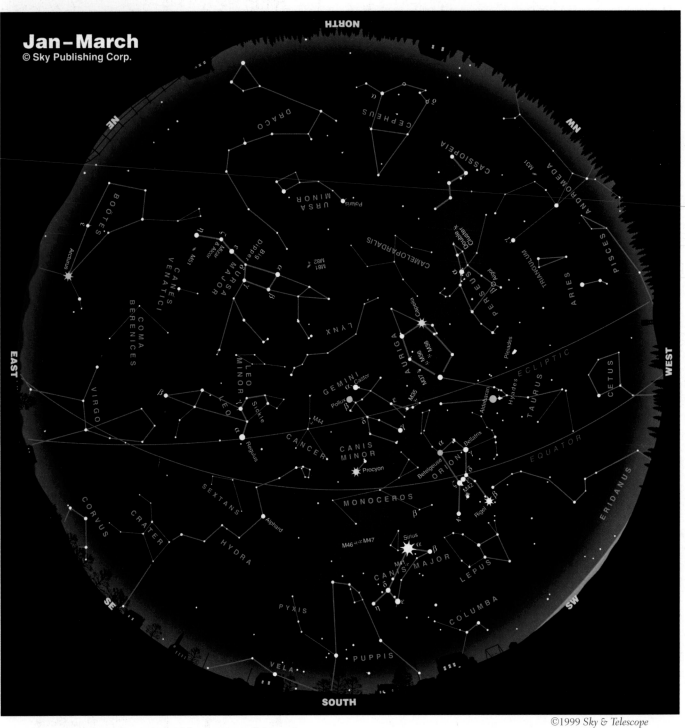

Jan–March
© Sky Publishing Corp.

©1999 *Sky & Telescope*

Use this chart January, February, and March.

Early January — 1 A.M.   Early February — 11 P.M.   Early March — 9 P.M.
Late January — Midnight   Late February — 10 P.M.   Late March — Dusk

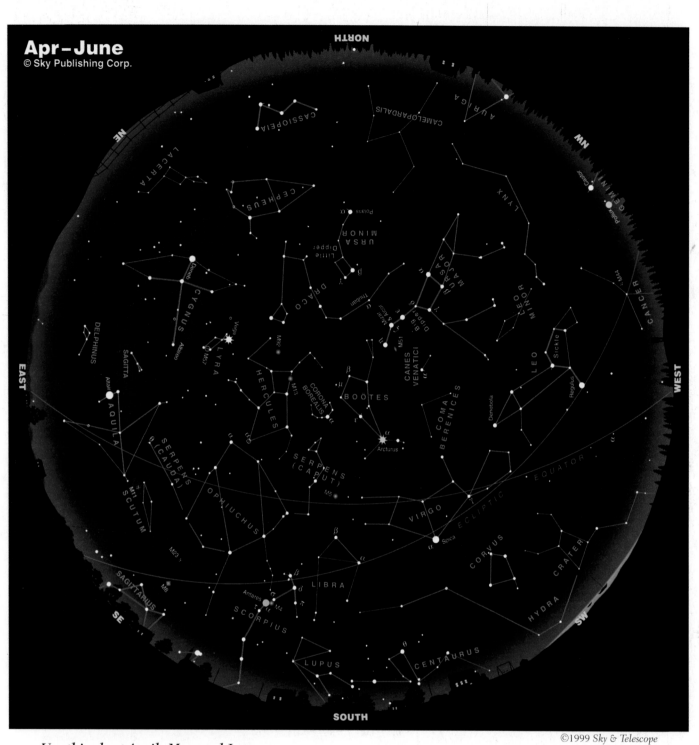

Apr–June
© Sky Publishing Corp.

©1999 Sky & Telescope

Use this chart April, May, and June.

Early April — 3 A.M.* Early May — 1 A.M.* Early June — 11 P.M.*
Late April — 2 A.M.* Late May — Midnight* Late June — Dusk

*Daylight Saving Time

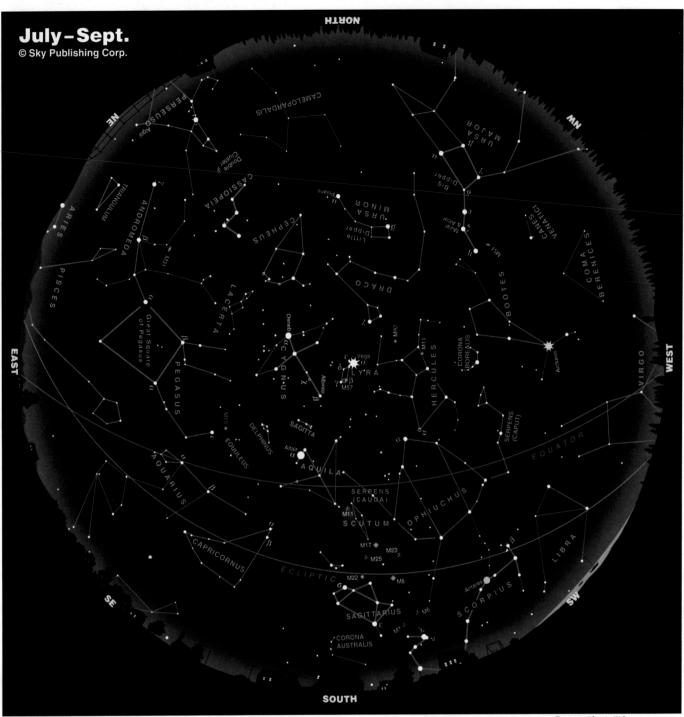

July–Sept.
© Sky Publishing Corp.

©1999 *Sky & Telescope*

Use this chart July, August, and September.

Early July — 1 A.M.*
Late July — Midnight*

Early August — 11 P.M.*
Late August — 10 P.M.*

Early September — 9 P.M.*
Late September — Dusk

*Daylight Saving Time

Oct.–Dec.
©Sky Publishing Corp.

©1999 *Sky & Telescope*

Use this chart October, November, and December.

Early October — 1 A.M.*   Early November — 10 P.M.   Early December — 8 P.M.
Late October — Midnight*   Late November — 9 P.M.   Late December — 7 P.M.

*Daylight Saving Time

GLOSSARY

21-cm line A spectral line from atomic hydrogen with wavelength 21 cm (in the radio portion of the spectrum).

absolute magnitude A measure of an object's luminosity; defined to be the apparent magnitude the object would have if it were located exactly 10 parsecs away.

absolute zero The coldest possible temperature, which is 0 K.

absorption (of light) The process by which matter absorbs radiative energy.

absorption-line spectrum A spectrum that contains absorption lines.

accelerating universe The possible fate of our universe in which a repulsive force (*see* cosmological constant) causes the expansion of the universe to accelerate with time. Its galaxies will recede from one another increasingly faster, and it will become cold and dark more quickly than a coasting universe.

acceleration The rate at which an object's velocity changes. Its standard units are m/s^2.

acceleration of gravity The acceleration of a falling object. On Earth, the acceleration of gravity, designated by g, is $9.8 \, m/s^2$.

accretion The process by which small objects gather together to make larger objects.

accretion disk A rapidly rotating disk of material that gradually falls inward as it orbits a starlike object (e.g., white dwarf, neutron star, or black hole).

active galactic nuclei The unusually luminous centers of some galaxies, thought to be powered by accretion onto supermassive black holes. Quasars are the brightest type of active galactic nuclei; radio galaxies also contain active galactic nuclei.

adaptive optics A technique in which telescope mirrors flex rapidly to compensate for the bending of starlight caused by atmospheric turbulence.

albedo Describes the fraction of sunlight reflected by a surface; albedo = 0 means no reflection at all (a perfectly black surface); albedo = 1 means all light is reflected (a perfectly white surface).

altitude (above horizon) The angular distance between the horizon and an object in the sky.

amino acids The building blocks of proteins.

analemma The figure-8 path traced by the Sun over the course of a year when viewed at the same place and the same time each

day; represents the discrepancies between apparent and mean solar time.

Andromeda Galaxy (M 31; the Great Galaxy in Andromeda) The nearest large spiral galaxy to the Milky Way.

angular momentum Momentum attributable to rotation or revolution. The angular momentum of an object moving in a circle of radius r is the product $m \times v \times r$.

angular resolution (of a telescope) The smallest angular separation that two point-like objects can have and still be seen as distinct points of light (rather than as a single point of light).

angular size (or **angular distance**) A measure of the angle formed by extending imaginary lines outward from our eyes to span an object (or between two objects).

annihilation *See* matter–antimatter annihilation

annular solar eclipse A solar eclipse during which the Moon is directly in front of the Sun but its angular size is not large enough to fully block the Sun; thus, a ring (or *annulus*) of sunlight is still visible around the Moon's disk.

Antarctic Circle The circle on the Earth with latitude 66.5°S.

antimatter Refers to any particle with the same mass as a particle of ordinary matter but whose other basic properties, such as electrical charge, are precisely opposite.

aphelion The point at which an object orbiting the Sun is farthest from the Sun.

apogee The point at which an object orbiting the Earth is farthest from the Earth.

apparent brightness The amount of light reaching us *per unit area* from a luminous object; often measured in units of $watts/m^2$.

apparent magnitude A measure of the apparent brightness of an object in the sky, based on the ancient system developed by Hipparchus.

apparent retrograde motion Refers to the apparent motion of a planet, as viewed from Earth, during the period of a few weeks or months when it moves westward relative to the stars in our sky.

apparent solar time Time measured by the actual position of the Sun in your local sky; defined so that noon is when the Sun is *on* the meridian.

arcminutes (or **minutes of arc**) One arcminute is 1/60 of 1°.

arcseconds (or **seconds of arc**) One arcsecond is 1/60 of an arcminute, or 1/3,600 of 1°.

Arctic Circle The circle on the Earth with latitude 66.5°N.

asteroid A relatively small and rocky object that orbits a star; asteroids are sometimes called *minor planets* because they are similar to planets but smaller.

asteroid belt The region of our solar system between the orbits of Mars and Jupiter in which asteroids are heavily concentrated.

astrobiology The study of life on Earth and beyond; emphasizes research into questions of the origin of life, the conditions under which life can survive, and the search for life beyond Earth.

astronomical unit (AU) The average distance (semimajor axis) of the Earth from the Sun, which is about 150 million km.

atmospheric pressure The surface pressure resulting from the overlying weight of an atmosphere.

atomic mass The combined number of protons and neutrons in an atom.

atomic number The number of protons in an atom.

atoms Consist of a nucleus made from protons and neutrons surrounded by a cloud of electrons.

aurora Dancing lights in the sky caused by charged particles entering our atmosphere; called the *aurora borealis* in the Northern Hemisphere and the *aurora australis* in the Southern Hemisphere.

azimuth (usually called *direction* in this book) Direction around the horizon from due north, measured clockwise in degrees. E.g., the azimuth of due north is 0°, due east is 90°, due south is 180°, and due west is 270°.

bar The standard unit of pressure, approximately equal to the Earth's atmospheric pressure at sea level.

baryonic matter Refers to ordinary matter made from atoms (because the nuclei of atoms contain protons and neutrons, which are both baryons).

baryons Particles, including protons and neutrons, that are made from three quarks.

basalt A type of volcanic rock that makes a low-viscosity lava when molten.

belts (on a jovian planet) Dark bands of sinking air that encircle a jovian planet at a particular set of latitudes.

Big Bang The event that gave birth to the universe.

Big Crunch If gravity ever reverses the universal expansion, the universe will someday begin to collapse and presumably end in a Big Crunch.

binary star system A star system that contains two stars.

biosphere Refers to the "layer" of life on Earth.

black hole A bottomless pit in spacetime. Nothing can escape from within a black hole, and we can never again detect or observe an object that falls into a black hole.

black smokers Structures around seafloor volcanic vents that support a wide variety of life.

blueshift A Doppler shift in which spectral features are shifted to shorter wavelengths, caused when an object is moving toward the observer.

bosons Particles, such as photons, to which the exclusion principle does not apply.

bound orbits Orbits on which an object travels repeatedly around another object; bound orbits are elliptical in shape.

brown dwarf An object too small to become an ordinary star because electron degeneracy pressure halts its gravitational collapse before fusion becomes self-sustaining; brown dwarfs have mass less than $0.08M_{Sun}$.

bubble (interstellar) The surface of a bubble is an expanding shell of hot, ionized gas driven by stellar winds or supernovae; inside the bubble, the gas is very hot and has very low density.

bulge (of a spiral galaxy) The central portion of a spiral galaxy that is roughly spherical (or football shaped) and bulges above and below the plane of the galactic disk.

Cambrian explosion The dramatic diversification of life on Earth that occurred between about 540 and 500 million years ago.

carbon stars Stars whose atmospheres are especially carbon-rich, thought to be near the ends of their lives; carbon stars are the primary sources of carbon in the universe.

carbonate rock A carbon-rich rock, such as limestone, that forms underwater from chemical reactions between sediments and carbon dioxide. On Earth, most of the outgassed carbon dioxide currently resides in carbonate rocks.

carbonate–silicate cycle The process that cycles carbon dioxide between the Earth's atmosphere and surface rocks.

CCD (charge coupled device) A type of electronic light detector that has largely replaced photographic film in astronomical research.

celestial coordinates The coordinates of right ascension and declination that fix an object's position on the celestial sphere.

celestial equator (CE) The extension of the Earth's equator onto the celestial sphere.

celestial navigation Navigation on the surface of the Earth accomplished by observations of the Sun and stars.

celestial sphere The imaginary sphere on which objects in the sky appear to reside when observed from Earth.

Celsius (temperature scale) The temperature scale commonly used in daily activity internationally. Defined so that, on Earth's surface, water freezes at 0°C and boils at 100°C.

central dominant galaxy A giant elliptical galaxy found at the center of a dense cluster of galaxies, apparently formed by the merger of several individual galaxies.

Cepheid See Cepheid variable

Cepheid variable A particularly luminous type of pulsating variable star that follows a period–luminosity relation and hence is very useful for measuring cosmic distances.

charged particle belts Zones in which ions and electrons accumulate and encircle a planet.

chemical enrichment The process by which the abundance of heavy elements (heavier than helium) in the interstellar medium gradually increases over time as these elements are produced by stars and released into space.

chromosphere The layer of the Sun's atmosphere below the corona; most of the Sun's ultraviolet light is emitted from this region, in which the temperature is about 10,000 K.

circulation cells (also called *Hadley cells*) Large-scale cells (similar to convection cells) in a planet's atmosphere that transport heat between the equator and the poles.

circumpolar star A star that always remains above the horizon for a particular latitude.

climate Describes the long-term average of weather.

close binary A binary star system in which the two stars are very close together.

closed universe The universe is closed if its average density is greater than the critical density, in which case spacetime must curve back on itself to the point where its overall shape is analogous to that of the surface of a sphere. In the absence of a repulsive force (*see* cosmological constant), a closed universe would someday stop expanding and begin to contract.

cluster of galaxies A collection of a few dozen or more galaxies bound together by gravity; smaller collections of galaxies are simply called *groups*.

cluster of stars A group of anywhere from several hundred to a million or so stars; star clusters come in two types—open clusters and globular clusters.

CNO cycle The cycle of reactions by which intermediate- and high-mass stars fuse hydrogen into helium.

coasting universe The possible fate of our universe in which the mass density of the universe is *smaller* than the critical density, so that the collective gravity of all

matter cannot halt the expansion. In the absence of a repulsive force (*see* cosmological constant), such a universe would keep expanding forever with little change in its rate of expansion.

coma (of a comet) The dusty atmosphere of a comet created by sublimation of ices in the nucleus when the comet is near the Sun.

comet A relatively small, icy object that orbits a star.

comparative planetology The study of the solar system by examining and understanding the similarities and differences among worlds.

compound (chemical) A substance made from molecules consisting of two or more atoms with different atomic number.

condensates Solid or liquid particles that condense from a cloud of gas.

condensation The formation of solid or liquid particles from a cloud of gas.

conduction (of energy) The process by which thermal energy is transferred by direct contact from warm material to cooler material.

conjunction (of a planet with the Sun) When a planet and the Sun line up in the sky.

conservation of angular momentum (law of) The principle that, in the absence of net torque (twisting force), the total angular momentum of a system remains constant.

conservation of energy (law of) The principle that energy (including mass-energy) can be neither created nor destroyed, but can only change from one form to another.

conservation of momentum (law of) The principle that, in the absence of net force, the total momentum of a system remains constant.

constellation A region of the sky; 88 official constellations cover the celestial sphere.

convection The energy transport process in which warm material expands and rises, while cooler material contracts and falls.

convection cell An individual small region of convecting material.

convection zone (of a star) A region in which energy is transported outward by convection.

Copernican revolution The dramatic change in human understanding of the cosmos in which the ancient belief in an Earth-centered (geocentric) universe was overturned by scientific evidence demonstrating that the Earth is one of many planets orbiting the Sun. The revolution began with the work of Nicholas Copernicus and played out over a period of more than a century. Other major players included Tycho Brahe, Kepler, Galileo, and Newton.

core (of a planet) The dense central region of a planet that has undergone differentiation.

core (of a star) The central region of a star, in which nuclear fusion can occur.

Coriolis effect Causes air or objects moving on a rotating planet to deviate from straight-line trajectories.

corona (solar) The tenuous uppermost layer of the Sun's atmosphere; most of the Sun's X rays are emitted from this region, in which the temperature is about 1 million K.

coronal holes Regions of the corona that barely show up in X-ray images because they are nearly devoid of hot coronal gas.

cosmic microwave background The remnant radiation from the Big Bang, which we detect using radio telescopes sensitive to microwaves (which are short-wavelength radio waves).

cosmic rays Particles such as electrons, protons, and atomic nuclei that zip through interstellar space at close to the speed of light.

cosmological constant The name given to a term in Einstein's equations of general relativity. If it is not zero, then it represents a repulsive force or a type of energy (sometimes called *dark energy* or *quintessence*) that might cause the expansion of the universe to accelerate with time.

cosmological horizon The boundary of our observable universe, which is where the lookback time is equal to the age of the universe. Beyond this boundary in spacetime, we cannot see anything at all.

cosmological redshift Refers to the redshifts we see from distant galaxies, caused by the fact that expansion of the universe stretches all the photons within it to longer, redder wavelengths.

critical density The precise average density for the entire universe that marks the dividing line between a recollapsing universe and one that will expand forever.

critical universe The possible fate of our universe in which the mass density of the universe *equals* the critical density. The universe will never collapse, but in the absence of a repulsive force it will expand more and more slowly as time progresses. *See* cosmological constant

crust (of a planet) The low-density surface layer of a planet that has undergone differentiation.

cycles per second Units of frequency for a wave; describes the number of peaks (or troughs) of a wave that pass by a given point each second. Equivalent to *hertz*.

dark energy Name sometimes given to energy that could be causing the expansion of the universe to accelerate. *See* cosmological constant

dark matter Matter that we infer to exist from its gravitational effects but from which we have not detected any light; dark matter apparently dominates the total mass of the universe.

daylight saving time Standard time plus 1 hour, so that the Sun appears on the meridian around 1 P.M. rather than around noon.

declination (dec) Analogous to latitude, but on the celestial sphere; it is the angular north-south distance between the celestial equator and a location on the celestial sphere.

degeneracy pressure A type of pressure unrelated to an object's temperature, which arises when electrons (electron degeneracy pressure) or neutrons (neutron degeneracy pressure) are packed so tightly that the exclusion and uncertainty principles come into play.

degenerate object An object in which degeneracy pressure is the primary pressure pushing back against gravity, such as a brown dwarf, white dwarf, or neutron star.

deuterium A form of hydrogen in which the nucleus contains a proton and a neutron, rather than only a proton (as is the case for most hydrogen nuclei).

differential rotation Describes the rotation of an object in which the equator rotates at a different rate than the poles.

differentiation The process in which gravity separates materials according to density, with high-density materials sinking and low-density materials rising.

diffraction grating A finely etched surface that can split light into a spectrum.

diffraction limit The angular resolution that a telescope could achieve if it were limited only by the interference of light waves; it is smaller (i.e., better angular resolution) for larger telescopes.

dimension (mathematical) Describes the number of independent directions in which movement is possible; e.g., the surface of the Earth is two-dimensional because only two independent directions of motion are possible (north-south and east-west).

direction (in local sky) One of the two coordinates (the other is altitude) needed to pinpoint an object in the local sky. It is the direction, such as north, south, east, or west, in which you must face to see the object. *See also* azimuth

disk component (of a galaxy) The portion of a spiral galaxy that looks like a disk and contains an interstellar medium with cool gas and dust; stars of many ages are found in the disk component.

Doppler effect (shift) The effect that shifts the wavelengths of spectral features in objects that are moving toward or away from the observer.

down quark One of the two quark types (the other is the up quark) found in ordinary protons and neutrons. Has a charge of $-\frac{1}{3}$.

dust (or **dust grains**) Tiny solid flecks of material; in astronomy, we often discuss interplanetary dust (found within a star system) or interstellar dust (found between the stars in a galaxy). *See also* interstellar dust grains

dust tail (of a comet) One of two tails seen when a comet passes near the Sun (the other is the plasma tail); composed of small solid particles pushed away from the Sun by the radiation pressure of sunlight.

dwarf elliptical galaxy A small elliptical galaxy with less than about a billion stars.

$E = mc^2$ Einstein's famous equation describes the equivalence between mass and energy, essentially telling us that mass is a form of energy (see *mass-energy*). In this equation, E stands for energy, m for mass, and c for the speed of light. If the mass m is measured in kilograms and the speed of light c is given in meters per second ($c = 3 \times 10^8$ m/s), the energy E will have units of joules.

Earth-orbiters (spacecraft) Spacecraft designed to study the Earth or the universe from Earth orbit.

eccentricity A measure of how much an ellipse deviates from a perfect circle; defined as the center-to-focus distance divided by the length of the semimajor axis.

eclipse Occurs when one astronomical object casts a shadow on another or crosses our line of sight to the other object.

eclipse seasons Periods during which lunar and solar eclipses can occur because the nodes of the Moon's orbit are nearly aligned with the Earth and Sun.

eclipsing binary A binary star system in which the two stars happen to be orbiting in the plane of our line of sight, so that each star will periodically eclipse the other.

ecliptic The Sun's apparent annual path among the constellations.

ecliptic plane The plane of the Earth's orbit around the Sun.

ejecta (from an impact) Debris ejected by the blast of an impact.

electromagnetic field An abstract concept used to describe how a charged particle would affect other charged particles at a distance.

electromagnetic force One of the four fundamental forces; it is the force that dominates atomic and molecular interactions.

electromagnetic spectrum The complete spectrum of light, including radio waves, infrared, visible light, ultraviolet light, X rays, and gamma rays.

electromagnetic wave A synonym for light, which consists of waves of electric and magnetic fields.

electron degeneracy pressure Degeneracy pressure exerted by electrons, as in brown dwarfs and white dwarfs.

electrons Fundamental particles with negative electric charge; the distribution of electrons in an atom gives the atom its size.

electron-volt (eV) A unit of energy equivalent to 1.60×10^{-19} joule.

electroweak era The era of the universe during which only three forces operated (gravity, strong force, and electroweak force), lasting from 10^{-38} second to 10^{-10} second after the Big Bang.

electroweak force The force that exists at high energies when the electromagnetic force and the weak force exist as a single force.

element (chemical) A substance made from individual atoms of a particular atomic number.

ellipse A type of oval that happens to be the shape of bound orbits. An ellipse can be drawn by moving a pencil along a string whose ends are tied to two tacks;

the locations of the tacks are the foci (singular, focus) of the ellipse.

elliptical galaxies Galaxies that appear rounded in shape, often longer in one direction, like a football. They have no disks and contain very little cool gas and dust compared to spiral galaxies, though they often contain very hot, ionized gas.

elongation (greatest) For Mercury or Venus, the point at which it appears farthest from the Sun in our sky.

emission (of light) The process by which matter emits energy in the form of light.

emission-line spectrum A spectrum that contains emission lines.

energy Broadly speaking, energy is what can make matter move. The three basic types of energy are kinetic, potential, and radiative.

equation of time Describes the discrepancies between apparent and mean solar time.

equivalence principle The fundamental starting point for general relativity, which states that the effects of gravity are exactly equivalent to the effects of acceleration.

era of atoms The era of the universe lasting from about 500,000 years to about 1 billion years after the Big Bang, during which it was cool enough for neutral atoms to form.

era of galaxies The present era of the universe, which began with the formation of galaxies when the universe was about 1 billion years old.

era of nuclei The era of the universe lasting from about 3 minutes to about 500,000 years after the Big Bang, during which matter in the universe was fully ionized and opaque to light. The cosmic background radiation was released at the end of this era.

era of nucleosynthesis The era of the universe lasting from about 0.001 second to about 3 minutes after the Big Bang, by the end of which virtually all of the neutrons and about one-seventh of the protons in the universe had fused into helium.

erosion The wearing down or building up of geological features by wind, water, ice, and other phenomena of planetary weather.

eruption The process of releasing hot lava on the planet's surface.

escape velocity The speed necessary for an object to completely escape the gravity of a large body such as a moon, planet, or star.

evaporation The process by which atoms or molecules escape into the gas phase from a liquid.

event Any particular point along a worldline represents a particular event; all observers will agree on the reality of an event but may disagree about its time and location.

event horizon The boundary that marks the "point of no return" between a black hole and the outside universe; events that occur within the event horizon can have no influence on our observable universe.

exchange particle According to the standard model of physics, each of the four fundamental forces is transmitted by the transfer of particular types of exchange particles.

excited state (of an atom) Any arrangement of electrons in an atom that has more energy than the ground state.

exclusion principle The law of quantum mechanics that states that two fermions cannot occupy the same quantum state at the same time.

exosphere The hot, outer layer of an atmosphere, where the atmosphere "fades away" to space.

exposure time The amount of time for which light is collected to make a single image.

extrasolar planet A planet orbiting a star other than our Sun.

Fahrenheit (temperature scale) The temperature scale commonly used in daily activity in the United States. Defined so that, on Earth's surface, water freezes at 32°F and boils at 212°F.

fall equinox (autumnal equinox) Refers both to the point in Virgo on the celestial sphere where the ecliptic crosses the celestial equator and to the moment in time when the Sun appears at that point each year (around September 21).

false-color image An image displayed in colors that are *not* the true, visible-light colors of an object.

fault (geological) A place where rocks slip sideways relative to one another.

feedback relationships Processes in which one property amplifies (positive feedback) or counteracts (negative feedback) the behavior of properties.

fermions Particles, such as electrons, neutrons, and protons, that obey the exclusion principle.

field An abstract concept used to describe how a particle would interact with a force. For example, the idea of a *gravitational field* describes how a particle would react to the local strength of gravity, and the idea of an *electromagnetic field* describes how a charged particle would respond to forces from other charged particles.

filter (for light) A material that transmits only particular wavelengths of light.

fireball A particularly bright meteor.

flare star A small, spectral type M star that displays particularly strong flares on its surface.

flat (or Euclidean) geometry Refers to any case in which the rules of geometry for a flat plane hold, such as that the shortest distance between two points is a straight line.

flat universe A universe in which the overall geometry of spacetime is flat (Euclidean), as would be the case if the density of the universe is equal to the critical density.

flybys (spacecraft) Spacecraft that fly past a target object (such as a planet), usually just once, as opposed to entering a bound orbit of the object.

focal plane The place where an image created by a lens or mirror is in focus.

focus (of a lens or mirror) The point at which rays of light that were initially parallel (such as light from a distant star) converge.

force Anything that can cause a change in momentum.

formation properties (of planets) In this book, for the purpose of understanding geological processes, planets are defined to be born with four formation properties: size (mass and radius), distance from the Sun, composition, and rotation rate.

frame of reference (in relativity) Two (or more) objects share the same frame of reference if they are *not* moving relative to each other.

free-fall Refers to conditions in which an object is falling without resistance; objects are weightless when in free-fall.

free-float frame A frame of reference in which all objects are weightless and hence float freely.

frequency Describes the rate at which peaks of a wave pass by a point; measured in units of 1/s, often called *cycles per second* or *hertz*.

frost line The boundary in the solar nebula beyond which ices could condense; only metals and rocks could condense within the frost line.

fundamental forces There are four known fundamental forces in nature: gravity, the electromagnetic force, the strong force, and the weak force.

fundamental particles Subatomic particles that cannot be divided into anything smaller.

galactic disk (of a spiral galaxy) *See* disk component

galactic fountain Refers to a model for the cycling of gas in the Milky Way Galaxy in which fountains of hot, ionized gas rise from the disk into the halo and then cool and form clouds as they sink back into the disk.

galactic wind A wind of low-density but extremely hot gas flowing out from a starburst galaxy, created by the combined energy of many supernovae.

galaxy A huge collection of anywhere from a few hundred million to more than a trillion stars, all bound together by gravity.

galaxy cluster *See* cluster of galaxies

galaxy evolution The formation and development of galaxies.

gamma-ray burst A sudden burst of gamma rays from deep space; such bursts apparently come from distant galaxies, but their precise mechanism is unknown.

gamma rays Light with very short wavelengths (and hence high frequencies)—shorter than those of X rays.

gas phase The phase of matter in which atoms or molecules can move essentially independently of one another.

gas pressure Describes the force (per unit area) pushing on any object due to surrounding gas. *See also* pressure

genetic code The "language" that living cells use to read the instructions chemically encoded in DNA.

geocentric universe (ancient belief in) The idea that the Earth is the center of the entire universe.

geological controlling factors In this book, for the purpose of understanding geological processes, geology is considered to be influenced primarily by four geological controlling factors: surface gravity, internal temperature, surface temperature, and the presence (and extent) of an atmosphere.

geological processes The four basic geological processes are impact cratering, volcanism, tectonics, and erosion.

geology The study of surface features (on a moon, planet, or asteroid) and the processes that create them.

giants (luminosity class III) Stars that appear just below the supergiants on the H–R diagram because they are somewhat smaller in radius and lower in luminosity.

global positioning system (GPS) A system of navigation by satellites orbiting the Earth.

global wind patterns (or **global circulation**) Wind patterns that remain fixed on a global scale, determined by the combination of surface heating and the planet's rotation.

globular cluster A spherically shaped cluster of up to a million or more stars; globular clusters are found primarily in the halos of galaxies and contain only very old stars.

gluons The exchange particles for the strong force.

grand unified theory (GUT) A theory that unifies three of the four fundamental forces—the strong force, the weak force, and the electromagnetic force (but not gravity)—in a single model.

granulation (on the Sun) The bubbling pattern visible in the photosphere, produced by the underlying convection.

gravitation (law of) *See* universal law of gravitation

gravitational constant The experimentally measured constant G that appears in the law of universal gravitation;

$$G = 6.67 \times 10^{-11} \frac{m^3}{kg \times s^2}.$$

gravitational contraction The process in which gravity causes an object to contract, thereby converting gravitational potential energy into thermal energy.

gravitational encounter Occurs when two (or more) objects pass near enough so that each can feel the effects of the other's gravity and can therefore exchange energy.

gravitational equilibrium Describes a state of balance in which the force of gravity pulling inward is precisely counteracted by pressure pushing outward.

gravitational lensing The magnification or distortion (into arcs, rings, or multiple images) of an image caused by light bending through a gravitational field, as predicted by Einstein's general theory of relativity.

gravitational redshift A redshift caused by the fact that time runs slow in gravitational fields.

gravitational time dilation The slowing of time that occurs in a gravitational field, as predicted by Einstein's general theory of relativity.

gravitational waves Predicted by Einstein's general theory of relativity, these waves travel at the speed of light and transmit distortions of space through the universe. Although not yet observed directly, we have strong indirect evidence that they exist.

gravitationally bound system Any system of objects, such as a star system or a galaxy, that is held together by gravity.

gravitons The exchange particles for the force of gravity.

gravity One of the four fundamental forces; it is the force that dominates on large scales.

grazing incidence (in telescopes) Reflections in which light grazes a mirror surface and is deflected at a small angle; commonly used to focus high-energy ultraviolet light and X rays.

great circle A circle on the surface of a sphere whose center is at the center of the sphere.

Great Red Spot A large, high-pressure storm on Jupiter.

greenhouse effect The process by which greenhouse gases in an atmosphere make a planet's surface temperature warmer than it would be in the absence of an atmosphere.

greenhouse gases Gases, such as carbon dioxide, water vapor, and methane, that are particularly good absorbers of infrared light but are transparent to visible light.

Gregorian calendar Our modern calendar, introduced by Pope Gregory in 1582.

ground state (of an atom) The lowest possible energy state of the electrons in an atom.

group (of galaxies) A few to a few dozen galaxies bound together by gravity. *See also* cluster of galaxies

GUT era The era of the universe during which only two forces operated (gravity and the grand-unified-theory or GUT force), lasting from 10^{-43} second to 10^{-38} second after the Big Bang.

GUT force The proposed force that exists at very high energies when the strong force, the weak force, and the electromagnetic force (but not gravity) all act as one.

habitable zone The region around a star in which planets could potentially have surface temperatures at which liquid water could exist.

half-life The time it takes for half of the nuclei in a given quantity of a radioactive substance to decay.

halo (of a galaxy) The spherical region surrounding the disk of a spiral galaxy.

Hawking radiation Radiation predicted to arise from the evaporation of black holes.

heavy elements In astronomy, *heavy elements* generally refers to all elements *except* hydrogen and helium.

helium-capture reactions Fusion reactions that fuse a helium nucleus into some other nucleus; such reactions can fuse carbon into oxygen, oxygen into neon, neon into magnesium, and so on.

helium flash The event that marks the sudden onset of helium fusion in the previously inert helium core of a low-mass star.

helium fusion The fusion of three helium nuclei into one carbon nucleus; also called the *triple-alpha reaction*.

hertz (Hz) The standard unit of frequency for light waves; equivalent to units of 1/s.

Hertzsprung–Russell (H–R) diagram A graph plotting individual stars as points, with stellar luminosity on the vertical axis and spectral type (or surface temperature) on the horizontal axis.

high-mass stars Stars born with masses above about $8M_{Sun}$; these stars will end their lives by exploding as supernovae.

horizon A boundary that divides what we can see from what we cannot see.

horizontal branch The horizontal line of stars that represents helium-burning stars on an H–R diagram for a cluster of stars.

horoscope A predictive chart made by an astrologer; in scientific studies, horoscopes have never been found to have any validity as predictive tools.

hot spot (geological) A place within a plate of the lithosphere where a localized plume of hot mantle material rises.

hour angle (HA) The angle or time (measured in hours) since an object was last on the meridian in the local sky. Defined to be 0 hours for objects that *are* on the meridian.

Hubble's constant A number that expresses the current rate of expansion of the universe; designated H_0, it is usually stated in units of km/s/Mpc. The reciprocal of Hubble's constant is the age the universe would have *if* the expansion rate had never changed.

Hubble's law Mathematically expresses the idea that more distant galaxies move away from us faster; its formula is $v = H_0 \times d$, where v is a galaxy's speed away from us, d is its distance, and H_0 is Hubble's constant.

hydrogen compounds Compounds that contain hydrogen and were common in the solar nebula, such as water (H_2O), ammonia (NH_3), and methane (CH_4).

hydrogen-shell burning Hydrogen fusion that occurs in a shell surrounding a stellar core.

hydrosphere Refers to the "layer" of water on the Earth consisting of oceans, lakes, rivers, ice caps, and other liquid water and ice.

hyperbola The precise mathematical shape of one type of unbound orbit (the other is a parabola) allowed under the force of gravity; at great distances from the attracting object, a hyperbolic path looks like a straight line.

hypernova A term sometimes used to describe a supernova (explosion) of a star so massive that it leaves a black hole behind.

hyperspace Any space with more than three dimensions.

hypothesis A tentative model proposed to explain some set of observed facts, but which has not yet been rigorously tested and confirmed.

ices (in solar system theory) Materials that are solid only at low temperatures, such as the hydrogen compounds water, ammonia, and methane.

image A picture of an object made by focusing light.

imaging (in astronomical research) The process of obtaining pictures of astronomical objects.

impact The collision of a small body (such as an asteroid or comet) with a larger object (such as a planet or moon).

impact basin A very large impact crater often filled by a lava flow.

impact crater A bowl-shaped depression left by the impact of an object that strikes a planetary surface (as opposed to burning up in the atmosphere).

impact cratering The excavation of bowl-shaped depressions (*impact craters*) by asteroids or comets striking a planet's surface.

impactor The object responsible for an impact.

inflation (of the universe) A sudden and dramatic expansion of the universe thought to have occurred at the end of the GUT era.

infrared light Light with wavelengths that fall in the portion of the electromagnetic spectrum between radio waves and visible light.

inner solar system Generally considered to encompass the region of our solar system out to about the orbit of Mars.

intensity (of light) A measure of the amount of energy coming from light of specific wavelength in the spectrum of an object.

interferometry A telescopic technique in which two or more telescopes are used in tandem to produce much better angular resolution than the telescopes could achieve individually.

intermediate-mass stars Stars born with masses between about 2–$8M_{Sun}$; these stars end their lives by ejecting a planetary nebula and becoming a white dwarf.

interstellar cloud A cloud of gas and dust between the stars.

interstellar dust grains Tiny solid flecks of carbon and silicon minerals found in cool interstellar clouds; they resemble particles of smoke and form in the winds of red giant stars.

interstellar medium Refers to gas and dust that fills the space between stars in a galaxy.

intracluster medium Hot, X-ray-emitting gas found between the galaxies within a cluster of galaxies.

inverse square law Any quantity that decreases with the square of the distance between two objects is said to follow an inverse square law.

inversion (atmospheric) A local weather condition in which air is colder near the surface than higher up in the troposphere—the opposite of the usual condition, in which the troposphere is warmer at the bottom.

Io torus A donut-shaped charged-particle belt around Jupiter that approximately traces Io's orbit.

ionization The process of stripping an electron from an atom.

ionization nebula A colorful, wispy cloud of gas that glows because neighboring hot stars irradiate it with ultraviolet photons that can ionize hydrogen atoms.

ionosphere A portion of the thermosphere in which ions are particularly common (due to ionization by X rays from the Sun).

ions Atoms with a positive or negative electrical charge.

irregular galaxies Galaxies that look neither spiral nor elliptical.

isotopes Each different isotope of an element has the *same* number of protons but a *different* number of neutrons.

jets High-speed streams of gas ejected from an object into space.

joule The international unit of energy, equivalent to about 1/4,000 of a Calorie.

jovian nebulae The clouds of gas that swirled around the jovian planets, from which the moons formed.

jovian planets Giant gaseous planets similar in overall composition to Jupiter.

Julian calendar The calendar introduced in 46 B.C. by Julius Caesar and used until it was replaced by the Gregorian calendar.

Kelvin (temperature scale) The most commonly used temperature scale in science, defined such that absolute zero is 0 K and water freezes at 273.15 K.

Kepler's first law States that the orbit of each planet about the Sun is an ellipse with the Sun at one focus.

Kepler's laws of planetary motion Three laws discovered by Kepler that describe the motion of the planets around the Sun.

Kepler's second law States that, as a planet moves around its orbit, it sweeps out equal areas in equal times. This tells us that a planet moves faster when it is closer to the Sun (near perihelion) than when it is farther from the Sun (near aphelion) in its orbit.

Kepler's third law States that the square of a planet's orbital period is proportional to the cube of its average distance from the Sun (semimajor axis), which tells us that more distant planets move more slowly in their orbits. In its original form, written $p^2 = a^3$. *See also* Newton's version of Kepler's third law

kinetic energy Energy of motion, given by the formula $\frac{1}{2}mv^2$.

Kuiper belt The comet-rich region of our solar system that spans distances of about 30–100 AU from the Sun; Kuiper belt comets have orbits that lie fairly close to the plane of planetary orbits and travel around the Sun in the same direction as the planets.

Large Magellanic Cloud One of two small, irregular galaxies (the other is the Small Magellanic Cloud) located about 150,000 light-years away; it probably orbits the Milky Way Galaxy.

large-scale structure (of the universe) Generally refers to structure of the universe on size scales larger than that of clusters of galaxies.

latitude The angular north-south distance between the Earth's equator and a location on the Earth's surface.

leap year A calendar year with 366 rather than 365 days; our current calendar (the Gregorian calendar) has a leap year every 4 years (by adding February 29) except in century years that are not divisible by 400.

length contraction Refers to the effect in which you observe lengths to be shortened in reference frames moving relative to you.

lenticular galaxies Galaxies that look lens-shaped when seen edge-on, resembling spiral galaxies without arms. They tend to have less cool gas than normal spiral galaxies but more gas than elliptical galaxies.

leptons Fermions *not* made from quarks, such as electrons and neutrinos.

life track A track drawn on an H–R diagram to represent the changes in a star's surface temperature and luminosity during its life; also called an *evolutionary track.*

light-collecting area (of a telescope) The area of the primary mirror or lens that collects light in a telescope.

light curve A graph of an object's intensity against time.

light gases (in solar system theory) Refers to hydrogen and helium, which never condense under solar nebula conditions.

light pollution Human-made light that hinders astronomical observations.

light-year The distance that light can travel in 1 year, which is 9.46 trillion km.

liquid phase The phase of matter in which atoms or molecules are held together but move relatively freely.

lithosphere The relatively rigid outer layer of a planet; generally encompasses the crust and the uppermost portion of the mantle.

Local Group The group of over 30 galaxies to which the Milky Way Galaxy belongs.

local sidereal time (LST) Sidereal time for a particular location, defined according to the position of the spring equinox in the local sky. More formally, the local sidereal time at any moment is defined to be the hour angle of the spring equinox.

local sky The sky as viewed from a particular location on Earth (or another solid object). Objects in the local sky are pinpointed by the coordinates of *altitude* and *direction* (or *azimuth*).

Local Supercluster The supercluster of galaxies to which the Local Group belongs.

longitude The angular east-west distance between the prime meridian (which passes through Greenwich) and a location on the Earth's surface.

lookback time Refers to the amount of time since the light we see from a distant object was emitted. I.e., if an object has a lookback time of 400 million years, we are seeing it as it looked 400 million years ago.

low-mass stars Stars born with masses less than about $2M_{Sun}$; these stars end their lives by ejecting a planetary nebula and becoming a white dwarf.

luminosity The total power output of an object, usually measured in watts or in units of solar luminosities ($L_{Sun} = 3.8 \times 10^{26}$ watts).

luminosity class Describes the region of the H–R diagram in which a star falls. Luminosity class I represents supergiants, III represents giants, and V represents main-sequence stars; luminosity classes II and IV are intermediate to the others.

luminosity–distance formula The formula that relates apparent brightness, luminosity, and distance:

$$\text{apparent brightness} = \frac{\text{luminosity}}{4\pi \times (\text{distance})^2}$$

lunar eclipse Occurs when the Moon passes through the Earth's shadow, which can occur only at full moon; may be total, partial, or penumbral.

lunar maria The regions of the Moon that look smooth from Earth and actually are impact basins.

lunar month *See* synodic month

lunar phase Describes the appearance of the Moon as seen from Earth.

MACHOs Stands for *massive compact halo objects* and represents one possible form of dark matter in which the dark objects are relatively large, like planets or brown dwarfs.

magma Underground molten rock.

magnetic braking The process by which a star's rotation slows as its magnetic field transfers its angular momentum to the surrounding nebula.

magnetic field Describes the region surrounding a magnet in which it can affect other magnets or charged particles in its vicinity.

magnetic field lines Lines that represent how the needles on a series of compasses would point if they were laid out in a magnetic field.

magnetosphere The region surrounding a planet in which charged particles are trapped by the planet's magnetic field.

magnitude system A system of describing stellar brightness by using numbers, called *magnitudes,* based on an ancient Greek way of describing the brightnesses of stars in the sky. This system uses *apparent magnitude* to describe a star's apparent brightness

and *absolute magnitude* to describe a star's luminosity.

main sequence (luminosity class V) The prominent line of points running from the upper left to the lower right on an H–R diagram; main-sequence stars shine by fusing hydrogen in their cores.

main-sequence fitting A method for measuring the distance to a cluster of stars by comparing the apparent brightness of the cluster's main sequence with the standard main sequence.

main-sequence lifetime The length of time for which a star of a particular mass can shine by fusing hydrogen into helium in its core.

main-sequence turnoff A method for measuring the age of a cluster of stars from the point on its H–R diagram where its stars turn off from the main sequence; the age of the cluster is equal to the main-sequence lifetime of stars at the main-sequence turn-off point.

mantle (of a planet) The rocky layer that lies between a planet's core and crust.

Martian meteorite This term is used to describe meteorites found on Earth that are thought to have originated on Mars.

mass A measure of the amount of matter in an object.

mass-energy The potential energy of mass, which has an amount $E = mc^2$.

mass exchange (in close binary star systems) The process in which tidal forces cause matter to spill from one star to a companion star in a close binary system.

mass extinction An event in which a large fraction of the species living on Earth go extinct, such as the event in which the dinosaurs died out about 65 million years ago.

mass increase (in relativity) Refers to the effect in which an object moving past you seems to have a mass greater than its rest mass.

mass-to-light ratio The mass of an object divided by its luminosity, usually stated in units of solar masses per solar luminosity. Objects with high mass-to-light ratios must contain substantial quantities of dark matter.

massive-star supernova A supernova that occurs when a massive star dies, initiated by the catastrophic collapse of its iron core; often called a Type II supernova.

matter–antimatter annihilation Occurs when a particle of matter and a particle of antimatter meet and convert all of their mass-energy to photons.

mean solar time Time measured by the average position of the Sun in your local sky over the course of the year.

meridian A half-circle extending from your horizon (altitude 0°) due south, through your zenith, to your horizon due north.

metals (in solar system theory) Elements, such as nickel, iron, and aluminum, that condense at fairly high temperatures.

meteor A flash of light caused when a particle from space burns up in our atmosphere.

meteor shower A period during which many more meteors than usual can be seen.

meteorite A rock from space that lands on Earth.

Metonic cycle The 19-year period, discovered by the Babylonian astronomer Meton, over which the lunar phases occur on the same dates.

Milky Way Used both as the name of our galaxy and to refer to the band of light we see in the sky when we look into the plane of the Milky Way Galaxy.

millisecond pulsars Pulsars with rotation periods of a few thousandths of a second.

model (scientific) A representation of some aspect of nature that can be used to explain and predict real phenomena without invoking myth, magic, or the supernatural.

molecular bands The tightly bunched lines in an object's spectrum that are produced by molecules.

molecular clouds Cool, dense interstellar clouds in which the low temperatures allow hydrogen atoms to pair up into hydrogen molecules (H_2).

molecular dissociation The process by which a molecule splits into its component atoms.

molecule Technically the smallest unit of a chemical element or compound; in this text, the term refers only to combinations of two or more atoms held together by chemical bonds.

momentum The product of an object's mass and velocity.

moon An object that orbits a planet.

mutations Errors in the copying process when a living cell replicates itself.

natural selection The process by which mutations that make an organism better able to survive get passed on to future generations.

nebula A cloud of gas in space, usually one that is glowing.

nebular capture The process by which icy planetesimals capture hydrogen and helium gas to form jovian planets.

nebular theory The detailed theory that describes how our solar system formed from a cloud of interstellar gas and dust.

net force The overall force to which an object responds; the net force is equal to the rate of change in the object's momentum, or equivalently to the object's mass × acceleration.

neutrino A type of fundamental particle that has extremely low mass and responds only to the weak force; neutrinos are leptons and come in three types—electron neutrinos, mu neutrinos, and tau neutrinos.

neutron degeneracy pressure Degeneracy pressure exerted by neutrons, as in neutron stars.

neutron star The compact corpse of a high-mass star left over after a supernova; typically contains a mass comparable to the mass of the Sun in a volume just a few kilometers in radius.

neutrons Particles with no electrical charge found in atomic nuclei, built from three quarks.

newton The standard unit of force in the metric system:

$$1 \text{ newton} = 1 \frac{\text{kg} \times \text{m}}{\text{s}^2}$$

Newton's first law of motion States that, in the absence of a net force, an object moves with constant velocity.

Newton's laws of motion Three basic laws that describe how objects respond to forces.

Newton's second law of motion States how a net force affects an object's motion. Specifically: force = rate of change in momentum, or force = mass × acceleration.

Newton's third law of motion States that, for any force, there is always an equal and opposite reaction force.

Newton's universal law of gravitation See universal law of gravitation

Newton's version of Kepler's third law This generalization of Kepler's third law can be used to calculate the masses of orbiting objects from measurements of orbital period and distance. Usually written as:

$$p^2 = \frac{4\pi^2}{G(M_1 + M_2)} a^3$$

nodes (of Moon's orbit) The two points in the Moon's orbit where it crosses the ecliptic plane.

nonbaryonic matter Refers to exotic matter that is not part of the normal composition of atoms, such as neutrinos or the hypothetical WIMPs.

nonscience As defined in this book, nonscience is any way of searching for knowledge that makes no claim to follow the scientific method, such as seeking knowledge through intuition, tradition, or faith.

north celestial pole (NCP) The point on the celestial sphere directly above the Earth's North Pole.

nova The dramatic brightening of a star that lasts for a few weeks and then subsides; occurs when a burst of hydrogen fusion ignites in a shell on the surface of an accreting white dwarf in a binary star system.

nuclear fission The process in which a larger nucleus splits into two (or more) smaller particles.

nuclear fusion The process in which two (or more) smaller nuclei slam together and make one larger nucleus.

nucleus (of an atom) The compact center of an atom made from protons and neutrons.

nucleus (of a comet) The solid portion of a comet, and the only portion that exists when the comet is far from the Sun.

observable universe The portion of the entire universe that, at least in principle, can be seen from Earth.

Olbers' paradox Asks the question of how the night sky can be dark if the universe is infinite and full of stars.

Oort cloud A huge, spherical region centered on the Sun, extending perhaps halfway to the nearest stars, in which trillions of comets orbit the Sun with random inclinations, orbital directions, and eccentricities.

opacity A measure of how much light a material absorbs compared to how much it transmits; materials with higher opacity absorb more light.

opaque (material) Describes a material that absorbs light.

open cluster A cluster of up to several thousand stars; open clusters are found only in the disks of galaxies and often contain young stars.

open universe The universe is open if its average density is less than the critical density, in which case spacetime has an overall shape analogous to the surface of a saddle.

opposition The point at which a planet appears opposite the Sun in our sky.

optical quality Describes the ability of a lens, mirror, or telescope to obtain clear and properly focused images.

orbital resonance Describes any situation in which one object's orbital period is a simple ratio of another object's period, such as 1/2, 1/4, or 5/3. In such cases, the two objects periodically line up with each other, and the extra gravitational attractions at these times can affect the objects' orbits.

orbiters (of other worlds) Spacecraft that go into orbit of another world for long-term study.

outer solar system Generally considered to encompass the region of our solar system beginning at about the orbit of Jupiter.

outgassing The process of releasing gases from a planetary interior, usually through volcanic eruptions.

oxidation Refers to chemical reactions, often with the surface of a planet, that remove oxygen from the atmosphere.

ozone The molecule O_3, which is a particularly good absorber of ultraviolet light.

ozone depletion Refers to the declining levels of atmospheric ozone found worldwide on Earth, especially in Antarctica, in recent years.

ozone hole A place where the concentration of ozone in the stratosphere is dramatically lower than is the norm.

pair production The process in which a concentration of energy spontaneously turns into a particle and its antiparticle.

parabola The precise mathematical shape of a special type of unbound orbit allowed under the force of gravity; if an object in a parabolic orbit loses only a tiny amount of energy, it will become bound.

paradigm (in science) Refers to general patterns of thought that tend to shape scientific beliefs during a particular time period.

paradox A situation that, at least at first, seems to violate common sense or contradict itself. Resolving paradoxes often leads to deeper understanding.

parallax The apparent shifting of an object against the background, due to viewing it from different positions. See also stellar parallax

parallax angle Half of a star's annual back-and-forth shift due to stellar parallax; related to the star's distance according to the formula

$$\text{distance in parsecs} = \frac{1}{p}$$

where p is the parallax angle in arcseconds.

parsec (pc) Approximately equal to 3.26 light-years; it is the distance to an object with a parallax angle of 1 arcsecond.

partial lunar eclipse A lunar eclipse in which the Moon becomes only partially covered by the Earth's umbral shadow.

partial solar eclipse A solar eclipse during which the Sun becomes only partially blocked by the disk of the Moon.

particle accelerator A machine designed to accelerate subatomic particles to high speeds in order to create new particles or to test fundamental theories of physics.

particle era The era of the universe lasting from 10^{-10} second to 0.001 second after the Big Bang, during which subatomic particles were continually created and destroyed and ending when matter annihilated antimatter.

peculiar velocity (of a galaxy) The component of a galaxy's velocity relative to the Milky Way that deviates from the velocity expected by Hubble's law.

penumbra The lighter, outlying regions of a shadow.

penumbral (lunar) eclipse A lunar eclipse in which the Moon passes only within the Earth's penumbral shadow and does not fall within the umbra.

perigee The point at which an object orbiting the Earth is nearest to the Earth.

perihelion The point at which an object orbiting the Sun is closest to the Sun.

period–luminosity relation The relation that describes how the luminosity of a Cepheid variable star is related to the period between peaks in its brightness; the longer the period, the more luminous the star.

phase (of matter) Describes the way in which atoms or molecules are held together; the common phases are solid, liquid, and gas.

photon An individual particle of light, characterized by a wavelength and a frequency.

photosphere The visible surface of the Sun, where the temperature averages just under 6,000 K.

pixel An individual "picture element" on a CCD.

Planck era The era of the universe prior to the Planck time.

Planck time The time when the universe was 10^{-43} second old, before which random energy fluctuations were so large that our current theories are powerless to describe what might have been happening.

Planck's constant A universal constant, abbreviated h, with value $h = 6.626 \times 10^{-34}$ joule $\times$ s.

planet An object that orbits a star and that, while much smaller than a star, is relatively large in size; there is no "official" minimum size for a planet, but the nine planets in our solar system all are at least 2,000 km in diameter.

planetary nebula The glowing cloud of gas ejected from a low-mass star at the end of its life.

planetesimals The building blocks of planets, formed by accretion in the solar nebula.

plasma A gas consisting of ions and electrons.

plasma tail (of a comet) One of two tails seen when a comet passes near the Sun (the other is the dust tail); composed of ionized gas blown away from the Sun by the solar wind.

plate tectonics The geological process in which plates are moved around by stresses in a planet's mantle.

plates (on a planet) Pieces of a lithosphere that apparently float upon the denser mantle below.

potential energy Energy stored for later conversion into kinetic energy; includes gravitational potential energy, electrical potential energy, and chemical potential energy.

power The rate of energy usage, usually measured in watts (1 watt = 1 joule/s).

precession The gradual wobble of the axis of a rotating object around a vertical line.

pressure Describes the force (per unit area) pushing on an object. In astronomy, we are generally interested in pressure applied by surrounding gas (or plasma). Ordinarily, such pressure is related to the temperature of the gas (*see* thermal pressure). In objects such as white dwarfs and neutron stars, pressure may arise from a quantum effect (*see* degeneracy pressure). Light can also exert pressure. (*See* radiation pressure.)

primary mirror The large, light-collecting mirror of a reflecting telescope.

prime focus (of a reflecting telescope) The first point at which light focuses after bouncing off the primary mirror; located in front of the primary mirror.

prime meridian The meridian of longitude that passes through Greenwich, England, defined to be longitude 0°.

primitive meteorites Meteorites that formed at the same time as the solar system itself, about 4.6 billion years ago.

processed meteorites Meteorites that apparently once were part of a larger object that "processed" the original material of the solar nebula into another form.

protogalactic cloud A huge, collapsing cloud of intergalactic gas from which an individual galaxy formed.

proton–proton chain The chain of reactions by which low-mass stars (including the Sun) fuse hydrogen into helium.

protons Particles found in atomic nuclei with positive electrical charge, built from three quarks.

protoplanetary disk A disk of material surrounding a young star (or protostar) that may eventually form planets.

protostar A forming star that has not yet reached the point where sustained fusion can occur in its core.

protostellar disk A disk of material surrounding a protostar; essentially the same as a protoplanetary disk, but may not necessarily lead to planet formation.

protostellar wind The relatively strong wind from a protostar.

protosun The central object in the forming solar system that eventually became the Sun.

pseudoscience Something that purports to be science or may appear to be scientific but that does not adhere to the testing and verification requirements of the scientific method.

pulsar A neutron star from which we see rapid pulses of radiation as it rotates.

pulsating variable stars Stars that alternately grow brighter and dimmer as their outer layers expand and contract in size.

quantum mechanics The branch of physics that deals with the very small, including molecules, atoms, and fundamental particles.

quantum state Refers to the complete description of the state of a subatomic particle, including its location, momentum, orbital angular momentum, and spin, to the extent allowed by the uncertainty principle.

quantum tunneling The process in which, thanks to the uncertainty principle, an electron or other subatomic particle appears on the other side of a barrier that it does not have the energy to overcome in a normal way.

quarks The building blocks of protons and neutrons, quarks are one of the two basic types of fermions (leptons are the other).

quasar The brightest type of active galactic nucleus.

radar ranging A method of measuring distances within the solar system by bouncing radio waves off planets.

radial motion The component of an object's motion directed toward or away from us.

radiation pressure Pressure exerted by photons of light.

radiation zone (of a star) A region of the interior in which energy is transported primarily by radiative diffusion.

radiative diffusion The process by which photons gradually migrate from a hot region (such as the solar core) to a cooler region (such as the solar surface).

radiative energy Energy carried by light; the energy of a photon is Planck's constant times its frequency, or $h \times f$.

radio galaxy A galaxy that emits unusually large quantities of radio waves; thought to contain an active galactic nucleus powered by a supermassive black hole.

radio waves Light with very long wavelengths (and hence low frequencies)—longer than those of infrared light.

radioactive dating The process of determining the age of a rock (i.e., the time since it solidified) by comparing the present amount of a radioactive substance to the amount of its decay product.

radioactive element (or **radioactive isotope**) A substance whose nucleus tends to fall apart spontaneously.

recession velocity (of a galaxy) The speed at which a distant galaxy is moving away from us due to the expansion of the universe.

recollapsing universe The possible fate of our universe in which the collective gravity of all its matter eventually halts and reverses the expansion. The galaxies will come crashing back together, and the universe will end in a fiery Big Crunch.

red giant A giant star that is red in color.

red-giant winds The relatively dense but slow winds from red giant stars.

redshift (Doppler) A Doppler shift in which spectral features are shifted to longer wavelengths, caused when an object is moving away from the observer.

reflecting telescope A telescope that uses mirrors to focus light.

reflection (of light) The process by which matter changes the direction of light.

refracting telescope A telescope that uses lenses to focus light.

rest wavelength The wavelength of a spectral feature in the absence of any Doppler shift or gravitational redshift.

retrograde motion Motion that is backward compared to the norm; e.g., we see Mars in apparent retrograde motion during the periods of time when it moves westward, rather than the more common eastward, relative to the stars.

revolution The orbital motion of one object around another.

right ascension (RA) Analogous to longitude, but on the celestial sphere; it is the angular east-west distance between the vernal equinox and a location on the celestial sphere.

rings (planetary) Consist of numerous small particles orbiting a planet within its Roche zone.

Roche tidal zone The region within two to three planetary radii (of any planet) in which the tidal forces tugging an object apart become comparable to the gravitational forces holding it together; planetary rings are always found within the Roche tidal zone.

rocks (in solar system theory) Material common on the surface of the Earth, such as silicon-based minerals, that are solid at temperatures and pressures found on Earth but typically melt or vaporize at temperatures of 500–1,300 K.

rotation The spinning of an object around its axis.

rotation curve A graph that plots rotational (or orbital) velocity against distance from the center for any object or set of objects.

runaway greenhouse effect A positive feedback cycle in which heating caused by the greenhouse effect causes more greenhouse gases to enter the atmosphere, which further enhances the greenhouse effect.

saddle-shaped (or hyperbolic) geometry Refers to any case in which the rules of geometry for a saddle-shaped surface hold, such as that two lines that begin parallel eventually diverge.

Sagittarius Dwarf A small, dwarf elliptical galaxy that is currently passing through the disk of the Milky Way Galaxy.

saros cycle The period over which the basic pattern of eclipses repeats, which is about 18 years $11\frac{1}{3}$ days.

satellite Any object orbiting another object.

scattered light Light that is reflected into random directions.

Schwarzschild radius A measure of the size of the event horizon of a black hole.

science The search for knowledge that can be used to explain or predict natural phenomena in a way that can be confirmed by rigorous observations or experiments.

scientific method An organized approach to explaining observed facts through science.

scientific theory A model of some aspect of nature that has been rigorously tested and has passed all tests to date.

secondary mirror A small mirror in a reflecting telescope, used to reflect light gathered by the primary mirror toward an eyepiece or instrument.

sedimentary rock A rock that formed from sediments created and deposited by erosional processes.

seismic waves Earthquake-induced vibrations that propagate through a planet.

semimajor axis Half the distance across the long axis of an ellipse; in this text, it is usually referred to as the *average* distance of an orbiting object, abbreviated *a* in the formula for Kepler's third law.

SETI (search for extraterrestrial intelligence) The name given to observing projects designed to search for signs of intelligent life beyond Earth.

shield volcano A shallow-sloped volcano made from the flow of low-viscosity basaltic lava.

shock wave A wave of pressure generated by gas moving faster than the speed of sound.

sidereal day The time of 23 hours 56 minutes 4.09 seconds between successive appearances of any particular star on the meridian; essentially the true rotation period of the Earth.

sidereal month About $27\frac{1}{4}$ days, the time required for the Moon to orbit the Earth once (as measured against the stars).

sidereal period (of a planet) A planet's actual orbital period around the Sun.

sidereal time Time measured according to the position of stars in the sky rather than the position of the Sun in the sky. *See also* local sidereal time

sidereal year The time required for the Earth to complete exactly one orbit as measured against the stars; about 20 minutes longer than the tropical year on which our calendar is based.

silicate rock A silicon-rich rock.

singularity The place at the center of a black hole where, in principle, gravity crushes all matter to an infinitely tiny and dense point.

Small Magellanic Cloud One of two small, irregular galaxies (the other is the Large Magellanic Cloud) located about 150,000 light-years away; it probably orbits the Milky Way Galaxy.

snowball Earth Name given to a hypothesis suggesting that, some 600–700 million years ago, the Earth experienced a period in which it became cold enough for glaciers to exist worldwide, even in equatorial regions.

solar activity Refers to short-lived phenomena on the Sun, including the emergence and disappearance of individual sunspots, prominences, and flares; sometimes called *solar weather*.

solar circle The Sun's orbital path around the galaxy, which has a radius of about 28,000 light-years.

solar day Twenty-four hours, which is the average time between appearances of the Sun on the meridian.

solar eclipse Occurs when the Moon's shadow falls on the Earth, which can occur only at new moon; may be total, partial, or annular.

solar flares Huge and sudden releases of energy on the solar surface, probably caused when energy stored in magnetic fields is suddenly released.

solar luminosity The luminosity of the Sun, which is approximately 4×10^{26} watts.

solar maximum The time during each sunspot cycle at which the number of sunspots is the greatest.

solar minimum The time during each sunspot cycle at which the number of sunspots is the smallest.

solar nebula The piece of interstellar cloud from which our own solar system formed.

solar neutrino problem Refers to the disagreement between the predicted and observed number of neutrinos coming from the Sun.

solar prominences Vaulted loops of hot gas that rise above the Sun's surface and follow magnetic field lines.

solar system (or **star system**) Consists of a star (sometimes more than one star) and all the objects that orbit it.

solar wind A stream of charged particles ejected from the Sun.

solid phase The phase of matter in which atoms or molecules are held rigidly in place.

sound wave A wave of alternately rising and falling pressure.

south celestial pole (SCP) The point on the celestial sphere directly above the Earth's South Pole.

spacetime The inseparable, four-dimensional combination of space and time.

spacetime diagram A graph that plots a spatial dimension on one axis and time on another axis.

spectral lines Bright or dark lines that appear in an object's spectrum, which we can see when we pass the object's light through a prismlike device that spreads out the light like a rainbow.

spectral resolution Describes the degree of detail that can be seen in a spectrum; the higher the spectral resolution, the more detail we can see.

spectral type A way of classifying a star by the lines that appear in its spectrum; it is related to surface temperature. The basic spectral types are designated by a letter (OBAFGKM, with O for the hottest stars and M for the coolest) and are subdivided with numbers from 0 through 9.

spectroscopic binary A binary star system whose binary nature is revealed because we detect the spectral lines of one or both stars alternately becoming blueshifted and redshifted as the stars orbit each other.

spectroscopy (in astronomical research) The process of obtaining spectra from astronomical objects

spectrum (of light) *See* electromagnetic spectrum

speed The rate at which an object moves. Its units are distance divided by time, such as m/s or km/hr.

speed of light The speed at which light travels, which is about 300,000 km/s.

spherical geometry Refers to any case in which the rules of geometry for the surface of a sphere hold, such as that lines that begin parallel eventually meet.

spheroidal component (of a galaxy) The portion of any galaxy that is spherical (or football-like) in shape and contains very little cool gas; generally contains only very old stars. Elliptical galaxies have only a spheroidal component, while spiral galaxies also have a disk component.

spin (quantum) *See* spin angular momentum

spin angular momentum Often simply called *spin*, it refers to the inherent angular momentum of a fundamental particle.

spiral arms The bright, prominent arms, usually in a spiral pattern, found in most spiral galaxies.

spiral density waves Gravitationally driven waves of enhanced density that move through a spiral galaxy and are responsible for maintaining its spiral arms.

spiral galaxies Galaxies that look like flat, white disks with yellowish bulges at their centers. The disks are filled with cool gas and dust, interspersed with hotter ionized gas, and usually display beautiful spiral arms.

spreading centers (geological) Places where hot mantle material rises upward between plates and then spreads sideways, creating new seafloor crust.

spring equinox (vernal equinox) Refers both to the point in Pisces on the celestial sphere where the ecliptic crosses the celestial equator and to the moment in time when the Sun appears at that point each year (around March 21).

standard candle An object for which we have some means of knowing its true luminosity, so that we can use its apparent brightness to determine its distance with the luminosity–distance formula.

standard model (of physics) The current theoretical model that describes the fundamental particles and forces in nature.

standard time Time measured according to the internationally recognized time zones.

star A large, glowing ball of gas that generates energy through nuclear fusion in its core. The term *star* is sometimes applied to objects that are in the process of becoming true stars (e.g., protostars) and to the remains of stars that have died (e.g., neutron stars).

star cluster *See* cluster of stars

starburst galaxy A galaxy in which stars are forming at an unusually high rate.

state (quantum) *See* quantum state

steady state theory A now-discredited theory that held that the universe had no beginning and looks about the same at all times.

Stefan–Boltzmann constant Constant that appears in the laws of thermal radiation, with value

$$\sigma = 5.7 \times 10^{-8} \frac{\text{watt}}{\text{m}^2 \times \text{Kelvin}^4}.$$

stellar evolution The formation and development of stars.

stellar parallax The apparent shift in the position of a nearby star (relative to distant objects) that occurs as we view the star from different positions in the Earth's orbit of the Sun each year.

stellar wind A stream of charged particles ejected from the surface of a star.

stratosphere An intermediate-altitude layer of the atmosphere that is warmed by the absorption of ultraviolet light from the Sun.

stratovolcano A steep-sided volcano made from viscous lavas that can't flow very far before solidifying.

stromatolites Large bacterial "colonies."

strong force One of the four fundamental forces; it is the force that holds atomic nuclei together.

subduction (of tectonic plates) The process in which one plate slides under another.

subduction zones Places where one plate slides under another.

subgiant A star that is between being a main-sequence star and being a giant; subgiants have inert helium cores and hydrogen-burning shells.

sublimation The process by which atoms or molecules escape into the gas phase from a solid.

summer solstice Refers both to the point on the celestial sphere where the ecliptic is farthest north of the celestial equator and to the moment in time when the Sun appears at that point each year (around June 21).

sunspot cycle The period of about 11 years over which the number of sunspots on the Sun rises and falls.

sunspots Blotches on the surface of the Sun that appear darker than surrounding regions.

superbubble Essentially a giant interstellar bubble, formed when the shock waves of many individual bubbles merge to form a single, giant shock wave.

supercluster Superclusters consist of many clusters of galaxies, groups of galaxies, and individual galaxies and are the largest known structures in the universe.

supergiants (luminosity class I) The very large and very bright stars that appear at the top of an H–R diagram.

supermassive black hole Giant black hole, with a mass millions to billions of times that of our Sun, thought to reside in the centers of many galaxies and to power active galactic nuclei.

supernova The explosion of a star.

Supernova 1987A A supernova witnessed on Earth in 1987; it was the nearest supernova seen in nearly 400 years and helped astronomers refine theories of supernovae.

supernova remnant A glowing, expanding cloud of debris from a supernova explosion.

synchronous rotation Describes the rotation of an object that always shows the same face to an object that it is orbiting because its rotation period and orbital period are equal.

synodic month (or **lunar month**) The time required for a complete cycle of lunar phases, which averages about $29\frac{1}{2}$ days.

synodic period (of a planet) The time between successive alignments of a planet and the Sun in our sky; measured from opposition to opposition for a planet beyond Earth's orbit, or from superior conjunction to superior conjunction for Mercury and Venus.

tangential motion The component of an object's motion directed across our line of sight.

tectonics The disruption of a planet's surface by internal stresses.

temperature A measure of the average kinetic energy of particles in a substance.

terrestrial planets Rocky planets similar in overall composition to Earth.

theories of relativity (*special* and *general*) Einstein's theories that describe the nature of space, time, and gravity.

thermal emitter An object that produces a thermal radiation spectrum; sometimes called a "blackbody."

thermal energy Represents the collective kinetic energy, as measured by temperature, of the many individual particles moving within a substance.

thermal escape The process in which atoms or molecules in a planet's exosphere move fast enough to escape into space.

thermal pressure The ordinary pressure in a gas arising from motions of particles that can be attributed to the object's temperature.

thermal pulses The predicted upward spikes in the rate of helium fusion, occurring every few thousand years, that occur near the end of a low-mass star's life.

thermal radiation The spectrum of radiation produced by an opaque object that depends only on the object's temperature; sometimes called "blackbody radiation."

thermosphere A high, hot X-ray-absorbing layer of an atmosphere, just below the exosphere.

tidal force A force that is caused when the gravity pulling on one side of an object is larger than that on the other side, causing the object to stretch.

tidal friction Friction within an object that is caused by a tidal force.

tidal heating A source of internal heating created by tidal friction. It is particularly important for satellites with eccentric orbits such as Io and Europa.

time dilation Refers to the effect in which you observe time running slower in reference frames moving relative to you.

timing (in astronomical research) The process of tracking how the light intensity from an astronomical object varies with time.

torque A twisting force that can cause a change in an object's angular momentum.

total apparent brightness *See* apparent brightness. We sometimes say "total apparent brightness" to distinguish it from wavelength-specific measures such as the apparent brightness measured in visible light.

total luminosity *See* luminosity. We sometimes say "total luminosity" to distinguish it from wavelength-specific measures such as the luminosity emitted in visible light or the X-ray luminosity.

total lunar eclipse A lunar eclipse in which the Moon becomes fully covered by the Earth's umbral shadow.

total solar eclipse A solar eclipse during which the Sun becomes fully blocked by the disk of the Moon.

totality (eclipse) The portion of either a total lunar eclipse, during which the Moon is fully within the Earth's umbral shadow, or a total solar eclipse, during which the Sun's disk is fully blocked by the Moon.

transmission (of light) The process in which light passes through matter without being absorbed.

transparent (material) Describes a material that transmits light.

triple-alpha reaction *See* helium fusion

Trojan asteroids Asteroids found within two stable zones that share Jupiter's orbit but lie 60° ahead of and behind Jupiter.

tropic of Cancer The circle on the Earth with latitude 23.5°N. It is the northernmost latitude at which the Sun ever passes directly overhead (at noon on the summer solstice).

tropic of Capricorn The circle on the Earth with latitude 23.5°S. It is the southernmost latitude at which the Sun ever passes directly overhead (at noon on the winter solstice).

tropical year The time from one spring equinox to the next, on which our calendar is based.

troposphere The lowest atmospheric layer, in which convection and weather occur.

Tully–Fisher relation A relationship among spiral galaxies showing that the faster a spiral galaxy's rotation speed, the more luminous it is; it is important because it allows us to determine the distance to a spiral galaxy once we measure its rotation rate and apply the luminosity–distance formula.

turbulence Rapid and random motion.

ultraviolet light Light with wavelengths that fall in the portion of the electromagnetic spectrum between visible light and X rays.

umbra The dark central region of a shadow.

unbound orbits Orbits on which an object comes in toward a large body only once, never to return; unbound orbits may be parabolic or hyperbolic in shape.

uncertainty principle The law of quantum mechanics that states that we can never know both a particle's position and its momentum, or both its energy and the time it has the energy, with absolute precision.

universal law of gravitation The law expressing the force of gravity (F_g) between two objects, given by the formula

$$F_g = G\frac{M_1 M_2}{d^2}$$
$$\left(G = 6.67 \times 10^{-11} \frac{m^3}{kg \times s^2}\right).$$

universal time (UT) Standard time in Greenwich (or anywhere on the prime meridian).

universe The sum total of all matter and energy.

up quark One of the two quark types (the other is the down quark) found in ordinary protons and neutrons. Has a charge of $+\frac{2}{3}$.

velocity The combination of speed and direction of motion; it can be stated as a speed in a particular direction, such as 100 km/hr due north.

virtual particles Particles that "pop" in and out of existence so rapidly that, according to the uncertainty principle, they cannot be directly detected.

viscosity Describes the "thickness" of a liquid in terms of how rapidly it flows; low-viscosity liquids flow quickly (e.g., water), while high-viscosity liquids flow slowly (e.g., molasses).

visible light The light our eyes can see, ranging in wavelength from about 400 to 700 nm.

visual binary A binary star system in which we can resolve both stars through a telescope.

voids Huge volumes of space between superclusters that appear to contain very little matter.

volatiles Refers to substances, such as water, carbon dioxide, and methane, that are usually found as gases, liquids, or surface ices on the terrestrial worlds.

volcanism The eruption of molten rock, or lava, from a planet's interior onto its surface.

wavelength The distance between adjacent peaks (or troughs) of a wave.

weak bosons The exchange particles for the weak force.

weak force One of the four fundamental forces; it is the force that mediates nuclear reactions; also the only force besides gravity felt by weakly interacting particles.

weakly interacting particles Particles, such as neutrinos and WIMPs, that respond only to the weak force and gravity; that is, they do not feel the strong force or the electromagnetic force.

weather Describes the ever-varying combination of winds, clouds, temperature, and pressure in a planet's troposphere.

weight The net force that an object applies to its surroundings; in the case of a stationary body on the surface of the Earth, weight = mass × acceleration of gravity.

weightless A weight of zero, as occurs during free-fall.

white dwarf limit (also called the *Chandrasekhar limit*) The maximum possible mass for a white dwarf, which is about $1.4M_{Sun}$.

white dwarf supernova A supernova that occurs when an accreting white dwarf reaches the white dwarf limit, ignites runaway carbon fusion, and explodes like a bomb; often called a *Type Ia supernova*.

white dwarfs The hot, compact corpses of low-mass stars, typically with a mass similar to the Sun compressed to a volume the size of the Earth.

WIMPs Stands for *weakly interacting massive particles* and represents a possible form of dark matter consisting of subatomic particles that are dark because they do not respond to the electromagnetic force.

winter solstice Refers both to the point on the celestial sphere where the ecliptic is farthest south of the celestial equator and to the moment in time when the Sun appears at that point each year (around December 21).

worldline A line that represents an object on a spacetime diagram.

wormholes The name given to hypothetical tunnels through hyperspace that might connect two distant places in our universe.

X rays Light with wavelengths that fall in the portion of the electromagnetic spectrum between ultraviolet light and gamma rays.

X-ray binary A binary star system that emits substantial amounts of X rays, thought to be from an accretion disk around a neutron star or black hole.

X-ray burster An object that emits a burst of X rays every few hours to every few days; each burst lasts a few seconds and is thought to be caused by helium fusion on the surface of an accreting neutron star in a binary system.

zenith The point directly overhead, which has an altitude of 90°.

zodiac The constellations on the celestial sphere through which the ecliptic passes.

zones (on a jovian planet) Bright bands of rising air that encircle a jovian planet at a particular set of latitudes.

ACKNOWLEDGMENTS

ILLUSTRATION BY JOE BERGERON: Figures 1.1, 1.3, 1.15, 1.16, 1.17, 2.6, 2.21, 7.3, 7.8, 7.9, 9.13, 10.12g, 14.3, 14.21, 14.22, 15.2, 15.8b, 15.12b, 16.1a, 16.15, 17.20, 17.32, 17.33, 19.2. Figures 20.2–20.5 ©Joe Bergeron

Part Opener I and Chapter Openers 1–2: ©Roger Ressmeyer/CORBIS

Part Opener II and Chapter Openers 3–6: "An Expanding Bubble in Space": NASA, Donald Walter (South Carolina State University), Paul Scowen and Brian Moore (Arizona State University

Part Opener III and Chapter Openers 7–11: "Valles Marineris—Point Perspective": Jody Swann, Tammy Becker, and Alfred McEwen of U.S. Geological Survey in Flagstaff, Arizona

Part Opener IV and Chapter Openers 12–15: "Hubble Peeks into a Stellar Nursery in a Nearby Galaxy": Hubble Heritage Team (AURA/STScI/NASA)

Part Opener V and Chapter Openers 16–20: "A Galaxy on the Edge": Hubble Heritage Team (AURA/STScI/NASA)

All Part Opener backgrounds, Visual Language Library Antique Celestial Charts and Illustrations, 1680–1880, Volume 3

CHAPTER 1 **1.2** NASA **1.4** Jerry Lodriguss **1.5** Photo by Jeff Bennett **1.6** Stan Maddock **Page 12** (**Sun**-left), Big Bear Solar Observatory/New Jersey Institute of Technology and NASA Marshall Space Flight Center; (**Sun**-right) courtesy of SOHO. SOHO is a project of international cooperation between ESA and NASA **Page 13** (**Mercury**-left) From the Voyage scale model solar system,

developed by Challenger Center for Space Science Education, the Smithsonian Institution, and NASA. Image created by ARC Science Simulations, ©2001; (**Mercury**-right) NASA/USGS **Page 14** (**Venus**-left) NASA, courtesy of NSSDC; (**Venus**-right) From the Voyage scale model solar system, developed by Challenger Center for Space Science Education, the Smithsonian Institution, and NASA. ©2001 David P. Anderson, Southern Methodist University **Page 15** (**Earth**-left) From the Voyage scale model solar system, developed by Challenger Center for Space Science Education, the Smithsonian Institution, and NASA. Image created by ARC Science Simulations, ©2001; (**Earth**-right) NASA **Page 16** (**Mars**-top) NASA/USGS, courtesy of NSSDC; (**Mars**-bottom) NASA **Page 17** (**Asteroid**) NEAR Project, Johns Hopkins University/APL, NASA **Page 18** (**Jupiter**) From the Voyage scale model solar system, developed by Challenger Center for Space Science Education, the Smithsonian Institution, and NASA. Image created by ARC Science Simulations, ©2001 **Page 19** (**Saturn**) From the Voyage scale model solar system, developed by Challenger Center for Space Science Education, the Smithsonian Institution, and NASA. Image created by ARC Science Simulations, ©2001 **Page 20** (**Uranus**) From the Voyage scale model solar system, developed by Challenger Center for Space Science Education, the Smithsonian Institution, and NASA. Image created by ARC Science Simulations, ©2001 **Page 21** (**Neptune**) From the Voyage scale model solar system, developed by Challenger Center for Space Science Education, the Smithsonian Institution, and NASA. Image created by ARC Science Simulations, ©2001 **Page 22** (**Pluto and Charon**) Dr. R. Albrecht (ESA/ESO) and NASA **Page 23** (**Comet Hale–Bopp**) Niescja Turner and Carter Emmart **1.7** Akira Fujii **1.8** Megan Donahue (STScI)

1.9 (**clockwise, starting far-top right**) ©Photo by Corel; ©Christian Jegou/Publiphoto/Photo Researchers, Inc.; ©David Gifford/Science Photo Library/Photo Researchers, Inc.; ©Blakeley Kim; ©Photo by Corel; NASA; Quade Paul/fiVth.com; ©John Eastcott and Yva Momatiuk/Photo Researchers, Inc. **1.19** Quade Paul/fiVth.com, PhotoDisc, and Hubble Space Telescope

CHAPTER 2 **2.1** Andrea Dupree (Harvard-Smithsonian CFA), Ronald Gilliland (STScI), ESA, and NASA **2.5** Gordon Garradd **2.10** (**c**) Richard Tauber Photography, San Francisco **2.13** ©David Nunuk **2.16** ©Dennis diCicco **2.17** ©Husmo-foto **2.19** (**Moons**) Akira Fujii; (**Earth**) NASA **2.23** (**top, bottom**) Akira Fujii; (**center**) Dennis diCicco **2.24, 2.25** Akira Fujii **2.26** Adapted from eclipse map by Fred Espenak, NASA/GSF: (http://sunearth.gsfc.nasa.gov/eclipse/eclipse.html) **2.27** ©Frank Zullo

CHAPTER 3 **3.2** ©Michael Yamashita/CORBIS **3.3** (a) ©N. Pecnik/Visuals Unlimited **3.4** ©Kenneth Garrett **3.5** ©Wm. E. Woolam/Southwest Parks; **3.6** ©1987 by Margaret R. Curtis **3.7** ©Richard A. Cooke III **3.8** ©Loren McIntyre/Woodfin Camp & Assoc. **3.9** ©Jeff Henry/Peter Arnold, Inc. **3.10** ©Oliver Strewe/Wave Productions Pty, Ltd. **3.11** Werner Forman Archive/Art Resource, NY **3.14** Courtesy of Carl Sagan Productions, Inc. From *Cosmos* (Random House), ©1980 Carl Sagan **3.15** ©Bettmann/CORBIS **Page 74** Giraudon/Art Resource, NY **Page 75** (**top**) Archive Photos; (**bottom**) The Granger Collection, NY **Page 76** ©Erich Lessing/Art Resource, NY **3.20** ©Jerry Lodriguss **Page 78** (**right**) ©Bettmann/CORBIS

CHAPTER 4 **4.1** ©2001 Stone/Dimitri Iundt **4.5** ©Science VU/Visuals Unlimited

CHAPTER 5 **Page 101** ©Bettmann/ CORBIS **5.8** NASA

CHAPTER 6 **6.1** ©Runk/Schoenberger from Grant Heilman Photography **6.15** (**b**) Yerkes Observatory **6.16** (**b**) Palomar Observatory/Caltech **6.18** Russ Underwood (W. M. Keck Observatory) **6.19** ©Richard Wainscoat **6.20** (**a, b**) Canada-France-Hawaii Telescope Corporation, Hawaii **6.21** NASA **6.22** Jodi Schoemer **6.24** David Parker, 1997/Science Library. *The Arecibo Observatory is part of the National Astronomy and Ionosphere Center, which is operated by Cornell Univ. under a cooperative agreement with the National Science Foundation* **6.25** ©Joel Gordon Photography **6.27** NASA/JPL/Caltech

CHAPTER 7 **7.2** C. R. O'Dell (Rice Univ.) and NASA **7.4** (**a**-left) NASA/STScI, courtesy of Alfred Schultz and Helen Hart; (**a**-right) JPL/Caltech, Franklin & Marshall College, and NASA at W. M. Keck Observatory; (**b**-left) M. J. McCaughrean (MPIA), C. R. O'Dell (Rice Univ.), and NASA; (**b**-right) J. Bally, D. Devine, and R. Sutherland **7.7** Quade Paul/fiVth.com **7.10** (**a, b**) NASA/JPL, courtesy NSSDC **7.11** ©William K. Hartmann **7.14** Data adapted from STARE project: (www.hao. ucar.edu/public/research/stare/hd209458.html)

CHAPTER 8 **8.1** (**Moon**) Akira Fujii; (**others**) NASA **8.2** (**a**) ©Don Davis; (**b**) NASA, courtesy NSSDC; (**c**) ©Barrie Rokeach; (**d**) Artis Planetarium/The Netherlands; (**e**) NASA/JPL **8.3** (**both**) ©Roger Ressmeyer/CORBIS **8.5** Don Gault, Peter Schmultz, and NASA Ames Research Center **8.6** (**a**) NASA, from the Apollo 16 crew's mapping camera; (**b**) NASA/USGS **8.7** NASA **8.8** ©2001 Stone/Paul Chesley **8.9** (**top**) NASA, courtesy of LPL; (**center**) NASA/JPL; (**bottom**) ©Forest Buchanan/Visuals Unlimited **8.10** (**left**) NASA/USGS; (**right**) NASA/JPL (Magellan Mission) **8.11** (**a, b**) NASA **8.15** GOES-10 satellite image. Colorization by ARC Science Simulations, ©1998 **8.16** (**b**) NASA **8.17** ©D. Cavagnaro/Visuals Unlimited **8.19** (**both**) NASA **8.20** (**a, c**) NASA/JPL; (**b**) NASA/USGS (Mariner 10 Mission) **8.21** (**a, c**) NASA/JPL; (**b**) Malin Space Science Systems and Caltech **8.22** (**all**) NASA/USGS **8.23** (**a–c**) NASA/JPL; (**d, f**) NASA/JPL/Malin Space Science Systems; (**e**) Dr. David E. Smith, NASA, and MOLA Science Team **8.24** NASA/JPL **8.25** (**all**) NASA/JPL

CHAPTER 9 **9.1** NASA **9.5** ©Michael Carroll **9.6** (**a**) STScI and R. Beebe and A. Simon (NMSU); (**b**) NASA Infrared Telescope Facility and Glenn Orton **9.7** (**Uranus**) E. Karkoschca (Univ. of Arizona and NASA); (**others**) NASA/JPL **9.8** John Spencer/Lowell Observatory **9.9** Courtesy Tim Parker/JPL **9.10** (**a**) PIRL/Lunar & Planetary Laboratory (Univ. of Arizona) and NASA; (**b**) Univ. of Arizona and NASA; (**c**-left) NASA/JPL; (**c**-right) Univ. of Arizona and NASA **9.12** DLR/NASA/JPL; (**b**) Arizona State Univ. and NASA; (**c**) NASA/JPL/Arizona State Univ. **9.14** (**a**) NASA/JPL; (**b**) NASA/JPL and Brown Univ. **9.15** (**a**) NASA/JPL; (**b**) NASA/JPL and Arizona State Univ. **9.16** (**a**) NASA/JPL/Caltech; (**b**) ©Don Davis; (**c**)

Seran Gibbard, Bruce Macintosh, Don Gavel, Claire Max (Lawrence Livermore National Laboratory) **9.17, 9.18** NASA/JPL **9.19** (**a**) NASA/USGS; (**b**) NASA/JPL **9.20** (**a**) S. Larson (Univ. of Arizona/LPL); (**b**) NASA/JPL; (**c**) ©William K. Hartmann **9.21** (**a, b**) NASA/JPL **9.22** (**Uranus**) Erich Karkoschka (Univ. of Arizona/LPL) and NASA; (**others**) NASA/JPL

CHAPTER 10 **10.1, 10.2** Data provided by Dave Tholen (Univ. of Hawaii) **10.3** Discovery photograph made January 7, 1976, by Eleanor F. Helin/JPL (Helin then associated with Caltech/Palomar Observatory) **10.4** (**a**) NASA/USGS; (**b**) NASA/JPL; (**c, d**) Johns Hopkins Univ./APL/NASA **10.5** Meteorite specimens courtesy of Robert Haag Meteorites **10.6** (**a**) ©Peter Ceravolo; (**b**) Tony and Daphne Hallas; (**c**) LASCO, SOHO Consortium, NRL, ESA, and NASA **10.7** (**left**) Astuo Kuboniwa, March 9, 1997, 19:25:00– 19:45:out, BISTAR Astronomical Observatory, Japan, Telescope: d=125mm, f=500mm refractor, Film: Ektracrome E100S; (**right/ inset**) U.S. Geological Survey/NASA **10.11** Eliot Young/Southwest Research Institute **10.12** (**a**) NASA/JPL/Caltech; (**b**) HST Jupiter Imaging Science Team; (**c**) MSSO, ANU/Science Library/Photo Researchers, Inc.; (**d**) Courtesy of Richard Wainscoat et al. (Univ. of Hawaii); (**e**) H. Hammel (MIT) and NASA; (**f**) HST Comet Team and NASA **10.13** (**c**) James W. Young/TMO/JPL/NASA **10.14** (**a**) Brad Snowder; (**b**) Kirk Johnson/ Denver Museum of Natural History **10.15** (**all**) Quade Paul/fiVth.com **10.16** TASS/ Sovfoto

CHAPTER 11 **11.1** NASA **11.2** (**Venus**) NASA/Magellan data, courtesy Peter Ford (MIT); (**Earth**) NOAA; (**Mars**) NASA/Viking data, courtesy Mike Mellon (Univ. of Colorado) **11.3** ©Tom Van Sant/The Geosphere Project/Corbis Stock Market **11.4** Digital image by Dr. Peter W. Sloss (NOAA/NES-DIS/NGDCO) **11.5** Mike Yamashita/ Woodfin Camp & Assoc. **11.8** (**a**) NASA; (**b**) Earth Satellite Corp./Science Photo Library/Photo Researchers, Inc. **11.10** (**a, c**) Biological Photo Services; (**b**) ©Kevin Collins/ Visuals Unlimited **11.12** (**a**) Woods Hole Oceanographic Institute; (**b**) ©Barrie Rokeach **11.14** Data obtained from C. D. Keeling and T. P. Whorf, Scripps Institution of Oceanography, University of California, La Jolla: (http://cdiac.esd.ornl.gov/ftp/maunaloa-co2/ maunaloa.co2) **11.15** NASA/GSFC (Science Visualization Studio) **11.16** (**a**) NASA; (**b**-both) NASA/Ames Research Center

CHAPTER 12 **12.1** Photo by Corel **12.3** National Optical Astronomy Observatories/ NSO, Sacramento Peak **12.7** National Optical Astronomy Observatories **12.8** Courtesy of Brookhaven National Laboratory **12.9** (**a, b**) ICRR (Institute for Cosmic Ray Research), Univ. of Tokyo **12.11** (**b**) Goram Scharmer/Stanford's Solar Observatories Group **12.12** (**a, b**) NOAO/National Solar Observatory **12.14** (**a**) Image from TRACE (Transition Region and Coronal Explorer), a mission of the Stanford-Lockheed Institute for Space Research (a joint program of the Lockheed-Martin Advanced Technology Center's

Solar and Astrophysics laboratory and Stanford's Solar Observatories Group), and part of the NASA Small Explorer program **12.15** Courtesy of SOHO. SOHO is a project of international cooperation between ESA and NASA **12.16** Image from TRACE (Transition Region and Coronal Explorer), a mission of the Stanford-Lockheed Institute for Space Research (a joint program of the Lockheed-Martin Advanced Technology Center's Solar and Astrophysics laboratory and Stanford's Solar Observatories Group), and part of the NASA Small Explorer program **12.17** Courtesy of B. Haisch and G. Slater (Lockheed Palo Alto Research Laboratory) **12.18** Data adapted from Marshall Space Flight Center: (http://science.msfc.nasa.gov/ssl/pad/solar)

CHAPTER 13 **Page 276** ©Bettmann/ CORBIS **13.3** Hubble Heritage Team (AURA/STScI/NASA) **Table 13.1** (**spectrum**) Roger Bell, Univ. of Maryland and Mike Briley, Univ. of Wisconsin–Oshkosh **13.4** (**photo**-center) Lowell Observatory **13.12** ©Anglo-Australian Observatory/Royal Observatory Edinburgh, photography by D. Malin **13.13** Hubble Heritage Team (AURA/ STScI/NASA)

CHAPTER 14 **14.1** ©Anglo-Australian Observatory, photograph by D. Malin **14.2** IPAC (Infrared Processing and Analysis Center) and Caltech/JPL **14.4** (**b, c**) C. Burrows and J. Morse (STScI), J. Hester (Arizona State Univ.), and NASA **14.13** (**a**) Nordic Optical Telescope, La Palma; (**b**) J. P. Harrington, K. J. Borkowski (Univ. of Maryland), and NASA; (**c**) B. Balick (Univ. of Washington), V. Icke (Leiden Univ.), G. Mellema (Stockholm Univ.), and NASA; (**d**) R. Sahai, J. Trauger (JPL), the WFPC2 Science Team, and NASA **14.19** ©ESO (European Southern Observatory) **14.20** ©Anglo-Australian Observatory, photograph by D. Malin

CHAPTER 15 **15.1** NASA/SAO/CXO **15.3** (**b**) M. Shara, B. Williams, and D. Zurek (STScI); R. Gilmozzi (ESO); D. Prialnik (Tel Aviv Univ.); and NASA **15.5** NASA and F. M. Walter (State Univ., New York at Stony Brook) **15.7** ©ESO (European Southern Observatory) **15.14** Andrew Fruchter (STScI) and NASA

CHAPTER 16 **16.2** (**counterclockwise from lower right**) D. Malin (Fig. 16.4); S. L. Snowden (Fig. 16.5); NASA/NSSDC (Fig. 16.11a); J. Bally (Fig. 16.8); J. Hester and P. Scowen (Fig. 16.10); D. Malin (Fig. 16.12) **16.3** H. Bond (STScI) and NASA **16.4** ©Anglo-Australian Observatory, photography by D. Malin **16.5** NASA/CXO/SAO/Rutgers and J. Hughes **16.6** (**a**) Tony and Daphne Hallas; (**b**) Jeff Hester (Arizona State Univ.) and NASA **16.7** Simulation by D. Strickland and I. Stevens **16.8, 16.9** John Bally (Univ. of Colorado) **16.10** Jeff Hester and Paul Scowen (Arizona State Univ.) and NASA **16.11** Courtesy of NASA's National Space Science Data Center at the Goddard Space Flight Center **16.12** ©Anglo-Australian Observatory/Royal Observatory Edinburgh, photography by D. Malin **16.13, 16.14** ©Anglo-Australian Observatory, photography by D. Malin **16.17** Hubble Heritage Team (AURA/STScI/NASA); Acknowledgments: N. Scoville (Caltech) and T. Rector (NOAO) **16.18** (**photo**) J. Trauger, JPL and NASA **16.19** (**a**) David Talent at the Cerro Tololo Inter-American Observatory, Chile; (**b**) AFRL/BMDO, courtesy of Dr. Mehradad Moshir; (**c**) NRAO/AUI, courtesy of Cornelia Lang (NRAO/UCLA) and Mark Morris (UCLA); (**d**) K. Y. Lo (Univ. of Illinois)

CHAPTER 17 **17.1** ©Sky Publishing Corporation, reproduced with permission **17.2** (**a**) ©ESO (European Southern Observatory); (**b**) ©IAC photo from plates taken with the Isaac Newton telescope, photo by D. Malin **17.3, 17.4** ©Anglo-Australian Observatory, photography by D. Malin **17.5** Hubble Heritage Team (AURA/STScI/NASA) **17.6** ©Anglo-Australian Observatory/Royal Observatory Edinburgh, photography by D. Malin **17.7** (**a, b**) ©Anglo-Australian Observatory, photography by D. Malin **17.8** ©Anglo-Australian Observatory/Royal Observatory Edinburgh, photography by D. Malin **17.11** The Observatories of the Carnegie Institution of Washington **17.12** (**a**) J. Trauger, JPL, and NASA; (**b**) Dr. W. L. Freedman (The Observatories of the Carnegie Institution of Washington) and NASA **17.13** Megan Donahue (STScI) **17.18** Robert Williams and the HDF Team (STScI) and NASA **17.21** Hyron Spinard, Univ. of California at Berkeley et al. **17.22** (**a**) Brad Whitmore (STScI) and NASA; (**b**) W. C. Keel (Univ. of Alabama) **17.23** Robert Williams and the HDF Team (STScI) and NASA **17.24** Frank Summers/American Museum of Natural History **17.25** ©Anglo-Australian Observatory, photography by D. Malin **17.26** Dr. Michael J. West/Univ. of Hawaii–Hilo **17.27** (**a**) MPE (Max Planck Institute); (**b**) Subaru Telescope, National Astronomical Observatory of Japan. ©2000 NAOJ **17.28** Hubble Heritage Team (AURA/STScI/NASA) **17.29** John Bahcall/Institute for Advanced Study and NASA **17.30** NRAO/AUI **17.31** John Biretta (STScI) **17.34** H. Ford (STScI/Johns Hopkins Univ.); L. Dressel, R. Harms, A. Kochhar (Applied Research Corp.); Z. Tsvetanov, A. Davidsen, G. Kriss (Johns Hopkins Univ.); R. Bohlin, G. Hartig (STScI); B. Margon (Univ. of Washington–Seattle); and NASA

CHAPTER 18 **18.3** (**photo**) ©Anglo-Australian Observatory, photography by D. Malin **18.5** Courtesy of Caltech **18.6** (**a**) Omar Lopez-Cruz and Iam Shelton; (**b**) Steve Snowden, NASA/GSFC/USRA **18.7** W. N. Colley and E. Turner (Princeton Univ.), J. A. Tyson (Bell Labs, Lucent Technologies), and NASA **18.9** A. Fruchter, the ERO Team (STScI, ST-ECF), and NASA **18.10** (**inset photos**) Charles Alcock/Lawrence Livermore National Laboratory **18.11** Michael Strauss, Princeton Univ. **18.12** (**a**) Harvard-Smithsonian Center for Astrophysics; (**b**) Emiko-Rose Koike/fiVth.com **18.13** Ben Moore, Dept. of Physics, Durham Univ., Durham City, England

CHAPTER 19 **19.5** ©Roger Ressmeyer/CORBIS **19.8** E. Bunn (St. Cloud State Univ.) **19.12** (**left**) John Kieffer/Peter Arnold, Inc.; (**right**) Joel Gordon Photography

CHAPTER 20 **20.1** (**a, b**) NASA/JPL **20.6** ©Seth Shostak

INDEX